Complex Multiplication

This is a self-contained account of the state of the art in classical complex multiplication that includes recent results on rings of integers and applications to cryptography using elliptic curves. The author is exhaustive in his treatment, giving a thorough development of the theories of elliptic functions, modular functions and quadratic number fields and providing a concise summary of the results from class field theory. The main results are accompanied by numerical examples, equipping any reader with all the tools and formulae they need.

Topics covered include: the construction of class fields over quadratic imaginary number fields by singular values of the modular invariant j and Weber's tau-function; explicit construction of rings of integers in ray class fields and Galois module structure; the construction of cryptographically relevant elliptic curves over finite fields; proof of Berwick's congruences using division values of the Weierstrass-p function; and relations between elliptic units and class numbers.

REINHARD SCHERTZ was Professor of Mathematics at the University of Augsburg in Germany until his retirement in 2008.

New Mathematical Monographs

All the titles listed below can be obtained from good booksellers or from Cambridge University Press. For a complete series listing, visit http://www.cambridge.org/uk/series/sSeries.asp?code=NMM

1 M. Cabanes and M. Enguehard *Representation Theory of Finite Reductive Groups*
2 J.B. Garnett and D.E. Marshall *Harmonic Measure*
3 P. Cohn *Free Ideal Rings and Localization in General Rings*
4 E. Bombieri and W. Gubler *Heights in Diophantine Geometry*
5 Y.J. Ionin and M.S. Shrikhande *Combinatorics of Symmetric Designs*
6 S. Berhanu, P.D. Cordaro and J. Hounie *An Introduction to Involutive Structures*
7 A. Shlapentokh *Hilbert's Tenth Problem*
8 G. Michler *Theory of Finite Simple Groups I*
9 A. Baker and G. Wüstholz *Logarithmic Forms and Diophantine Geometry*
10 P. Kronheimer and T. Mrowka *Monopoles and Three-Manifolds*
11 B. Bekka, P. de la Harpe and A. Valette *Kazhdan's Property (T)*
12 J. Neisendorfer *Algebraic Methods in Unstable Homotopy Theory*
13 M. Grandis *Directed Algebraic Topology*
14 G. Michler *Theory of Finite Simple Groups II*

Complex Multiplication

REINHARD SCHERTZ
Universität Augsburg

CAMBRIDGE
UNIVERSITY PRESS

CAMBRIDGE UNIVERSITY PRESS
Cambridge, New York, Melbourne, Madrid, Cape Town,
Singapore, São Paulo, Delhi, Tokyo, Mexico City

Cambridge University Press
The Edinburgh Building, Cambridge CB2 8RU, UK

Published in the United States of America by Cambridge University Press, New York

www.cambridge.org
Information on this title: www.cambridge.org/9780521766685

First published 2010

A catalogue record for this publication is available from the British Library

Library of Congress Cataloguing in Publication Data

Schertz, Reinhard, 1943–
Complex multiplication / Reinhard Schertz.
p. cm. – (New mathematical monographs; 15)
ISBN 978-0-521-76668-5 (Hardback)
1. Multiplication, Complex. I. Title. II. Series.
QA564.S294 2010
516.3′52–dc22

2009051874

ISBN 978-0-521-76668-5 Hardback

to Marlies

Contents

Preface

The aim of this book is to give an account of the state of the art in classical complex multiplication including, in particular, recent results on rings of integers and applications to cryptography using elliptic curves. All requisites needed about elliptic functions, modular functions and quadratic number fields are developed in this book and the results from class field theory are summarised in compact form. Further, most of the main results presented in the following chapters are accompanied by a plethora of numerical examples.† The reader interested in the application of the various explicit results will therefore find all the necessary tools in this book.

After the early results of Abel and Kronecker at the beginning of and mid nineteenth century, Weber at the start of the twentieth century gave the first systematic account of complex multiplication in his "Lehrbuch der Algebra III". The aim of this theory is to generate abelian extensions of quadratic imaginary number fields by values of elliptic functions and modular functions. Up until 1931 further accounts of the theory were given by Fricke (1916, 1922) and Fueter (1924, 1927). Finally, Hasse (1927) using class field theory that had developed in the meantime, presented a very short and elegant version of complex multiplication. His work contains the generation of ray class fields over a quadratic imaginary number field by singular values of the modular invariant j and Weber's τ function, using in the proof, besides class field theory, only the discriminant from the theory of elliptic functions. A more detailed exposition of the theory including a proof of the principal ideal theorem was provided by Deuring (1958). However, results on rings of integers had not been thus far obtained.

† I would like to thank the KANT group of TU Berlin headed by Michael Pohst for their help in computation by KASH (www.math.tu-berlin.de/kant).

Geometrically, in complex multiplication the generation of ray class fields over imaginary quadratic number fields is quite analogous to the construction of cyclotomic fields by roots of unity, which are torsion points of the unit circle. In complex multiplication the role of the unit circle is taken by a suitable elliptic curve E: the coefficients of E generate the Hilbert class field Ω, and the ray class fields are obtained by adjoining to Ω the x-coordinate of some torsion point of E. In view of this analogy one may pose the question whether it is possible to find explicit construction not only for the fields but also for their rings of integers as algebra or as Galois modules. Further, analogous to cyclotomic theory, one may ask for constructions of unit groups together with formulae for the class number. In fact, all this has been shown in the works of Leopoldt (1954, 1962) to be possible for cyclotomic fields. The solutions to these problems in complex multiplication form the central topics of this book. The following problems are treated in detail:

- Classical and simple generators for ring class fields and ray class fields
- Construction of rings of integers in ray class fields by explicit basis
- Galois module structure of these rings of integers including explicit construction of Galois generators and associated orders
- Construction of unit groups of maximal rank including their relation to class numbers
- Proof and generalisation of Berwick's congruences for the singular values of the modular invariant j

A recent application of complex multiplication described in this book is concerned with

- the construction cryptographically relevant elliptic curves over finite fields.

As shown in Chapter 9, the problem behind this construction is to find generating polynomials with small coefficients for abelian extensions of a quadratic imaginary number field. In contrast to cyclotomic theory, this is a non-trivial task, because the singular values of the modular invariant and the Weber τ function have minimal polynomials with astronomic coefficients.

Compared to cyclotomic theory the results obtained in complex multiplication, so far, seem complete. On the other hand there are numerous interesting questions concerning the generalisation to abelian varieties of higher dimension, for which a thorough understanding of complex multiplication is essential. These questions are in fact of very real interest because varieties of higher dimension can also be applied in cryptography. Such results can, for example, be found in the works of Weng (2001, 2003) for curves of genus 2 and 3. Conversely, the generalisation of the construction of class fields including results on the structure of rings of integers and class numbers remain unsolved. To obtain such results it may be useful to look at geometric analogies and analytic relations between cyclotomic fields with the unit circle of genus 0 and class fields of complex multiplication with their elliptic curves of genus 1 as described in Chapters 6 and 7 and to find such relations between curves of genus 1 and a higher genus.

1
Elliptic functions

In complex multiplication the abelian extensions of an imaginary quadratic number field are generated by the coefficients of suitable elliptic curves together with the coordinates of torsion points in terms of values of elliptic functions and modular functions, as explained in Chapter 6. Moreover, using elliptic functions, we will study rings of integers and unit groups, and we will find explicit constructions for them later in Chapters 7 and 8. Therefore, besides the parametrisation of elliptic curves by the Weierstrass $\wp$ function and its derivative, we need to study more closely the σ function and the η function that are involved in, for example, the calculations of discriminants in Chapter 7. The resolvent formula of section 1.7 and the p-adic limits will be needed for the Galois module structure of Chapter 8. For the same reason we will study Weierstrass equations that will also be used in the applications to cryptography in Chapter 10. The division polynomials in section 1.5 will be crucial for the proof of Berwick's congruences in Chapter 9.

1.1 Values of elliptic functions

In the complex plane we fix a lattice, i.e. a free abelian group of rank 2,

$$\mathfrak{L} = \mathbb{Z}\omega_1 + \mathbb{Z}\omega_2,$$

generated by two over $\mathbb{R}$ linearly independent numbers ω_1, ω_2. Any two such numbers are called a basis of $\mathfrak{L}$, and we write

$$\mathfrak{L} = [\omega_1, \omega_2].$$

A function $f : \mathbb{C} \to \mathbb{C} \cup \{\infty\}$ is called **elliptic** with respect to $\mathfrak{L}$ if f is meromorphic and if the points of $\mathfrak{L}$ are periods of f:

$$f(z + \omega) = f(z) \text{ for all } \omega \in \mathfrak{L}.$$

For a fixed number $\gamma \in \mathbb{C}$ and a basis ω_1, ω_2 of $\mathfrak{L}$ the set

$$F = F_\gamma = \{\gamma + x_1\omega_1 + x_2\omega_2 \mid 0 \le x_j < 1\}$$

is a systems of repesentatives of $\mathbb{C}/\mathfrak{L}$ and is called a **fundamental parallelogram** for $\mathfrak{L}$. It is therefore sufficient to study the values of an elliptic function on a fundamental parallelogram.

Theorem 1.1.1 *Let f be a non-constant elliptic function without poles on the boundary of F. Then the sum of its residues in F is equal to zero:*

$$\sum_{z \in F} Res(f, z) = 0.$$

Proof By periodicity of f we have the equalities

$$\int_{\gamma}^{\gamma+\omega_j} f(z)dz = \int_{\gamma+\omega}^{\gamma+\omega_j+\omega} f(z)dz = -\int_{\gamma+\omega_j+\omega}^{\gamma+\omega} f(z)dz$$

for every $\omega \in \mathfrak{L}$. So in the integral of f along the boundary of F_γ the integrals along opposite edges cancel out. The assertion of Theorem 1.1.1 now follows from the theorem of residues. □

Applying Theorem 1.1.1 to $\frac{f}{f-w}$, $w \in \mathbb{C}$, for an elliptic function f, the argument principle tells us:

Theorem 1.1.2 *Let f be a non-constant elliptic function. Then for every $w \in \mathbb{C} \cup \{\infty\}$ the number of solutions $a \in F$ of the equation $f(a) = w$, counted according to multiplicity, is equal to the number of poles of f in F. This number is called the order of f.*

Let f be a non-constant elliptic function. Then for $a \in \mathbb{C}$ we define the **order** of f at a to be the unique number $m \in \mathbb{Z}$ such that $g_a(z) := \frac{f(z)}{(z-a)^m}$ is holomorphic in a and $g_a(a) \neq 0$.

Theorem 1.1.3 *Let a_j be the points in F, where the non-constant elliptic function f has non-vanishing order m_j. Then*

$$\sum_j m_j = 0 \qquad \textit{and} \qquad \sum_j m_j a_j \in \mathfrak{L}.$$

Proof The first assertion is a special case of Theorem 1.1.2. We prove the second assertion. Since there are only finitely many zeros and poles in F, we can shift F such that all poles and zeros of f are in the interior of F and not on the boundary. According to the residue theorem we then have

$$2\pi i \sum_j m_j a_j = \int_{\partial F} z \frac{f'(z)}{f(z)} dz.$$

Now, combining, as in the proof of Theorem 1.1.1, the integrals along opposite edges, we obtain on the right-hand side an expression of the form

$$\int_{\gamma}^{\gamma+\omega_1} z\frac{f'(z)}{f(z)}dz - \int_{\gamma+\omega_2}^{\gamma+\omega_1+\omega_2} z\frac{f'(z)}{f(z)}dz = \int_{\gamma}^{\gamma+\omega_1} \omega_2\frac{f'(z)}{f(z)}dz$$

$$= \omega_2 \int_{f(\overline{\gamma,\gamma+\omega_1})} \frac{du}{u} = \omega_2 2\pi i k, \quad k \in \mathbb{Z},$$

where $\overline{\gamma, \gamma+\omega_1}$ denotes the path from γ to $\gamma+\omega_1$. This makes the assertion clear. □

From the second assertion of Theorem 1.1.3 we further obtain

Theorem 1.1.4 *An elliptic function with at most one pole of order 1 in F is constant.*

1.2 The functions $\sigma(z|\mathfrak{L})$, $\zeta(z|\mathfrak{L})$ and $\wp(z|\mathfrak{L})$

We start by constructing the most important functions used in the sequel. They can all be derived from the **Weierstrass σ function** of a lattice $\mathfrak{L}$:

$$\sigma(z) = z \cdot \prod_{\omega\in\mathfrak{L}\setminus\{0\}} (1 - \tfrac{z}{\omega})e^{\frac{z}{\omega}+\frac{1}{2}\frac{z^2}{\omega^2}}.$$

Absolute convergence and holomorphy of the defining product for $\sigma(z)$ follow from:

Lemma 1.2.1 *For $k \geq 3$ the series $\sum_{\omega\in\mathfrak{L}\setminus\{0\}} \frac{1}{\omega^k}$ is absolutely convergent.*

Proof Observing that for $n \in \mathbb{N}$ there are $8n$ lattice points in

$$\mathfrak{L}_n = \{x_1\omega_1 + x_2\omega_2 \mid (x_1, x_2) \in [-n, n] \times \{\pm n\} \cup \{\pm n\} \times [-n, n]\}, \quad n \in \mathbb{N},$$

we obtain

$$\sum_{n=1}^{N}\sum_{\omega\in\mathfrak{L}_n}\frac{1}{|\omega|^k}\le\frac{8}{\delta^k}\sum_{n=1}^{N}\frac{1}{n^{k-1}}\text{ with }\delta=\min\{|\omega|\mid\omega\in\mathfrak{L}_1\},$$

which implies the assertion of Lemma 1.2.1. □

All elliptic functions can be derived from σ though σ itself is not elliptic. Taking a logarithmic derivative we first obtain the **elliptic zeta function**

$$\zeta(z)=\frac{d}{dz}\log(\sigma(z))=\frac{\sigma'(z)}{\sigma(z)}=\frac{1}{z}+\sum_{\omega\in\mathfrak{L}\setminus\{0\}}\left[\frac{1}{z-\omega}+\frac{1}{\omega}+\frac{z}{\omega^2}\right],$$

then, by differentiation we obtain the **Weierstrass $\wp$ function**

$$\wp(z)=-\zeta'(z)=\frac{1}{z^2}+\sum_{\omega\in\mathfrak{L}\setminus\{0\}}\left[\frac{1}{(z-\omega)^2}-\frac{1}{\omega^2}\right] \tag{1.1}$$

and its derivative

$$\wp'(z)=-2\sum_{\omega\in\mathfrak{L}}\frac{1}{(z-\omega)^3}.$$

$\wp'$ is meromorphic by definition and clearly elliptic because the series for $\wp'$ is absolutely convergent. Moreover we have:

Theorem 1.2.2 *$\wp$ is an elliptic function.*

Proof $\wp'$ being elliptic implies that $\wp(z+\omega_j)=\wp(z)+c_j$ with a constant c_j. Setting $z=-\frac{\omega_j}{2}$ and, keeping in mind that $\wp$ is an even function without a pole at $-\frac{\omega_j}{2}$, we find that $c_j=0$. Therefore, $\wp$ is elliptic. □

On the other hand σ and ζ are not elliptic, as can be seen by the following theorem in combination with Theorem 1.2.5. By integration we obtain from Theorem 1.2.2 the transformation formulae:

Theorem 1.2.3 *For $\omega\in\mathfrak{L}$ we have*

$$\zeta(z+\omega)=\zeta(z)+\eta(\omega),$$
$$\sigma(z+\omega)=\psi(\omega)e^{\eta(\omega)(z+\frac{\omega}{2})}\sigma(z)$$

with a constant $\eta(\omega)$, the so-called quasi-period, and the factor

$$\psi(\omega)=\begin{cases}1 & \text{for } \omega\in 2\mathfrak{L},\\ -1 & \text{for } \omega\in\mathfrak{L}\setminus 2\mathfrak{L}.\end{cases}$$

Clearly, the map $\eta : \mathfrak{L} \to \mathbb{C}$, defined by the transformation formula of the zeta function, is an additive homomorphism. Further, by the unique representation

$$z = z_1\omega_1 + z_2\omega_2, \quad z_j \in \mathbb{R},$$

for $z \in \mathbb{C}$ we can extend η to $\mathbb{C}$ by

$$\eta(z) := z_1\eta_1 + z_2\eta_2$$

with $\eta_j = \eta(\omega_j), j = 1, 2$. Now we define:

$$l(u, w) := u\eta(w) - \eta(u)w \qquad \text{for } u, w \in \mathbb{C}.$$

Further, to simplify notation, we set

$$z^* := \eta(z)$$

and

$$\zeta^*(z) := \zeta(z) - z^*,$$
$$\sigma^*(z) = e^{-\frac{zz^*}{2}}\sigma(z).$$

The transformation formulae of Theorem 1.2.3 can then be written as:

Theorem 1.2.4 *For $\tau \in \mathfrak{L}$ we have*

$$\begin{aligned} \zeta^*(z+\tau) &= \zeta^*(z), \\ \sigma^*(z+\tau) &= \psi(\tau)e^{\frac{1}{2}l(z,\tau)}\sigma^*(z). \end{aligned}$$

Therefore, ζ^* is periodic with respect to $\mathfrak{L}$ but not elliptic. For a better understanding of the factor $e^{\frac{1}{2}l(z,\tau)}$ in the transformation formula of σ^* we need:

Theorem 1.2.5 (Legendre Relation) *Let ω_1, ω_2 be a basis of the lattice $\mathfrak{L}$ with $\Im(\frac{\omega_1}{\omega_2}) > 0$. Then the quasi-periods $\eta_j = \eta(\omega_j)$ of the zeta function satisfy*

$$\omega_1\eta_2 - \omega_2\eta_1 = 2\pi i.$$

Proof We assume 0 to be an interior point of F. Then, $\zeta(z)$ has exactly one pole of order 1 in F. So, by the theorem of residues

$$2\pi i = \int_{\partial F} \zeta(z)dz.$$

Using the transformation formula of the zeta function and adding up integrals along opposite edges as in the proof of Theorem 1.1.1, this becomes

$$-\int_{\gamma+\omega_1}^{\gamma} \eta_2 dz - \int_{\gamma}^{\gamma+\omega_2} \eta_1 dz = \omega_1\eta_2 - \omega_2\eta_1,$$

which proves Theorem 1.2.5. □

From Theorem 1.2.5 we now obtain:

Lemma 1.2.6 *For $u = u_1\omega_1 + u_2\omega_2,\ w = w_1\omega_1 + w_2\omega_2$ we have*

$$l(u,w) = 2\pi i(u_1w_2 - u_2w_1),$$

hence

$$e^{l(u,w)} = 1,\ \text{for } u, w \in \mathfrak{L}.$$

Further, we have the rules

$$\begin{aligned} l(au,w) &= al(u,w),\ \text{for } a \in \mathbb{R}, \\ l(au,w) &= l(u,\bar{a}w)\ \text{for } a \in \mathbb{C}. \end{aligned}$$

Proof Let ω_1, ω_2 be a basis of $\mathfrak{L}$ with $\Im(\frac{\omega_1}{\omega_2}) > 0$. Then, by definition of l and Legendre's relation in Theorem 1.2.5

$$l(u,w) = 2\pi i(u_1w_2 - u_2w_1) = det\begin{pmatrix} u_1 & u_2 \\ w_1 & w_2 \end{pmatrix} 2\pi i,$$

with the real coordinates u_j, w_j in the representations $u = u_1\omega_1 + u_2\omega_2, w = w_1\omega_1 + w_2\omega_2$. The first two assertions now follow immediately. To prove the third, let A be the representing matrix of a with respect to the basis ω_1, ω_2,

$$a\begin{pmatrix} \omega_1 \\ \omega_2 \end{pmatrix} = A\begin{pmatrix} \omega_1 \\ \omega_2 \end{pmatrix}.$$

Then, $det(A)A^{-1}$ is the matrix belonging to $\bar{a}$, and the coordinates of au and $\bar{a}w$ are given by

$$(u_1,u_2)A \quad \text{and} \quad (w_1,w_2)det(A)A^{-1}.$$

Hereafter, the third assertion of the lemma follows from the above representation of l as a determinant, observing that $N(a) = det(A)$. □

1.3 Construction of elliptic functions

A non-constant elliptic function f has only a finite number of points modulo $\mathfrak{L}$, where its order is different from 0 (zeros or poles). Let these points be:

$$a_1, \ldots, a_s \text{ with orders } m_1, \ldots, m_s.$$

Then, according to Theorem 1.1.3,

$$\sum_{j=1}^{s} m_j = 0 \text{ and } \sum_{j=1}^{s} m_j a_j \in \mathfrak{L}. \tag{1.2}$$

Conversely, for $a_j \in \mathbb{C}$ and $m_j \in \mathbb{Z}$ satisfying (1.2), an elliptic function is defined by

$$g(z) = \prod_{j=1}^{s} \left(e^{z a_j^*} \sigma(z - a_j) \right)^{m_j}, \tag{1.3}$$

having the same zeros and poles including orders as f. To see this, we use the transformation formula of the σ function for $\omega \in \mathfrak{L}$:

$$g(z + \omega) = \psi(\omega)^{\sum m_j} e^{\omega(\sum m_j a_j)^* - \omega^*(\sum m_i a_j)} g(z).$$

Herein the exponent of e is a multiple of $2\pi i$ due to Legendre's relation and the relation $\sum m_j a_j \in \mathfrak{L}$. Further, by asumption $\sum m_j = 0$. So we have $g(z + \omega) = g(z)$. Hence g is elliptic.

The function $\frac{f}{g}$ must be constant by Theorem 1.1.4, and we have proved the following theorem:

Theorem 1.3.1 (Abel–Jacobi) *Let $a_1, \ldots, a_s \in \mathbb{C}$ and $m_1, \ldots, m_s \in \mathbb{Z} \setminus \{0\}$. There exists an elliptic function f, such that the a_i are mod $\mathfrak{L}$ the only points, where the order of f is non-zero and equal to m_i iff condition (1.2) is satisfied. Every such function is, up to a constant factor, equal to the product in (1.3).*

As an example we consider the function $\wp(z) - \wp(a)$ for $a \in \mathbb{C} \setminus \mathfrak{L}$. Here we have $a_1 = a$, $a_2 = -a$, $a_3 = 0$; $m_1 = m_2 = 1$, $m_3 = -2$. Therefore, by Theorem 1.3.1

$$\wp(z) - \wp(a) = C \frac{\sigma(z - a)\sigma(z + a)}{\sigma(z)^2}$$

with a constant C. To determine C, we multiply both sides of the equation by $\sigma(z)^2$ and take the limit for $z \to 0$. This shows that $C = \frac{1}{\sigma(a)^2}$, and the first assertion of the following theorem is proved. To prove the second assertion, we divide the first formula by $z - a$ and take the limit for $a \to z$.

Theorem 1.3.2

(i) $\wp(z) - \wp(a) = -\frac{\sigma(z-a)\sigma(z+a)}{\sigma(a)^2\sigma(z)^2}$ *for* $a \in \mathbb{C} \setminus \mathfrak{L}$.

(ii) $\wp'(z) = -\frac{\sigma(2z)}{\sigma(z)^4}$.

Clearly, the set of elliptic functions with respect to a lattice is a field under addition and multiplication. In the sequel it will be denoted by $\mathbb{C}_{\mathfrak{L}}$. Using the results of section 1.1, it follows:

Theorem 1.3.3 *$\mathbb{C}_{\mathfrak{L}}$ is generated over $\mathbb{C}$ by $\wp$ and $\wp'$: $\mathbb{C}_{\mathfrak{L}} = \mathbb{C}(\wp, \wp')$.*

Proof First, we show every even elliptic function to be a rational function of $\wp$. Therefore, we need:

Lemma 1.3.4 *Let f be an even function from $\mathbb{C}_{\mathfrak{L}} \setminus \mathbb{C}$ and $a \in \mathbb{C}$ with $2a \in \mathfrak{L}$. Then, the order of f at a is divisible by 2.*

Proof Writing down the Taylor expansion of f at a,

$$f(z) = c_m(z-a)^m + c_{m+1}(z-a)^{m+1} + \ldots, \qquad c_m \neq 0,$$

we find

$$f(z) = f(-z+2a) = f(a + (a-z)) = (-1)^m c_m (z-a)^m + \ldots$$

Hence m must be even. □

Lemma 1.3.5 *$\wp'(z)$ is of order 3 and has three simple zeros at the half-periods*

$$w_1 = \frac{\omega_1}{2}, \; w_2 = \frac{\omega_1 + \omega_2}{2}, \; w_3 = \frac{\omega_2}{2}.$$

For $a \in \mathbb{C} \setminus \mathfrak{L}$ the function $\wp(z) - \wp(a)$ has a simple zero at each of the points $\pm a$ if $2a \notin \mathfrak{L}$ and a zero of order 2 at a if $2a \in \mathfrak{L}$.

Proof $\wp'$ being an odd function, we can conclude, using the Taylor expansion as in the proof of Lemma 1.3.4, that the half-periods are zeros of $\wp'(z)$. Moreover, since $\wp'$ has order 3, these are modulo $\mathfrak{L}$ the only zeros. The assertion of Lemma 1.3.5 about $\wp$ now follows because $\wp$ has order 2. □

Proof of Theorem 1.3.3. First, we let f be a non-constant *even* elliptic function with $a_1, .., a_s$ being modulo $\mathfrak{L}$ all points where f has an order $m_j \neq 0$. We set $m'_j = m_j$ resp. $m'_j = \frac{m_j}{2}$ if $2a_j \notin \mathfrak{L}$ resp. $2a_j \in \mathfrak{L}$. Then, by Lemma 1.3.5 the function

$$g(z) = f(z) \prod_{j=1}^{s} (\wp(z) - \wp(a_j))^{-m'_j}$$

can only have zeros or poles in $\mathfrak{L}$ and Theorem 1.1.3 tells us that g must be constant, so f is a rational function of $\wp$. □

Now, let f be an arbitrary elliptic function. We write f as the sum of an even and an odd function,

$$f(z) = \tfrac{f(z)+f(-z)}{2} + \tfrac{f(z)-f(-z)}{2\wp'(z)} \wp'(z),$$

which is a linear combination of 1 and $\wp'$ with coefficients, that are even elliptic functions and thus rational functions in $\wp$. This proves the assertion of Theorem 1.3.3.

1.4 Algebraic and geometric properties of elliptic functions

The **Eisenstein series** of a lattice $\mathfrak{L}$,

$$G_m(\mathfrak{L}) := \sum_{\omega \in \mathfrak{L}}{}' \frac{1}{\omega^{2m}}, \quad m \geq 2,$$

are absolutely convergent by Lemma 1.2.1. We set

$$g_2 = g_2(\mathfrak{L}) = 60 G_2(\mathfrak{L}),$$

$$g_3 = g_3(\mathfrak{L}) = 140 G_3(\mathfrak{L}).$$

Theorem 1.4.1 *$\wp$ and $\wp'$ satisfy the algebraic equation*

$$\wp'^2 = 4\wp^3 - g_2\wp - g_3.$$

The polynomial $4X^3 - g_2X - g_3$ has three pairwise different zeros

$$e_j = \wp(w_j), \quad j = 1, 2, 3,$$

with the half-periods w_j. Its discriminant is

$$\Delta(\mathfrak{L}) = 16(e_1 - e_2)^2(e_1 - e_3)^2(e_2 - e_3)^2.$$

For every lattice

$$\Delta(\mathfrak{L}) \neq 0,$$

and by known formulae, we find

$$\Delta(\mathfrak{L}) = g_2^3 - 27g_3^2.$$

For the proof we need:

Lemma 1.4.2 *In a neighbourhood of zero $\wp$ has the Laurent expansion*

$$\wp(z) = \frac{1}{z^2} + \sum_{m=1}^{\infty} (2m+1) G_{m+1}(\mathfrak{L}) z^{2m}.$$

Proof In the series (1.1) of $\wp$ we write

$$\frac{1}{(z-\omega)^2} - \frac{1}{\omega^2} = \frac{1}{\omega^2}\left(\frac{1}{(1-\frac{z}{\omega})^2} - 1\right) = \sum_{n=2}^{\infty} n \frac{z^{n-1}}{\omega^{n+1}}.$$

Then, because of absolute convergence, we can interchange summation over ω with summation over n. Further, observing that $\sum_{\omega \in \mathfrak{L}}{}' \frac{1}{\omega^k} = 0$ for every odd $k \geq 3$, we obtain the Laurent expansion of our assertion. □

Proof of Theorem 1.4.1. Writing the function

$$g(z) = \wp'^2 - (4\wp^3 - g_2\wp - g_3)$$

in terms of the Laurent expansion of Lemma 1.4.2 we find that g has no poles and $g(0) = 0$. Therefore, by Theorem 1.1.2, $g = 0$, which proves the first assertion. To prove the remaining assertions, we observe that $\wp'(w_i) = 0$ according to Lemma 1.3.5 for every half-period w_i. The polynomial $4X^3 - g_2X - g_3$ therefore has the zeros $\wp(w_j), j = 1, 2, 3$. Moreover, these are pairwise different because $\wp'(w_j) = 0$ and thus $\wp$ has order 2 at w_j. This finishes the proof of Theorem 1.4.1. □

Theorem 1.4.1 gives rise to a bijection

$$z + \mathfrak{L} \mapsto Q(z) := \begin{cases} (1 : \wp(z) : \wp'(z)) \text{ if } z \notin \mathfrak{L}, \\ \left(\frac{1}{\wp'(z)} : \frac{\wp(z)}{\wp'(z)} : 1\right) \text{ if } \wp'(z) \neq 0 \end{cases} \tag{1.4}$$

between $\mathbb{C}/\mathfrak{L}$ and the projective curve

$$E := \{(t : x : y) \in \mathbb{P}^2(\mathbb{C}) \mid y^2 t = 4x^3 - g_2xt^2 - g_3t^3\}.$$

By the definition

$$Q(z_1) + Q(z_2) := Q(z_1 + z_2)$$

E is endowed with a group structure with the point $Q(0) = (0 : 0 : 1)$ at infinity as a neutral element. The algebraic properties of this group structure follow from the addition formula for $\wp$ that we are now going to derive.

The analytic source of the addition formula is the first formula in Theorem 1.3.2:

$$\wp(z) - \wp(z') = -\frac{\sigma(z+z')\sigma(z-z')}{\sigma(z)^2\sigma(z')^2}.$$

Taking both sides of the logarithmic derivative with respect to z and z', we find

$$\begin{aligned}\frac{\wp'(z)}{\wp(z)-\wp(z')} &= \zeta(z+z') + \zeta(z-z') - 2\zeta(z),\\ -\frac{\wp'(z')}{\wp(z)-\wp(z')} &= \zeta(z+z') - \zeta(z-z') - 2\zeta(z'),\end{aligned}$$

and adding the two formulae yields:

Theorem 1.4.3 (addition theorem of the ζ function) *For $z, z' \in \mathbb{C} \setminus \mathfrak{L}$ with $z \not\equiv \pm z'$ mod $\mathfrak{L}$ we have*

$$\zeta(z+z') = \zeta(z) + \zeta(z') + \frac{1}{2}\frac{\wp'(z) - \wp'(z')}{\wp(z) - \wp(z')}.$$

By differentiation of the formula in Theorem 1.4.3 with respect to z and z', we obtain

$$\wp(z+z') = \wp(z) - \frac{1}{2}\frac{\wp''(z)(\wp(z)-\wp(z')) - (\wp'(z)-\wp'(z'))\wp'(z)}{(\wp(z)-\wp(z'))^2},$$

$$\wp(z+z') = \wp(z') - \frac{1}{2}\frac{-\wp''(z')(\wp(z)-\wp(z')) - (\wp'(z)-\wp'(z'))(-\wp'(z'))}{(\wp(z)-\wp(z'))^2}.$$

Further, by differentiation of the algebraic equation in Theorem 1.4.1 we find

$$\wp''(z) = 6\wp(z)^2 - \frac{g_2}{2}.$$

Adding the two formulae for $\wp(z+z')$ and expressing $\wp''$ by $\wp$ using the last formula shows:

Theorem 1.4.4 (addition theorem of the $\wp$ function) *For $z, z' \in \mathbb{C} \setminus \mathfrak{L}$ with $z \not\equiv \pm z'$ mod $\mathfrak{L}$ we have*

$$\wp(z+z') = -\wp(z) - \wp(z') + \frac{1}{4}\left(\frac{\wp'(z)-\wp'(z')}{\wp(z)-\wp(z')}\right)^2 \quad \textit{if } z \not\equiv \pm z' \textit{ mod } \mathfrak{L},$$

$$\wp(2z) = -2\wp(z) + \frac{1}{4}\left(\frac{\wp''(z)}{\wp'(z)}\right)^2 \quad \textit{if } 2z \not\equiv 0 \textit{ mod } \mathfrak{L}.$$

The formulae of Theorem 1.4.4 tell us that E is an algebraic group with respect to the defined addition, i.e. the coordinates of $Q(z_1)+Q(z_2)$ are rational functions in the coordinates of $Q(z_1)$ and $Q(z_2)$ with coefficients in $\mathbb{Q}[g_2, g_3]$.

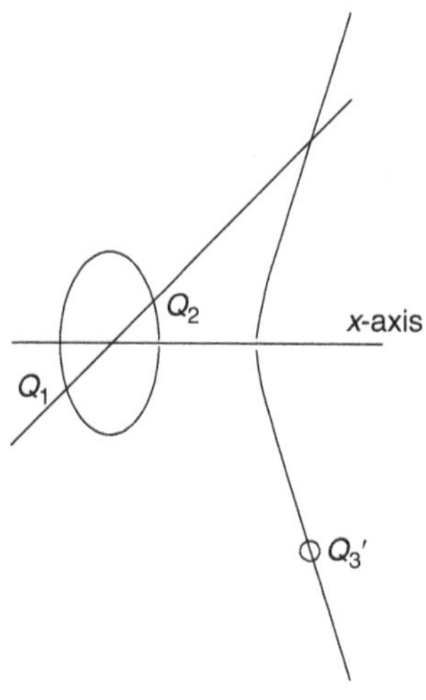

Furthermore, the above formulae allow the following geometric interpretation: for two points $Q_1, Q_2 \in E$ the line through Q_1 and Q_2 has exactly one more point of intersection $Q_3 = (t:x:y)$ with E and from the equation of intersection one finds (e.g. Lang (1970), page 12)

$$Q_1 + Q_2 + Q_3 = 0.$$

Therefore, we can conclude

$$Q_1 + Q_2 = Q_3' \text{ mit } Q_3' = (t:x:-y).$$

This relation shows us again that the coordinates of Q_1+Q_2 are rational functions of the coordinates of Q_1 and Q_2 with coefficients in $\mathbb{Q}[g_2, g_3]$.

1.5 Division polynomials

The aim of this section is to compute algebraic equations for the x-coordinates of torsion points of E. Our presentation is based on Cassels (1949), Lang (1978) and Meyer (1995). For a natural number n, we consider the function

$$f_n(z) = (-1)^{n+1}\frac{\sigma(nz)}{\sigma(z)^{n^2}},$$

which is elliptic according to Theorem 1.3.1. f_n satisfies the following product relation:

Theorem 1.5.1

$$f_n^2(z) = n^2 \prod_{\substack{u\in\frac{1}{n}\mathfrak{L}\setminus\mathfrak{L} \\ u \bmod \mathfrak{L}}} (\wp(z) - \wp(u)).$$

Proof Given an elliptic function $f \neq 0$ having non-zero order μ_j exactly at points u_j modulo $\mathfrak{L}$ we call the formal linear combination

$$(f) = \sum_j \mu_j(u_j)$$

of the points u_j the **divisor** of f. According to Liouville's theorems the quotient of two elliptic functions having the same divisor is a constant. Hence the quotient of both the left- and the right-hand sides in Theorem 1.5.1 are a constant because both have the same divisor

$$\sum_{\substack{u\in\frac{1}{n}\mathfrak{L} \\ u \bmod \mathfrak{L}}} 2(u) \quad - 2(n^2-1)(0).$$

The constant is computed via the Laurent expansions at zero of the functions involved in the formula. □

Theorem 1.5.2 *Let $m > n$ be natural numbers. We set $\wp_m(z) = \wp(mz)$ and $\wp_n(z) = \wp(nz)$. Then*

$$\wp_m - \wp_n = -\frac{f_{m+n}f_{m-n}}{f_m^2 f_n^2}.$$

Proof First, we determine the divisor of each side. It is given by:

$$\sum_{\substack{u\in\frac{1}{m}\mathfrak{L} \\ u \bmod \mathfrak{L}}} -2(u) + \sum_{\substack{u\in\frac{1}{n}\mathfrak{L} \\ u \bmod \mathfrak{L}}} -2(u) + \sum_{\substack{u\in\frac{1}{m+n}\mathfrak{L} \\ u \bmod \mathfrak{L}}} (u) + \sum_{\substack{u\in\frac{1}{m-n}\mathfrak{L} \\ u \bmod \mathfrak{L}}} (u).$$

For the right-hand side this follows from defining the property of f_m and f_n. For the left, it is obtained by looking at the formula

$$\wp(mz) - \wp(nz) = -\frac{\sigma((m+n)z)\sigma((m-n)z)}{\sigma(mz)^2\sigma(nz)^2}.$$

Therefore, the quotient of both sides must be a constant C and, comparing Laurent expansions, we find $C = 1$. □

The functions f_m satisfy the **identity:**

Theorem 1.5.3

$$f_{m+1}f_{m-1}f_n^2 - f_{n+1}f_{n-1}f_m^2 = f_{m+n}f_{m-n} \textit{ for } \quad m > n > 1.$$

Proof The proof is obtained by applying Theorem 1.5.2 twice:

$$-\frac{f_{m+n}f_{m-n}}{f_m^2 f_n^2} = \wp_m - \wp_n = (\wp_m - \wp) - (\wp_n - \wp)$$

$$= -\frac{f_{m+1}f_{m-1}}{f_m^2 f_1^2} + \frac{f_{n+1}f_{n-1}}{f_n^2 f_1^2} = -\frac{f_{m+1}f_{m-1}f_n^2 - f_{n+1}f_{n-1}f_m^2}{f_m^2 f_n^2}.$$

□

Using Theorem 1.5.3 for the pairs $(n+1, n)$ and $(n+1, n-1)$ and, keeping in mind $f_2 = \wp'$, we obtain the following **recursion formulae:**

Theorem 1.5.4

(i) $f_{2n+1} = f_{n+2}f_n^3 - f_{n+1}^3 f_{n-1}$,

(ii) $f_{2n}f_2 = f_n(f_{n+2}f_{n-1}^2 - f_{n-2}f_{n+1}^2)$.

Using these formulae, the f_n can be computed recursively if f_1, f_2, f_3, f_4 are known. The latter are determined via the addition theorem and Theorem 1.5.2:

$$\begin{aligned} f_1 &= 1, \\ f_2 &= \wp', \\ f_3 &= 3\wp^4 - \tfrac{3}{2}g_2\wp^2 - 3g_3\wp - \tfrac{1}{16}g_2^2, \\ f_4 &= \tfrac{1}{2}\wp'(4\wp^6 - 5g_2\wp^4 - 20g_3\wp^3 - \tfrac{5}{4}g_2^2\wp^2 - g_2g_3\wp - 2g_3^2 \\ &\qquad + \tfrac{1}{16}g_2^3). \end{aligned} \tag{1.5}$$

In particular, the recursion formulae show

$$f_n \in \mathbb{C}[\wp] \text{ for } n \text{ odd},$$
$$f_n \in \wp'\mathbb{C}[\wp] \text{ for } n \text{ even}.$$

Writing

$$f_n = \begin{cases} \Psi_n(\wp) \text{ for } n \text{ odd}, \\ \frac{\wp'}{2}\Psi_n(\wp) \text{ for } n \text{ even}, \end{cases}$$

with the polynomials

$$\Psi_n(\wp) = n \prod_{\substack{u \in \frac{1}{n}\mathfrak{L} \setminus \frac{1}{2}\mathfrak{L} \\ \pm u \ mod\ \mathfrak{L}}} (\wp - \wp(u)) \in \mathbb{C}[\wp],$$

the above formulae yield

$$\begin{aligned} \Psi_1 &= 1, \\ \Psi_2 &= 2, \\ \Psi_3 &= 3\wp^4 - \frac{3}{2}g_2\wp^2 - 3g_3\wp - \frac{1}{16}g_2^2, \\ \Psi_4 &= 4\wp^6 - 5g_2\wp^4 - 20g_3\wp^3 - \frac{5}{4}g_2^2\wp^2 - g_2g_3\wp - 2g_3^2 + \frac{1}{16}g_2^3, \end{aligned}$$

and we have the recursion formulae:

Theorem 1.5.5

$$\Psi_{2n+1} = \begin{cases} \Psi_{n+2}\Psi_n^3 - \Psi_{n+1}^3\Psi_{n-1}16\wp'^4 \ \textit{if}\ 2\nmid n, \\ 16\wp'^4\Psi_{n+2}\Psi_n^3 - \Psi_{n+1}^3\Psi_{n-1} \ \textit{if}\ 2|n, \end{cases}$$

$$\Psi_{2n} = \Psi_n(\Psi_{n+2}\Psi_{n-1}^2 - \Psi_{n-2}\Psi_{n+1}^2).$$

For illustration and for later purposes we apply Theorem 1.5.5 to derive the representation for Ψ_5:

$$\begin{aligned} \Psi_5 &= 5\,\wp^{12} - \frac{31}{2}\,g_2\,\wp^{10} - 95\,g_3\,\wp^9 - \frac{105}{16}\,g_2{}^2\wp^8 + 15\,g_2\,g_3\,\wp^7 \\ &+ \left(\frac{75}{16}\,g_2{}^3 - 15\,g_3{}^2\right)\wp^6 + \frac{87}{8}\,g_2{}^2 g_3\,\wp^5 \\ &+ \left(-\frac{133}{256}\,g_2{}^4 + \frac{17}{2}\,g_2\,g_3{}^2 - 8\left(\frac{1}{32}\,g_2{}^3 - g_3{}^2\right)g_2 + \frac{3}{2}\left(\frac{3}{16}\,g_2{}^3 + 9\,g_3{}^2\right)g_2\right)\wp^4 \\ &+ \left(-\frac{5}{8}\,g_2{}^3 g_3 - 10\,g_3{}^3 - 8\left(\frac{1}{32}\,g_2{}^3 - g_3{}^2\right)g_3 + 3\left(\frac{3}{16}\,g_2{}^3 + 9\,g_3{}^2\right)g_3\right)\wp^3 \\ &+ \left(\frac{3}{512}\,g_2{}^5 - \frac{1}{2}\,g_2{}^2 g_3{}^2 + \left(\frac{1}{32}\,g_2{}^3 - g_3{}^2\right)g_2{}^2 + \frac{1}{16}\left(\frac{3}{16}\,g_2{}^3 + 9\,g_3{}^2\right)g_2{}^2\right)\wp^2 \\ &+ \left(\frac{9}{256}\,g_2{}^4 g_3 - \frac{1}{2}\,g_2\,g_3{}^3 + 2\left(\frac{1}{32}\,g_2{}^3 - g_3{}^2\right)g_2\,g_3\right)\wp \\ &+ \frac{1}{4096}\,g_2{}^6 + \left(\frac{1}{32}\,g_2{}^3 - g_3{}^2\right)g_3{}^2 \end{aligned}$$

Proof of the formulae (1.5):
$f_1 = 1$ follows immediately from the definition of f_1. The equality $f_2 = \wp'$ means

$$-\frac{\sigma(2z)}{\sigma(z)^4} = \wp',$$

and this follows from Theorem 1.3.2 or from the fact that the divisor of $\wp'$ is

$$(\wp') = -3(0) + \sum_{\substack{u \in \frac{1}{2}\mathfrak{L} \\ u \bmod \mathfrak{L}}} (u).$$

To compute f_3 we use Theorem 1.5.2. We obtain

$$\wp_2 - \wp = -\frac{f_3 f_1}{f_2^2} = -\frac{f_3}{\wp'^2}$$

and further, using the addition formula:

$$\wp_2 = -2\wp + \frac{1}{4}\left(\frac{\wp''}{\wp'}\right)^2.$$

Differentiation of $\wp'^2 = 4\wp^3 - g_2\wp - g_3$ leads to

$$2\wp'\wp'' = 12\wp^2\wp' - g_2\wp', \text{ hence } \frac{\wp''}{2\wp'} = \frac{3\wp^2 - \frac{1}{4}g_2}{\wp'}.$$

Putting this into the previous equation we obtain

$$\wp_2 - \wp = \frac{-3\wp\wp'^2 + 9\wp^4 - \frac{3}{2}\wp^2 g_2 + \frac{1}{16}g_2^2}{\wp'^2},$$

and, comparing this with the first formula gives us the asserted form for f_3. The formula for f_4 is deduced analogously. □

1.6 Weierstrass functions

In the sequel let $x, y \in \mathbb{C}_{\mathfrak{L}}$ be two functions having exactly one pole mod $\mathfrak{L}$ at 0 of order 2 resp. 3. We call two such functions a **pair of Weierstrass functions for $\mathfrak{L}$**. Liouville's theorem tells us that this condition is equivalent to x and y being of the form

$$x = a + b\wp, \quad y = c + d\wp + e\wp' \text{ with } a, b, c, d, e \in \mathbb{C} \text{ and } b, e \neq 0.$$

This implies that every pair x, y of Weierstrass functions after multiplication by non-zero constants satisfies a **Weierstrass equation**, i.e. an irreducible equation of the form

$$y^2 + a_1 xy + a_3 y = x^3 + a_2 x^2 + a_4 x + a_6 \tag{1.6}$$

with coefficients $a_1, \ldots, a_6 \in \mathbb{C}$. So the map

$$z + \mathfrak{L} \mapsto Q(z) := \begin{cases} (1 : x(z) : y(z)) \text{ for } z \notin \mathfrak{L}, \\ \left(\frac{1}{y(z)} : \frac{x(z)}{y(z)} : 1\right) \text{ for } y(z) \neq 0. \end{cases} \tag{1.7}$$

defines a bijection between $\mathbb{C}/\mathfrak{L}$ and the projective curve E defined by

$$y^2 t + a_1 xyt + a_3 yt^2 = x^3 + a_2 x^2 t + a_4 xt^2 + a_6 t^3.$$

Setting

$$Q(z_1) + Q(z_2) := Q(z_1 + z_2)$$

E is endowed with a group structure with the point $Q(0) = (0 : 0 : 1)$ at infinity as neutral element. Further, x and y being obtained from $\wp$ and $\wp'$ by an affine linear transformation, this group structure again allows the following geometric interpretation: for $Q_1, Q_2, Q_3 \in E$ with $Q_1 + Q_2 + Q_3 = 0$, the point Q_3 is the third intersection point of the line through Q_1 and Q_2 with the curve E. Moreover, this implies that the coordinates of $Q_1 + Q_2$ are rational functions of the coordinates of Q_1 and Q_2 with coefficients in $\mathbb{Q}[a_1, a_2, a_3, a_4, a_6]$. Therefore, given a field extension L of $\mathbb{Q}(a_1, a_2, a_3, a_4, a_6)$, the set $E(L)$ of all points in E having coordinates in L is a subgroup of E.

For a field extension L of $\mathbb{Q}(a_1, a_2, a_3, a_4, a_6)$ we will consider the subfield $L(x, y)$ of $\mathbb{C}(x, y)$ later. In this situation it is often necessary to know whether a given function $f \in \mathbb{C}(x, y) = \mathbb{C}_{\mathfrak{L}}$ is an element of $L(x, y)$. Here, the following theorem is useful.

Theorem 1.6.1 *Let x, y be a pair of Weierstrass functions with respect to $\mathfrak{L}$ and $f \in \mathbb{C}(x, y)$. Further, let $(z_n)_{n \geq 1}$ be an injective sequence, $\lim_{n\to\infty} z_n = 0$, containing no poles of f. Then*

$$f \in L(x, y)$$

with $L := \mathbb{Q}(f(z_1), \ldots, f(z_N), x(z_1), \ldots, x(z_N), y(z_1), \ldots, y(z_N))$ for some $N \in \mathbb{N}$.

Proof f has a representation of the form

$$n(x) f = a(x) + b(x) y \tag{1.8}$$

with polynomials

$$n(X), a(X), b(X) \in \mathbb{C}[X], \quad \gcd(n(X), a(X), b(X)) = 1.$$

Assuming that $n(X)$ is normalised, these polynomials are unique. Further, by the identity theorem for meromorphic functions the coefficients can also be characterised as the unique solutions of a system of linear equations arising from (1.8) by inserting the values $f(z_n), x(z_n)$ and $y(z_n)$. Thus, the coefficients of $a(X), b(X)$ and $n(X)$ for normalised $n(X)$ are in the field generated over $\mathbb{Q}$ by all values $f(z_n), x(z_n)$ and $y(z_n)$. More precisely the finitely many coefficients of the polynomials $a(X), b(X)$ and $n(X)$ must lie in a subfield generated over $\mathbb{Q}$ by a finite number of values $f(z_n), x(z_n)$ and $y(z_n)$. □

1.6.1 *Expansions at zero*

In the following, let x, y be a pair of Weierstrass functions. Then

$$W := -\frac{x}{y} \tag{1.9}$$

has a zero at 0 of order 1, thus being a **uniformising parameter** at 0 of the Riemann surface defined by equation (1.6) and the parametrisation (1.7). Hence, all functions from $\mathbb{C}_{\mathfrak{L}}$ and, in particular, x have a Laurent expansion of powers of W converging in a neighbourhood of 0. The coefficients of this expansion are related to $a_1, \ldots, a_6$ as follows:

Theorem 1.6.2 *In a neighbourhood of 0 we have Laurent expansions of the form*

$$\begin{aligned} -\tfrac{1}{y} &= W^3 + \sum_{n\geq 4} A_n W^n \\ x &= W^{-2} + \sum_{n\geq -1} B_n W^n \\ y &= -W^{-3} + \sum_{n\geq -2} C_n W^n \end{aligned}$$

with coefficients $A_n, B_n, C_n \in \mathbb{Z}[a_1, \ldots, a_6]$.

Proof Dividing the above Weierstrass equation by $-y^3$ gives us the equation

$$s = W^3 + (a_1 W + a_2 W^2)s + (a_3 + a_4 W)s^2 + a_6 s^3 =: f(W, s) \tag{1.10}$$

with

$$s := -\frac{1}{y}.$$

Substituting this equation into itself recursively,

$$s = f(W, f(W, f(W, \dots, f(W, s)))),$$

we obtain for every $N \in \mathbb{N}$

$$s = \sum_{n=3}^{N-1} A_n W^n + W^N g_N$$

with an elliptic function g_N holomorphic at 0 and coefficients $A_3 = 1$ and $A_n \in \mathbb{Z}[a_1, \dots, a_6]$ for $n \geq 3$. More precisely, g_N is a polynomial of s and W with coefficients in $\mathbb{Z}[a_1, \dots, a_6]$. On the other hand, s having a zero at 0 has a Laurent series expansion in a neighbourhood of 0:

$$s = \sum_{n \geq 3} d_n W^n.$$

From

$$\sum_{n=3}^{N-1} (A_n - d_n) W^n = O(W^N) \text{ for } W \to 0$$

we can conclude that

$$d_n = A_n \text{ for } n = 3, \dots, N-1$$

for every $N \in \mathbb{N}$, thereby proving the first assertion of our theorem. The third assertion follows immediately from the first and implies the second using the relation $x = -Wy$. □

For the expansion of an arbitrary function in $\mathbb{C}_{\mathfrak{L}}$ we have:

Theorem 1.6.3 *For $f \in \mathbb{C}_{\mathfrak{L}}$ let $\xi_1, .., \xi_m$ mod $\mathfrak{L}$ be all poles $\not\equiv 0$ mod $\mathfrak{L}$ of f. Further, let $(z_k)_{k \geq 1}$, $z_k \notin \mathfrak{L}$, be a sequence converging to 0 with $f(z_k), W(z_k), s(z_k)$ being algebraic numbers.*
Then, there exist $l \in \mathbb{N}$ and $r \in R_0$,

$$R_0 := \mathbb{Z}[a_1, \dots, a_6, x(\xi_1), \dots, x(\xi_m), W(z_1), \dots, W(z_l), s(z_1), \dots, s(z_l), f(z_1), \dots, f(z_l)],$$

such that the coefficients in the Laurent expansion

$$f = \sum_{n \geq n_0} d_n W^n$$

are all in $\frac{1}{r} R_0$.

Proof For sufficiently large $N \in \mathbb{N}$

$$g := \left\{ \prod_{j=1}^{m} (x - x(\xi_j)) \right\}^N f$$

has no pole outside $\mathfrak{L}$. Therefore,

$$g = p(x) + q(x)y$$

with polynomials $p, q \in \mathbb{C}[X]$, which implies that for a sufficiently large $M \in \mathbb{N}$

$$s^M g = h(W, s)$$

with a polynomial h, that is unique if we assume its degree with respect to s to be minimal. The coefficients of h satisfy the linear equations

$$s(z_k)^M g(z_k) = h(W(z_k), s(z_k)), \quad k \in \mathbb{N},$$

and they are even uniquely determined by these equations because by the identity theorem $s^M g$ is uniquely determined by its values $s(z_k)^M g(z_k)$, $k \in \mathbb{N}$. The coefficients of the solution are in a field generated by finitely many of the values $g(z_k), x(z_k), y(z_k)$. Inserting the W-expansions of x and s from Theorem 1.6.2 into

$$f = \left\{ \prod_{j=1}^{m} (x - x(\xi_j)) \right\}^{-N} s^{-M} h(W, s),$$

gives us the assertion of Theorem 1.6.3. □

Next, we will express the addition formula 1.4.3 in terms of a power series. The starting point is the geometric interpretation according to which for $(z_1, z_2, z_3) \in \mathbb{C}^3$ with $z_1 + z_2 + z_3 \equiv 0 \bmod \mathfrak{L}$ the points $Q(z_i)$ are the three intersection points of a line and the projective curve E defined by $Y^2 Z = 4X^3 - g_2 X Z^2 - g_3 Z^3$. By a linear transformation T,

$$T : (p_1 : p_2 : p_3) \mapsto (ap_1 + bp_2 : cp_1 + dp_2 + ep_3 : p_3),$$

E is mapped to the curve E' defined by

$$Y^2 Z + a_1 XYZ + a_3 YZ^2 = X^3 + a_2 X^2 Z + a_4 XZ^2 + a_6 Z^3.$$

Here again the images $T(p(z_i))$ are the three intersection points of a line with E'. This way, using Theorem 1.6.2, we find the following result:

Theorem 1.6.4 *There exist two power series*

$$G(X,Y), I(X) \in \mathbb{Z}[a_1,\ldots,a_6][[X,Y]]$$

and a neighbourhood U of 0 such that for all $z, z_1, z_2 \in U$ the power series are convergent for $X, Y = W(z), W(z_i)$ and

$$\begin{aligned} W(z_1+z_2) &= W(z_1)+W(z_2)+W(z_1)W(z_2)G(W(z_1),W(z_2)), \\ W(-z) &= -W(z)(1+W(z)I(W(z)). \end{aligned}$$

Proof We consider three points $z_1, z_2, z_3 \neq 0$ in $\mathbb{C}$ with

$$z_1+z_2+z_3=0.$$

For z_i sufficiently near to 0 the values $W(z_i), s(z_i)$ are finite. Hence the corresponding points on E' have projective coordinates

$$(W_j : 1 : s_j) \text{ with } W_j = W(z_j), s_j = s(z_j), j = 1,2,3.$$

To show that W_3 is a power series in W_1 and W_2 we observe that $(W_3 : 1 : s_3)$ is the third intersection point with E' of the line through $(W_1 : 1 : s_1)$ and $(W_2 : 1 : s_2)$. We first assume that $W_1 \neq W_2$ and consider the equation

$$s = mW + b$$

of the line through (W_1, s_1) and (W_2, s_2) in the (W, s)-plane. Inserting this into the equation $s = f(W, s)$ for E' from (1.10), we obtain a cubic equation for W:

$$\begin{aligned} mW+b &= W^3+(a_1W+a_2W^2)(mW+b)+(a_3+a_4W)(mW+b)^2 \\ &\quad +a_6(mW+b)^3. \end{aligned}$$

Herein the coefficient of W^3 is

$$1+a_2m+a_4m^2+a_6m^3.$$

Therefore, looking at the coefficients of W^2, we obtain

$$W_1+W_2+W_3 = -\frac{a_1m+a_2ba_3m^2+2a_4bm+3a_6m^2}{1+a_2m+a_4m^2+a_6m^3}. \tag{1.11}$$

Now we substitute the expansion of m and b that we obtain from Theorem 1.6.2:

$$m = \frac{s_1-s_2}{W_1-W_2} = \sum_{n\geq 3} A_n \frac{W_1^n-W_2^n}{W_1-W_2} = W_1^2+W_1W_2+W_2^2+\cdots,$$

$$b = s_1 - mW_1 = \sum_{n\geq 3} A_n W_1^n \quad - mW_1.$$

Then (1.11) shows that W_3 is of the form

$$W_3 = F(W_1, W_2)$$

with a power series

$$F(X,Y) \in \mathbb{Z}[a_1, \ldots, a_6][[X,Y]]$$

that is independent of W_1, W_2. Next, we derive an expansion for $W(z_1 + z_2)$ as a power series of W_3. We observe that

$$W(z_1 + z_2) = W(-z_3) = -\frac{x(-z_3)}{y(-z_3)},$$

and from $\wp(-z_3) = \wp(z_3)$ it follows that

$$x(-z_3) = x(z_3).$$

Hence, the numbers $y(\pm z_3)$ satisfy the quadratic equation

$$Y^2 + a_1 x(z_3) Y + a_3 Y = x(z_3)^3 + a_2 x(z_3)^2 + a_4 x(z_3) + a_6.$$

For the following we first assume that $z_3 \notin \frac{1}{2}\mathfrak{L}$. Then $\wp'(-z_3) \neq \wp'(z_3)$, which implies that $y(-z_3) \neq y(z_3)$ and, looking at the coefficient of Y in the quadratic equation, we obtain the relation:

$$y(-z_3) = -y(z_3) - a_1 x(z_3) - a_3,$$

which by continuity of y also holds for $z_3 \in \frac{1}{2}\mathfrak{L}$. Thus, we obtain

$$W(-z_3) = -\frac{x(z_3)}{-y(z_3) - a_1 x(z_3) - a_3} = -\frac{W_3}{1 - a_1 W_3 - a_3 s_3}.$$

Inserting the expansion of s_3 as a power series of W_3, we find that

$$W(z_1 + z_2) = H(W_3)$$

with

$$H(X) \in \mathbb{Z}[a_1, \ldots, a_6][[X]].$$

This formula also holds for $z_3 \in \frac{1}{2}\mathfrak{L}$, because both sides are continuous functions of $z_3 = -(z_1 + z_2)$ and, taking the limit $z_1, z_2 \to 0$, it follows that $F(0,0) = 0$. Therefore, inserting F in H again yields a power series

$$H(F(X,Y)) \in \mathbb{Z}[a_1, \ldots, a_6][[X,Y]].$$

We then have

$$W(z_1 + z_2) = H(F(W(z_1), W(z_2))),$$

and by continuity this formula also holds for $z_1, z_2, z_1 + z_2 \neq 0$ in a neighbourhood of 0. Finally, $H(F(X, Y))$ has the form asserted in the theorem because the limits of

$$W(z_1 + z_2) - W(z_1) - W(z_2)$$

for $z_1 \to 0$ and $z_2 \to 0$ are equal to zero.

To prove the formula for $W(-z)$, we use the above relation

$$W(-z) = -\frac{x(z)}{-y(z) - a_1 x(z) - a_3} = -\frac{W(z)}{1 - a_1 W(z) - a_3 s}$$

for $z = z_3$ and insert the expansion of W as a power series in s. This yields the formula for $W(-z)$. □

1.6.2 p-adic limits

The series in Theorems 1.6.2 to 1.6.4 are understood as convergent with respect to the usual metric of $\mathbb{C}$. Under certain conditions we will now show that they are also convergent **to the same limit** with respect to the metric of a discrete valuation ring. We let R be a discrete valuation ring in $\mathbb{C}$ with maximal ideal $\mathfrak{P} = R\pi$. Then we have:

Theorem 1.6.5 *Let $a_1, \dots, a_6 \in R$ and $\xi \in \mathbb{C}$ with*

$$W(\xi) \in \mathfrak{P}.$$

Then with the coefficients A_n, B_n, C_n from Theorem 1.6.2 we have the $\mathfrak{P}$-adic expansions

$$-\tfrac{1}{y(\xi)} = W(\xi)^3 + \sum_{n \geq 4} A_n W(\xi)^n,$$
$$x(\xi) = W(\xi)^{-2} + \sum_{n \geq -1} B_n W(\xi)^n,$$
$$y(\xi) = -W(\xi)^{-3} + \sum_{n \geq -2} C_n W(\xi)^n,$$

where for the last two formulae $W(\xi) \neq 0$ has to be required.

Proof Following the proof of Theorem 1.6.3, we start by showing the first assertion. For $W(\xi) = 0$ the algebraic equation between x and y implies that $s(\xi) = 0$. Therefore the expansion in our assertion is trivial. In the case of $W(\xi) \in \mathfrak{P} \setminus \{0\}$ the algebraic equation yields $s(\xi) \in \mathfrak{P}$ and, applying the equation between s and W recursively as at the beginning of the proof of Theorem 1.6.2, leads to

$$s - \sum_{n\geq 3}^{N-1} A_n W^n \in W^N \mathbb{Z}[a_1, \dots, a_6, s, W]$$

for every $N \in \mathbb{N}$. This implies that

$$s(\xi) - \sum_{n\geq 3}^{N-1} A_n W(\xi)^n \in \mathfrak{P}^N,$$

thus proving the first assertion for $s(\xi) = -\frac{1}{y(\xi)}$. The two remaining assertions of Theorem 1.6.5 can be deduced from the first as in the proof of Theorem 1.6.2. □

Using the $\mathfrak{P}$-adic expansion of $-\frac{1}{y(\xi)}$ in Theorem 1.6.5 we obtain, as described in the proof of Theorem 1.6.4:

Theorem 1.6.6 *Let $a_1, \dots, a_6 \in R$ and $\xi, \xi_1, \xi_2 \in \mathbb{C}$ with*

$$W(\xi), W(\xi_1), W(\xi_2) \in \mathfrak{P}.$$

Then we have the $\mathfrak{P}$-adic expansions

$$W(\xi_1 + \xi_2) = W(\xi_1) + W(\xi_2) + W(\xi_1)W(\xi_2)G(W(\xi_1), W(\xi_2)),$$
$$W(-\xi) = -W(\xi)(1 + W(\xi)I(W(\xi))$$

with the power series $G(X, Y)$, $I(X)$ from Theorem 1.6.4.

The following theorem is a p-adic version of Theorem 1.6.3.

Theorem 1.6.7 *Let f be as in Theorem 1.6.3 and R a valuation ring in $\mathbb{C}$ with maximal ideal $\mathfrak{P}$ containing the ring R_0 of Theorem 1.6.3. Then for $W(\xi) \in \mathfrak{P}$ we have*

$$f(\xi) = \sum_{n\geq n_0} d_n W(\xi)^n$$

with the series from Theorem 1.6.3 as a $\mathfrak{P}$-adic limit.
Furthermore, let $R = R_1 \subseteq R_2 \subseteq \dots$ be a sequence of valuation rings with prime elements π_i satisfying

$$\pi_{k+1}^{e_k} \sim \pi_k \text{ with } e_k \in \mathbb{N},\ e_k \geq 2 \quad \text{for almost all } k.$$

For a suitable sequence $\eta_k \in \mathbb{C}$ let

$$W(\eta_k) \in \pi_k R_k \text{ for all } k.$$

Then the coefficients d_n are all in R if all values $f(\eta_k)$ are integral algebraic over R for $\mathfrak{P}$.

Proof The $\mathfrak{P}$-adic covergence follows immediately by inserting the series for x and y into the representation of f in the proof of Theorem 1.6.3.

It remains to prove that the coefficients d_n are in R under the above conditions on the $W(\eta_k)$. By Theorem 1.6.3 we have

$$d_n \in \frac{1}{r}R, \quad n \geq 0,$$

with an $r \in R \setminus \{0\}$ that is independent of n. Now, if π_1 does not divide any denominator of the coefficients d_n, we are done. Otherwise let π_1^l, $l > 0$, be the maximal power of π_1 dividing the denominator of some d_n. We chose the minimal $n = n_1$ with the denominator of d_{n_1} being associated with π_1^l. From $e_k \geq 2$ for almost all k it follows that for sufficiently large k the denominator of $d_n W(\eta_k)^n$ with $n \neq n_1$ and $d_n \neq 0$ is associated with a decidedly smaller power of π_k than the denominator of $d_{n_1} W(\eta_k)^{n_1}$, so by the strict triangular inequality, $f(\eta_k)$ cannot be integral over R – contrary to our assuming $f(\eta_k)$. This proves that all d_n are in R. □

Finally, we consider the relations between different pairs of Weierstrass functions and present some known formulae that can be found, for instance, in Silverman, "The Arithmic of Elliptic Curves" (1985). Let x, y be a pair of Weierstrass functions for the lattice $\mathfrak{L}$ satisfying the irreducible equation

$$\Phi(X,Y) = Y^2 + a_1XY + a_3Y - (X^3 + a_2X^2 + a_4X + a_6) = 0.$$

Then

$$\mathbb{C}(\wp, \wp') = \mathbb{C}(x,y) \cong Quot(\mathbb{C}[X,Y]/(\Phi(X,Y)))$$

is an algebraic function field over $\mathbb{C}$ of genus one. Let R be a valuation ring with maximal ideal $\mathfrak{p}$ and residue field k. We assume $\Phi(X,Y) \in R[X,Y]$, and by $\overline{\Phi}(X,Y)$ we denote the image of $\Phi(X,Y)$ under the natural map $\overline{} : R[X,Y] \to k[X,Y]$. We say that $\Phi(X,Y)$ has **good reduction modulo $\mathfrak{p}$** if

$$Quot(k[X,Y]/(\overline{\Phi}(X,Y)))$$

is again a function field of genus one over k, which is equivalent to the indivisibility of the discriminant Δ of $\Phi(X,Y)$ by $\mathfrak{p}$. For $a_1 = a_3 = 0$ the discriminant Δ is the discriminant of the cubic polynomial $x^3 + a_2X^2 + a_4X + a_6$ and in the general case Δ is defined as follows:

$$\Delta = -b_2^2 b_8 - 8b_4^3 - 27b_6^2 + 9b_2 b_4 b_6$$

with

$$\begin{aligned} b_2 &= a_1^2 + 4a_2, \\ b_4 &= 2a_4 + a_1 a_3, \\ b_6 &= a_3^2 + 4a_6, \\ b_8 &= a_1^2 a_6 + 4a_2 a_6 - a_1 a_3 a_4 + a_2 a_3^2 - a_4^2. \end{aligned}$$

The quantity

$$j := \frac{(b_2^2 - 24b_4)^3}{\Delta}$$

only depends on the underlying lattice $\mathfrak{L} = \mathbb{Z}\omega_1 + \mathbb{Z}\omega_2$, $\Im\left(\frac{\omega_1}{\omega_2}\right) > 0$, and is equal to the modular invariant $j\left(\frac{\omega_1}{\omega_2}\right)$ defined later in section 2.4.

Now let x, y and $\tilde{x}, \tilde{y}$ be two pairs of Weierstrass functions with respect to the lattice $\mathfrak{L}$. Then they are related by two equations of the form

$$\begin{aligned} x &= u^2\tilde{x} + r, \\ y &= u^3\tilde{y} + u^2 v\tilde{x} + t \end{aligned}$$

with coefficients $r, t, u, v \in \mathbb{C}, u \neq 0$. For the discriminants of the corresponding equations $\Phi(X, Y)$ and $\tilde{\Phi}(X, Y)$, we have

$$\Delta = u^{12}\tilde{\Delta}. \tag{1.12}$$

If, in particular, $\Phi(X, Y), \tilde{\Phi}(X, Y) \in R[X, Y]$ have good reduction modulo $\mathfrak{p}$, it follows that

$$u \sim 1 \text{ and } r, t, v \in R.$$

This implies:

Theorem 1.6.8 *Let $\Phi(X, Y), \tilde{\Phi}(X, Y) \in R[X, Y]$ be equations of two pairs x, y and $\tilde{x}, \tilde{y}$ of Weierstrass functions with good reduction modulo $\mathfrak{p}$. Then, in a neighbourhood of 0 the uniforming parameters $W = -\frac{x}{y}$, $\tilde{W} = -\frac{\tilde{x}}{\tilde{y}}$ are related by the series*

$$W = \sum_{n \geq 1} r_n \tilde{W}^n,$$

with coefficients $r_n \in R$ *and* $r_1 \sim 1$.
Further, for every $\xi \in \mathbb{C}$ *with* $\tilde{W}(\xi) \in \mathfrak{p}$ *the expansion*

$$W(\xi) = \sum_{n \geq 1} r_n \tilde{W}(\xi)^n$$

is also valid as a $\mathfrak{p}$*-adic limit.*

Proof According to the above relation we have

$$W = \frac{u^2\tilde{W} - \frac{r}{\tilde{y}}}{u^3 - u^2 v \tilde{W} + \frac{t}{\tilde{y}}},$$

with a unit u. Inserting the expansion of $\frac{1}{\tilde{y}}$ from Theorems 1.6.2 and 1.6.5, gives us the desired expansion. □

1.7 Elliptic resolvents

We follow the exposition in Schertz (1999). Let $\hat{\mathfrak{L}} \supseteq \mathfrak{L}$ be two complex lattices. Then for $\xi \in \hat{\mathfrak{L}}$ the map

$$\tau_\xi : g(z) \mapsto g(z + \xi)$$

defines an automorphism of $\mathbb{C}_{\mathfrak{L}}/\mathbb{C}_{\hat{\mathfrak{L}}}$ depending on ξ only modulo $\hat{\mathfrak{L}}$. Clearly, $\mathbb{C}_{\hat{\mathfrak{L}}}$ is the fixed field of all these automorphisms. Hence, $\mathbb{C}_{\mathfrak{L}}/\mathbb{C}_{\hat{\mathfrak{L}}}$ is a Galois extension with Galois group

$$\mathfrak{G} = \{\tau_\xi | \xi \in \hat{\mathfrak{L}}\} \cong \hat{\mathfrak{L}}/\mathfrak{L}.$$

We are now going to construct suitable resolvents for the extension $\mathbb{C}_{\mathfrak{L}}/\mathbb{C}_{\hat{\mathfrak{L}}}$. Therefore, in order to distinguish between elliptic functions for different lattices, we will use the notation $f(z|\mathfrak{L})$ and $l_{\mathfrak{L}}(u, w)$ rather than $f(z)$ for $f = \sigma, \zeta, \wp, \ldots$ and $l(u, w)$.

For a complex lattice we consider Klein's normalisation of the Weierstrass σ function

$$\varphi(z|\mathfrak{L}) := \sigma^*(z|\mathfrak{L}) \sqrt[12]{\Delta(\mathfrak{L})}, \tag{1.13}$$

with some 12-th root of $\Delta(\mathfrak{L})$. We chose an arbitrary $\gamma \in \mathbb{C} \setminus \mathfrak{L}$ and define

$$h_\gamma(z|\mathfrak{L}) := e^{-\frac{1}{2} l_{\mathfrak{L}}(z,\gamma)} \frac{\varphi(z + \gamma|\mathfrak{L})}{\varphi(z|\mathfrak{L})} \quad \text{with} \quad l_{\mathfrak{L}}(z, \gamma) = z\gamma^* - z^*\gamma. \tag{1.14}$$

Here the 12-th root of $\Delta(\mathfrak{L})$ is of no importance. Later in the applications it will come into play. The function h_γ has the following properties:

$$h_\gamma(z|\mathfrak{L}) \quad \text{is meromorphic in } \mathbb{C}, \tag{1.15}$$

$$e^{l_{\mathfrak{L}}(z,\gamma)}h_\gamma(z|\mathfrak{L}) \;\; \text{is periodic with respect to } \mathfrak{L}, \tag{1.16}$$

$$\begin{gathered} h_\gamma(z|\mathfrak{L}\bar{\nu}^{-1}) = h_{\gamma\nu}(z\nu|\mathfrak{L}) = \epsilon(\nu)h_\gamma(z\nu|\mathfrak{L}) \text{ with} \\ \epsilon(\nu) = \psi_{\mathfrak{L}}(\gamma(\nu-1))e^{\frac{1}{2}N(\gamma)l_{\mathfrak{L}}(1,\nu)} \text{ for } \nu \equiv 1 \bmod \tfrac{1}{\gamma}\mathfrak{L}, \end{gathered} \tag{1.17}$$

where the bar denotes complex conjugation and $N(.)$ the complex norm. The proof of (1.15) to (1.17) follows immediately from the transformation formula of Theorem 1.2.4 using Lemma 1.2.6. For the following we need:

Lemma 1.7.1 *Let $\mathfrak{L} \subset \hat{\mathfrak{L}}$ be complex lattices and $u, w \in \mathbb{C}$. Then*

$$l_{\hat{\mathfrak{L}}}(u,w) = [\hat{\mathfrak{L}} : \mathfrak{L}]l_{\mathfrak{L}}(u,w).$$

Proof The assertion follows from Lemma 1.2.6, keeping in mind that the coordinates of u and w with respect to the basis $\hat{\omega}_1, \hat{\omega}_2$ with $\Im(\frac{\hat{\omega}_1}{\hat{\omega}_2}) > 0$ are given by

$$(u_1,u_2)B \text{ and } (w_1,w_2)B,$$

where B as a matrix transforming a basis of $\hat{\mathfrak{L}}$ into a basis of $\mathfrak{L}$ has the determinant $[\hat{\mathfrak{L}} : \mathfrak{L}]$. □

For the following, let $\hat{\mathfrak{L}} \supset \mathfrak{L}$ be two fixed complex lattices with index

$$[\hat{\mathfrak{L}} : \mathfrak{L}] = n,$$

and let χ be a character of $\hat{\mathfrak{L}}/\mathfrak{L}$. By the following theorem we will obtain suitable resolvents for the extensions $\mathbb{C}_{\mathfrak{L}}/\mathbb{C}_{\hat{\mathfrak{L}}}$.

Theorem 1.7.2

$$\sum_{\substack{\xi \in \hat{\mathfrak{L}} \\ \xi \bmod \mathfrak{L}}} e^{l_{\mathfrak{L}}(\xi,\gamma)}h_\gamma(z+\xi|\mathfrak{L})\bar{\chi}(\xi) = \sqrt[12]{\frac{\Delta(\hat{\mathfrak{L}})}{\Delta(\mathfrak{L})}}C_\chi(z) \text{ with}$$

$$C_\chi(z) = e^{-\frac{1}{2}l_{\hat{\mathfrak{L}}}(z,\frac{\gamma}{n}+\mu_\chi)}\frac{\varphi(\gamma|\mathfrak{L})\varphi(z+\frac{\gamma}{n}+\mu_\chi|\hat{\mathfrak{L}})}{\varphi(z|\hat{\mathfrak{L}})\varphi(\frac{\gamma}{n}+\mu_\chi|\hat{\mathfrak{L}})},$$

and an isomorphism

$$\mu: \; X(\hat{\mathfrak{L}}/\mathfrak{L}) \to (\tfrac{1}{n}\mathfrak{L})/\hat{\mathfrak{L}}, \quad \chi \mapsto \mu_\chi.$$

$X(\hat{\mathfrak{L}}/\mathfrak{L})$ denotes the group of characters of $\hat{\mathfrak{L}}/\mathfrak{L}$.

Note that $C_\chi(z)$ depends only on the class of μ_χ modulo $\hat{\mathfrak{L}}$.

Proof We set

$$R_\chi(z) := \sum_{\substack{\xi \in \hat{\mathfrak{L}} \\ \xi \bmod \mathfrak{L}}} e^{l_{\mathfrak{L}}(\xi,\gamma)} h_\gamma(z+\xi|\mathfrak{L})\bar{\chi}(\xi).$$

For $\omega \in \hat{\mathfrak{L}}$ we obtain from (1.16) the transformation formula

$$R_\chi(z+\omega) = \chi(\xi) e^{-l_{\mathfrak{L}}(\omega,\gamma)} R_\chi(z).$$

By the determinant formula in the proof of Lemma 1.2.6 we see that there exists a $\mu_\chi \in \frac{1}{n}\mathfrak{L}$, uniquely determined modulo $\hat{\mathfrak{L}}$, such that

$$\chi(\xi) = e^{l_{\hat{\mathfrak{L}}}(\xi,-\mu_\chi)} \text{ for all } \xi \in \hat{\mathfrak{L}}.$$

Using the rules of Lemma 1.2.6, the transformaton formula for R_χ then becomes:

$$R_\chi(z+\omega) = e^{-l_{\hat{\mathfrak{L}}}(\omega,\delta_\chi)} R_\chi(z) \text{ with } \delta_\chi = \mu_\chi + \frac{\gamma}{n}.$$

According to Lemma 1.7.1 the same transformation formula holds for $h_{\delta_\chi}(z|\hat{\mathfrak{L}})$. Hence $R_\chi(z) h_{\delta_\chi}(z|\hat{\mathfrak{L}})^{-1}$ is an elliptic function with respect to $\hat{\mathfrak{L}}$. Moreover, this function has at most one pole of order 1 at $z \equiv -\delta_\chi$. Therefore, by Theorem 1.1.4 it must be a constant c that is equal to the limit:

$$c = \lim_{z\to 0} \frac{R_\chi(z)}{h_{\delta_\chi}(z|\hat{\mathfrak{L}})} = \frac{\varphi(\gamma|\mathfrak{L})}{\varphi(\delta_\chi|\hat{\mathfrak{L}})} \sqrt[12]{\frac{\Delta(\hat{\mathfrak{L}})}{\Delta(\mathfrak{L})}}.$$

This proves the formula of our theorem. □

For later applications the formula of Theorem 1.7.2 needs a modification, for in general neither $h_\gamma(z|\mathfrak{L})$ is in $\mathbb{C}_{\mathfrak{L}}$ nor is the factor $e^{l_{\mathfrak{L}}(\xi,\gamma)}$ a character of $\hat{\mathfrak{L}}/\mathfrak{L}$. Therefore, R_χ cannot be a resolvent for the extension $\mathbb{C}_{\mathfrak{L}}/\mathbb{C}_{\hat{\mathfrak{L}}}$. We choose one of the various modifications described in Schertz (1999): Let $\omega \in \mathbb{C}$ have the properties

$$(\overline{\omega}-1)\gamma \in \mathfrak{L}$$

and

$$\omega\hat{\mathfrak{L}} \subseteq \mathfrak{L}.$$

We define the function

$$G(z) := G(z|\mathfrak{L}) := \frac{h_\gamma(z|\mathfrak{L})}{h_\gamma(\omega z|\mathfrak{L})},$$

and by (1.15) to (1.17) we see that G is elliptic with respect to $\mathfrak{L}$. Further,

$$\chi_0(\xi) = e^{l_{\mathfrak{L}}(\xi(1-\omega),\gamma)}$$

defines a character of $\hat{\mathfrak{L}}/\mathfrak{L}$, and by (1.16) we obtain

$$G(z+\xi|\mathfrak{L}) = \chi_0(\xi)e^{l_{\mathfrak{L}}(\xi,\gamma)}\frac{h_\gamma(z+\xi|\mathfrak{L})}{h_\gamma(\omega z|\mathfrak{L})}.$$

Theorem 1.7.2 now yields:

Theorem 1.7.3

$$(G(z|\mathfrak{L}),\chi) := \sum_{\substack{\xi\in\hat{\mathfrak{L}} \\ \xi \bmod \mathfrak{L}}} G(z+\xi|\mathfrak{L})\bar{\chi}(\xi) = \sqrt[12]{\frac{\Delta(\hat{\mathfrak{L}})}{\Delta(\mathfrak{L})}}C_{\chi\bar{\chi}_0}(z)h_\gamma^{-1}(\omega z|\mathfrak{L}).$$

For the analytic interpretation of Theorem 1.7.3 we observe that $\mathbb{C}_{\mathfrak{L}}$ is a $\mathbb{C}_{\hat{\mathfrak{L}}}[\mathfrak{G}]$ module with respect to

$$G(z|\mathfrak{L}) \circ \sum_{[\xi]\in\hat{\mathfrak{L}}/\mathfrak{L}} a_\xi(z)[\xi] := \sum_{[\xi]\in\hat{\mathfrak{L}}/\mathfrak{L}} a_\xi(z)G(z+\xi|\mathfrak{L}).$$

Therefore, the resolvent in Theorem 1.7.3 is a Galois resolvent for the extension $\mathbb{C}_{\mathfrak{L}}/\mathbb{C}_{\hat{\mathfrak{L}}}$. For a finite subset $S \subset \mathbb{C}/\mathfrak{L}$ we let $\mathbb{O}_{\hat{\mathfrak{L}}}$ and $\mathbb{O}_{\mathfrak{L}}$ be the subrings of functions in $\mathbb{C}_{\hat{\mathfrak{L}}}$ and $\mathbb{C}_{\mathfrak{L}}$ having no poles outside S. For S that is large enough the functions C_χ are all units in $\mathbb{O}_{\hat{\mathfrak{L}}}$ and it follows that

$$\mathbb{O}_{\mathfrak{L}} = G \circ \mathbb{O}_{\hat{\mathfrak{L}}}[\mathfrak{G}],$$

because the discriminant of the set $G \circ \mathfrak{G}$ is a unit in $\mathbb{O}_{\hat{\mathfrak{L}}}$.

To finish, we will establish a kind of resolvent formula for the ζ^* function, too. For this purpose we must distinguish between the principal and the non-principal characters of $\hat{\mathfrak{L}}/\mathfrak{L}$. First, we normalise ζ^* in analogy to σ^* by

$$Z(z|\mathfrak{L}) := \frac{\zeta^*(z|\mathfrak{L})}{\sqrt[12]{\Delta(\mathfrak{L})}}.$$

Theorem 1.7.4

$$\sum_{\substack{\xi\in\hat{\mathfrak{L}}\\ \xi \bmod \mathfrak{L}}} Z(z+\xi|\mathfrak{L}) = \sqrt[12]{\frac{\Delta(\hat{\mathfrak{L}})}{\Delta(\mathfrak{L})}}\, Z(z|\hat{\mathfrak{L}}).$$

Proof The Laurent expansion of $\wp'$ gives us directly

$$\sum_{\substack{\xi\in\hat{\mathfrak{L}}\\ \xi \bmod \mathfrak{L}}} \wp'(z+\xi|\mathfrak{L}) = \wp'(z|\hat{\mathfrak{L}}),$$

and, integrating twice,

$$\sum_{\substack{\xi\in\hat{\mathfrak{L}}\\ \xi \bmod \mathfrak{L}}} \zeta(z+\xi|\mathfrak{L}) = \zeta(z|\hat{\mathfrak{L}}) + Az + B$$

with constants A, B. Replacing ζ by ζ^* this relation becomes

$$\sum_{\substack{\xi\in\hat{\mathfrak{L}}\\ \xi \bmod \mathfrak{L}}} \zeta^*(z+\xi|\mathfrak{L}) = \zeta^*(z|\hat{\mathfrak{L}}) + g(z)$$

with $g(z) = \eta(z|\hat{\mathfrak{L}}) - \sum_{\xi} \eta(z+\xi|\mathfrak{L}) - (Az+B)$.

Herein g is an affine linear function that is periodic with respect to $\mathfrak{L}$. Thus, g must be a constant that is equal to zero because ζ^* is an odd function. This implies the assertion of Theorem 1.7.4. □

Theorem 1.7.5 *Let χ be a non-principal character of $\hat{\mathfrak{L}}/\mathfrak{L}$. Then there exists a $\delta_\chi \in \frac{1}{n}\mathfrak{L}\setminus\hat{\mathfrak{L}}$ such that $\chi(\xi) = e^{-l_{\hat{\mathfrak{L}}}(\xi,\delta_\chi)}$ for all $\xi \in \hat{\mathfrak{L}}$ and*

$$\sum_{\substack{\xi\in\hat{\mathfrak{L}}\\ \xi \bmod \mathfrak{L}}} Z(z+\xi|\mathfrak{L})\bar{\chi}(\xi) = e^{-\frac{1}{2}l_{\hat{\mathfrak{L}}}(z,\delta_\chi)}\sqrt[12]{\frac{\Delta(\hat{\mathfrak{L}})}{\Delta(\mathfrak{L})}\frac{\varphi(z+\delta_\chi|\hat{\mathfrak{L}})}{\varphi(z|\hat{\mathfrak{L}})\varphi(\delta_\chi|\hat{\mathfrak{L}})}}.$$

Proof Using the relation $\sum_{\xi} \chi(\xi) = 0$, we find

$$R_\chi(z) := \sum_{\substack{\xi\in\hat{\mathfrak{L}}\\ \xi \bmod \mathfrak{L}}} \zeta^*(z+\xi|\mathfrak{L})\bar{\chi}(\xi)$$

$$= \sum_{\substack{\xi\in\hat{\mathfrak{L}}\\ \xi \bmod \mathfrak{L}}} \zeta(z+\xi|\mathfrak{L})\bar{\chi}(\xi) - \sum_{\substack{\xi\in\hat{\mathfrak{L}}\\ \xi \bmod \mathfrak{L}}} \xi^*\chi(\xi).$$

So $R_\chi(z)$ is a meromorphic function satsifying the transformation formula $R_\chi(z+\omega) = R_\chi(z)\chi(\omega)$ for all $\omega \in \hat{\mathfrak{L}}$. The proof is now completed in analogy to the proof of Theorem 1.7.4. □

1.8 *q*-expansions

In view of homogeneity,

$$\begin{aligned}
\sigma(z\lambda|\mathfrak{L}\lambda) &= \sigma(z|\mathfrak{L})\lambda, \\
\zeta(z\lambda|\mathfrak{L}\lambda) &= \zeta(z|\mathfrak{L})\lambda^{-1}, \\
\wp(z\lambda|\mathfrak{L}\lambda) &= \wp(z|\mathfrak{L})\lambda^{-2}, \\
\wp'(z\lambda|\mathfrak{L}\lambda) &= \wp'(z|\mathfrak{L})\lambda^{-3}, \\
G_m(\mathfrak{L}\lambda) &= G_m(\mathfrak{L})\lambda^{-2m},
\end{aligned}$$

it suffices to consider lattices of the form $\mathfrak{L} = \mathbb{Z}\omega + \mathbb{Z},\ \Im(\omega) > 0$ in the following. Here, for simplification, we write

$$f(z|\omega) := f(z|\mathbb{Z}\omega + \mathbb{Z}).$$

The source of all formulae derived in this section is the following theorem:

Theorem 1.8.1 (q-expansion of $\sigma(z|\omega)$) *For $\omega \in \mathbb{C}, \Im(\omega) > 0$, and $z \in \mathbb{C}$ we set $q = e^{2\pi i\omega}$, $Q = e^{2\pi iz}, Q^{\pm\frac{1}{2}} = e^{\pm\pi iz}$. Further, let $\eta_2 = \eta(1|\omega)$ be the quasi-period belonging to 1. Then we have*

$$\sigma(z|\omega) = \frac{1}{2\pi i}e^{\frac{1}{2}\eta_2 z^2}(Q^{\frac{1}{2}} - Q^{-\frac{1}{2}})\prod_{n=1}^{\infty}\frac{(1-q^nQ)(1-q^nQ^{-1})}{(1-q^n)^2}.$$

Proof The function

$$g(z) = e^{\frac{1}{2}\eta_2 z^2}(Q^{\frac{1}{2}} - Q^{-\frac{1}{2}})\prod_{n=1}^{\infty}(1-q^nQ)(1-q^nQ^{-1})$$

is holomorphic in $\mathbb{C}$ having the same zeros as $\sigma(z|\omega)$. Substituing $z+1$ resp. $z+\omega$ for z we obtain $-Q^{\pm\frac{1}{2}}$ resp. $q^{\pm\frac{1}{2}}Q^{\pm\frac{1}{2}}$ for $Q^{\pm\frac{1}{2}}$ and, using the Legendre relation $\omega\eta_2 - \eta_1 = 2\pi i$, we obtain the transformation formulae

$$\begin{aligned}
g(z+1) &= -e^{\eta_2(z+\frac{1}{2})}g(z), \\
g(z+\omega) &= -e^{-\eta_1(z+\frac{\omega}{2})}g(z).
\end{aligned}$$

The same transformation formulae holding for $\sigma(z|\omega)$ according to Theorem 1.2.3, we can conclude that the quotient $\frac{\sigma(z)}{g(z)}$ is a holomorphic elliptic function, hence equal to a constant by Theorem 1.1.2. The limit for $z \to 0$ determines the constant and finishes the proof of Theorem 1.8.1. □

Taking logarithmic derivatives in the formula of Theorem 1.8.1, we obtain the q-expansion of the ζ function :

$$\begin{aligned}\zeta(z|\omega) &= \eta_2 z + \pi i \frac{Q^{\frac{1}{2}} + Q^{-\frac{1}{2}}}{Q^{\frac{1}{2}} - Q^{-\frac{1}{2}}} + 2\pi i \sum_{n=1}^{\infty} \left[\frac{q^n Q^{-1}}{1 - q^n Q^{-1}} - \frac{q^n Q}{1 - q^n Q} \right] \\ &= \eta_2 z + \pi i + 2\pi i \frac{1}{Q-1} + 2\pi i \sum_{m,n \geq 1} q^{nm} (Q^{-m} - Q^m) \quad (1.18)\end{aligned}$$

To determine η_2, we use the generating relation for the Bernoulli numbers B_n

$$\frac{1}{e^u - 1} = \sum_{n=1}^{\infty} \frac{B_n}{n!} u^{n-1} = \frac{1}{u} - \frac{1}{2} + \sum_{l=1}^{\infty} \frac{B_{2l}}{(2l)!} u^{2l-1} = \frac{1}{u} - \frac{1}{2} + \frac{1}{12} u - \frac{1}{720} u^3 \ldots \quad (1.19)$$

to replace the term $\frac{1}{Q-1}$. Thus (1.18) yields the expansion into powers of z:

$$\begin{aligned}\zeta(z|\omega) = \frac{1}{z} &+ \left(\eta_2 + (2\pi i)^2 \left(\frac{1}{12} - 2 \sum_{m=1}^{\infty} \frac{mq^m}{1 - q^m} \right) \right) z \\ &+ \sum_{l=2}^{\infty} \frac{(2\pi i)^{2l}}{(2l-1)!} \left(\frac{B_{2l}}{2l} - 2 \sum_{m=1}^{\infty} \frac{m^{2l-1} q^m}{1 - q^m} \right) z^{2l-1}.\end{aligned}$$

On the other hand, analogous to Lemma 1.4.2,

$$\zeta(z|\omega) = \frac{1}{z} - G_2(\omega) z^3 - G_3(\omega) z^5 - \cdots \quad (1.20)$$

Now, comparing coefficients leads us to the q-expasion of η_2:

$$\eta_2 = \frac{(2\pi i)^2}{12} \left(-1 + 24 \sum_{m=1}^{\infty} \frac{mq^m}{1 - q^m} \right).$$

By inserting this formula into (1.18), we obtain the following q-expansion for $\zeta(z|\omega)$. Finally, by twice differentiating this formula we obtain the q-expansion for $\wp(z|\omega)$ and $\wp'(z|\omega)$:

Theorem 1.8.2 (q-expansion of $\zeta(z|\omega), \wp(z|\omega), \wp'(z|\omega)$)

$$\begin{aligned}
\zeta(z|\omega) &= \frac{(2\pi i)^2}{12}\left(-1+24\sum_{m=1}^{\infty}\frac{mq^m}{1-q^m}\right)z+\pi i+2\pi i\frac{1}{Q-1} \\
&\quad +2\pi i\sum_{m=1}^{\infty}\left(\frac{q^mQ}{1-q^mQ}-\frac{q^mQ^{-1}}{1-q^mQ^{-1}}\right), \\
\wp(z|\omega) &= (2\pi i)^2\left[\frac{1}{12}+\frac{Q}{(1-Q)^2}+\sum_{m=1}^{\infty}\left(\frac{q^mQ}{(1-q^mQ)^2}+\frac{q^mQ^{-1}}{(1-q^mQ^{-1})^2}\right.\right. \\
&\qquad \left.\left. -2\frac{mq^m}{1-q^m}\right)\right], \\
\wp'(z|\omega) &= (2\pi i)^3\left[\frac{Q(1+Q)}{(1-Q)^3}+\sum_{n=1}^{\infty}\left(\frac{q^mQ(1+q^mQ)}{(1-q^mQ)^3}+\frac{q^mQ^{-1}(1+q^mQ^{-1})}{(1-q^mQ^{-1})^3}\right)\right].
\end{aligned}$$

The q-expansions of the ζ^* function the σ^* function follow from Theorem 1.8.2 using Legendre's relation:

Theorem 1.8.3 (q-expansion of $\zeta^*(z|\omega), \sigma^*(z|\omega)$) *For $z = z_1\omega + z_2$ we have*

$$\begin{aligned}
\zeta^*(z|\omega) &= 2\pi i\left[z_1+\frac{1}{2}\frac{Q^{\frac{1}{2}}+Q^{-\frac{1}{2}}}{Q^{\frac{1}{2}}-Q^{-\frac{1}{2}}}+\sum_{n=1}^{\infty}\left(\frac{q^nQ^{-1}}{1-q^nQ^{-1}}-\frac{q^nQ}{1-q^nQ}\right)\right], \\
\sigma^*(z|\omega) &= \frac{1}{2\pi i}\left[Q^{\frac{1}{2}z_1}(Q^{\frac{1}{2}}-Q^{-\frac{1}{2}})\prod_{n=1}^{\infty}\frac{(1-q^nQ)(1-q^nQ^{-1})}{(1-q^n)^2}\right].
\end{aligned}$$

Further, by comparing (1.19) and (1.20), we find the q-expansion of the Eisenstein series:

Theorem 1.8.4 (q-expansion of the Eisenstein series) *For $l \geq 2$ we have*

$$\begin{aligned}
G_l(\omega) &= -\frac{(2\pi i)^{2l}}{(2l-1)!}\left(\frac{B_{2l}}{2l}-2\sum_{m=1}^{\infty}\frac{m^{2l-1}q^m}{1-q^m}\right) \\
&= -\frac{(2\pi i)^{2l}}{(2l-1)!}\left(\frac{B_{2l}}{2l}-2\sum_{n=1}^{\infty}\sigma_{2l-1}(n)q^n\right).
\end{aligned}$$

with

$$\sigma_k(n) := \sum_{0<d|n} d^k.$$

In particular, we have thus derived the q-expansion of g_2, g_3 and the discriminant Δ in Theorem 1.4.1:

Theorem 1.8.5 (q-expansion of g_2, g_3)

$$g_2(\omega) = \frac{(2\pi i)^4}{2^2 3}\left[1 + 240T_3\right],$$
$$g_3(\omega) = \frac{(2\pi i)^6}{2^3 3^3}\left[-1 + 504T_5\right]$$

with

$$T_k := \sum_{n=1}^{\infty} \sigma_k(n) q^n.$$

Theorem 1.8.5 implies

Theorem 1.8.6 (q-expansion of Δ)

$$\Delta(\omega) = (2\pi i)^{12} \sum_{n=1}^{\infty} \tau(n) q^n$$

with

$$\tau(n) \in \mathbb{Z} \text{ for all } n \in \mathbb{N} \text{ and } \tau(1) = 1.$$

Proof According to Theorem 1.8.5 we have

$$\begin{aligned} \tfrac{1}{(2\pi i)^{12}}\left(g_2(\omega)^3 - 27 g_3(\omega)^2\right) &= \tfrac{1}{2^6 3^3}\left\{[1 + 240T_3]^3 - [-1 + 504T_5]^2\right\} \\ &\equiv \tfrac{1}{12}\left\{5T_3 + 7T_5\right\} \mod q^2\mathbb{Z}[[q]]. \end{aligned}$$

Herein

$$5T_3 + 7T_5 = 12q + \sum_{n=2}^{\infty}\left(\sum_{0<d|n} d^3(5 + 7d^2)\right) q^n,$$

and because $d^3(5 + 7d^2)$ is divisible by 12 for all $d \in \mathbb{N}$, this implies the assertion of Theorem 1.8.6. A further proof of Theorem 1.8.6 can be found in the next section. □

1.9 Dedekind's η function and σ-product formula

The function defined in the upper half plane $\Im(\omega) > 0$ by the infinite product

$$\eta(\omega) := q^{\frac{1}{24}} \prod_{n=1}^{\infty}(1 - q^n) \qquad \text{with } q = e^{2\pi i \omega}, q^{\frac{1}{24}} = e^{\frac{2\pi i \omega}{24}},$$

is called the **Dedekind** η **function**. We emphasise that this function has nothing to do with the quasi-periods of the elliptic ζ function in Theorem 1.2.3.

Theorem 1.9.1 (product formula of the σ^* function) *Let* $\mathfrak{L} = [\omega, 1], \hat{\mathfrak{L}} = [\frac{\omega}{n_1}, \frac{1}{n_2}]$ *be complex lattices,* $\Im(\omega) > 0,\ n_1, n_2 \in \mathbb{N}$. *We fix the following system of representatives*

$$\xi = \frac{x\omega}{n_1} + \frac{y}{n_2}, \quad x = 0, \ldots, n_1 - 1, \ y = 0, \ldots, n_2 - 1,$$

for the residue classes of $\hat{\mathfrak{L}}/\mathfrak{L}$. *Then*

$$\prod_{\xi} (2\pi i) e^{-\frac{1}{2} l_{\mathfrak{L}}(z,\xi)} \sigma^*(z + \xi|\mathfrak{L}) \eta(\omega)^2 = \zeta \cdot (2\pi i) \sigma^*(z|\hat{\mathfrak{L}}) \eta \left(\frac{n_2 \omega}{n_1} \right)^2 n_2$$

with

$$\zeta = -\zeta_4^{n_1 n_2 + n_1} \zeta_8^{(n_1 - 1)(n_2 - 1)}.$$

Dividing the left-hand side by the right-hand side and taking the limit for $z \to 0$ *yields*

$$\prod_{\xi \neq 0} (2\pi i) \sigma^*(\xi|\mathfrak{L}) = \zeta \frac{\eta(\frac{n_1 \omega}{n_2})^2 n_2}{\eta(\omega)^{2 n_1 n_2}}.$$

Proof The formula is derived via the product expansions of the functions involved, using the following notation:

$$Q_w := e^{2\pi i w}, \ Q_w^{\frac{1}{2}} = e^{\pi i w}, \ q = Q_\omega, \ \hat{q} = Q_{\frac{n_2 \omega}{n_1}}.$$

The q-expansion of $(2\pi i)\sigma^*(w|\mathfrak{L})\eta(\omega)^2$ being given by

$$(2\pi i)\sigma^*(w|\mathfrak{L})\eta(\omega)^2 = Q_w^{\frac{1}{2} w_1} (Q_w^{\frac{1}{2}} - Q_w^{-\frac{1}{2}}) q^{\frac{1}{12}} \prod_{n=1}^{\infty} (1 - q^n Q_w)(1 - q^n Q_w^{-1}),$$

the product in Theorem 1.6.4 is of the form

$$\prod_{\xi} e^{-\frac{1}{2} l_{\mathfrak{L}}(z,\xi)} (2\pi i) \sigma^*(z + \xi|\mathfrak{L}) \eta(\omega)^2 = f_1 f_2 f_3$$

with

$$\begin{aligned} f_1 &= e^{\frac{2\pi i}{2}(\sum_{x,y} (z+\xi)(z_1+\xi_1) - z_1 \xi_2 + z_2 \xi_1)}, \\ f_2 &= \prod_{x,y} q^{\frac{1}{12}} (Q_{z+\xi}^{\frac{1}{2}} - Q_{z+\xi}^{-\frac{1}{2}}), \\ f_3 &= \prod_{n=1}^{\infty} \prod_{x,y} (1 - q^n Q_{z+\xi})(1 - q^n Q_{z+\xi}^{-1}). \end{aligned}$$

Using the formulae $\sum_{k=1}^{m-1} k = \frac{m(m-1)}{2}$ and $\sum_{k=1}^{m-1} k^2 = \frac{m(m-1)(2m-1)}{6}$, we now compute

$$f_1 = \zeta_8^{(n_1-1)(n_2-1)} Q_{n_2 z}^{\frac{n_1 z_1}{2}} Q_{n_2 z}^{\frac{n_1-1}{2}} \hat{q}^{\frac{(n_1-1)(2n_1-1)}{12}}, \quad \hat{q} = q^{\frac{n_1}{n_2}}.$$

Further, in view of the identity $\prod_{y=1}^{n_2} (a - b\zeta_{n_2}^y) = a^{n_2} - b^{n_2}$, we find that

$$f_2 = -\zeta_4^{n_1 n_2 + n_1} Q_{n_2 z}^{-\frac{n_1-1}{2}} \hat{q}^{n_1 n_2 - \frac{n_1(n_1-1)}{4}} \prod_{x=1}^{n_1-1} (1 - \hat{q}^x Q_{n_2 z})$$

and in the same way

$$f_3 = \left(\prod_{k=n_1}^{\infty} (1 - \hat{q}^k Q_{n_2 z}) \right) \left(\prod_{k=1}^{\infty} (1 - \hat{q}^k Q_{n_2 z}^{-1}) \right).$$

Putting together the identities for f_1, f_2, f_3 and, appealing to the identity $\sigma^*(n_2 z | \frac{n_2 \omega}{n_1}) = \sigma^*(z|\hat{\mathfrak{L}}) n_2$ following from the homogeneity of the σ function, we end up with the formula asserted in Theorem 1.9.1. □

An application of Theorem 1.9.1 is:

Theorem 1.9.2 *For $\Im(\omega) > 0$ we have*

$$\Delta(\omega) = (2\pi i)^{12} \eta(\omega)^{24}.$$

Proof According to Theorem 1.4.1 we have

$$\Delta(\omega) = 16(\wp(\xi_1) - \wp(\xi_2))^2 (\wp(\xi_1) - \wp(\xi_3))^2 (\wp(\xi_2) - \wp(\xi_3))^2$$

with the half periods

$$\xi_1 = \frac{\omega}{2}, \; \xi_2 = \frac{\omega + 1}{2}, \; \xi_3 = \frac{1}{2}$$

and the $\wp$ function for $\mathbb{Z}\omega + \mathbb{Z}$. By Theorem 1.3.2 and the definition of σ^* the differences of $\wp$ values can be written as

$$\wp(\xi_j) - \wp(\xi_k) = -\frac{\sigma^*(\xi_j + \xi_k)\sigma^*(\xi_j - \xi_k)}{\sigma^*(\xi_j)^2 \sigma^*(\xi_k)^2}$$

and, using the transformation formula for σ^*, we can write

$$\wp(\xi_1) - \wp(\xi_2) = i \frac{\sigma^*(\xi_3)^2}{\sigma^*(\xi_1)^2 \sigma^*(\xi_2)^2},$$

$$\wp(\xi_1) - \wp(\xi_3) = i\frac{\sigma^*(\xi_2)^2}{\sigma^*(\xi_1)^2\sigma^*(\xi_3)^2},$$
$$\wp(\xi_2) - \wp(\xi_3) = \frac{\sigma^*(\xi_1)^2}{\sigma^*(\xi_1)^2\sigma^*(\xi_3)^2}.$$

Thus

$$\Delta(\omega) = \frac{-16}{(\sigma^*(\xi_1)\sigma^*(\xi_2)\sigma^*(\xi_3))^4}.$$

The identity of Theorem 1.9.2 now follows from Theorem 1.9.1 with $n_1 = n_2 = 2$. □

Using Theorem 1.9.2, the formula of Theorem 1.9.1 can be rewritten in an invariant form by taking 12-th powers:

Theorem 1.9.3 *Let $\mathfrak{L} \subseteq \hat{\mathfrak{L}}$ be complex lattices. Then for any system $\{\xi\}$ of representatives for the residues of $\hat{\mathfrak{L}}/\mathfrak{L}$ we have*

$$\prod_{\xi} e^{-6l_{\mathfrak{L}}(z,\xi)}\varphi(z+\xi|\mathfrak{L})^{12} = \zeta\varphi(z|\hat{\mathfrak{L}})^{12}$$

and

$$\prod_{\xi \not\equiv 0 \text{ mod } \mathfrak{L}} \varphi(\xi|\mathfrak{L})^{12} = \zeta\frac{\Delta(\hat{\mathfrak{L}})}{\Delta(\mathfrak{L})}$$

with some $[\hat{\mathfrak{L}} : \mathfrak{L}]$-th root of unity ζ dependent on the system $\{\xi\}$.

1.10 The transformation formula of the Dedekind η function

In the following let $M = \left(\begin{smallmatrix} a & b \\ c & d \end{smallmatrix}\right)$ be a unimodular matrix, i.e.

$$a, b, c, d \in \mathbb{Z} \text{ and } \det(M) = ad - bc = 1.$$

We abuse notation by taking the same letter for the linear transformation

$$Mz = \frac{az+b}{cz+d}.$$

We recall that M transforms a basis ω_1, ω_2 of a lattice $\mathfrak{L} = [\omega_1, \omega_2] = [\omega, 1]\omega_2$, $\omega = \frac{\omega_1}{\omega_2}$, with $\Im(\omega) > 0$ into another such basis. Therefore, by homogeneity of Δ we obtain the relation

$$\Delta(M\omega)(c\omega + d)^{-12} = \Delta(\mathfrak{L}) = \Delta(\omega).$$

Taking 24-th roots and referring to Theorem 1.9.2, this implies that

$$\eta(M\omega) = \epsilon(M)\sqrt{c\omega + d}\ \eta(\omega)$$

with a 24-th root of unity $\epsilon(M)$ and $\Re(\sqrt{c\omega + d}) > 0$. According to Theorem 2.1.3 the group of unimodular matrices is generated by

$$S = \begin{pmatrix} 0 & -1 \\ 1 & 0 \end{pmatrix}, \quad T = \begin{pmatrix} 1 & 1 \\ 0 & 1 \end{pmatrix}.$$

It then suffices to determine $\epsilon(M)$ for these two matrices. $\epsilon(T) = \zeta_{24}$ follows directly from the definition of the η function. $\epsilon(S)$ can be deduced from the above formula, because for $\omega \in i\mathbb{R}^+$ the two values $\eta(\omega)$ and $\eta(\frac{-1}{\omega})$ are real and positive:

$$\epsilon(S) = \zeta_{24}^{-3}, \quad \epsilon(T) = \zeta_{24}. \tag{1.21}$$

C. Meyer (1957) has derived the following explicit formula for $\epsilon(M)$. Let M be normalised by

$$c \geq 0 \text{ and } d > 0 \text{ if } c = 0. \tag{1.22}$$

Then

$$\epsilon(M) = \begin{cases} (\frac{a}{c})\zeta_{24}^{ab+2ac-3c+cd(1-a^2)} & \text{if } c \equiv 1 \bmod 2, \\ (\frac{c}{|a|})\zeta_{24}^{ab-ac+3a-3+cd(1-a^2)} & \text{if } a \equiv 1 \bmod 2 \text{ and } c \neq 0, \\ \zeta_{24}^{b} & \text{if } c = 0, \end{cases} \tag{1.23}$$

where $(\frac{a}{c})$ and $(\frac{0}{c}) = 1$ denotes the Legendre symbol. To write $\epsilon(M)$ in all three cases by one formula, we define c_1 and $\lambda \in \mathbb{Z}$ by

$$c = c_1 2^\lambda \text{ with } c_1 \equiv 1 \bmod 2 \text{ if } c \neq 0,$$
$$c_1 = \lambda = 1 \text{ if } c = 0.$$

For $c \neq 0$ we have, according to the quadratic reciprocity law,

$$\left(\frac{c}{|a|}\right) = (-1)^{\frac{a-1}{2}\frac{c_1-1}{2} + \lambda\frac{a^2-1}{8}} \left(\frac{a}{c_1}\right).$$

Putting this into the second formula in (1.23), we obtain the following formula for $\epsilon(M)$, holding in all cases:

Theorem 1.10.1 *Let $M = \begin{pmatrix} a & b \\ c & d \end{pmatrix}$ be a unimodular matrix, normalised by (1.22). Then, with the above definition of c_1 and λ we have*

$$\epsilon(M) = (\tfrac{a}{c_1})\zeta_{24}^{ba+c(d(1-a^2)-a)+3(a-1)c_1+\lambda\frac{3}{2}(a^2-1)}.$$

Since all unimodular matrices are products of S and T, it is of course possible to deduce Theorem 1.10.1 from the formulae for $\epsilon(S)$ and $\epsilon(T)$ by showing that the formula holds for $\epsilon(SM)$ and $\epsilon(TM)$ if it is valid for $\epsilon(M)$. We omit the tedious verification.

2
Modular functions

Modular functions naturally come into play, when coefficients of elliptic curves and torsion points are studied. Let $\mathfrak{L} = \mathbb{Z}\omega_1 + \mathbb{Z}\omega_2$, $\Im\left(\frac{\omega_1}{\omega_2}\right) > 0$, be a lattice. Then, due to homogeneity, the j-invariant of the corresponding Weierstrass equation,

$$j = 12^3 \frac{g_2(\mathfrak{L})^3}{\Delta(\mathfrak{L})},$$

can be viewed as a function of $\omega := \frac{\omega_1}{\omega_2}$:

$$j = j(\omega).$$

Since j only depends on the lattice, j is invariant under all unimodular transformations $\omega \mapsto \frac{a\omega+b}{c\omega+d}, a, b, c, d \in \mathbb{Z}; ad - bc = 1$, which is the essential property of a modular function for the full modular group.

Similar functions are defined by the division values of the Weierstrass $\wp$ function and σ function, as, for example

$$f(\omega) := \wp\left(\frac{x\omega + y}{N}\,\middle|\,\mathbb{Z}\omega + \mathbb{Z}\right) \qquad \text{with fixed } x, y \in \mathbb{Z},\ N \in \mathbb{N}.$$

They are considered when studying the torsion points of elliptic curves.

In the following, we will explain all the things we need from modular functions. In particular, we will derive algebraic equations between different modular functions that are crucial for the algebraic properties of certain "singular values".

2.1 The modular group

The subgroup

$$\Gamma := \left\{ M = \begin{pmatrix} a & b \\ c & d \end{pmatrix} \middle| \; a, b, c, d \in \mathbb{Z}, \det(M) = 1 \right\}$$

of $Gl_2(\mathbb{C})$ is called the **modular group**. For $M = \left(\begin{smallmatrix} a & b \\ c & d \end{smallmatrix}\right) \in \Gamma$ the linear transformation

$$M(z) = \frac{az+b}{cz+d}$$

is a bijection both of the upper-half plane,

$$\mathbb{H} = \{z \in \mathbb{C} \mid \Im(z) > 0\}$$

and the extended upper-half plane

$$\mathbb{H}^* = \mathbb{H} \cup \mathbb{Q} \cup \{i\infty\},$$

which becomes evident by the formula

$$\Im(M(z)) = \frac{\Im(z) \det(M)}{|cz+d|^2}.$$

Further, for the action of Γ on $\mathbb{H}^*$ we have the rule

$$(MN)(z) = M(N(z)) \text{ for } M, N \in \Gamma.$$

So, for every subgroup U of Γ an equivalence relation is defined by

$$z' \underset{U}{\sim} z :\iff z' = M(z) \text{ for a } M \in U.$$

This leads us to the definition:

Definition 2.1.1 Let U be a subgroup of Γ. Then a set F_U of $\mathbb{H}^*$ is called a fundamental domain of U if

(i) every $z \in \mathbb{H}^*$ is equivalent modulo U to a point in F_U and
(ii) different inner points of F_U are not equivalent modulo U.

Clearly, every system of representatives of $\mathbb{H}^*$ modulo U is a fundamental domain, and, conversely, a fundamental domain becomes a system of representatives by taking away some points of the boundary. For $U = \Gamma$ a fundamental domain is given by:

Theorem 2.1.2 *The set*
$F := \{z \in \mathbb{H} \mid -\frac{1}{2} \le \Re(z) \le 0, |z| \ge 1\} \cup \{z \in \mathbb{H} \mid 0 < \Re(z) < \frac{1}{2}, |z| > 1\} \cup \{i\infty\}.$
is a fundamental domain for Γ*, and it is even a system of representatives.*

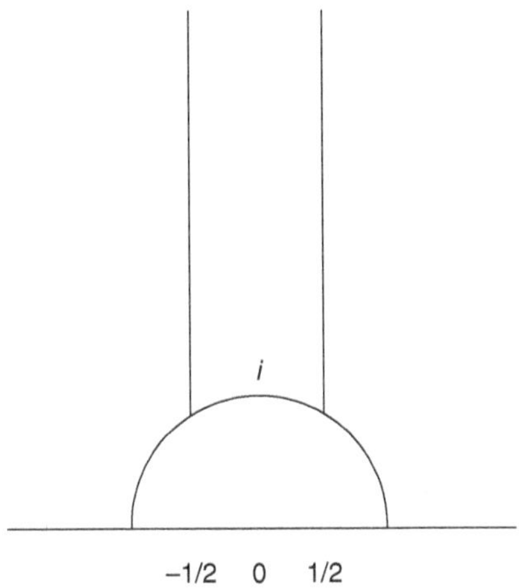

Proof First, we show that every point $z \in \mathbb{H}^* \setminus F$ is equivalent to a point in F. For $z = \frac{a}{b} \in \mathbb{Q}$ with $\gcd(a,b) = 1$ we can find $c, d \in \mathbb{Z}$ with $ac + bd = 1$. Then, $M = \left(\begin{smallmatrix} c & d \\ -b & a \end{smallmatrix}\right)$ is in Γ and $M(z) = i\infty \in F$. For $z \in \mathbb{H} \setminus F$ we show that z is mapped to F by a product of

$$S = \begin{pmatrix} 0 & -1 \\ 1 & 0 \end{pmatrix} \text{ and } T = \begin{pmatrix} 1 & 1 \\ 0 & 1 \end{pmatrix}.$$

Therefore, we consider the set

$$I_z := \{\Im(M(z)) \mid M \in < S, T >, \ \Im(M(z)) \ge \Im(z)\}.$$

I_z is finite because for $M = \left(\begin{smallmatrix} a & b \\ c & d \end{smallmatrix}\right) \in \Gamma$ the inequality $\Im(M(z)) \ge \Im(z)$ leads to $|cz + d| \le 1$, which for a fixed z can only be satisfied by a finite number of $c, d \in \mathbb{Z}$. Thus, there exists a $M \in < S, T >$ such that $\Im(M(z))$ is maximal. Further, by applying T we can achieve

$$-\frac{1}{2} \le \Re(M(z)) < \frac{1}{2}.$$

We contend that $|M(z)| \geq 1$, which then implies

$$M(z) \in F \text{ or } SM(z) \in F.$$

The proof of $|M(z)| \geq 1$ follows from the maximality of $\Im(M(z))$, because $|M(z)| < 1$ would imply $\Im(SM(z)) = \frac{\Im(M(z))}{|M(z)|^2} > \Im(M(z))$ in contradiction to the maximality of $\Im(M(z))$.

Now we show that the points of F are pairwise inequivalent, whereby in particular the second property of Definition 2.1.1 is proved for F. Let $z, z_1 \in F$ be two different points and $M = \left(\begin{smallmatrix} a & b \\ c & d \end{smallmatrix}\right) \in \Gamma$ with $z_1 = M(z)$. Clearly $z \neq i\infty$ and M is no power of T. Hence

$$|cz+d|^2 = c^2|z|^2 + d^2 + 2cd\Re(z) \geq c^2 + d^2 - |cd| \geq (|c| - |d|)^2 + |cd| \geq 1,$$

where for $|z| > 1$ one of the inequalities is strict. Therefore, $\Im(z_1) \leq \Im(z)$. In the same way we conclude that $\Im(z) \leq \Im(z_1)$ and thus $\Im(z_1) = \Im(z)$. This implies that $|cz + d| = 1$, $|z| = 1$ and analogously $|z_1| = 1$. Looking at the definition of F, we then find $z = z_1$ in contradiction to $z \neq z_1$. □

From the first part of the proof of Theorem 2.1.2 we can derive the following theorem:

Theorem 2.1.3

$$\Gamma = < S, T > .$$

Proof We fix an element $z \in F$ that is transcendental over $\mathbb{Q}$. For $M \in \Gamma$ the first part of the proof of Theorem 2.1.2 shows the existence of an $N \in < S, T >$ with $N(M(z)) \in F$. This implies that $NM(z) = z$. Writing $NM = \left(\begin{smallmatrix} a & b \\ c & d \end{smallmatrix}\right)$, the trancendence of z implies the equalities $c = b = 0, a = d$ and thus $NM = \pm \left(\begin{smallmatrix} 1 & 0 \\ 0 & 1 \end{smallmatrix}\right) \in < S, T >$. □

Using Theorem 2.1.2, we are able to construct fundamental domains for any subgroup of Γ:

Theorem 2.1.4 *Let U be a subgroup of Γ and M_j a system of representatives for Γ modulo $U \cup (-U)$. Then*

$$F_U = \bigcup_j M_j(F)$$

is a fundamental domain for U.

The sets $M_j(F)$ in Theorem 2.1.4 are called **fundamental triangles**. Two fundamental triangles are called adjacent if the intersection of their closures is unimodular equivalent to one of the edges $\{ e^{i\varphi} \mid \frac{2\pi}{6} \leq \varphi \leq \frac{2\pi}{3}\}$, $e^{\frac{\pi i}{3}} + i\mathbb{R}_{\geq 0}$ of F. For a subgroup U of finite index in Γ it is shown in Schöneberg (1974) that the system M_j can always be chosen such that F_U is obtained from F by successively adding adjacent fundamental triangles.

Example 2.1.5 Let $U = \Gamma_0(2) := \{\left(\begin{smallmatrix} a & b \\ c & d \end{smallmatrix}\right) \in \Gamma \mid b \equiv 0 \bmod 2\}$. Here

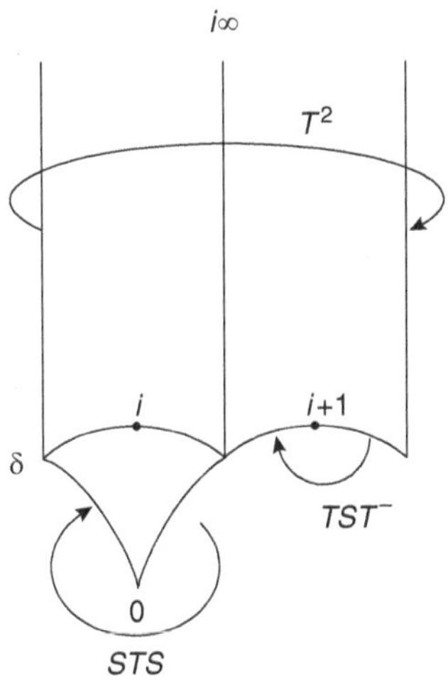

$U = -U$ and $[\Gamma : U] = 3$, as we will see later. Choosing $I = \left(\begin{smallmatrix} 1 & 0 \\ 0 & 1 \end{smallmatrix}\right), S, T$ as system of representatives, we obtain the above fundamental domain.

2.2 Congruence subgroups

Given a natural number N, we call

$$\Gamma(N) := \left\{ M \in \Gamma \mid M \equiv \pm \begin{pmatrix} 1 & 0 \\ 0 & 1 \end{pmatrix} \mod N \right\}$$

the **principal congruence group of level** N. Every subgroup U of Γ with

$$\Gamma \supseteq U \supseteq \Gamma(N)$$

for some $N \in \mathbb{N}$ is called a congruence subgroup modulo N. Being the kernel of the reduction map

$$\kappa : \Gamma \to SL_2(\mathbb{Z}/N\mathbb{Z}), \quad \begin{pmatrix} a & b \\ c & d \end{pmatrix} \mapsto \begin{pmatrix} \overline{a} & \overline{b} \\ \overline{c} & \overline{d} \end{pmatrix},$$

$\Gamma(N)$ has a finite index in Γ. By computation of the cardinality of $SL_2(\mathbb{Z}/N\mathbb{Z})$ and by proving the surjectivity of κ, one finds the formula

$$[\Gamma : \Gamma(N)] = N^3 \prod_{p|N} \left(1 - \frac{1}{p^2}\right),$$

that will not be needed in the sequel.

For the special type of congruence subgroup that we are going to consider, we need some facts about primitive matrices. By $\mathbb{P}_r$ we denote the set of all primitive 2×2 matrices of determinant $r \in \mathbb{N}$, i.e. $\mathbb{P}_r$ consists of all matrices

$$A = \begin{pmatrix} a & b \\ c & d \end{pmatrix} \in \mathbb{Z}^{2\times 2} \text{ with } ad - bc = r \text{ and } \gcd(a, b, c, d) = 1.$$

Proposition 2.2.1 *For every matrix $R_0 \in \mathbb{P}_r$ we have $\mathbb{P}_r = \Gamma R_0 \Gamma$.*

Proof Multiplying from both sides successively by S and T every matrix $R \in \mathbb{P}_r$ can be transformed to $\left(\begin{smallmatrix} r & 0 \\ 0 & 1 \end{smallmatrix}\right)$. Therefore, with suitable unimodular matrices $M_i, N_i,\ i = 1, 2$ we have

$$R = M_1 \begin{pmatrix} r & 0 \\ 0 & 1 \end{pmatrix} M_2 \text{ and } R_0 = N_1 \begin{pmatrix} r & 0 \\ 0 & 1 \end{pmatrix} N_2.$$

It follows that

$$R = M_1 N_1^{-1} R_0 N_2^{-1} M_2 \in \Gamma R_0 \Gamma.$$

This implies the assertion of proposition 2.2.1 because, conversely, every matrix in $\Gamma R_0 \Gamma$ is primitive and has the determinant r. □

For a given matrix $R \in \mathbb{P}_r$ we now consider the subgroup

$$\Gamma_R := \Gamma \cap R^{-1} \Gamma R$$

of Γ. If R is equal to one of the special matrices $\left(\begin{smallmatrix} r & 0 \\ 0 & 1 \end{smallmatrix}\right)$ and $\left(\begin{smallmatrix} 1 & 0 \\ 0 & r \end{smallmatrix}\right)$ we have the description

$$\Gamma_{\left(\begin{smallmatrix} r & 0 \\ 0 & 1 \end{smallmatrix}\right)} = \Gamma^0(r) := \{\left(\begin{smallmatrix} a & b \\ c & d \end{smallmatrix}\right) \in \Gamma \mid b \equiv 0 \bmod r\},$$
$$\Gamma_{\left(\begin{smallmatrix} 1 & 0 \\ 0 & r \end{smallmatrix}\right)} = \Gamma_0(r) := \{\left(\begin{smallmatrix} a & b \\ c & d \end{smallmatrix}\right) \in \Gamma \mid c \equiv 0 \bmod r\},$$

which implies that $\Gamma_{\left(\begin{smallmatrix} r & 0 \\ 0 & 1 \end{smallmatrix}\right)}$ is a congruence subgroup modulo r. We even have:

Theorem 2.2.2 *Γ_R is a congruence subgroup modulo r for every $R \in \mathbb{P}_r$.*

Proof By the next theorem we know that all Γ_R, $R \in \mathbb{P}_r$, are conjugate to $\Gamma_{\left(\begin{smallmatrix} r & 0 \\ 0 & 1 \end{smallmatrix}\right)}$ and by definition $\Gamma(r)$ is a normal subgroup of Γ, so the inclusion $\Gamma_R \supseteq \Gamma(r)$ holds for every $R \in \mathbb{P}_r$. □

Theorem 2.2.3 *For $R, R' \in \mathbb{P}_r$ and $M, M' \in \Gamma$ we have*

(i) $\Gamma R = \Gamma R' \Longrightarrow \Gamma_R = \Gamma_{R'}$,
(ii) $M^{-1}\Gamma_R M = \Gamma_{RM}$,
(iii) $\Gamma_R M = \Gamma_R M' \iff \Gamma RM = \Gamma RM'$.

Proof (i) and (ii) are easily verified. For (iii), we assume that $\Gamma_R M = \Gamma_R M'$. Multiplying by R from the left, we obtain $R\Gamma \cap \Gamma RM = R\Gamma \cap \Gamma RM'$. This implies that $RM \in R\Gamma \cap \Gamma RM' \subseteq \Gamma RM'$, hence $\Gamma RM \subseteq \Gamma RM'$. In the same way we see that $\Gamma RM' \subseteq \Gamma RM$, and it follows that $\Gamma RM = \Gamma RM'$. The opposite implication is evident by $\Gamma_R M = \Gamma \cap R^{-1}\Gamma RM$. □

Applying the third assertion of Theorem 2.2.3, we have a description of the right cosets of Γ_R by primitive matrices. Therefore, we define on $\mathbb{P}_r$ the **equivalence relation**

$$R \sim R' :\iff \Gamma R = \Gamma R'.$$

Theorem 2.2.4 *The number of equivalence classes modulo $\sim$ is finite. A system of representatives is given by the triangular matrices*

$$\begin{pmatrix} a & b \\ 0 & d \end{pmatrix}, \quad a > 0, ad = r, \ \ \gcd(a, b, d) = 1, \ \ b \ mod \ d.$$

Proof Multiplying by S and T from the left, every matrix R clearly can be transformed into a triangular matrix of the above form, so R is equivalent to one of the above matrices. If two such triangular matrices $R = \left(\begin{smallmatrix} a & b \\ 0 & d \end{smallmatrix}\right)$, $R' = \left(\begin{smallmatrix} a' & b' \\ 0 & d' \end{smallmatrix}\right)$ are equivalent, it follows that

$$\frac{1}{r}\begin{pmatrix} a'd & ab' - a'b \\ 0 & ad' \end{pmatrix} = R'R^{-1} \in \Gamma$$

and thus $a'd = ad' = r, ab' - a'b \equiv 0 \bmod r$. This implies that $a = a', d = d'$ and $b \equiv b' \bmod d$ because of $ad = a'd' = r$. Hence $R = R'$. □

From the third assertion of Theorem 2.2.3 and Theorem 2.2.4 we obtain:

Theorem 2.2.5 *Let $R_1, \dots, R_{\psi(r)}$ be a system of representatives of $\mathbb{P}_r$ modulo $\sim$ and R an arbitrary matrix in $\mathbb{P}_r$. Then by Theorem 2.2.1 there exist unimodular matrices M_μ with*

$$\Gamma R_\mu = \Gamma R M_\mu, \mu = 1, \dots, \psi(r).$$

So by Theorem 2.2.3

$$\Gamma = \bigcup_{\mu=1}^{\psi(r)} \Gamma_R M_\mu$$

and, in particular,

$$[\Gamma : \Gamma_R] = \psi(r).$$

Remark 2.2.6 The following assertions are easy to prove
Let r, s be in $\mathbb{N}$ with $\gcd(r, s) = 1$. Then

(i) $\mathbb{P}_{rs} = \mathbb{P}_r \mathbb{P}_s$,
(ii) for $R, R' \in \mathbb{P}_r$, $S, S' \in \mathbb{P}_s$ we have the equivalence

$$RS \sim R'S' \iff (R \sim R' \text{ and } S \sim S'),$$

(iii) $\psi(r) = r \prod_{p|r} (1 + \frac{1}{p})$.

2.3 Definition of modular forms

In what follows we consider meromorphic functions f on $\mathbb{H}$ that behave nicely under unimodular transformations. Therefore, given a number $k \in \mathbb{Z}$ and a matrix $M = \left(\begin{smallmatrix} a & b \\ c & d \end{smallmatrix}\right)$ of positive determinant, we define the operation

$$[f|_k M](\omega) := det(M)^{\frac{k}{2}} (c\omega + d)^{-k} f(M(\omega)).$$

Then $[f|_k M]$ is again a meromorphic function on $\mathbb{H}$, and we have the rule

$$[f|_k (MN)] = [[f|_k M]|_k N]. \tag{2.1}$$

Now we define:

Definition 2.3.1 Let f be a meromorphic function on $\mathbb{H}$ and U a subgroup of Γ with $[\Gamma : U] < \infty$. Then f is called a modular form for U of weight $k \in \mathbb{Z}$ if the following conditions are satisfied:

(i) $[f|_k M] = f$ for all $M \in U$.

(ii) For every $M \in \Gamma$ there exist $c_0 \in \mathbb{R}$, $l \in \mathbb{N}$ and $n_0 \in \mathbb{Z}$, so that for $\Im(\omega) \geq c_0$ we have a series expansion

$$[f|_k M](\omega) = \sum_{n \geq n_0} a_n q^{\frac{n}{l}}, \qquad q^{\frac{1}{l}} = e^{\frac{2\pi i \omega}{l}},$$

with coefficients $a_n \in \mathbb{C}$.

In the case $k = 0$ we call f a modular function.

To understand the second condition, we consider the topology $\mathbb{O}^*$ in $\mathbb{H}^*$ generated by the open disks in $\mathbb{H}$ and the images of the sets

$$U_c = \{\omega \in \mathbb{H} \mid \Im(\omega) > c\} \cup \{i\infty\}, \quad c \in \mathbb{R}^+,$$

under unimodular transformations. For an $M \in \Gamma \backslash < T >$ the image $M(U_c)$ is the union of $\{M(i\infty)\}$ and the inside of a circle in $\mathbb{H}^*$ with $\mathbb{R}$ being the tangent line in the point $M(i\infty)$. Hence, condition (ii) describes the behaviour of f near the points in $\mathbb{Q} \cup \{i\infty\}$, which are called the cusps.

Now we consider a function f satisfying condition (i). Using the above rule (2.1) it follows that $[f|_k M]$ is invariant under all $N \in M^{-1}UM$. As $M^{-1}UM$ has a finite index in Γ as well, this implies that a power T^l, $l \in \mathbb{N}$, is in $M^{-1}UM$. Thus, $[f|_k M]$ has period l, which implies that $[f|_k M]$ can be represented as a Laurent series in $q^{\frac{1}{l}}$ in every strip

$$c_0 < \Im(\omega) < c_1,$$

where it is holomorphic. Hence, condition (ii) means that $[f|_k M]$ is holomorphic for $\Im(\omega) \geq c_0$ having at most a pole for $q^{\frac{1}{l}} \to 0$.

Further, for $k = 0$ the defining conditions tell us that a modular function can be extended to $\mathbb{H}^*$ as a continuous map to $\mathbb{C} \cup \{\infty\}$ with respect to the topology $\mathbb{O}^*$.

For later convenience we define the **homogeneous modular form** for a given modular form f of weight k:

$$f\begin{pmatrix} \omega_1 \\ \omega_2 \end{pmatrix} := f\left(\frac{\omega_1}{\omega_2}\right) \omega_2^{-k} \text{ for } \omega_1, \omega_2 \in \mathbb{C} \setminus \{0\} \text{ with } \Im\left(\frac{\omega_1}{\omega_2}\right) > 0.$$

In this notation condition (i) in the above definition is equivalent to

$$f\left(M\begin{pmatrix} \omega_1 \\ \omega_2 \end{pmatrix}\right) = f\begin{pmatrix} \omega_1 \\ \omega_2 \end{pmatrix} \quad \text{for all } M \in U.$$

Conversely, the modular form f can be written in terms of its homogeneous modular form by

$$f(\omega) = f\begin{pmatrix} \omega \\ 1 \end{pmatrix}.$$

2.4 Examples of modular forms and modular functions

2.4.1 The functions g_2, g_3 and Δ

Using the lattice functions g_2, g_3 and Δ from 1.4, we obtain modular forms of weight 4, 6 and 12 for Γ by

$$\begin{aligned} g_2(\omega) &= 60G_2(\mathbb{Z}\omega + \mathbb{Z}), \\ g_3(\omega) &= 140G_3(\mathbb{Z}\omega + \mathbb{Z}), \\ \Delta(\omega) &= \Delta(\mathbb{Z}\omega + \mathbb{Z}) \end{aligned}$$

with corresponding homogeneous modular forms

$$\begin{aligned} g_2\left(\begin{smallmatrix}\omega_1\\ \omega_2\end{smallmatrix}\right) &= 60G_2(\mathbb{Z}\omega_1 + \mathbb{Z}\omega_2), \\ g_3\left(\begin{smallmatrix}\omega_1\\ \omega_2\end{smallmatrix}\right) &= 140G_3(\mathbb{Z}\omega_1 + \mathbb{Z}\omega_2), \\ \Delta\left(\begin{smallmatrix}\omega_1\\ \omega_2\end{smallmatrix}\right) &= \Delta(\mathbb{Z}\omega_1 + \mathbb{Z}\omega_2). \end{aligned}$$

Since these functions are invariant by all modular substitutions, it suffices to derive the q-expansion required in Definition 2.3.1 only for the unity matrix, which has already been done in 1.8.5 and 1.8.6.

An important modular form of weight 1 is defined by the η function:

$$\sqrt[12]{\Delta(\omega)} := 2\pi\ \eta(\omega)^2, \quad \sqrt[12]{\Delta\left(\begin{smallmatrix}\omega_1\\ \omega_2\end{smallmatrix}\right)} := 2\pi\ \eta\left(\tfrac{\omega_1}{\omega_2}\right)^2 \omega_2^{-1}. \tag{2.2}$$

It is modular for $\Gamma(12)$, as can be seen by the transformation formula of the η function.

2.4.2 The functions j, $\sqrt[3]{j}$, $\sqrt[2]{j - 12^3}$, j_R, φ_R

Quotients of modular forms of the same weight are modular functions. In this way we obtain the following modular functions:

$$j(\omega) := 12^3 \tfrac{g_2(\omega)^3}{\Delta(\omega)}$$

is called the **modular invariant**. It is holomorphic in $\mathbb{H}$ and modular for Γ. Further, we can define the following roots as holomorphic functions defined on the upper-half plane:

$$\begin{aligned} \sqrt[3]{j(\omega)} &:= \gamma_2(\omega) := 12 \tfrac{g_2(\omega)}{\sqrt[3]{\Delta(\omega)}}, \\ \sqrt[2]{j(\omega) - 12^3} &:= \gamma_3(\omega) := 6^3 \tfrac{g_3(\omega)}{\sqrt[2]{\Delta(\omega)}}. \end{aligned}$$

They are modular for $\Gamma(3)$ resp. $\Gamma(2)$, as we will see later.

Modular functions for Γ_R with a primitive matrix R of determinant r are, for instance, given by

$$j_R(\omega) := j(R(\omega)) \text{ and } \varphi_R(\omega) := r^{12} \tfrac{\Delta(R\left(\begin{smallmatrix}\omega\\ 1\end{smallmatrix}\right))}{\Delta\left(\begin{smallmatrix}\omega\\ 1\end{smallmatrix}\right)}.$$

For a unimodular matrix M we find that

$$j_R(M(\omega)) = j_{RM}(\omega) \text{ and } \varphi_R(M(\omega)) = \varphi_{RM}(\omega),$$

and obviously j_R and φ_R only depend on the equivalence class ΓR, so in view of Theorem 2.2.4 we have

$$j_R = j_{\left(\begin{smallmatrix} a & b \\ 0 & d \end{smallmatrix}\right)} \text{ and } \varphi_R = \varphi_{\left(\begin{smallmatrix} a & b \\ 0 & d \end{smallmatrix}\right)}$$

with a triangular matrix $\left(\begin{smallmatrix} a & b \\ 0 & d \end{smallmatrix}\right)$ that is equivalent to R. Hence the q-expansions required in Definition 2.3.1 are obtained from the expansions for j and Δ.

2.4.3 η-quotients

In many cases roots of φ_R are modular functions for Γ_R. By the transformation formula of the η function, we find, for instance, that the following functions are modular for $\Gamma_{\left(\begin{smallmatrix} 1 & 0 \\ 0 & n \end{smallmatrix}\right)}$, $n \in \mathbb{N}$:

$$\begin{array}{ll}
\left(\frac{\eta(\frac{\omega}{n})}{\eta(\omega)}\right)^8 \gamma_2(\omega)^{n-1} & \text{for } 3 \nmid n, \\
\left(\frac{\eta(\frac{\omega}{n})}{\eta(\omega)}\right)^6 \gamma_3(\omega)^{\frac{n-1}{2}} & \text{for } 2 \nmid n, \\
\left(\frac{\eta(\frac{\omega}{n})}{\eta(\omega)}\right)^m, \ m = \gcd(3, n) & \text{for } n = t^2, \ t \in \mathbb{N} \text{ and } 2 \nmid n, \\
\frac{\eta(\frac{\omega}{p})\eta(\frac{\omega}{q})}{\eta(\frac{\omega}{pq})\eta(\omega)} (\gamma_2(\omega)\gamma_3(\omega))^{\frac{p-1}{2}\frac{q-1}{2}} & \text{for } n = pq, \ p, q \in \mathbb{N}, \gcd(6, n) = 1.
\end{array}$$

For n not prime to 6 the last statement is no longer true. Looking closely at the transformation formula of the η function we find that the following functions are modular functions for $\Gamma_{\left(\begin{smallmatrix} 1 & 0 \\ 0 & rn \end{smallmatrix}\right)}$ with a suitable r.

$$\frac{\eta(\frac{\omega}{p})\eta(\frac{\omega}{q})}{\eta(\frac{\omega}{pq})\eta(\omega)} (\gamma_2(\omega)\gamma_3(\omega))^{\frac{p-1}{2}\frac{q-1}{2}}$$

$$\text{if } n = pq, \ p, q \in \mathbb{N}, \ p, q \text{ odd},$$

$$\text{with } r = \begin{cases} 1 & \text{if} \quad 3 \nmid n \text{ or } 3|p \text{ and } 3|(q-1) \\ 3 & \text{if} \quad 3|p, \ 3 \nmid (q-1). \end{cases}$$

$$\frac{\eta(\frac{\omega}{p})\eta(\frac{\omega}{q})}{\eta(\frac{\omega}{pq})\eta(\omega)} \gamma_2(\omega)^{(p-1)(q-1)}$$

$$\text{if } n = pq, \ p, q \in \mathbb{N}, \ p \text{ even}, \ q \text{ odd},$$

$$\text{with } r = \begin{cases} 4 & \text{if} \quad 3 \nmid n \text{ or } 3|p \text{ and } 3|(q-1) \\ 12 & \text{if} \quad 3|p, \ 3 \nmid (q-1). \end{cases}$$

$$\frac{\eta(\frac{\omega}{p})\eta(\frac{\omega}{q})}{\eta(\frac{\omega}{pq})\eta(\omega)} (\gamma_2(\omega)\gamma_3(\omega))^{(p-1)(q-1)}$$

$$\text{if } n = pq, \ p, q \in \mathbb{N}, \ p, q \text{ even},$$

$$\text{with } r = \begin{cases} 8 & \text{if} \quad 3 \nmid n \text{ or } 3|p \text{ and } 3|(q-1) \\ 24 & \text{if} \quad 3|p, \ 3 \nmid (q-1). \end{cases}$$

To prove the last assertions, let p, q be natural numbers and $g_{p,q}$ the function

$$g_{p,q}(\omega) = \frac{\eta(\frac{\omega}{p})\eta(\frac{\omega}{q})}{\eta(\frac{\omega}{pq})\eta(\omega)}$$

By Theorem 1.10.1 we find for $M = \left(\begin{smallmatrix} a & b \\ c & d \end{smallmatrix}\right) \in \Gamma$ with $n|b$

$$g_{p,q}(M(\omega)) = \zeta_{24}^{-e} g_{p,q}(\omega),$$

$$e = \left(a\tfrac{b}{pq} + c(d(1-a^2)-a)\right)(p-1)(q-1) \; + \; 3(a-1)c_1(p_1-1)(q_1-1),$$

c_1 being defined as in Theorem 1.10.1 and p_1, q_1 denoting the odd parts of p, q. Further, we find in the same way for γ_2 and γ_3:

$$\gamma_2(M(\omega)) = \zeta_{24}^{-8e'}\gamma_2(\omega), \quad \gamma_3(M(\omega)) = \zeta_{24}^{-12e'}\gamma_3(\omega),$$

$$e' = ab + c(d(1-a^2)-a) + 3(a-1)c_1.$$

Herein the condition $n|b$ is of course not necessary. Using these formulae, we find that all the above functions satisfy the invariance under unimodular transformations required in the definition of modular functions for $\Gamma_{\left(\begin{smallmatrix} 1 & 0 \\ 0 & n \end{smallmatrix}\right)}$ resp. for $\Gamma_{\left(\begin{smallmatrix} 1 & 0 \\ 0 & rn \end{smallmatrix}\right)}$. The q-expansions needed are obtained in the same way as for j_R und φ_R. Further, the last formulae imply the above assertion about γ_2, γ_3.

To construct functions in $\mathbb{C}_{\Gamma(N)}$ for an arbitrary $N \neq 1$ we begin by defining normalisations of Weber's $\wp$ function and the σ^* function using $\sqrt[12]{\Delta}$ and g_2, g_3:

2.4.4 Weber's τ function

Weber's τ function is defined by

$$\tau^{(e)}(z|{}^{\omega_1}_{\omega_2}) := g^{(e)}({}^{\omega_1}_{\omega_2})\wp(z|{}^{\omega_1}_{\omega_2})^{\frac{e}{2}}, \quad e = 2, 4, 6,$$

with the normalising factor

$$g^{(2)} = -2^7 3^5 \tfrac{g_2 g_3}{\Delta},$$

$$g^{(4)} = 2^8 3^4 \tfrac{g_2^2}{\Delta},$$

$$g^{(6)} = -2^9 3^6 \tfrac{g_3}{\Delta}.$$

In the case $e = 2$ we simply write

$$\tau(z|{}^{\omega_1}_{\omega_2}) = \tau^{(2)}(z|{}^{\omega_1}_{\omega_2}).$$

Obviously, τ is homogeneous of degree 0 and only depending on the lattice generated by ω_1, ω_2.

2.4.5 The natural normalisation of the $\wp$ function

For various arithmetic and numeric applications, we need, instead of Weber's τ function, the following "natural" normalisation of the $\wp$ function:

$$p(z|{}^{\omega_1}_{\omega_2}) := \frac{\wp(z|{}^{\omega_1}_{\omega_2})}{\sqrt[6]{\Delta({}^{\omega_1}_{\omega_2})}}.$$

However, according to (2.2) the 6-th root of Δ is no modular form for the full modular group, and hence the p function depends on the basis ω_1, ω_2.

2.4.6 Klein's normalisation of the σ function

Klein's normalisation of the σ function is defined by

$$\varphi(z|{}^{\omega_1}_{\omega_2}) := \sigma^*(z|{}^{\omega_1}_{\omega_2}) \sqrt[12]{\Delta({}^{\omega_1}_{\omega_2})},$$

where the 12-th root of Δ is chosen according to (2.2).

2.4.7 Transformation of $\tau^{(e)}, p, \varphi$

Let f be one of the functions $\tau^{(e)}, p, \varphi$. Then f is homogeneous of degree zero, i.e.

$$f(\lambda z|{}^{\lambda\omega_1}_{\lambda\omega_2}) = f(z|{}^{\omega_1}_{\omega_2}) \text{ for } \lambda \in \mathbb{C}^*.$$

For every

$$\underline{x} = (x_1, x_2) \in \frac{1}{N}(\mathbb{Z} \times \mathbb{Z}) \setminus \mathbb{Z} \times \mathbb{Z}$$

we define an N-th division value by

$$f_{\underline{x}}(\omega) := f(\underline{x}({}^{\omega}_{1})|{}^{\omega}_{1}) = f(x_1\omega + x_2|{}^{\omega}_{1}).$$

This is a modular function for $\Gamma(N)$ resp. $\Gamma(6N)$ resp. $\Gamma(12N^2)$, which follows from the transformation formulae

$$\tau^{(e)}_{\underline{x}}(M(\omega)) = \tau^{(e)}_{\underline{x}M}(\omega),$$

$$p_{\underline{x}}(M(\omega)) = \epsilon(M)^{-4} p_{\underline{x}M}(\omega),$$

$$\varphi_{\underline{x}}(M(\omega)) = \epsilon(M)^{2} \varphi_{\underline{x}M}(\omega).$$

Herein M is an arbitrary unimodular matrix. $\epsilon(M)$ denotes the root of unity from the η-transformation formula having the property $\epsilon(M)^4 = 1$

resp. $\epsilon(M)^2 = 1$ for $M \in \Gamma(6)$ resp. $M \in \Gamma(12)$. Further, by periodicity of the $\wp$ function we have

$$\wp\left(\underline{x}M\left(\begin{smallmatrix}\omega\\1\end{smallmatrix}\right)\middle|\begin{smallmatrix}\omega\\1\end{smallmatrix}\right) = \wp\left(\underline{x}\left(\begin{smallmatrix}\omega\\1\end{smallmatrix}\right)\middle|\begin{smallmatrix}\omega\\1\end{smallmatrix}\right) \text{ for } M \in \Gamma(N)$$

and by the transformation formula of the σ^* function:

$$\sigma^*\left(\underline{x}M\left(\begin{smallmatrix}\omega\\1\end{smallmatrix}\right)\middle|\begin{smallmatrix}\omega\\1\end{smallmatrix}\right) = \sigma^*\left(\underline{x}\left(\begin{smallmatrix}\omega\\1\end{smallmatrix}\right)\middle|\begin{smallmatrix}\omega\\1\end{smallmatrix}\right) \text{ for } M \in \Gamma(2N^2).$$

The meromorphy of $\varphi_{\underline{x}}$ is not immediate, because σ^* is not meromorphic. However, we observe that according to Legendre's relation we have the equality $\omega^* = \omega\eta_2 - 2\pi i$, by which it is easy to see that $\varphi_{\underline{x}}$ is meromorphic.

Finally, the q-expansions required in Definition 2.3.1 for $f_{\underline{x}}(M(\omega))$, $M \in \Gamma$, follow, using the above relations from Theorems 1.8.2 and 1.8.3.

2.5 Modular functions for Γ

2.5.1 Construction of modular functions for Γ

Clearly the set of modular functions for Γ is a field extension of $\mathbb{C}$, which in the sequel will be denoted by $\mathbb{C}_\Gamma$. The main result of this section is:

Theorem 2.5.1 $$\mathbb{C}_\Gamma = \mathbb{C}(j).$$

The proof of Theorem 2.5.1 needs some preparations. For the following let F be the fundamental domain for Γ from Theorem 2.1.2. Let f be a non-constant modular form for Γ of weight k. Then for $\omega \in F \setminus \{i\infty\}$ we denote by

$n_w(f)$ the index of the first coefficient in the Laurent expansion of f at $w \in \mathbb{H}$

and

$n_w(f)$ the index of the first coefficient in the q-expansion of f at w if $w \sim i\infty$.

Proposition 2.5.2 *Let f be a non-constant modular form for Γ and let $w, w' \in \mathbb{H}^*$ be the equivalent modulo Γ. Then $n_w(f) = n_{w'}(f)$.*

Proof For $w \sim i\infty$ the assertion is immediate by definition of $n_w(f)$. For $w \not\sim i\infty$ let $M = \left(\begin{smallmatrix}a & b\\ c & d\end{smallmatrix}\right) \in \Gamma$ with $w' = M(w)$. M maps a small disc D_w with centre w onto a small disc containing w' as an interior point.

By the argument principle, we then find that

$$\begin{aligned} n_{w'}(f) &= \tfrac{1}{2\pi i} \oint_{M(D_w)} d\log f(\omega) = \tfrac{1}{2\pi i} \oint_{D_w} d\log f(M(\xi)) \\ &= \tfrac{1}{2\pi i} \oint_{D_w} d\log(c\xi + d)^k f(\xi) \\ &= \tfrac{1}{2\pi i} \oint_{D_w} d\log(c\xi + d)^k + \tfrac{1}{2\pi i} \oint_{D_w} d\log f(\xi) \\ &= \tfrac{1}{2\pi i} \oint_{D_w} d\log f(\xi) = n_w(f). \end{aligned}$$

Herein $\frac{1}{2\pi i} \oint_{D_w} d\log(c\xi + d)^k = 0$, because $c\xi + d \neq 0$ for $\xi \in \mathbb{H}$. □

Theorem 2.5.3 *Let f be a non-constant modular form of weight k for Γ. Then*

$$\left(\sum_{w \in F \setminus \{i,\rho\}} n_w(f) \right) + \frac{n_i(f)}{2} + \frac{n_\rho(f)}{3} = \frac{k}{12}.$$

Proof By the argument principle we have

$$2\pi i \sum_{w \in F \setminus \{i,\rho,\infty\}} n_w(f) = \int_\gamma \frac{f'(\omega)}{f(\omega)} d\omega = \int_\gamma d\log f(\omega),$$

where γ denotes the path along the boundary of a lower section of F as described in the figure below. The section is chosen such that in the upper part of F there are no poles and zeros other than possibly at $i\infty$. This may be due to condition (ii) in Definition 2.3.1. Further, at $i, \rho, -\overline{\rho}$ and at possible poles and zeros on the boundary of F the path is modified by small arcs of radius ϵ. Since $f(\omega+1) = f(\omega)$ we first have

$$\int_{D'}^{A} d\log f(\omega) + \int_{C'}^{D} d\log f(\omega) = 0.$$

The integrals from A' to B and from B' to C are calculated in the same way. Here, the substitution $\omega \mapsto \frac{-1}{\omega}$ maps the path from A' to B onto the path from C to B'. We use the transformation formula

$$f\left(\tfrac{-1}{\omega}\right) = \omega^k f(\omega),$$

by which we obtain

$$\log f\left(\tfrac{-1}{\omega}\right) = k \log \omega + \log f(\omega) + c$$

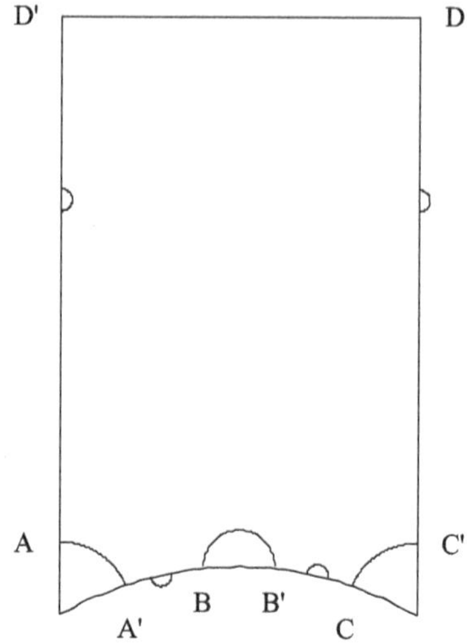

with a constant c and then

$$-\int_{B'}^{C} d\log f(\omega) = \int_{A'}^{B} d\log f\left(\tfrac{-1}{\omega}\right) = k\int_{A'}^{B} d\log\omega + \int_{A'}^{B} d\log f(\omega).$$

Hence

$$\lim_{\epsilon\to 0}\left(\int_{A'}^{B} d\log f(\omega) + \int_{B'}^{C} d\log f(\omega)\right) = -\lim_{\epsilon\to 0} k\int_{A'}^{B} d\log\omega = k\frac{2\pi i}{12}.$$

To calculate the integral from D to D' we use the q-expansion of f:

$$f(\omega) = \hat{f}(q(\omega)),\quad q(\omega) = e^{2\pi i\omega},$$

$$\hat{f}(q) = c_n q^n + c_{n+1}q^{n+1} + \cdots,\quad c_n \neq 0, n = n_{i\infty}(f),$$

and we observe that by $\omega \mapsto e^{2\pi i\omega}$ the path from D to D' is mapped onto the boundary of a circle around 0. This implies that

$$\int_{D}^{D'} d\log f(\omega) = -\oint d\log \hat{f}(q) = -2\pi i\; n_{i\infty}(f).$$

Next, we evaluate the integral along the arc from A to A'. Let A'' be the intersection point of the arc and the tangent of the unit circle at $\rho = e^{\frac{2\pi i}{3}}$. Then, with the distance $\delta(A'', A')$ and the radius ϵ, it follows from the tangent property that

$$\lim_{\epsilon \to 0} \frac{\delta(A', A'')}{\epsilon} = 0.$$

Further, in a neighbourhood of ρ

$$\frac{f'(\omega)}{f(\omega)} = O(\tfrac{1}{|\omega-\rho|}) \text{ for } \omega \to \rho.$$

Hence

$$\lim_{\epsilon \to 0} \int\limits_{A''}^{A'} d\log f(\omega) = 0.$$

and, observing

$$\frac{f'(\omega)}{f(\omega)} = n_\rho(f)\frac{1}{\omega - \rho} + O(1) \text{ for } \omega \to \rho,$$

it follows that

$$\lim_{\epsilon \to 0} \int\limits_{A}^{A'} d\log f(\omega) = -n_\rho(f) \lim_{\epsilon \to 0} \int\limits_{A}^{A''} \frac{d\omega}{\omega - \rho} = -n_\rho(f)\frac{2\pi i}{6}.$$

In the same way it can be shown that

$$\lim_{\epsilon \to 0} \int\limits_{C}^{C'} d\log f(\omega) = -n_{-\overline{\rho}}(f)\frac{2\pi i}{6} = -n_\rho(f)\frac{2\pi i}{6}$$

and

$$\lim_{\epsilon \to 0} \int\limits_{B}^{B'} d\log f(\omega) = -n_i(f)\frac{2\pi i}{2}.$$

Combining all the results, we obtain

$$\int\limits_{\gamma} d\log f(\omega) = 2\pi i \left(\frac{k}{12} - \frac{1}{3}n_\rho(f) - \frac{1}{2}n_i(f) - n_{i\infty}(f) \right),$$

which proves the asserted formula of Theorem 2.5.3. □

Let f be a non-constant modular function for Γ. Then, by Theorem 2.5.3 we can conclude that

$$n_w(f) \equiv 0 \bmod 2 \text{ if } w \sim i,$$
$$n_w(f) \equiv 0 \bmod 3 \text{ if } w \sim \rho.$$

Now we define

$$n_w^*(f) := \begin{cases} n_w(f) \text{ if } w \not\sim i, \rho, \\ \frac{1}{2} n_w(f) \text{ if } w \sim i, \\ \frac{1}{3} n_w(f) \text{ if } w \sim \rho \end{cases}$$

and for $c \in \mathbb{C}$

$$N_c(f) := \sum_{w \in F} \max(0, n_w^*(f - c)),$$
$$N_\infty(f) := - \sum_{w \in F} \min(0, n_w^*(f)).$$

By Theorem 2.5.3 we then obtain:

Theorem 2.5.4 *Let f be a non-constant modular function for Γ. Then*

$$N_c(f) = N_\infty(f)$$

for all $c \in \mathbb{C}$. The common value $N_c(f)$ is called the order of f.

As an example we consider the modular invariant j, which is holomorphic in $\mathbb{H}$. By Theorems 1.8.5 and 1.8.6 we find out that the q-expansion starts with q^{-1}. Hence:

Theorem 2.5.5 *We have*

$$j(\mathbb{H}) = \mathbb{C}$$

and

$$j(\omega_1) = j(\omega_2) \iff \omega_1 \sim \omega_2 \text{ modulo } \Gamma$$

for all $\omega_1, \omega_2 \in \mathbb{H}$.

As an application of this theorem we will now show that every elliptic curve over $\mathbb{C}$, defined by an equation of the form

$$y^2 = 4x^3 - a_2 x - a_3 \quad \text{with} \quad a_2, a_3 \in \mathbb{C}, \;\; a_2^3 - 27a_3^2 \neq 0,$$

can be parametrised as in (1.4) by the Weierstrass $\wp$ function for a lattice and its derivative.

Remark 2.5.6 Using Theorem 2.5.3 it is easy to prove again that $\Delta(\omega) \neq 0$ for $\omega \in \mathbb{H}$.

Theorem 2.5.7 *Let $a_2, a_3 \in \mathbb{C}$ have the property $a_2^3 - 27a_3^2 \neq 0$. Then there exists a complex lattice $\mathfrak{L}$ such that*

$$a_2 = g_2(\mathfrak{L}) \text{ and } a_3 = g_3(\mathfrak{L}).$$

Proof By 2.5.5 there exists an $\omega \in \mathbb{H}$ with

$$12^3 \frac{a_2^3}{a_2^3 - 27a_3^2} = j(\omega) = 12^3 \frac{g_2(\mathfrak{L}_1)^3}{g_2(\mathfrak{L}_1)^3 - 27g_3(\mathfrak{L}_1)^2}, \quad \mathfrak{L}_1 = [\omega, 1].$$

Hence

$$a_2^3 = \lambda g_2(\mathfrak{L}_1)^3 \text{ and } a_2^3 - 27a_3^2 = \lambda g_2(\mathfrak{L}_1)^3 - 27g_3(\mathfrak{L}_1)^2$$

with some $\lambda \in \mathbb{C}, \lambda \neq 0$, and further, with a suitable 12-th root ξ of λ:

$$a_2 = \xi^4 g_2(\mathfrak{L}_1) = g_2(\xi^{-1}\mathfrak{L}_1), \quad a_3 = \xi^3 g_3(\mathfrak{L}_1) = g_2(\xi^{-1}\mathfrak{L}_1).$$

Therefore, $\mathfrak{L} = \xi^{-1}\mathfrak{L}_1$ is a lattice having the required property. □

Proof of Theorem 2.5.1: Our conclusions are similar to the proof of Theorem 1.3.3. Let f be a non-constant function in $\mathbb{C}_\Gamma$ and $\omega_1, \ldots, \omega_m$ the different points $\omega \in F \setminus \{i\infty\}$, where $n_\omega(f) \neq 0$. j having order 1, it follows that

$$g = f \prod_{\mu=1}^{m} (j - j(\omega_\mu))^{-n^*_{\omega_\mu}(f)}$$

has no poles or zeros other than possibly at $i\infty$. Hence by Theorem 2.5.4, g must be a constant, which proves that f is a rational function of j. □

This proof of Theorem 2.5.1 also shows that:

Theorem 2.5.8 *The ring of modular functions for Γ that are holomorphic in $\mathbb{H}$ is given by $\mathbb{C}[j]$.*

2.5.2 The q-expansion principle

For the later construction of algebraic numbers the field of constants $\mathbb{C}$ is unsuitable. We will, rather, need functions in $\mathbb{Q}(j)$ or polynomials of j having coefficients in an additive subgroup of $\mathbb{C}$. These functions will be characterised by the q-**expansion principle**.

By Theorems 1.8.5 and 1.8.6 we obtain the q-expansion of j:

$$j(\omega) = \frac{\left(1 + 240 \sum_{n=1}^{\infty} \sum_{d|n} d^3 q^n\right)^3}{q \prod_{n=1}^{\infty} (1-q^n)^{24}} = q^{-1} + \sum_{n=0}^{\infty} c_n q^n \in \frac{1}{q}\mathbb{Z}[[q]].$$

This leads us to:

Theorem 2.5.9 (q-expansion principle)

(i) *For* $f = \sum_{n \geq n_0} a_n q^n \in \mathbb{C}_\Gamma = \mathbb{C}(j)$ *and a subfield* Λ *of* $\mathbb{C}$ *we have*

$$f \in \Lambda(j) \iff f \in q^{n_0}\Lambda[[q]].$$

(ii) *For* $f = \sum_{n \geq n_0} a_n q^n \in \mathbb{C}[j]$ *and an additive subgroup* $\mathfrak{a}$ *of* $\mathbb{C}$ *we have*

$$f \in \mathfrak{a}[j] \iff f \in q^{n_0}\mathfrak{a}[[q]].$$

Proof The implication from the left to the right is trivial for both assertions. To prove the opposite implication for the first assertion, we write f as a rational function of j according to Theorem 2.5.1:

$$f = \frac{A_M j^M + \cdots + A_0}{B_N j^N + \cdots + B_0}.$$

Assuming the polynomials in the numerator and the denominator to be coprime, the coefficients A_k, B_k are uniquely determined up to a common factor. Multiplying the equation by the denominator of the right-hand side and inserting the q-expansion of j and its powers,

$$j^k f = \sum_{n=-\infty}^{\infty} d_\nu^{(k)} q^n, \quad j^k = \sum_{n=-\infty}^{\infty} c_\nu^{(k)} q^n,$$

we obtain a linear equation for the A_k, B_k's:

$$\sum_{n=-\infty}^{\infty} (B_N d_n^{(N)} + \cdots + B_0 d_n^{(0)} - (A_M c_n^{(M)} + \cdots + A_0 c_n^{(0)}))q^n = 0.$$

By the identity theorem for power series this is equivalent to the infinite system of linear equations

$$B_N d_n^{(N)} + \cdots + B_0 d_n^{(0)} - (A_M c_n^{(M)} + \cdots + A_0 c_n^{(0)}), \quad n \in \mathbb{Z}.$$

This system having coefficients in Λ and the space of solutions being one-dimensional, we can conclude that there exists a non-zero solution $(A_M, \ldots, A_0, B_N, \ldots, B_0) \in \Lambda^{M+N+2}$. Hence $f \in \Lambda(j)$.

To prove the missing implication in (ii), we write

$$f = A_M j^M + \cdots + A_0$$

with coefficients $A_k \in \mathbb{C}$. We then use the fact that the q-expansion of

j^k starts with q^{-k}:

$$j^k = q^{-k} + \sum_{n=-k+1}^{\infty} c_n^{(k)} q^n \in q^{-k}\mathbb{Z}[[q]].$$

Inserting these expansions into the above equation, we obtain

$$\sum_{n=n_0}^{\infty} a_n q^n =$$

$$\begin{array}{lll} & A_M & (q^{-M} + c_{-(M-1)}^{(M)} q^{-(N-1)} + \cdots + c_0^{(M)} + \cdots) \\ + & A_{M-1} & (\qquad\qquad q^{-(M-1)} + \cdots + c_0^{(M-1)} + \cdots) \\ & \vdots & \\ + & A_0 & (\qquad\qquad\qquad\qquad 1 + \cdots). \end{array}$$

Herein by assumption the a_n are in $\mathfrak{a}$, and because the $c_n^{(k)}$ are in $\mathbb{Z}$ we find recursively that $A_M \in \mathfrak{a}, A_{m-1} \in \mathfrak{a}, \ldots, A_0 \in \mathfrak{a}$ by comparing coefficients. □

2.6 Modular functions for subgroups of Γ

2.6.1 The isomorphisms of $\mathbb{C}_U/\mathbb{C}_\Gamma$

Clearly, the set of modular functions for a subgroup U of Γ is a field extension of $\mathbb{C}_\Gamma$ that we will denote by C_U in the sequel. More precisely we have:

Theorem 2.6.1 *Let U be a subgroup of Γ with $-\left(\begin{smallmatrix}1&0\\0&1\end{smallmatrix}\right) \in U$ and $[\Gamma : U] < \infty$. Then:*

(i) *$\mathbb{C}_U/\mathbb{C}_\Gamma$ is algebraic of degree $[\mathbb{C}_U : \mathbb{C}_\Gamma] = [\Gamma : U]$.*

(ii) *Every $M \in \Gamma$ defines an isomorphism*

$$(f^{\lambda_M})(\omega) = f(M(\omega))$$

of $\mathbb{C}_U$ onto $\mathbb{C}_{M^{-1}UM}$ with $\lambda_M|\mathbb{C}_\Gamma = id_{\mathbb{C}_\Gamma}$, which only depends on the coset UM.

(iii) *For a decomposition $\Gamma = \bigcup_{\nu=1}^{n} UM_\nu$ into right cosets of Γ modulo U we obtain by $\lambda_{M_\nu}, \nu = 1, \ldots, n$, all different embeddings of $\mathbb{C}_U/\mathbb{C}_\Gamma$ into the algebraic closure of $\mathbb{C}_\Gamma$.*

(iv) *The* **main-polynomial** *of a function* $f \in \mathbb{C}_U$,

$$F(X,j) := \prod_{\nu=1}^{n} (X - f^{\lambda_{M_\nu}}),$$

has coefficients in $\mathbb{C}_\Gamma$ *and, for* f *holomorphic in* $\mathbb{H}$ *we even have*

$$F(X,j) \in \mathbb{C}[X,j].$$

Proof The assertions (ii)–(iv), apart from the fact that the λ_{M_ν} are all different, are obtained by direct verification, recalling Theorem 2.5.8 for the second part of (iv). To prove that all λ_{M_ν} are different, which implies (i), we must prove the existence of a function in $\mathbb{C}_U$ such that all conjugates $f^{\lambda_{M_\nu}}$ are different. Such functions will explicitly be constructed for all subgroups to be considered in the sequel. In the general case the existence of these functions is proved by applying the Riemann–Roch Theorem to the compact Riemann surface associated with $U\backslash H^*$. □

2.6.2 The extended q-expansion principle

For the construction of algebraic numbers we need modular functions with a main-polynomial having coefficients in $\Lambda(j)$ for a subfield Λ of $\mathbb{C}$, as for instance $\Lambda = \mathbb{Q}$. To characterise such functions we generalise the q-expansion principle:

Theorem 2.6.2 (extended q-expansion principle) *Let* U *be a subgroup of* Γ, $[\Gamma : U] < \infty$, *and let* $f \in \mathbb{C}_U$ *be holomorphic in* $\mathbb{H}$ *with main-polynomial* $F(X,j)$.

(i) *If ONE conjugate* f^{λ_M} *of* f *has* q*-coefficients in* Λ, *then*

$$F(X,j) \in \Lambda[X,j].$$

(ii) *If ALL conjugates* f^{λ_M} *of* f *have* q*-coefficients in a subring* $\mathfrak{a}$ *of* $\mathbb{C}$, *then*

$$F(X,j) \in \mathfrak{a}[X,j].$$

Proof The holomorphy of f implies the holomorphy of its conjugates. Hence, the coefficients of the minimal polynomial $p(X,j)$ of f over $\mathbb{C}_\Gamma$ are in $\mathbb{C}[j]$:

$$p(X,j) = \sum_{\nu=0}^{n} \left(\sum_{\mu=0}^{m} c_{\nu,\mu} j^\mu \right) X^\nu, \qquad c_{\nu,\mu} \in \mathbb{C}.$$

The coefficients $c_{\nu,\mu}$ are uniquely determined by

$$\sum_{\mu=0}^{m} c_{n,\mu} j^\mu = 1 \text{ and } p(f^{\lambda_M}, j) = 0.$$

Inserting the q-expansions

$$j^\mu (f^{\lambda_M})^\nu = \sum_{k=-\infty}^{\infty} a_k^{(\nu,\mu)} q^{\frac{k}{t}}$$

and by comparing coefficients, we obtain a system of linear equations equivalent to $p(f^{\lambda_M}, j) = 0$:

$$c_{n,0} = 1, \qquad \sum_{\nu=0}^{n} \sum_{\mu=0}^{m} c_{\nu,\mu} a_n^{(\nu,\mu)} = 0, \quad \nu \in \mathbb{Z}.$$

By assumption about the q-expansion of f^{λ_M}, the coefficients $a_n^{(\nu,\mu)}$ are in Λ. Hence, by linear algebra there exists a non-trivial solution $(c_{\nu,\mu}) \in \Lambda^{n+m+2}$. This implies that the minimal polynomial over $\mathbb{C}_\Gamma$ has coefficients in $\Lambda[j]$. The same holds for the polynomial $F(X, j)$, which is a power of $p(X, j)$.

Under the assumptions made in (ii) the coefficients

$$p_\nu(j) = \sum_{\mu=0}^{m} c_{\nu,\mu} j^\mu$$

of the minimal polynomial are in $\mathfrak{a}[j]$ by Theorem 2.5.9. Again this implies that $F(X, j) \in \mathfrak{a}[X, j]$, because $F(X, j)$ is a power of $p(X, j)$. □

2.7 Modular functions for Γ_R

In 2.4.2 we have seen that j_R and φ_R are in $\mathbb{C}_{\Gamma_R}$ for a primitive matrix R. In this section we will show that these functions are generators of $\mathbb{C}_{\Gamma_R}/\mathbb{C}_\Gamma$. Furthermore, our aim is to prove $\mathbb{Q}(j, j_R) = \mathbb{Q}(j, \varphi_R)$ and to characterise the functions in this field.

Theorem 2.7.1 *Let g_R be one of the functions j_R and φ_R. Then:*

(i) $\mathbb{C}_{\Gamma_R} = \mathbb{C}(j, g_R)$.

(ii) *For a complete system R_μ of inequivalent matrices of determinant r the functions $g_{R_\mu}, \mu = 1, \ldots, \psi(r)$, are all different and constitute a complete system of conjugates of g_R over $\mathbb{C}_\Gamma$.*

Proof By Theorem 2.2.5 we see that by g_{R_μ} all conjugates of g_R over $\mathbb{C}_\Gamma$ are given. It remains to prove that the g_{R_μ} are pairwise different, which will be shown by looking at their q-expansions. We recall that

$$j = q^{-1} + \sum_{n=0}^{\infty} c_n q^n, \qquad \Delta\binom{z}{1} = (2\pi)^{12} q \prod_{n=1}^{\infty} (1-q^n)^{24}.$$

The g_{R_μ} only depend on the equivalence class of R_μ, so we appeal to Theorem 2.2.5, and we assume that $R_\mu = \left(\begin{smallmatrix} a & b \\ 0 & d \end{smallmatrix}\right)$. Then

$$j_{\left(\begin{smallmatrix} a & b \\ 0 & d \end{smallmatrix}\right)} = \zeta_d^b q^{\frac{a}{d}} + \sum_{n=0}^{\infty} c_n \zeta_d^{bn} q^{\frac{an}{d}},$$

$$\varphi_{\left(\begin{smallmatrix} a & b \\ 0 & d \end{smallmatrix}\right)} = a^{12} \frac{\zeta_d^a q^{\frac{an}{d}} \prod_{n=1}^{\infty} (1 - \zeta_d^{an} q^{\frac{an}{d}})^{24}}{q \prod_{n=1}^{\infty} (1-q^n)^{24}}.$$

Herein the leading coefficients are all different if R_μ runs through the system of representatives in Theorem 2.2.5. □

In view of later applications we consider the fields

$$\mathbb{Q}_\Gamma := \mathbb{Q}(j) \quad \text{and} \quad \mathbb{Q}_{\Gamma_R} := \mathbb{Q}(j, j_R).$$

According to Theorem 2.6.2 the minimal polynomial of j_R over $\mathbb{C}_\Gamma$ has coefficients in $\mathbb{Q}_\Gamma$, so by Theorem 2.7.1 we obtain:

Theorem 2.7.2 *We have*

$$[\mathbb{Q}_{\Gamma_R} : \mathbb{Q}_\Gamma] = \psi(r),$$

and the different isomorphisms of $\mathbb{Q}_{\Gamma_R}/\mathbb{Q}_\Gamma$ *are given by*

$$\lambda_{M_\mu}, \; \mu = 1, \ldots, \psi(r),$$

if RM_μ, $\mu = 1, \ldots, \psi(r)$, $M_\mu \in \Gamma$, *is a system of representatives of primitive matrices of determinant* r.

To characterise the functions in $\mathbb{Q}_{\Gamma_R}$, we set up the following notation for the conjugates of $f \in \mathbb{C}_{\Gamma_R}$ over $\mathbb{C}_\Gamma$. We set:

$$f_{R'} := f^{\lambda_{M'}}, \qquad \text{for } \Gamma R M' = \Gamma R', \; M' \in \Gamma.$$

In particular, using this notation, we can write $f = f_R$.

Theorem 2.7.3 *Let R, R_0 be in $\mathbb{P}_r$ and $f_R \in \mathbb{C}_{\Gamma_R}$. We assume j_{R_0} to have q-coefficients in $\mathbb{Q}$. Then*

$$f_R \in \mathbb{Q}_{\Gamma_R} \iff (f_{R_0} \text{ has } q\text{-coefficients in } \mathbb{Q}).$$

Proof The implication from the left to the right is trivial. To prove the opposite implication, we observe that j_{R_0} is a generating element for the extension $\mathbb{C}_{\Gamma_R}/\mathbb{C}_\Gamma$. Hence

$$f_{R_0} = \sum_{\mu=1}^{\psi(r)} a_\mu(j) j_{R_0}^\mu$$

with coefficients $a_\mu(j) \in \mathbb{C}(j)$ that are uniquely determined. Inserting the q-expansion of the functions involved, we obtain a system of linear equations over $\mathbb{Q}$ for the q-coefficients of $a_\mu(j)$. Hence, the $a_\mu(j)$ have q-coefficients in $\mathbb{Q}$, which implies that they are in $\mathbb{Q}(j)$, so the above representation shows that $f_{R_0} \in \mathbb{Q}_{\Gamma_{R_0}}$, and it follows that $f_R \in \mathbb{Q}_{\Gamma_R}$. □

For the following we define the **modular polynomial** of order r to be the main-polynomial of j_R over $\mathbb{Q}_\Gamma$,

$$I_r(X, j) := \prod_{\mu=1}^{\psi(r)} (X - j_{R_\mu}),$$

which is also the minimal polynomial of j_R over $\mathbb{Q}_\Gamma$. Herein R_μ runs through a system of representatives of primitive matrices of determinant r.

Theorem 2.7.4 *We have:*

(i) $I_r(X, j) \in \mathbb{Z}[X, j]$,
(ii) $I_r(X, X) \neq 0$,
(iii) *The leading coefficient of $I_r(X, X)$ is ± 1 for $r \in \mathbb{N} \setminus \mathbb{N}^2$.*

Proof $j_{\left(\begin{smallmatrix} r & 0 \\ 0 & 1 \end{smallmatrix}\right)}$ having q-coefficients in $\mathbb{Q}$ and j_R being holomorphic in $\mathbb{H}$, the first part of Theorem 2.6.2 tells us that

$$I_r(X, j) \in \mathbb{Q}[X, j],$$

and from the second part we obtain

$$I_r(X, j) \in \mathfrak{O}[X, j]$$

because the q-coefficients of the functions j_R are in the ring $\mathfrak{O}$ of algebraic integers. This proves the first assertion of Theorem 2.7.4.

To prove the remaining assertions, we consider the q-expansion of $I_r(j,j)$, and we observe that the lowest coefficient in this q-expansion is equal to the leading coefficient of the polynomial $I_r(X,X)$. Using the representatives of Theorem 2.2.4 we find that:

$$I_r(j,j) = \prod_{\left(\begin{smallmatrix} a & b \\ 0 & d \end{smallmatrix}\right)} \left([q^{-1} + \cdots] - [\zeta_r^{-ab} q^{-\frac{a}{d}} + \cdots]\right).$$

Since $\frac{a}{d} \neq 1$ if r is not a square, the leading coefficient of $I_r(X,X)$ is a product of roots of unity, hence equal to ± 1, because $I_r(X,X)$ is in $\mathbb{Q}[X,X]$. For $r = t^2$ the leading coefficient is given by

$$\pm \prod_{b \bmod t} (1 - \zeta_t^b) = \pm \begin{cases} t & \text{if } t \text{ is a prime power,} \\ 1 & \text{otherwise.} \end{cases}$$

In particular, this shows again that $I_r(X,X) \neq 0$.

□

Now we consider the function φ_R with the main-polynomial

$$\Phi_r(X,j) := \prod_{\mu=1}^{\psi(r)} (X - \varphi_{R_\mu}).$$

We have:

Theorem 2.7.5 *Let R be a primitive matrix of determinant r. Then:*

(i) $\mathbb{Q}_{\Gamma_R} = \mathbb{Q}_\Gamma(\varphi_R)$,

(ii) $\Phi_r(X,j) = X^{\psi(r)} + B^{(r)}_{\psi(r)-1}(j) X^{\psi(r)-1}(j) + \ldots + B_0^{(r)}(j) \in \mathbb{Z}[X,j]$.

Herein $B_0^{(r)}(j) = \pm \prod_{\mu=1}^{\psi(r)} a_\mu^{12}$ *with the representatives* $R_\mu = \left(\begin{smallmatrix} a_\mu & b_\mu \\ 0 & d_\mu \end{smallmatrix}\right)$ *from Theorem 2.2.4.*

(iii) *For a prime p we have* $B_0^{(p)}(j) = p^{12}$.

Proof The first assertion of Theorem 2.7.5 follows from Theorem 2.7.3 because $\varphi_{\left(\begin{smallmatrix} r & 0 \\ 0 & 1 \end{smallmatrix}\right)}$ has q-coefficients in $\mathbb{Q}$. For arbitrary $R \in \mathbb{P}_r$ the q-coefficients of φ_R are in $\mathbb{Z}[\zeta_r]$, and φ_R is holomorphic, so by 2.5.9 the coefficients of $\Phi_r(X,j)$ are in $\mathbb{Z}[j]$. Furthermore, φ_R having no zeros in

$\mathbb{H}$ implies that

$$B_0^{(r)}(j) = (-1)^{\psi(r)} \prod_{\mu=1}^{\psi(r)} \varphi_{R_\mu}$$

as a polynomial in $\mathbb{Z}[j]$ must be a constant. Hence, it is equal to the product of the coefficients of the lowest terms in the q-expansions:

$$B_0^{(r)}(j) = (-1)^{\psi(r)} \prod_{\mu=1}^{\psi(r)} \zeta_r^{a_\mu b_\mu} a_\mu^{12}.$$

If $r = p$ is a prime, this product is equal to p^{12}. □

In the later applications we will sometimes need the representation of a function f as a rational function of j and another function. This is obtained by:

Theorem 2.7.6 *Let P be a primitive matrix, whose determinant is a prime p. Let $f_P \in \mathbb{Q}_{\Gamma_P}$ be holomorphic in $\mathbb{H}$ with the property that the conjugates $f_{P'}$, $P' \in \mathbb{P}_p$, have q-coefficients that are algebraic integers. Then:*

$$f_P \Phi_p'(\varphi_P, j) = a_0(j) + a_1(j)\varphi_P + \cdots + a_p(j)\varphi_P^p$$

$$\textit{with } a_\mu(j) \in \mathbb{Z}[j], \ \mu = 0, \ldots, p.$$

Further, $a_0(j) \in p\mathbb{Z}[j]$ if $f_{\left(\begin{smallmatrix} p & 0 \\ 0 & 1 \end{smallmatrix}\right)}$ has q-coefficients divisible by p.

Proof The proof follows from the q-expansion principle, using the following algebraic lemma:

Lemma 2.7.7 *Let $k(\varphi)/k$ be a separable field extension of degree n, and let*

$$\Phi(X) = X^n + B_{n-1}X^{n-1} + \cdots + B_0$$

be the minimal polynomial of φ over k. Then:

(i) $\frac{1}{\Phi'(\varphi)}, \ldots, \frac{\varphi^{n-1}}{\Phi'(\varphi)}$ *is a basis of $k(\varphi)/k$.*

(ii) *Every $f \in k(\varphi)$ has a representation*

$$f = \frac{a_0 + a_1\varphi + \cdots + a_{n-1}\varphi^{n-1}}{\Phi'(\varphi)},$$

where the coefficients $a_\mu \in k$ are given by the formula

$$a_\mu = \sum_{\nu=\mu+1}^{n} B_\nu tr_{k(\varphi)/k}(f\varphi^{\nu-(\mu+1)}).$$

Before proving the lemma, we first use it to prove Theorem 2.7.6. Therefore we apply the lemma to the extension $\mathbb{Q}_\Gamma(\varphi_P)/\mathbb{Q}_\Gamma$ and the function f_P. The formula of the lemma then shows that the coefficients $a_\mu(j)$ are polynomials of j, because the $B_\nu^{(p)}(j)$ and the functions f_P and φ_P with all their conjugates are holomorphic. Further, the hypothesis on f_P implies that the q-coefficients of the $a_\mu(j)$ are algebraic integers. Therefore, by the q-expansion principle, the $a_\mu(j)$ are in $\mathbb{Z}[j]$.

Applying the isomorphism λ_{M_0} with $\Gamma P M_0 = \Gamma\left(\begin{smallmatrix} p & 0 \\ 0 & 1 \end{smallmatrix}\right)$ to the representation, we obtain

$$f_{\left(\begin{smallmatrix} p & 0 \\ 0 & 1 \end{smallmatrix}\right)}\Phi_p'(\varphi_{\left(\begin{smallmatrix} p & 0 \\ 0 & 1 \end{smallmatrix}\right)}, j) = a_0(j) + a_1(j)\varphi_{\left(\begin{smallmatrix} p & 0 \\ 0 & 1 \end{smallmatrix}\right)} + \cdots + a_p(j)\varphi^p_{\left(\begin{smallmatrix} p & 0 \\ 0 & 1 \end{smallmatrix}\right)}.$$

Herein $f_{\left(\begin{smallmatrix} p & 0 \\ 0 & 1 \end{smallmatrix}\right)}$ as well as $\varphi_{\left(\begin{smallmatrix} p & 0 \\ 0 & 1 \end{smallmatrix}\right)}$ have q-coefficients divisible by p and, since $\Phi'(\varphi_{\left(\begin{smallmatrix} p & 0 \\ 0 & 1 \end{smallmatrix}\right)})$ has integral q-coefficients, the representation implies that the q-coefficients of $a_0(j)$ are divisible by p. Applying again the q-expansion principle, we find that $a_0(j) \in p\mathbb{Z}[j]$. □

Proof of Lemma 2.7.7: Writing

$$\Phi(X) = \prod_{\mu=1}^{n}(X - \varphi^{\sigma_\mu})$$

with the different isomorphisms $\sigma_1, \ldots, \sigma_n$ of $k(\varphi)/k$, we obtain

$$\left(\frac{\Phi(X) - \Phi(\varphi)}{X - \varphi}\right)^{\sigma_\mu}\Bigg|_{X=\varphi^{\sigma_\mu}} = \Phi'(\varphi^{\sigma_\mu})\delta_{\mu,\nu},$$

where $\delta_{\mu,\nu}$ denotes the Kronecker symbol. By summation over ν this becomes

$$\left[tr_{K(\varphi)/k}\left(f\frac{\Phi(X) - \Phi(\varphi)}{X - \varphi}\right)\right]_{X=\varphi} = f\Phi'(\varphi),$$

where the trace is applied to the coefficients of the polynomial $f\frac{\Phi(X)-\Phi(\varphi)}{X-\varphi}$. Writing

$$\frac{\Phi(X) - \Phi(\varphi)}{X - \varphi} = \sum_{\mu=1}^{n} B_\mu \frac{X^\mu - \varphi^\mu}{X - \varphi} = \sum_{\mu=1}^{n} B_\mu \sum_{\nu=0}^{\mu-1} \varphi^{\mu-(\nu+1)} X^\nu$$
$$= \sum_{\nu=0}^{n-1}\left(\sum_{\mu=\nu+1}^{n} B_\mu \varphi^{\mu-(\nu+1)}\right) X^\nu,$$

we end up with the formula

$$\sum_{\nu=0}^{n-1}\left(\sum_{\mu=\nu+1}^{n} B_\mu tr_{k(\varphi)/k}\left(f\varphi^{\mu-(\nu+1)}\right)\right)\varphi^\nu = f\Phi'(\varphi).$$

This implies the assertion of the lemma because by separability $\Phi'(\varphi) \neq 0$. □

2.8 Modular functions for $\Gamma(N)$

Let N a natural number,

$$\underline{x} = \left(\frac{a_1}{N}, \frac{a_2}{N}\right) \in \frac{1}{N}(\mathbb{Z}\times\mathbb{Z}) \setminus \mathbb{Z}\times\mathbb{Z}$$

and

$$\tau_{\underline{x}}(\omega) := g^{(2)}(\omega)\wp\left(\underline{x}\begin{pmatrix}\omega\\1\end{pmatrix}\middle|\begin{matrix}\omega\\1\end{matrix}\right)$$

an N**-th division value of Weber's** τ **function** as defined in 2.4.7. With these division values we have

Theorem 2.8.1 *$\mathbb{C}_{\Gamma(N)}$ is Galois over $\mathbb{C}_\Gamma$ with Galois group*

$$G(\mathbb{C}_{\Gamma(N)}/\mathbb{C}_\Gamma) = \{\lambda_M \mid M \in \Gamma,\ M \ mod\, \Gamma(N)\} \cong \Gamma/\Gamma(N).$$

$\mathbb{C}_{\Gamma(N)}$ is generated over $\mathbb{C}_\Gamma$ by the set of functions

$$\tau_{\underline{x}}, \quad \underline{x} \in \frac{1}{N}(\mathbb{Z}\times\mathbb{Z}) \setminus \mathbb{Z}\times\mathbb{Z}.$$

Proof We know that the N-th division values of the τ function are in $\mathbb{C}_{\Gamma(N)}$. Let $\mathbb{K}$ be the subfield generated by them. We consider the corresponding subgroup of Γ,

$$U = \{M \in \Gamma \mid \lambda_M|\mathbb{K} = id_{\mathbb{K}}\},$$

which contains $\Gamma(N)$. The assertion of Theorem 2.8.1 is proved by showing that $U = \Gamma(N)$. For $M \in U$ we have

$$\tau_{\underline{x}} = \tau_{\underline{x}}^{\lambda_M} = \tau_{\underline{x}M} \text{ for all } \underline{x} \in \frac{1}{N}(\mathbb{Z}\times\mathbb{Z}) \setminus \mathbb{Z}\times\mathbb{Z}.$$

This implies that $M \in \Gamma(N)$ because, as we will show,

$$\tau_{\underline{x}} = \tau_{\underline{x}'} \iff \underline{x} \equiv \pm\underline{x}' \bmod \mathbb{Z}\times\mathbb{Z}.$$

To prove this equivalence, we look at the q-expansion of $\tau_{\underline{x}}$. Using the identity

$$\frac{q^m Q^{\pm 1}}{(1-q^m Q^{\pm 1})} = \sum_{n=1}^{\infty} n q^{mn} Q^{\pm n}$$

we obtain by modifying the series in Theorem 1.8.2:

$$\tau_{\underline{x}}(\omega) = [q^{-1} + \cdots] \left[1 + \frac{12Q}{(1-Q)^2} + 12 \sum_{n,m=1}^{\infty} n q^{nm} (Q^n + Q^{-n} - 2) \right]$$

$$\text{with } q = e^{2\pi i \omega} \text{ and } Q = \zeta_N^{a_2} q^{\frac{a_1}{N}}.$$

Herein, however, for reason of convergence, we have to assume that $|qQ^{\pm 1}| < 1$, which can be achieved by changing $\underline{x}$ modulo $\mathbb{Z} \times \mathbb{Z}$. Further, the q-coefficients of the factor $[q^{-1} + \cdots]$ due to $g^{(2)}$ are in $\mathbb{Z}$. This formula implies the above equivalence. So we have proved Theorem 2.8.1. □

According to the above formula, $\tau_{(\frac{a}{N},0)}$ has a rational q-expansion. Further, an arbitrary N-th division value is conjugate to $\tau_{(\frac{a}{N},0)}$ over $\mathbb{Q}_\Gamma$ for a suitable $a \in \mathbb{Z}$. Therefore, by the extended q-expansion principle the main-polynomial of an N-th division value is in $\mathbb{Q}(j)[X]$. Analogously to 2.7 we therefore consider the extension of $\mathbb{Q}(j)$ generated by all N-th division values. However, for the later applications we should take the N-th cyclotomic field $\mathbb{Q}_N = \mathbb{Q}(\zeta_N)$ as a field of constants rather than $\mathbb{Q}$. To characterise these functions we define the **modular function field of level** N,

$$F_N := \{ f \in \mathbb{C}_{\Gamma(N)} \mid f \textit{ having } q\textit{-coefficients in } \mathbb{Q}_N \textit{ in all cusps} \},$$

first introduced by Söhngen (1935). We have:

Theorem 2.8.2 *F_N is generated over $\mathbb{Q}_N(j)$ by all N-th division values of the τ function .*

Proof By the primitive element theorem we know there is a linear combination

$$g = \sum_{\underline{x}} n_{\underline{x}} \tau_{\underline{x}}, \quad n_{\underline{x}} \in \mathbb{Z},$$

of N-th division values generating $\mathbb{C}_{\Gamma(N)}$ over $\mathbb{C}_\Gamma$. Observing now that the q-coefficients of g are in $\mathbb{Q}_N$, we can show as in the proof of Theorem 2.7.3 that:

$$F_N \subseteq \mathbb{Q}_N(j, g).$$

Hence, F_N is contained in the field generated over $\mathbb{Q}_N(j)$ by the N-th division values. The opposite inclusion is trivial. $\square$

Remark 2.8.3 In the proof of Theorem 2.8.2 we only used the functions in F_N as having q-coefficients in $\mathbb{Q}_N$. Therefore, the proof shows that a function in $\mathbb{C}_{\Gamma(N)}$ having q-coefficients in $\mathbb{Q}_N$ necessarily also has q-coefficients in $\mathbb{Q}_N$ in ALL cusps.

By Theorems 2.8.1 and 2.8.2 the extension $F_N/\mathbb{Q}_N(j)$ is Galois and

$$G(F_N/\mathbb{Q}_N(j)) \cong \Gamma/\Gamma(N).$$

To determine the Galois group of $F_N/\mathbb{Q}_\Gamma$, we have to extend the automorphisms

$$\lambda_a = (\zeta_N \mapsto \zeta_N^a), \quad a \in (\mathbb{Z}/N\mathbb{Z})^*,$$

of $\mathbb{Q}_N$ to F_N. For this purpose we set

$$\left(\sum_{n=-\infty}^{\infty} a_n q^{\frac{n}{N}}\right)^{\lambda_a} := \sum_{n=-\infty}^{\infty} (a_n^{\lambda_a}) q^{\frac{n}{N}}$$

for a Laurent series $\sum\limits_{n=-\infty}^{\infty} a_n q^{\frac{n}{N}}$ in $q^{\frac{1}{N}}$ with coefficients in $\mathbb{Q}_N$, thereby defining an automorphism of the field of Laurent series over $\mathbb{Q}_N$. The action of λ_a on a division value $\tau_{\underline{x}}$ can also be described by

$$\tau_{\underline{x}}^{\lambda_a} = \tau_{\underline{x}\left(\begin{smallmatrix}1 & 0\\ 0 & a\end{smallmatrix}\right)}.$$

In view of Theorem 2.8.2 this implies that the restriction of λ_a to F_N is an automorphism of F_N. Hence the Galois group $F_N/\mathbb{Q}_\Gamma$ is generated by the two subgroups

$$\{\lambda_M \mid M \in \Gamma\} \cong \Gamma/\Gamma(N) \text{ and } \{\lambda_a \mid a \in (\mathbb{Z}/N\mathbb{Z})^*\} \cong (\mathbb{Z}/N\mathbb{Z})^*.$$

Further, the action of $\lambda_a, a \in (\mathbb{Z}/N\mathbb{Z})^*$, and $\lambda_M, M \in \Gamma$, on the N-th division values shows that for every $\lambda \in G(F_N/\mathbb{Q}_\Gamma)$ there is a matrix $A \in Gl_2(Z/N\mathbb{Z})$ with

$$\zeta_N^\lambda = \zeta_N^{\det A} \text{ and } \tau_{\underline{x}}^\lambda = \tau_{\underline{x}A}$$

for all N-th partial values $\tau_{\underline{x}}$. A is a uniquely determined modulo N up to a factor ± 1 and, since every matrix $A \in Gl_2(\mathbb{Z}/N\mathbb{Z})$ has a decomposition of the form

$$A \equiv M_1\left(\begin{smallmatrix}1 & 0\\ 0 & a\end{smallmatrix}\right)M_2 \bmod N \text{ with } M_1, M_2 \in \Gamma \text{ and } a \in \mathbb{Z} \text{ prime to } N,$$

it defines an element $\lambda \in G(F_N/\mathbb{Q}_\Gamma)$.

Theorem 2.8.4 *F_N is Galois over $\mathbb{Q}_\Gamma$ and*

$$G(F_N/\mathbb{Q}_\Gamma) \cong Gl_2(\mathbb{Z}/N\mathbb{Z})/\{\pm 1\}.$$

The action of the automorphism λ_A corresponding to a matrix $A \in Gl_2(\mathbb{Z}/N\mathbb{Z})$ on the generating elements ζ_N and $\tau_{\underline{x}}$ is given by

$$\zeta_N^{\lambda_A} = \zeta_N^{\det(A)} \text{ and } \tau_{\underline{x}}^{\lambda_A} = \tau_{\underline{x}A}.$$

For an arbitrary function $f \in F_N$ the action is calculated via a decomposition of the above form of A:

$$f^{\lambda_A} = [[f^{\lambda_{M_1}}]^{\lambda_{\left(\begin{smallmatrix}1&0\\0&a\end{smallmatrix}\right)}}]^{\lambda_{M_2}},$$

observing that by definition $\lambda_{\left(\begin{smallmatrix}1&0\\0&a\end{smallmatrix}\right)}$ is the application of automorphism λ_a of $\mathbb{Q}_N$ to the q-coefficients of a function in F_N. To simplify notation, we often write the action of λ_A as

$$f \circ A := f^{\lambda_A}.$$

2.9 The field $\mathbb{Q}(\gamma_2, \gamma_3)$

In order to find a simple algebraic equation for modular functions, it is sometimes useful to replace $\mathbb{Q}(j)$ as the base field by one of the extensions

$$\mathbb{Q}(\gamma_2), \quad \mathbb{Q}(\gamma_3) \quad \text{or} \quad \mathbb{Q}(\gamma_2, \gamma_3).$$

of $\mathbb{Q}(j)$. We set

$$\begin{aligned} U_2 &:= \{M \in \Gamma \mid \gamma_2^{\lambda_M} = \gamma_2\}, \\ U_3 &:= \{M \in \Gamma \mid \gamma_3^{\lambda_M} = \gamma_3\}, \\ U_6 &:=< U_2, U_3 > . \end{aligned}$$

By the η-transformation formula, we see that $\Gamma(6)$ is a subgroup of U_6. Therefore, we have the inclusion

$$\mathbb{C}_{U_6} \subseteq \mathbb{C}_{\Gamma(N)}, \text{ for } 6|N.$$

In particular, the q-expansions of the functions in $\mathbb{C}_{U_6} = \mathbb{C}(\gamma_2, \gamma_3)$ are power series of $q^{\frac{1}{6}}$. We are now going to find a suitable generalisation of Theorems 2.5.2 and 2.6.2.

Theorem 2.9.1 *The ring of holomorphic functions in $\mathbb{C}(\gamma_2, \gamma_3)$ is given by $\mathbb{C}[\gamma_2, \gamma_3]$.*

Theorem 2.9.2

(i) *Let* $f = \sum_{n \geq n_0} a_n q^{\frac{n}{6}}$ *be in* $\mathbb{C}(\gamma_2, \gamma_3)$*, and let* Λ *be a subfield of* $\mathbb{C}$*. Then:*

$$f \in \Lambda(\gamma_2, \gamma_3) \iff f \in q^{\frac{n_0}{6}} \Lambda[[q^{\frac{1}{6}}]].$$

(ii) *Let* $f = \sum_{n \geq n_0} a_n q^{\frac{n}{6}}$ *be in* $\mathbb{C}[\gamma_2, \gamma_3]$ *and let* $\mathfrak{a}$ *be an additive subgroup of* $\mathbb{C}$*. Then*

$$f \in \mathfrak{a}[\gamma_2, \gamma_3] \iff f \in q^{\frac{n_0}{6}} \mathfrak{a}[[q^{\frac{1}{6}}]].$$

Proof of Theorem 2.9.1 We show that in every $\mathbb{H}$ holomorphic function $f \in \mathbb{C}(\gamma_2, \gamma_3)$ is a polynomial in γ_2 and γ_3. For the functions in $\mathbb{C}(\gamma_2)$ the proof is completely analogous to the proof of Theorem 2.5.8 because $\mathbb{C}(\gamma_2)$ is a rational function field and γ_2 takes every value in $\mathbb{H}$.

$\mathbb{C}(\gamma_2, \gamma_3)/\mathbb{C}(\gamma_2)$ being of degree 2, every function in $f \in \mathbb{C}(\gamma_2, \gamma_3)$ can be written in the form

$$f = a(\gamma_2) + b(\gamma_2)\gamma_3, \qquad a(\gamma_2), b(\gamma_2) \in \mathbb{C}(\gamma_2).$$

Therefore, for f holomorphic in $\mathbb{H}$ we have to show that $a(\gamma_2)$ and $b(\gamma_2)$ are polynomials. Let λ_M be the non-trivial automorphism of $\mathbb{C}(\gamma_2, \gamma_3)/\mathbb{C}(\gamma_2)$ then

$$f + f^{\lambda_M} = 2a(\gamma_2) \text{ and } (f - f^{\lambda_M})^2 = (2b(\gamma_2))^2(\gamma_2^3 - 12^3)$$

are holomorphic in $\mathbb{H}$ and thus polynomials in γ_2. $X^3 - 12^3$ having only simple zeros in $\mathbb{C}$, this implies that $a(\gamma_2)$ and also $b(\gamma_2)$ are polynomials in γ_2. Hence f is in $\mathbb{C}[\gamma_2, \gamma_3]$. □

Proof of Theorem 2.9.2 The q-expansions of γ_2 and γ_3 are of the form

$$\gamma_2 = q^{-\frac{1}{3}} \left[1 + \sum_{n=1}^{\infty} c_n q^n\right],$$
$$\gamma_3 = q^{-\frac{1}{2}} \left[1 + \sum_{n=1}^{\infty} d_n q^n\right],$$

with $c_n, d_n \in \mathbb{Z}$. The implication from the right to the left in the first part of Theorem 2.9.2 is now proved analogously to the proof of the first part of Theorem 2.5.9 by inserting the q-expansion of the functions involved into the representation $f = a(\gamma_2) + b(\gamma_2)\gamma_3$. For f having q-coefficients in Λ, this shows that the coefficients of $a(X)$ and $b(X)$ are solutions to a system of linear equations with coefficients in Λ.

To prove the second assertion we write

$$f = \sum_{(\mu,\nu)\in\{0,1,2\}\times\{0,1\}} \gamma_2^\mu \gamma_3^\nu p_{\mu,\nu}(j)$$

with polynomials $p_{\mu,\nu}$. Herein

$$\gamma_2^\mu \gamma_3^\nu p_{\mu,\nu}(j) = \sum_{n\equiv 2\mu+3\nu \bmod 6} b_n q^{\frac{n}{6}}.$$

The q-expansions of different summands containing no common power of q, we can conclude that the $p_{\mu,\nu}(j)$ have q-coefficients in $\mathfrak{a}$ if this is true for the q-coefficients of f. By Theorem 2.5.9 we now obtain the implication from the right to the left in the second part of Theorem 2.9.2. The converse implication is trivial. □

By Theorems 2.9.1 and 2.9.2 we come analogously to Theorem 2.9.3:

Theorem 2.9.3 *Let s be one of the numbers 2,3,6, and let U be a subgroup of Γ with $[\Gamma : U] < \infty$. Let $f \in \mathbb{C}_U$ be holomorphic in $\mathbb{H}$. Then, in particular, f is a function in $\mathbb{C}_{U\cap U_s}$. By $m(X)$ we denote the minimal polynomial of f over $\mathbb{C}_{U_s}$. Then:*

(i) *If ONE conjugate f^{λ_M} of f over $\mathbb{C}_{U_s}$ has q-coefficients in the field Λ, then*

$$\begin{aligned} m(X) &\in \Lambda[X,\gamma_s] \quad &\text{for} \quad s = 2,3,\\ m(X) &\in \Lambda[X,\gamma_2,\gamma_3] \quad &\text{for} \quad s = 6. \end{aligned}$$

(ii) *If ALL conjugates of f over $\mathbb{C}_{U_s}$ have q have q-coefficients in a subring $\mathfrak{a}$ of $\mathbb{C}$, then*

$$\begin{aligned} m(X) &\in \mathfrak{a}[X,\gamma_s] \quad &\text{for} \quad s = 2,3,\\ m(X) &\in \mathfrak{a}[X,\gamma_2,\gamma_3] \quad &\text{for} \quad s = 6. \end{aligned}$$

2.10 Lower powers of η-quotients

The lower powers of η-quotients considered in section 2.4.3 satisfy rather simple algebraic equations over the field $\mathbb{Q}(\gamma_2,\gamma_3)$. Similarly to Weber (1908), §72, for $n \in \mathbb{N}$ we study the functions

$$\eta_n(\omega) := \sqrt{n}\frac{\eta(n\omega)}{\eta(\omega)}.$$

In view of later applications we also consider the functions

$$\eta_{p,q}(\omega) := \frac{\eta(p\omega)\eta(q\omega)}{\eta(pq\omega)\eta(\omega)}.$$

with odd primes p, q. More precisely, we will need the functions

$$g_n(\omega) := \frac{\eta\left(\frac{\omega}{n}\right)}{\eta(\omega)}. \tag{2.3}$$

and

$$g_{p,q}(\omega) := \frac{\eta\left(\frac{\omega}{p}\right)\eta\left(\frac{\omega}{q}\right)}{\eta\left(\frac{\omega}{pq}\right)\eta(\omega)}, \tag{2.4}$$

which, up to a constant factor, are conjugate to η_n resp. to $\eta_{p,q}$ over $\mathbb{C}(\gamma_2, \gamma_3)$. For instance, as can be derived from the next two theorems,

$$g_n^6 \text{ is conjugate over } \mathbb{C}(\gamma_2, \gamma_3) \text{ to } (-1)^{\frac{n-1}{2}}\eta_n^6,$$

and for odd primes p, q

$$g_{p,q}^m \text{ is conjugate over } \mathbb{C}(\gamma_2, \gamma_3) \text{ to } (-1)^{\frac{p-1}{2}\frac{q-1}{2}}\eta_{p,q}^m \text{ with } m = \gcd(pq, 3).$$

To compute the equations of some powers of η_n and $\eta_{p,q}$ over $\mathbb{Q}(\gamma_2, \gamma_3)$ we need explicit formulae for their conjugates, so let s again be one of the numbers $2, 3, 6$. Then, if the natural number n is coprime to s, we have

$$[U_s : \Gamma_{\left(\begin{smallmatrix} n & 0 \\ 0 & 1 \end{smallmatrix}\right)} \cap U_s] = [\Gamma : \Gamma_{\left(\begin{smallmatrix} n & 0 \\ 0 & 1 \end{smallmatrix}\right)}]$$

with the above definition of U_s. Setting $\mathbb{Q}_2 := \mathbb{Q}(\gamma_2), \mathbb{Q}_3 := \mathbb{Q}(\gamma_3)$ and $\mathbb{Q}_6 := \mathbb{Q}(\gamma_2, \gamma_3)$, this implies that isomorphisms of $\mathbb{Q}_{\Gamma_{\left(\begin{smallmatrix} n & 0 \\ 0 & 1 \end{smallmatrix}\right)}}\mathbb{Q}_s$ over $\mathbb{Q}_s$ are obtained by determining the matrices M_μ in Theorem 2.6.1 in U_s. Therefore, we choose a system of representatives of primitive matrices of determinant n of the form

$$R_\mu = \begin{pmatrix} a & b \\ 0 & d \end{pmatrix} \text{ with } b \equiv 0 \bmod s$$

and then matrices $M_\mu \in U_s$ so that

$$\begin{pmatrix} n & 0 \\ 0 & 1 \end{pmatrix} M_\mu = N_\mu \begin{pmatrix} a & b \\ 0 & d \end{pmatrix}$$

with a matrix $N_\mu \in \Gamma$. Expressing M_μ by

$$N_\mu = \begin{pmatrix} \alpha & \beta \\ \gamma & \delta \end{pmatrix}$$

yields

$$M_\mu = \begin{pmatrix} \frac{\alpha a}{n} & \frac{\alpha b + \beta d}{n} \\ \gamma a & \gamma b + \delta d \end{pmatrix}.$$

Now we choose

$$\alpha = d,$$
$$\beta = -b + \nu a, \ \nu \in \mathbb{Z}, \text{ so that } \gcd(\alpha, \beta) = 1$$
$$\text{and then}$$
$$\gamma, \delta \in \mathbb{Z} \text{ with } \alpha\delta - \beta\gamma = 1.$$

Thus, for a given primitive matrix $R_\mu = \left(\begin{smallmatrix} a & b \\ 0 & d \end{smallmatrix}\right)$ we have determined unimodular matrices

$$M_\mu = \begin{pmatrix} 1 & \nu \\ \gamma a & \gamma b + \delta d \end{pmatrix}, N_\mu = \begin{pmatrix} d & -b + \nu a \\ \gamma & \delta \end{pmatrix}$$

with $\left(\begin{smallmatrix} n & 0 \\ 0 & 1 \end{smallmatrix}\right) M_\mu = N_\mu \left(\begin{smallmatrix} a & b \\ 0 & d \end{smallmatrix}\right)$. As mentioned above for n coprime to s we can find a system of representatives R_μ such that

$$b \equiv 0 \bmod 8 \text{ if } s = 2,$$
$$b \equiv 0 \bmod 3 \text{ if } s = 3,$$
$$b \equiv 0 \bmod 24 \text{ if } s = 6.$$

The above construction shows that in these cases also ν and γ can be found divisible by 3, 8, 24 with $\gamma > 0$. For M_μ and N_μ we then have the congruences

$$M_\mu \equiv \begin{pmatrix} 1 & 0 \\ 0 & 1 \end{pmatrix} \bmod t \text{ and } N_\mu \equiv \begin{pmatrix} d & 0 \\ 0 & d \end{pmatrix} \bmod t$$

with $t = 3, 8, 24$.

For $3 \nmid n$ we know by section 2.4.3 that:

$$\eta_n^8 \gamma_2^{n-1} \in \mathbb{Q}_{\Gamma_{\left(\begin{smallmatrix} n & 0 \\ 0 & 1 \end{smallmatrix}\right)}}.$$

According to Theorem 2.7.1 η_n^{24} is a generating element for $\mathbb{Q}_{\Gamma_{\left(\begin{smallmatrix} n & 0 \\ 0 & 1 \end{smallmatrix}\right)}}$ over $\mathbb{Q}_\Gamma$ and it is also a generator for $\mathbb{Q}_{\Gamma_{\left(\begin{smallmatrix} n & 0 \\ 0 & 1 \end{smallmatrix}\right)}} \mathbb{Q}_2$ because $\mathbb{Q}_{\Gamma_{\left(\begin{smallmatrix} n & 0 \\ 0 & 1 \end{smallmatrix}\right)}} \cap \mathbb{Q}_2 = \mathbb{Q}_\Gamma$. The minimal polynomial of η_n^8 over $\mathbb{Q}_2$ is given by

$$\prod_\mu (X - \eta_n^{8\lambda_{M_\mu}})$$

with the above system $\{M_\mu\}$. Using the transformation formula of the η function as in 2.4.3 and, bearing in mind the congruences for M_μ and N_μ, we obtain explicit formulae for the conjugates. Similar results are obtained for η_n^6 and η_2^n if n is prime to 2 resp. 6. This leads to:

Theorem 2.10.1

(i) *Let n be a natural number not divisible by 3. Then η_n^8 is a generating function for $\mathbb{Q}_{\Gamma_{\left(\begin{smallmatrix} n & 0 \\ 0 & 1\end{smallmatrix}\right)}}\mathbb{Q}_2$ over $\mathbb{Q}_2$, and the conjugates of η_n^8 over $\mathbb{Q}_2$ are given by*

$$a^4 \left(\frac{\eta(\frac{a\omega+b}{d})}{\eta(\omega)}\right)^8,$$

where $\left(\begin{smallmatrix} a & b \\ 0 & d\end{smallmatrix}\right)$ is a system of representatives of primitive matrices of determinant n with $b \equiv 0 \bmod 3$.

(ii) *Let n be a natural number not divisible by 2. Then η_n^6 is a generating function for $\mathbb{Q}_{\Gamma_{\left(\begin{smallmatrix} n & 0 \\ 0 & 1\end{smallmatrix}\right)}}\mathbb{Q}_3$ over $\mathbb{Q}_3$ and the conjugates of η_n^6 over $\mathbb{Q}_3$ are given by*

$$(-1)^{\frac{d-1}{2}} a^3 \left(\frac{\eta(\frac{a\omega+b}{d})}{\eta(\omega)}\right)^6,$$

where $\left(\begin{smallmatrix} a & b \\ 0 & d\end{smallmatrix}\right)$ is a system of representatives of primitive matrices of determinant n with $b \equiv 0 \bmod 4$.

(iii) *Let n be a natural number prime to 6. Then η_n^2 is a generating function for $\mathbb{Q}_{\Gamma_{\left(\begin{smallmatrix} n & 0 \\ 0 & 1\end{smallmatrix}\right)}}\mathbb{Q}_6$ over $\mathbb{Q}_6$, and the conjugates of η_n^2 over $\mathbb{Q}_6$ are given by*

$$(-1)^{\frac{d-1}{2}} a \left(\frac{\eta(\frac{a\omega+b}{d})}{\eta(\omega)}\right)^2,$$

where $\left(\begin{smallmatrix} a & b \\ 0 & d\end{smallmatrix}\right)$ is a system of representatives of primitive matrices of determinant n with $b \equiv 0 \bmod 12$.

(iv) *Let n be an odd natural number which is a perfect square. We set $m = 1$ if n is divisible by 3 and $m = 3$ otherwise. Then η_n^m is a generating function of $\mathbb{Q}_{\Gamma_{\left(\begin{smallmatrix} n & 0 \\ 0 & 1\end{smallmatrix}\right)}}$ over $\mathbb{Q}_\Gamma$, and the conjugates of η_n^m over $\mathbb{Q}_\Gamma$ are given by*

$$\left(\left(\tfrac{b}{e}\right) i^{\frac{d-1}{2}} \sqrt{a} \frac{\eta(\frac{a\omega+b}{d})}{\eta(\omega)}\right)^m,$$

where $\left(\begin{smallmatrix} a & b \\ 0 & d\end{smallmatrix}\right)$ is a system of representatives of primitive matrices of determinant n with $b \equiv 0 \bmod 24$ and $e = \gcd(a, d)$.

Proof The results in the cases (i)–(iii) are obtained as described above. To prove (iv), we use the transformation formula of the η function, which yields the above formula with the Legendre symbol $(\frac{\gamma}{d})$ instead of the factor $(\frac{b}{e})$. In view of $1 = \alpha\delta - \beta\gamma \equiv -\beta\gamma \bmod d$ and $\beta = -b + \nu a$, the symbol $(\frac{\gamma}{d})$ becomes $(\frac{b-\nu a}{d})$. Herein $b - \nu a$ for all possible choices of ν is running through the residue classes modulo d that are congruent to b modulo a. The images $\eta_n^{\lambda M_\mu}$ only depend on the matrices $\left(\begin{smallmatrix} a & b \\ 0 & d \end{smallmatrix}\right)$, so the Legendre symbols $(\frac{b-\nu a}{d})$ are all equal, and the common value is $(\frac{b}{e})$. □

A similar result can be derived for the functions $\eta_{p,q}(\omega)$ with $p, q \in \mathbb{N}$ prime to 6.

Theorem 2.10.2 *Let p, q be primes other than 2 and 3. Then $\eta_{p,q}$ is a generating function of $\mathbb{Q}_{\Gamma_{\left(\begin{smallmatrix} pq & 0 \\ 0 & 1 \end{smallmatrix}\right)}}\mathbb{Q}_6$ over $\mathbb{Q}_6$, and the conjugates over $\mathbb{Q}_6$ are of the following form:*

(i) *In the case $p = q$:*

$$\frac{\eta(p\omega)\eta(p\omega)}{\eta(p^2\omega)\eta(\omega)},$$

$$(-1)^{\frac{p-1}{2}\frac{q-1}{2}} \frac{\eta(\frac{\omega+b}{p})^2}{\eta(\frac{\omega+b}{p^2})\eta(\omega)}, \qquad b \bmod p, b \equiv 0 \bmod 24,$$

$$\sqrt{p}\left(\frac{b}{p}\right) i^{\frac{1-p}{2}} \frac{\eta(p\omega)^2}{\eta(\frac{p\omega+b}{p})\eta(\omega)}, \qquad b \bmod p, b \equiv 0 \bmod 24, p \nmid b.$$

(ii) *In the case $p \neq q$:*

$$\frac{\eta(p\omega)\eta(q\omega)}{\eta(pq\omega)\eta(\omega)},$$

$$\left(\frac{p}{q}\right) \frac{\eta(p\omega)\eta(\frac{\omega+\frac{b}{p}}{q})}{\eta(\frac{p\omega+b}{q})\eta(\omega)}, \qquad b \bmod q, b \equiv 0 \bmod 24p,$$

$$\left(\frac{q}{p}\right) \frac{\eta(\frac{\omega+\frac{b}{q}}{p})\eta(q\omega)}{\eta(\frac{q\omega+b}{p})\eta(\omega)}, \qquad b \bmod p, b \equiv 0 \bmod 24q,$$

$$(-1)^{\frac{p-1}{2}\frac{q-1}{2}} \frac{\eta(\frac{\omega+b}{p})\eta(\frac{\omega+b}{q})}{\eta(\frac{\omega+b}{pq})\eta(\omega)}, \qquad b \bmod pq, b \equiv 0 \bmod 24.$$

Proof In the proof of Theorem 2.10.2 we use the same arguments as in the proof of Theorem 2.10.1. In addition, we have to determine the action of the M_μ on the cosets $\Gamma\left(\begin{smallmatrix} p & 0 \\ 0 & 1 \end{smallmatrix}\right)$ and $\Gamma\left(\begin{smallmatrix} q & 0 \\ 0 & 1 \end{smallmatrix}\right)$, where the M_μ are defined by the relation

$$\begin{pmatrix} pq & 0 \\ 0 & 1 \end{pmatrix} M_\mu = N_\mu \begin{pmatrix} a & b \\ 0 & d \end{pmatrix}.$$

Therefore, we observe that an equation

$$\begin{pmatrix} ad\,0 \\ 0\;\,1 \end{pmatrix} M = \begin{pmatrix} \alpha & \beta \\ \gamma & \delta \end{pmatrix} \begin{pmatrix} a\,b \\ 0\,d \end{pmatrix}$$

implies that

$$\begin{pmatrix} a\,0 \\ 0\,1 \end{pmatrix} M = \begin{pmatrix} \frac{\alpha}{d} & \beta \\ \gamma & \delta d \end{pmatrix} \begin{pmatrix} a\,b \\ 0\,1 \end{pmatrix}$$

and

$$\begin{pmatrix} d\,0 \\ 0\,1 \end{pmatrix} M = \begin{pmatrix} \alpha & \frac{\beta}{a} \\ \gamma a & \delta \end{pmatrix} \begin{pmatrix} 1 & \frac{b}{a} \\ 0 & d \end{pmatrix}.$$

In the same way

$$\begin{pmatrix} ad\,0 \\ 0\;\,1 \end{pmatrix} M = \begin{pmatrix} \alpha & \beta \\ \gamma & \delta \end{pmatrix} \begin{pmatrix} 1 & b \\ 0 & ad \end{pmatrix}$$

implies that

$$\begin{pmatrix} a\,0 \\ 0\,1 \end{pmatrix} M = \begin{pmatrix} \frac{\alpha}{d} & \beta \\ \gamma & \delta d \end{pmatrix} \begin{pmatrix} 1\,b \\ 0\,a \end{pmatrix}.$$

Using these relations, the conjugates are obtained by the transformation formula of the η function. For the third formula in the case $p = q$ the transformation formula of the η function first yields $(\frac{\gamma}{p})$ instead of $(\frac{b}{p})$. Then, using the congruence

$$\gamma b \equiv -\gamma(-b + \nu p) \equiv -\gamma\beta \equiv \alpha\delta - \gamma\beta \equiv 1 \bmod p,$$

we find that the two factors are equal. Finally, by looking at their q-expansions, the conjugates all turn out to be different, which proves that $\eta_{p,q}$ is a generator for $\mathbb{Q}_{\Gamma_{\left(\begin{smallmatrix} pq & 0 \\ 0 & 1 \end{smallmatrix}\right)}}\mathbb{Q}_6$ over $\mathbb{Q}_6$. □

Example 2.10.3 We are now able to determine the minimal polynomials for the powers of g_n and for $g_{p,q}$ by solving the system of equation in the proof of Theorem 2.9.3 (see also Weber (1908), pp. 251, 253, for the first five equations):

$$m_{g_2^8}(X) = X^3 - \gamma_2 X + 16$$

$$m_{g_3^6} = X^4 + 18X^2 - \gamma_3 X - 27$$

$$m_{g_5^2}(X) = X^6 + 10X^3 - \gamma_2 X + 5$$
$$m_{g_7^2}(X) = X^8 + 7\cdot 2X^6 + 7\cdot 9X^4 + 7\cdot 10X^2 - \gamma_3 X - 7$$
$$m_{g_{11}^2} = X^{12} - 11\cdot 90X^6 + 11\cdot 40\gamma_2 X^4 - 11\cdot 15\gamma_3 X^3 + 11\cdot 2\gamma_2^2 X^2 - \gamma_2\gamma_3 X - 11$$

$$\begin{aligned}
m_{g_{5,7}}(X) &= X^{48} + (-j + 708)\, X^{47} + (35\, j + 171402)\, X^{46} \\
&+ (-525\, j + 15185504)\, X^{45} + (4340\, j + 248865015)\, X^{44} \\
&+ (-20825\, j + 1763984952)\, X^{43} + (52507\, j + 6992359702)\, X^{42} \\
&+ (-22260\, j + 19325688804)\, X^{41} + (-243035\, j + 42055238451)\, X^{40} \\
&+ (596085\, j + 70108209360)\, X^{39} + (-272090\, j + 108345969504)\, X^{38} \\
&+ (-671132\, j + 121198179480)\, X^{37} + (969290\, j + 155029457048)\, X^{36} \\
&+ (-1612065\, j + 97918126080)\, X^{35} + (2493785\, j + 141722714700)\, X^{34} \\
&+ (647290\, j - 1509796288)\, X^{33} + (-3217739\, j + 108236157813)\, X^{32} \\
&+ (3033590\, j - 93954247716)\, X^{31} + (-5781615\, j + 91135898154)\, X^{30} \\
&+ (1744085\, j - 108382009680)\, X^{29} + (1645840\, j + 66862445601)\, X^{28} \\
&+ (-2260650\, j - 66642524048)\, X^{27} + (6807810\, j + 38019611082)\, X^{26} \\
&+ (-2737140\, j - 28638526644)\, X^{25} + (2182740\, j + 17438539150)\, X^{24} \\
&+ (-125335\, j - 8820058716)\, X^{23} + (-1729889\, j + 5404139562)\, X^{22} \\
&+ (1024275\, j - 1967888032)\, X^{21} + (-1121960\, j + 1183191681)\, X^{20} \\
&+ (395675\, j - 370697040)\, X^{19} + (-54915\, j + 103145994)\, X^{18} \\
&+ (15582\, j - 42145404)\, X^{17} + (34755\, j - 15703947)\, X^{16} \\
&+ (-6475\, j - 3186512)\, X^{15} + (1120\, j - 4585140)\, X^{14} \\
&+ (-176\, j + 1313040)\, X^{13} + \left(j^2 - 1486\, j - 38632\right) X^{12} \\
&+ (-7\, j + 399000)\, X^{11} + (-19\, j + 211104)\, X^{10} + (-9\, j + 6771)\, X^{8} \\
&+ (8\, j - 6084)\, X^{7} + (7\, j - 5258)\, X^{6} \\
&+ (j - 792)\, X^{5} - 105\, X^{4} + 16\, X^{3} + 42\, X^{2} + 12\, X + 1
\end{aligned}$$

For higher p, q the coefficients of $m_{g_{p,q}}$ are growing very fast. Without skipping most of the summands the following example would fill three pages.

$$m_{g_{5,17}}(X) = X^{108} + \left(-{\gamma_2}^2 j^2 + 1983\, {\gamma_2}^2 j - 703818\, {\gamma_2}^2\right) X^{107}$$

....

$$\begin{aligned}
&+ (358985068495\, j^4 - 3090344731374757377745\, j^3 \\
&-17585251716130674348255782\, j^2 - 7911838808006536150814586061\, j \\
&+157207014552439419850187630184 4)\, X^{54}
\end{aligned}$$

....
$+\left(\gamma_2{}^2 j^2 - 1983\,\gamma_2{}^2 j + 728366\,\gamma_2{}^2\right) X^5 + 14076\,\gamma_2\,X^4 + (-5\,j - 21824)\,X^3$
$+\,5\,\gamma_2\,X + 1$

3
Basic facts from number theory

To investigate how singular values of elliptic and modular functions are linked to class fields over quadratic imaginary number fields, we need some facts about orders and ideals in quadratic number fields. Furthermore, we will collect the basic definitions and results from class field theory. Here we will use the classical language because this marries well with singular values of elliptic and modular functions that are essentially dependent on ideals.

3.1 Ideal theory of suborders in a quadratic number field

3.1.1 Fractional ideals, integral ideals, proper ideals, regular ideals

Let K be a quadratic imaginary number field. Let d denote the discriminant and $\mathfrak{O}_K$ the maximal order of K. $\mathfrak{O}_K$ is a free $\mathbb{Z}$-module of rank 2, described by a basis

$$\mathfrak{O}_K = [\omega, 1] \quad \text{with} \quad \omega = \frac{d + \sqrt{d}}{2},$$

where we are using the notation

$$[a, b] = \mathbb{Z}a + \mathbb{Z}b$$

for a lattice. An order in K is defined as a subring containing the unit element 1 of K, which is also a rank two $\mathbb{Z}$-module. Every order is a subring of $\mathfrak{O}_K$ and for every $t \in \mathbb{N}$ there exists a unique order $\mathfrak{O}$, with $[\mathfrak{O}_K : \mathfrak{O}] = t$, given by

$$\mathfrak{O} = \mathfrak{O}_t := [t\omega, 1].$$

With these notations $\mathfrak{O}_K = \mathfrak{O}_1$. The conductor of $\mathfrak{O} = \mathfrak{O}_t$ is defined as the greatest integral ideal of $\mathfrak{O}$ that is also an ideal of $\mathfrak{O}_1$:

$$\mathfrak{t} = \{\xi \in \mathfrak{O} \mid \xi\mathfrak{O}_1 \subseteq \mathfrak{O}\}.$$

The conductor of $\mathfrak{O}_t$ is given by

$$\mathfrak{t} = \mathfrak{O}_1 t.$$

Therefore, $\mathfrak{O}_t$ is also called the order of conductor $t\mathfrak{O}_1$ or "order of conductor t".

Definition 3.1.1 Let $\mathfrak{O}$ be an order in K and $\mathfrak{a}$ a free $\mathbb{Z}$-module of rank 2 in K. Then $\mathfrak{a}$ is called

(i) an ideal of $\mathfrak{O}$ if $\mathfrak{O}\mathfrak{a} \subseteq \mathfrak{a}$,
(ii) an integral ideal of $\mathfrak{O}$ if $\mathfrak{O}\mathfrak{a} \subseteq \mathfrak{a} \subseteq \mathfrak{O}$,
(iii) a proper ideal of $\mathfrak{O}$ if $\mathfrak{O} = \{\xi \in K \mid \xi\mathfrak{a} \subseteq \mathfrak{a}\}$.

The next theorem shows that every free $\mathbb{Z}$-module of rank 2 in K is a proper ideal of some order in K.

Theorem 3.1.2 *Let $\mathfrak{a} = [\alpha_1, \alpha_2]$ be a free $\mathbb{Z}$-module of rank 2 in K and $\alpha := \frac{\alpha_1}{\alpha_2}$. Then:*

(i) *$\mathfrak{O} := \{\xi \in K \mid \xi\mathfrak{a} \subseteq \mathfrak{a}\}$ is an order in K.*
(ii) *Let $AX^2 + BX + C = 0$, $A, B, C \in \mathbb{Z}, A > 0$, and $\gcd(A, B, C) = 1$, be the primitive quadratic equation satisfied by α. Then*

$$\mathfrak{O} = [A\alpha, 1] = \mathfrak{O}_t.$$

Denote by

$$D(\alpha) := B^2 - 4AC$$

the discriminant of $AX^2 + BX + C$. Then the conductor t of $\mathfrak{O}$ is obtained from the relation

$$D(\alpha) = t^2 d$$

with the discriminant d of K. Thus, the proper ideals of $\mathfrak{O}$ are exactly the free $\mathbb{Z}$-modules of rank 2 in K, with a ratio of basis satisfying a primitive quadratic equation of discriminant $t^2 d$.

Notation: Later we will consider the values of modular functions f depending on the ratio α of basis of a proper ideal $\mathfrak{a}$ of some order $\mathfrak{O}_t$. Since the assertions to be stated about $f(\alpha)$ mostly depend only on t, it will be convenient not only to define the discriminant $D(\alpha)$ of a quadratic imaginary number $\alpha \in \mathbb{H}$ but also the **conductor of** α to be the conductor of the order of $[\alpha, 1]$.

Proof Writing the elements $\xi \in K$ as $\xi = m\alpha + n$ with $m, n \in \mathbb{Q}$, we find that

$$\xi[\alpha,1] \subseteq [\alpha,1] \iff \left\{ \begin{array}{r} m\alpha^2 + n\alpha = n\alpha - \frac{m}{A}(B\alpha + C) \in [\alpha,1] \\ m\alpha + n \in [\alpha,1] \end{array} \right\}$$

$$\iff \left\{ \begin{array}{c} A|\gcd(mB, mC) \\ m, n \in \mathbb{Z} \end{array} \right\} \iff \left\{ \begin{array}{c} A|m \\ m, n \in \mathbb{Z} \end{array} \right\} \iff \xi \in [A\alpha, 1].$$

This implies that $\mathfrak{O} = [A\alpha, 1]$. In particular, $\mathfrak{O}$ is an order in K. Since the conductor t of $\mathfrak{O}$ is equal to the index

$$t = [\mathfrak{O}_1 : \mathfrak{O}]$$

it can also be expressed by the absolute value of the determinant of a rational matrix transforming a $\mathbb{Z}$-basis of $\mathfrak{O}_1$ into a basis of $\mathfrak{O}$. Therefore, in view of

$$A\alpha = \frac{-B \pm \sqrt{B^2 - 4AC}}{2} = \frac{-B \pm t\sqrt{d}}{2}$$

we find the conductor to be t. □

The **product** $\mathfrak{a}\mathfrak{b}$ of two free $\mathbb{Z}$-modules $\mathfrak{a}, \mathfrak{b}$ of rank 2 in K is defined as usual as the $\mathbb{Z}$-module generated by the products ab of elements $a \in \mathfrak{a}$ and $b \in \mathfrak{b}$. The **norm** $N(\mathfrak{a})$ of a proper $\mathfrak{O}$-ideal $\mathfrak{a}$ is defined as the absolute value of a determinant of a rational matrix transforming a $\mathbb{Z}$-basis of $\mathfrak{O}$ into a $\mathbb{Z}$-basis of $\mathfrak{a}$.

Theorem 3.1.3 *The set of proper ideals of $\mathfrak{O} = \mathfrak{O}_t$ is a group $\mathfrak{J}_t$ under multiplication with $\mathfrak{O}$ as neutral element. For $\mathfrak{a} \in \mathfrak{J}_t$ we have*

$$\mathfrak{a}\mathfrak{a}^\tau = N(\mathfrak{a})\mathfrak{O},$$

where τ denotes the generating automorphism of $K/\mathbb{Q}$.

Proof Clearly, it suffices to prove the equality of the product of $\mathfrak{a}$ and $\mathfrak{a}^\tau$. So let

$$\mathfrak{a} = [\alpha_1, \alpha_2] = \alpha_2[\alpha, 1] \text{ with } \alpha = \frac{\alpha_1}{\alpha_2}.$$

α satisfies a primitive quadratic equation

$$AX^2 + BX + C = 0.$$

Denoting the trace of $K/\mathbb{Q}$ by $tr(.)$, we find that

$$\begin{aligned}\mathfrak{a}\mathfrak{a}^{\tau} &= N(\alpha_2)[\alpha,\alpha^{\tau},N(\alpha),1] = [N(\alpha_2)[\alpha,tr(\alpha),N(\alpha),1] \\ &= \left[\alpha,\frac{B}{A},\frac{C}{A},1\right] = \frac{N(\alpha_2)}{A}[A\alpha,1] = \frac{N(\alpha_2)}{A}\mathfrak{O}.\end{aligned}$$

Now

$$\begin{pmatrix}\alpha_1\\ \alpha_2\end{pmatrix} = \alpha_2\begin{pmatrix}\alpha\\ 1\end{pmatrix} = \begin{pmatrix}\frac{1}{A} & 0\\ 0 & 1\end{pmatrix} D\begin{pmatrix}A\alpha\\ 1\end{pmatrix}$$

with the representing matrix D of α_2. Thus

$$N(\mathfrak{a}) = \left|det\left(\begin{pmatrix}\frac{1}{A} & 0\\ 0 & 1\end{pmatrix} D\right)\right| = \frac{|N(\alpha_2)|}{|A|},$$

which is the asserted identity. □

Examples of proper ideals other than principal ideals are given by regular ideals:

Definition 3.1.4 Let $\mathfrak{O}$ be the order of conductor $\mathfrak{t}$ in K and let $\mathfrak{a}$ be an ideal of $\mathfrak{O}$. Then $\mathfrak{a}$ is called a regular ideal of $\mathfrak{O}$ if there exists $\xi \in \mathfrak{O}$, prime to $\mathfrak{t}$, such that

$$\xi\mathfrak{a} + \mathfrak{t} = \mathfrak{O}.$$

Theorem 3.1.5 *Every regular ideal of $\mathfrak{O}$ is a proper ideal of $\mathfrak{O}$.*

Proof It suffices to show that every ideal $\mathfrak{a}$ of $\mathfrak{O}$ with

$$\mathfrak{a} + \mathfrak{t} = \mathfrak{O}$$

is invertible. So let $\tilde{\mathfrak{a}} := \mathfrak{O}_1\mathfrak{a}$ the ideal of $\mathfrak{O}_1$ above $\mathfrak{a}$ and $\tilde{\mathfrak{a}}^{-1}$ its inverse. We set

$$\mathfrak{a}' := \mathfrak{O} + \mathfrak{t}\tilde{\mathfrak{a}}^{-1}.$$

Clearly, this is an ideal of $\mathfrak{O}$, and we find that

$$\mathfrak{a}\mathfrak{a}' = \mathfrak{a}(\mathfrak{O} + \mathfrak{t}\tilde{\mathfrak{a}}^{-1}) = \mathfrak{a} + \mathfrak{t}\mathfrak{a}(\mathfrak{O}_1\tilde{\mathfrak{a}}^{-1}) = \mathfrak{a} + \mathfrak{t}\tilde{\mathfrak{a}}\tilde{\mathfrak{a}}^{-1} = \mathfrak{a} + \mathfrak{t} = \mathfrak{O}.$$

Hence, $\mathfrak{a}\mathfrak{a}' = \mathfrak{O}$. Now it is easy to see that $\mathfrak{a}$ is a proper ideal of $\mathfrak{O}$. □

3.1.2 Ideal groups

Theorem 3.1.6 *Let $\mathfrak{O}$ be the order of conductor $\mathfrak{t}$ in K. Then the set $\mathfrak{J}_t^{(0)}$ of regular ideals of $\mathfrak{O}$ is a subgroup of the group of proper ideals of $\mathfrak{O}$. Let $\mathfrak{A}^{\mathfrak{t}}$ denote the group of fractional ideals of $\mathfrak{O}_1$ prime to $\mathfrak{t}$. Then the map*

$$\kappa : \mathfrak{J}_t^{(0)} \to \mathfrak{A}^{\mathfrak{t}}, \qquad \mathfrak{a}_t \mapsto \mathfrak{O}_1 \mathfrak{a}_t,$$

defines a bijection. For a given $\mathfrak{a} \in \mathfrak{A}^{\mathfrak{t}}$ the corresponding proper ideal $\mathfrak{a}_t$ of $\mathfrak{O}$ is obtained by:

$$\mathfrak{a}_t = \mathfrak{a} \cap S_t,$$

where S_t denotes the quotient ring of $\mathfrak{O}$ defined by the semi-group of elements prime to t. For an integral ideal $\mathfrak{a}$ in $\mathfrak{A}^{\mathfrak{t}}$ we have

$$\mathfrak{a}_t = \mathfrak{a} \cap \mathfrak{O},$$

and

$$\mathfrak{O}/\mathfrak{a}_t \cong \mathfrak{O}_1/\mathfrak{a}.$$

Proof To start, let $\mathfrak{a}$ be an integral ideal of $\mathfrak{O}_1$ prime to $\mathfrak{t}$:

$$\mathfrak{a} + \mathfrak{t} = \mathfrak{O}_1.$$

We contend that

$$\mathfrak{a} \cap \mathfrak{O} + \mathfrak{t} = \mathfrak{O}, \quad \mathfrak{O}_1(\mathfrak{a} \cap \mathfrak{O}) = \mathfrak{a} \quad \text{and} \quad \mathfrak{O}/(\mathfrak{a} \cap \mathfrak{O}) \cong \mathfrak{O}_1/\mathfrak{a}. \tag{3.1}$$

The first equality is immediate, since $\mathfrak{t} \subseteq \mathfrak{O}$. The second is obtained as follows: We have

$$\mathfrak{O}_1(\mathfrak{a} \cap \mathfrak{O}) = (\mathfrak{a} + \mathfrak{t})(\mathfrak{a} \cap \mathfrak{O}) = \mathfrak{a}(\mathfrak{a} \cap \mathfrak{t}) + \mathfrak{t}(\mathfrak{a} \cap \mathfrak{O}),$$

and herein the second summand can be modified because $\mathfrak{t}$ is an integral ideal of $\mathfrak{O}_1$ prime to $\mathfrak{a} \cap \mathfrak{O}$:

$$\mathfrak{t} \cap (\mathfrak{a} \cap \mathfrak{O}) = (\mathfrak{t} \cap \mathfrak{a}) \cap \mathfrak{O} = (\mathfrak{t}\mathfrak{a}) \cap \mathfrak{O} = \mathfrak{t}\mathfrak{a}.$$

Therefore, we obtain

$$\mathfrak{O}_1(\mathfrak{a} \cap \mathfrak{O}) = \mathfrak{a}(\mathfrak{a} \cap \mathfrak{O}) + \mathfrak{t}\mathfrak{a} = \mathfrak{a}(\mathfrak{a} \cap \mathfrak{O} + \mathfrak{t}) = \mathfrak{a}\mathfrak{O} = \mathfrak{a}.$$

To prove the third assertion in (3.1) we derive from

$$\mathfrak{O}_1 = \mathfrak{a} + \mathfrak{t} \subseteq \mathfrak{a} + \mathfrak{O} \subseteq \mathfrak{O}_1$$

the equality $\mathfrak{O}_1 = \mathfrak{a} + \mathfrak{O}$, which implies that

$$\mathfrak{O}_1/\mathfrak{a} = (\mathfrak{O} + \mathfrak{a})/\mathfrak{a} \cong \mathfrak{O}/(\mathfrak{a} \cap \mathfrak{O}).$$

Now we consider an integral regular ideal $\mathfrak{a}$ of $\mathfrak{O}$, and we show analogously to (3.1)

$$\mathfrak{a}\mathfrak{O}_1 + \mathfrak{t} = \mathfrak{O}_1 \text{ and } (\mathfrak{a}\mathfrak{O}_1) \cap \mathfrak{O} = \mathfrak{a}.$$

Herein the first equality follows from $\mathfrak{a} + \mathfrak{t} = \mathfrak{O}$ by multiplication with $\mathfrak{O}_1$. The second is obtained as the second equality in (3.1). We have

$$(\mathfrak{a}\mathfrak{O}_1) \cap \mathfrak{O} = (\mathfrak{a}\mathfrak{O}_1) \cap (\mathfrak{a} + \mathfrak{t}).$$

Recalling the inclusion $\mathfrak{a}\mathfrak{O}_1 \supseteq \mathfrak{a}$, the right-hand side can be modified:

$$\mathfrak{a}\mathfrak{O}_1 \cap \mathfrak{O} = \mathfrak{a} + (\mathfrak{a}\mathfrak{O}_1) \cap \mathfrak{t} = \mathfrak{a} + (\mathfrak{a}\mathfrak{O}_1)\mathfrak{t} = \mathfrak{a} + \mathfrak{a}\mathfrak{t} = \mathfrak{a}.$$

This completes the proof of the asserted bijection for integral ideals in $\mathfrak{J}_t^{(0)}$ and $\mathfrak{A}^{\mathfrak{t}}$. For arbitrary ideals in $\mathfrak{J}_t^{(0)}$ resp. in $\mathfrak{A}^{\mathfrak{t}}$ the assertions of the theorem can be reduced to the "integral case": therefore let $\mathfrak{a}$ be an arbitrary ideal in $\mathfrak{A}^{\mathfrak{t}}$. Then there exists some ξ in $\mathfrak{O}_1$ prime to t such that

$$\xi\mathfrak{a} + \mathfrak{t} = \mathfrak{O}_1.$$

Replacing ξ by a power of ξ, we can further achieve that $\xi \equiv 1 \bmod \mathfrak{t}$ so that ξ is in $\mathfrak{O}$. ξ being a unit in S_t, we obtain by intersecting

$$\mathfrak{O} = \mathfrak{O}_1 \cap S_t = (\xi\mathfrak{a} + \mathfrak{t}) \cap S_t = \xi(\mathfrak{a} \cap S_t) + \mathfrak{t}$$

and then

$$\mathfrak{O}_1(\mathfrak{a} \cap S_t) = \frac{1}{\xi}\mathfrak{O}_1(\xi\mathfrak{a} \cap S_t) = \frac{1}{\xi}(\xi\mathfrak{a} \cap \mathfrak{O}) = \frac{1}{\xi}\xi\mathfrak{a} = \mathfrak{a}.$$

In the same way we can show that:

$$\xi\mathfrak{O}_1\mathfrak{a} + \mathfrak{t} = \mathfrak{O}_1 \text{ and } (\mathfrak{a}\mathfrak{O}_1) \cap S_t = \mathfrak{a}$$

for a regular ideal $\mathfrak{a}$ of $\mathfrak{O}$ and a $\xi \in \mathfrak{O}$ that is prime to t with $\xi\mathfrak{a}+\mathfrak{t} = \mathfrak{O}$. This completes the proof of Theorem 3.1.6. □

The **ring ideal class group** of $\mathfrak{O} = \mathfrak{O}_t$ is defined by

$$\mathfrak{R}_t := \mathfrak{J}_t/\mathfrak{H}_t,$$

where

$$\mathfrak{H}_t := \{\gamma\mathfrak{O} \mid \gamma \in K, \gamma \neq 0\}$$

denotes the subgroup of principal ideals in $\mathfrak{J}_t$. We have:

Theorem 3.1.7 *Every ring ideal class in $\mathfrak{R}_t$ contains an integral regular ideal. In view of Theorem 3.1.6 this implies that*

$$\mathfrak{A}^{\mathfrak{t}}/\mathfrak{U}_{\mathfrak{t}} \cong \mathfrak{R}_t \quad by \quad \mathfrak{a}\mathfrak{U}_t \mapsto (\mathfrak{a} \cap S_t)\mathfrak{H}_t,$$

where, as above, $\mathfrak{A}^{\mathfrak{t}}$ denotes the group of fractional ideals of $\mathfrak{O}_1$ that are prime to $\mathfrak{t}$. $\mathfrak{U}_{\mathfrak{t}}$ denotes the subgroup

$$\mathfrak{A}_{\mathfrak{t}} := \{\tfrac{\alpha_1}{\alpha_2}\mathfrak{O}_1 \mid \alpha_i\mathfrak{O}_t + \mathfrak{t} = \mathfrak{O}_t\},$$

and $\mathfrak{A}^{\mathfrak{t}}/\mathfrak{U}_{\mathfrak{t}}$ is called the **ring divisor class group** *modulo $\mathfrak{t}$.*

In the later applications we will need a further isomorphism analogous to that in Theorem 3.1.7. We fix an integral ideal $\mathfrak{f}$ of $\mathfrak{O}_t$ that is not necessarily a proper ideal of $\mathfrak{O}_t$ and define the subgroups

$$\mathfrak{J}_{t,\mathfrak{f}} := \{\mathfrak{a} \in \mathfrak{J}_t \mid \mathfrak{a} = \tfrac{\mathfrak{a}_1}{\mathfrak{a}_2}, \mathfrak{a}_i \in \mathfrak{J}_t, \mathfrak{a}_i + \mathfrak{f} = \mathfrak{O}_t\},$$
$$\mathfrak{S}_{t,\mathfrak{f}} := \{\mathfrak{a} = \tfrac{\alpha_1}{\alpha_2}\mathfrak{O}_t \mid \alpha_i\mathfrak{O}_t + \mathfrak{f} = \mathfrak{O}_t, \alpha_1 \equiv \alpha_2 \bmod \mathfrak{f}\}.$$

We call $\mathfrak{S}_{t,\mathfrak{f}}$ the **ray modulo** $\mathfrak{f}$ of $\mathfrak{O}_t$ and

$$\mathfrak{R}_{t,\mathfrak{f}} := \mathfrak{J}_{t,\mathfrak{f}}/\mathfrak{S}_{t,\mathfrak{f}}$$

the **ray class group modulo** $\mathfrak{f}$ of $\mathfrak{O}_t$. In particular, for $t = 1$ we denote by

$$\mathfrak{K}_{\mathfrak{f}} := \mathfrak{J}_{1,\mathfrak{f}}/\mathfrak{S}_{1,\mathfrak{f}}$$

the ray class group modulo $\mathfrak{f}$. Then, analogously to Theorem 3.1.7 we have:

Theorem 3.1.8 *In every ray class modulo $\mathfrak{f}$ of $\mathfrak{O}_t$ there exists an integral regular ideal. Thus, we have the isomorphism*

$$\mathfrak{A}^{\mathfrak{t}\mathfrak{f}}/\mathfrak{U}_{\mathfrak{t},\mathfrak{f}} \cong \mathfrak{J}_{t,\mathfrak{f}}/\mathfrak{S}_{t,\mathfrak{f}} \quad by \quad \mathfrak{a}\mathfrak{U}_{\mathfrak{t},\mathfrak{f}} \mapsto (\mathfrak{a} \cap S_t)\mathfrak{S}_{t,\mathfrak{f}},$$

where, as above, $\mathfrak{A}^{\mathfrak{t}\mathfrak{f}}$ denotes the group of fractional ideals of $\mathfrak{O}_1$ that are prime to $\mathfrak{t}\mathfrak{f}$, and by $\mathfrak{U}_{\mathfrak{t},\mathfrak{f}}$ we denote the subgroup

$$\mathfrak{U}_{\mathfrak{t},\mathfrak{f}} := \{\tfrac{\alpha_1}{\alpha_2}\mathfrak{O}_1 \mid \alpha_i\mathfrak{O}_t + \mathfrak{t}\mathfrak{f} = \mathfrak{O}_t,\ \alpha_1 \equiv \alpha_2 \ mod\ \mathfrak{f}\}.$$

Proof of Theorems 3.1.7 and 3.1.8 In view of Theorem 3.1.6 it suffices to show that there exists a regular ideal in every ideal class of $\mathfrak{J}_t$ modulo $\mathfrak{H}_t$ resp. of $\mathfrak{J}_{t,\mathfrak{f}}$ modulo $\mathfrak{S}_{t,\mathfrak{f}}$. For the classes modulo $\mathfrak{H}_t$ this follows by Theorem 3.1.10, as will be explained after this theorem. For the classes modulo $\mathfrak{S}_{t,\mathfrak{f}}$ we need some further facts about the ring $\mathfrak{O}_t$: Let $\mathfrak{a}$ be in

$\mathfrak{J}_{t,\mathfrak{f}}$. Adapting the conclusions after the proof of Theorem 3.1.10 we can find an element $\lambda \in K$ and an ideal $\mathfrak{b}$ of $\mathfrak{O}_t$, prime to $t\mathfrak{f}$, such that

$$\mathfrak{a} = \lambda \mathfrak{b}.$$

Herein we have by definition

$$\mathfrak{a} = \tfrac{\mathfrak{a}_1}{\mathfrak{a}_2} \text{ with integral ideals } \mathfrak{a}_i \in \mathfrak{J}_{t,\mathfrak{f}},$$

and we can write

$$\lambda \mathfrak{O}_t = \frac{\mathfrak{a}_1}{\mathfrak{a}_2 \mathfrak{b}} = \frac{\mathfrak{a}_1 (\mathfrak{a}_2 \mathfrak{b})^{e-1}}{(\mathfrak{a}_2 \mathfrak{b})^e}.$$

Choosing an exponent $e \in \mathbb{N}$ so that $(\mathfrak{a}_2 \mathfrak{b})^e$ is principal, we obtain

$$\lambda = \tfrac{\lambda_1}{\lambda_2} \text{ with } \lambda_i \mathfrak{O}_t + \mathfrak{f} = \mathfrak{O}_t.$$

As we will show, we can find two elements $\lambda_i' \in \mathfrak{O}_t$, $i = 1, 2$, prime to $t\mathfrak{f}$ such that

$$\lambda_i' \equiv \lambda_i \bmod \mathfrak{f}.$$

Then $\lambda \mathfrak{b}$ and $\mathfrak{b}' := \frac{\lambda_1'}{\lambda_2'} \mathfrak{b}$ are in the same class modulo $\mathfrak{S}_{t,\mathfrak{f}}$. Further, $\mathfrak{b}'$ is regular and is in the ray class of $\mathfrak{a}$. Multiplying $\mathfrak{b}'$ by a power of $\lambda_2'^e$, satisfying the congruence $\lambda_2'^e \equiv 1 \bmod \mathfrak{f}$, we find that $\mathfrak{b}'' := \lambda_2'^e \mathfrak{b}'$ is a regular ideal in the class of $\mathfrak{a}$.

To construct λ_i' we observe that $\mathfrak{O}_t$ is a noetherian ring, in which every ideal $\neq (0)$ is maximal. This follows from the fact that every ideal $\neq (0)$ is a free rank two module over $\mathbb{Z}$, which implies that $\mathfrak{O}_t / \mathfrak{a}$ is finite. Hence, the two ideals $\mathfrak{f}$ and $\mathfrak{f}\mathfrak{t}$ have a decomposition as a product of primary ideals:

$$\mathfrak{f} = \mathfrak{q}_1 \cdot \ldots \cdot \mathfrak{q}_m \quad \text{resp.} \quad \mathfrak{f}\mathfrak{t} = \mathfrak{q}_1' \cdot \ldots \cdot \mathfrak{q}_n',$$

where the radicals

$$\mathfrak{p}_i = \sqrt{\mathfrak{q}_i} \text{ resp. } \mathfrak{p}_i' = \sqrt{\mathfrak{q}_i'}$$

are pairwise coprime.

$$\mathfrak{p}_i + \mathfrak{p}_j = \mathfrak{O}_t \text{ and } \mathfrak{p}_i' + \mathfrak{p}_j' = \mathfrak{O}_t \text{ if } i \neq j.$$

Since $\mathfrak{f}$ divides $\mathfrak{f}\mathfrak{t}$, the $\mathfrak{p}_i$ must be some of the $\mathfrak{p}_i'$, which implies $m \leq n$. Therefore, we may assume that

$$\mathfrak{p}_i = \mathfrak{p}_i', \quad i = 1, \ldots, m.$$

$\mathfrak{O}_t$ being noetherian we further have the inclusions

$$\mathfrak{p}_i^e \subseteq \mathfrak{q}_i, \quad \mathfrak{p}_i'^e \subseteq \mathfrak{q}_i'$$

with a suitable common exponent e. For λ with $\lambda\mathfrak{O}_t + \mathfrak{f} = \mathfrak{O}_t$ we pick a solution of the system of simultaneous congruences

$$\begin{aligned} \lambda' &\equiv \lambda \bmod \mathfrak{f}, \\ \lambda' &\equiv 1 \bmod \mathfrak{p}_i'^e, \qquad i = m+1, \dots, n, \end{aligned}$$

which is possible because the ideals involved are pairwise coprime. Such a solution λ' satisfies

$$\lambda' \equiv \lambda \bmod \mathfrak{f} \text{ and } \lambda\mathfrak{O}_t + \mathfrak{f}\mathfrak{t} = \mathfrak{O}_t.$$

This finishes the proof. □

By Theorems 3.1.7 and 3.1.8 we find the following formulae:

Theorem 3.1.9 *The indices of $\mathfrak{R}_t$ and $\mathfrak{R}_{t,\mathfrak{f}}$ are given by:*

$$h_t := |\mathfrak{R}_t| = h_K \frac{w_t(\mathfrak{O}_t)\Phi_1(\mathfrak{t})}{w_1(\mathfrak{O}_1)\Phi_t(\mathfrak{t})} = h_K \frac{w_t(\mathfrak{O}_t)}{w_1(\mathfrak{O}_1)}\, t \prod_{p|t} \left(1 - \left(\frac{d}{p}\right)\frac{1}{p}\right),$$

$$|\mathfrak{R}_{t,\mathfrak{f}}| = h_t \frac{w_t(\mathfrak{f})}{w_t(\mathfrak{O}_t)} \Phi_t(\mathfrak{f}).$$

Herein h_K denotes the class number of K, and for an integral ideal $\mathfrak{a}$ of $\mathfrak{O}_t$ we denote by $w_t(\mathfrak{a})$ the number of roots of unity $\xi \in \mathfrak{O}_t$ with $\xi \equiv 1$ mod $\mathfrak{a}$. Φ_t is the Euler function of $\mathfrak{O}_t$.

Proof In view of Theorem 3.1.7, to determine $|\mathfrak{R}_t|$, we must find a formula for $[\mathfrak{A}^{\mathfrak{t}} : \mathfrak{U}_{\mathfrak{t}}]$. Let $\mathfrak{H}_1^{\mathfrak{t}}$ denote the subgoup of principal ideals of $\mathfrak{O}_1$, that are prime to t. Since every ideal of $\mathfrak{O}_1$ is equivalent to an ideal prime to $\mathfrak{t}$, the groups $\mathfrak{J}_1/\mathfrak{H}_1$ and $\mathfrak{A}^{\mathfrak{t}}/\mathfrak{H}_1^{\mathfrak{t}}$ are isomorphic, hence

$$[\mathfrak{A}^{\mathfrak{t}} : \mathfrak{H}_1^{\mathfrak{t}}] = h_K.$$

To compute $[\mathfrak{H}_1^{\mathfrak{t}} : \mathfrak{U}_{\mathfrak{t}}]$, we consider the epimorphism

$$\begin{aligned} \kappa : \overline{\mathfrak{O}}_1^{\,*} \times \overline{\mathfrak{O}}_1^{\,*} &\to \mathfrak{H}_1^{\mathfrak{t}}/\mathfrak{U}_{\mathfrak{t}} \\ (\overline{\nu}, \overline{\nu'}) &\mapsto (\tfrac{\nu}{\nu'})\mathfrak{U}_{\mathfrak{t}} \end{aligned}$$

with the group of prime residues $\overline{\mathfrak{O}}_1^{\,*}$ of $\mathfrak{O}_1$ modulo $\mathfrak{t}$. It is easy to see that this map is well defined, its kernel consisting of all pairs

$$(\overline{\nu}, \overline{\nu'}) \text{ with } \overline{\nu} = \overline{\nu'\alpha\epsilon}$$

with some $\alpha \in S_t$ prime to $\mathfrak{t}$ and some unit $\epsilon \in \mathfrak{O}_1$. Now we have $S_t/S_t\mathfrak{t} \cong \mathfrak{O}_t/\mathfrak{t}$. Therefore, the elements of the kernel of κ can also be characterised by

$$(\overline{\nu}, \overline{\nu'}) \text{ with } \overline{\nu} = \overline{\nu' \alpha \epsilon}, \quad \overline{\alpha} \in U, \epsilon \in W,$$

where U denotes the group of prime residues modulo $\mathfrak{t}$ in $\mathfrak{O}_t$ and W the subgroup of prime residue classes modulo $\mathfrak{t}$ represented by units of $\mathfrak{O}_1$. Hence

$$|ker(\kappa)| = \Phi_1(\mathfrak{t})|UW|.$$

Using $|UW| = \frac{|U||W|}{|U \cap W|}$, this implies that

$$|ker(\kappa)| = \frac{\Phi_1(\mathfrak{t})\Phi_t(\mathfrak{t})w_1(\mathfrak{O}_1)}{w_t(\mathfrak{O}_t)}$$

and the formula for $|\mathfrak{R}_t|$ is proved.

To prove the asserted formula for $|\mathfrak{R}_{t,\mathfrak{f}}|$, we consider the inclusions

$$\mathfrak{J}_{t,\mathfrak{f}} \supseteq \mathfrak{H}_t^{\mathfrak{f}} \supseteq \mathfrak{S}_{t,\mathfrak{f}},$$

where

$$\mathfrak{H}_t^{\mathfrak{f}} := \{\tfrac{\nu_1}{\nu_2}\mathfrak{O}_t \mid \nu_i \mathfrak{O}_t + \mathfrak{f} = \mathfrak{O}_t\}$$

denotes the subgroup of principal ideals of $\mathfrak{J}_{t,\mathfrak{f}}$. Since, as will be explained after Theorem 3.1.10, in every ring ideal class there exists an integral ideal prime to $\mathfrak{f}\mathfrak{t}$, this implies that

$$[\mathfrak{J}_{t,\mathfrak{f}} : \mathfrak{H}_t^{\mathfrak{f}}] = h_t.$$

Therefore, it remains to be shown that

$$[\mathfrak{H}_t^{\mathfrak{f}} : \mathfrak{S}_{t,\mathfrak{f}}] = \frac{1}{w_t(\mathfrak{O}_t)}\Phi_t(\mathfrak{f}).$$

Therefore, we consider the epimorphism

$$\kappa : \begin{array}{ccc} \overline{\mathfrak{O}}_t^* \times \overline{\mathfrak{O}}_t^* & \to & \mathfrak{H}_t^{\mathfrak{f}}/\mathfrak{S}_{t,\mathfrak{f}} \\ (\overline{\nu}, \overline{\nu'}) & \mapsto & (\frac{\nu}{\nu'}\mathfrak{O}_t)\mathfrak{S}_{t,\mathfrak{f}} \end{array},$$

where $\overline{\mathfrak{O}}_t^*$ denotes the group of prime residues of $\mathfrak{O}_t$ modulo $\mathfrak{f}$. The kernel consists of all pairs

$$(\overline{\nu_1}, \overline{\nu_2}) \text{ with } \overline{\nu_2} = \overline{\nu_1 \epsilon}$$

for a unit ϵ in $\mathfrak{O}_t$. Hence

$$|ker(\kappa)| = \Phi_t(\mathfrak{f})w_t(\mathfrak{O}_t),$$

and this implies the asserted formula for $|\mathfrak{R}_{t,\mathfrak{f}}|$. □

Theorem 3.1.10 (n-system) *Let n be a natural number. Let $\alpha_0 \in \mathbb{H}$ be the ratio of a basis of a proper ideal of $\mathfrak{O}_t$ and let*

$$A_0X^2 + B_0X + C_0 = 0 \quad with \quad \gcd(A_0, n) = 1, A_0 > 0.$$

be the primitive quadratic equation satisfied by α_0. Then, in every ring ideal class modulo t there exists an ideal $\mathfrak{a} = [\alpha_1, \alpha_2]$, such that the ratio of the basis $\alpha = \frac{\alpha_1}{\alpha_2} \in \mathbb{H}$ satisfies a primitive equation of the form

$$AX^2 + BX + C = 0$$

with

$$\gcd(A, n) = 1, A > 0 \quad and \quad B \equiv B_0 \ mod\ n.$$

Choosing such an ideal in every ring ideal class, we call the system of corresponding ratios of basis an n-system.

Remark 3.1.11 In particular, by choosing $n = t$ in Theorem 3.1.10 it follows that in every class there exists an ideal

$$\mathfrak{a} = A[\alpha, 1] = \left[A, \frac{-B + t\sqrt{d}}{2}\right] \quad \text{with} \quad \gcd(A, t) = 1,$$

hence a regular ideal.

Theorem 3.1.12 *Let n be a natural number, and let $\alpha \in \mathbb{H}$ be the ratio of a basis of a proper ideal of $\mathfrak{O}_t$ satisfying a primitive equation*

$$AX^2 + BX + C = 0 \quad with \quad \gcd(A, n) = 1, A > 0.$$

Then there exists a unimodular transformation $M \in \Gamma(n)$, so that $\alpha' := M(\alpha)$ is a root of a primitive equation

$$A'X^2 + B'X + C' = 0$$

with

$$\gcd(A', nt) = 1, A' > 0 \quad and \quad B' \equiv B \ mod\ n.$$

Proof of Theorem 3.1.10 let α be the ratio of a basis of a proper ideal $\mathfrak{a}$ of $\mathfrak{O}_t$, and let n be a natural number. Since the ratios of basis in $\mathbb{H}$ of ideals equivalent to $\mathfrak{a}$ are exactly the images of α under unimodular transformations, we have to determine $M \in \Gamma$ such that $M(\alpha)$ satisfies an equation having the desired properties. We proceed by induction on the number of primes dividing n. For $n = 1$ nothing is to be shown, so we assume the assertion to be proved for n, and we let $n' = np^s, s \in \mathbb{N}$, with a prime p not dividing n. To prove the assertion for n', we consider, for a given $\mu \in \mathbb{Z}$, the unimodular transformation

$$M_\mu := \begin{pmatrix} 1 & -\mu \\ 0 & 1 \end{pmatrix}, \quad N_\mu := \begin{pmatrix} 1 & 0 \\ -\mu & 1 \end{pmatrix}.$$

For the root $\omega \in \mathbb{H}$ of a primitive quadratic equation

$$AX^2 + BX + C = 0,$$

we find that $M_\mu(\omega)$ resp. $N_\mu(\omega)$ satisfy

$$AX^2 + (B + 2\mu A)X + (C + \mu B + \mu^2 A) = 0 \text{ resp.}$$

$$(A + \mu B + \mu^2 C)X^2 + (B + 2\mu C)X + C = 0.$$

Now we assume α to be a root of the primitive equation

$$AX^2 + BX + C = 0 \text{ with } \gcd(A, n) = 1, A > 0 \text{ and } B \equiv B_0 \bmod n.$$

Then, transforming α by some product of M_μ and N_μ with $\mu \in \mathbb{N}$ divisible by n, these conditions will not be lost if the μ's are sufficiently large. Note that $B^2 - 4AC < 0$ and $A > 0$ imply $C > 0$. In this way we can further achieve the new A as prime to np. In the cases $p \neq 2$ or ($p = 2$ and $B \equiv 0 \bmod 2$) this becomes clear by writing $A + \mu B + \mu^2 C = A + \mu(B + \mu C)$. For $p = 2$ and $B \equiv 1 \bmod 2$ one has first to transform α by a suitable M_μ so that $C \equiv 0 \bmod 2$. Hereafter, we obtain an equation with $A \equiv 1 \bmod 2$ by applying a suitable N_μ. In order for B to satisfy the desired congruence modulo np^s, we apply a suitable M_μ. Thus, we end up with an equation, in which $B \equiv B_0 \bmod np^s$. If $p \neq 2$, this is immediate. If $p = 2$, we have to observe that the coefficients of X of two primitive quadratic equations having the same discriminant have the same parity. □

Proof of Theorem 3.1.12 Proceeding as in the last proof, we can achieve the new coefficient A to be also prime to t. □

3.1.3 Primitive matrices and bases of ideals

In what follows we will consider the action of primitive matrices on $\mathbb{Z}$-basis of ideals. Therefore, we set up some notation: for $\alpha_1, \alpha_2 \in \mathbb{C}$, linearly independent over $\mathbb{R}$, we define

$$\alpha := \frac{\alpha_1}{\alpha_2}, \qquad \underline{\alpha} := \begin{pmatrix} \alpha_1 \\ \alpha_2 \end{pmatrix} \quad \text{and} \quad [\underline{\alpha}] := \mathbb{Z}\alpha_1 + \mathbb{Z}\alpha_2.$$

For a rational 2×2-matrix $S = \left(\begin{smallmatrix} a & b \\ c & d \end{smallmatrix}\right)$ of determinant $s \neq 0$, the ratio of the basis $S\underline{\alpha}$ is then given by $S(\alpha) = \frac{a\alpha+b}{c\alpha+d}$.

Theorem 3.1.13 *Let S be a primitive matrix of determinant s and $[\underline{\alpha}] \in \mathfrak{I}_t$. Then:*

(i) *$[S\underline{\alpha}] \in \mathfrak{I}_{t'}$ with $t'|ts$ and $t|t's$.*

(ii) *If $t' = st$, then $\mathfrak{O}_t[S\underline{\alpha}] = [\underline{\alpha}]$, and if S' is another primitive matrix of determinant s with $[S'\underline{\alpha}] \in \mathfrak{I}_{ts}$, then $S' \sim SD_\epsilon$, where D_ϵ is the matrix of a unit in $\mathfrak{O}_t$ with respect to the basis $\underline{\alpha}$.*

(iii) *If $s|t$ and $t' = \frac{t}{s}$, then $s\mathfrak{O}_{\frac{t}{s}}[\underline{\alpha}] = [S\underline{\alpha}]$. In particular, the equivalence class of S is uniquely determined by $[S\underline{\alpha}] \in \mathfrak{I}_{\frac{t}{s}}$.*

Proof (i) By changing the basis in $[\underline{\alpha}]$ and $[S\underline{\alpha}]$ we can arrive at $S = \left(\begin{smallmatrix} 1 & 0 \\ 0 & s \end{smallmatrix}\right)$. Assume $AX^2 + BX + C = 0$ to be the primitive equation of α. Then $S(\alpha)$ satisfies

$$s^2AX^2 + sBX + C = 0.$$

The primitive equation for $S(\alpha)$ is then obtained by dividing by the $\gcd(s^2A, sB, C)$. Recalling $\gcd(A, B, C) = 1$, it follows that

$$D(S(\alpha))|s^2D(\alpha) \text{ and } D(\alpha)|s^2D(S(\alpha)).$$

The relation between $D(\alpha)$ and t from Theorem 3.1.2 now proves (i) of Theorem 3.1.13.

(ii) In the case of $t' = ts$ the above equation is necessarily primitive and thus C prime to s. According to Theorem 3.1.2 we have $\mathfrak{O}_t = [A\alpha, 1]$ and computation shows that

$$\begin{aligned} \mathfrak{O}_t[S\underline{\alpha}] &= \alpha_2[A\alpha, 1][\alpha, s] = \alpha_2[A\alpha^2, \alpha, s] \\ &= \alpha_2[B\alpha + C, \alpha, s] = [\alpha, 1] = \alpha_2[\alpha, 1] = [\underline{\alpha}]. \end{aligned}$$

If S' is a further primitive matrix of determinant s satisfying

$$[S\underline{\alpha}] = \epsilon[S\underline{\alpha}] \tag{3.2}$$

with an element $\epsilon \in K$, then by multiplication with $\mathfrak{O}_t$ we obtain

$$[\underline{\alpha}] = \epsilon[\underline{\alpha}].$$

This implies ϵ to be a unit in $\mathfrak{O}_t$ and, from (3.2) it follows that

$$MS = S'D_\epsilon$$

with a unimodular matrix M and the matrix D_ϵ of ϵ with respect to $\underline{\alpha}$. Hence, S is equivalent to $S'D_\epsilon$.

(iii) If $[S\underline{\alpha}]$ is in $\mathfrak{I}_{\frac{t}{s}}$, then

$$AX^2 + \frac{B}{s}X + \frac{C}{s^2} = 0$$

is the primitive equation of $S(\alpha) = \frac{\alpha}{s}$. Hence B is divisible by s and C by s^2. A must be prime to s. Further, by Theorem 3.1.2

$$\mathfrak{O}_{\frac{t}{s}} = [A\frac{\alpha}{s}, 1].$$

Now we compute

$$\begin{aligned}
\mathfrak{O}_{\frac{t}{s}}[\underline{\alpha}] &= \alpha_2[A\frac{\alpha}{s}, 1][\alpha, 1] = \alpha_2[A\frac{\alpha^2}{s}, \alpha, A\frac{\alpha}{s}, 1] \\
&= \alpha_2[\frac{B\alpha + C}{s}, \frac{\alpha}{s}, 1] = \alpha_2[\frac{\alpha}{s}, 1] = \frac{1}{s}[S\underline{\alpha}].
\end{aligned}$$

This completes the proof of 3. □

The next theorem makes the statement of Theorem 3.1.13 more precise if s is a prime.

Theorem 3.1.14 *Let P be a primitive matrix, whose determinant is a prime p, and we let $[\underline{\alpha}] \in \mathfrak{I}_t$. Then:*

(i) *If $p|t$, there exists a primitive matrix P_1 of determinant p with*

$$[P\underline{\alpha}] \in \mathfrak{I}_{tp} \text{ if } P \not\sim P_1,$$
$$[P\underline{\alpha}] \in \mathfrak{I}_{tp^{-1}} \text{ if } P \sim P_1.$$

(ii) *In the case $p\nmid t$ the result depends on the decomposition of p in K:*

(a) *If $p = \mathfrak{p}$, then we always have $[P\underline{\alpha}] \in \mathfrak{I}_{tp}$.*

(b) *If $p = \mathfrak{p}^2$, then there exists a primitive matrix $P_\mathfrak{p}$, such that $[P_\mathfrak{p}\underline{\alpha}]$ is a basis of $[\underline{\alpha}]\mathfrak{p}_t$ with $\mathfrak{p}_t = \mathfrak{p} \cap \mathfrak{O}_t$. And then for P we have*

$$[P\underline{\alpha}] = [\underline{\alpha}]\mathfrak{p}_t \in \mathfrak{I}_t \text{ if } P \sim P_\mathfrak{p},$$
$$[P\underline{\alpha}] \in \mathfrak{I}_{tp} \text{ if } P \not\sim P_\mathfrak{p}.$$

(c) *If $p = \mathfrak{p}\overline{\mathfrak{p}}$, then there exist two inequivalent primitive matrices $P_{\mathfrak{p}}, P_{\overline{\mathfrak{p}}}$, such that $[P_{\mathfrak{p}}\underline{\alpha}]$ resp. $[P_{\overline{\mathfrak{p}}}\underline{\alpha}]$ is a basis of $[\underline{\alpha}]\mathfrak{p}_t$ resp. $[\underline{\alpha}]\overline{\mathfrak{p}_t}$ with $\mathfrak{p}_t = \mathfrak{p} \cap \mathfrak{O}_t$. And then for P we have*

$$[P\underline{\alpha}] = [\underline{\alpha}]\mathfrak{p}_t \in \mathfrak{J}_t \text{ if } P \sim P_{\mathfrak{p}},$$
$$[P\underline{\alpha}] = [\underline{\alpha}]\overline{\mathfrak{p}_t} \in \mathfrak{J}_t \text{ if } P \sim P_{\overline{\mathfrak{p}}},$$
$$[P\underline{\alpha}] \in \mathfrak{J}_{tp} \text{ if } P \not\sim P_{\mathfrak{p}}, P_{\overline{\mathfrak{p}}}.$$

Proof (i) In view of Theorem 3.1.13 we only have to show that for $p \mid t$ there is no primitive matrix of determinant p with $D(P(\alpha)) = D(\alpha)$. As in the proof of Theorem 3.1.13 we may assume that $P = \left(\begin{smallmatrix} 1 & 0 \\ 0 & p \end{smallmatrix}\right)$. Let $AX^2 + BX + C = 0$ be the primitive equation of α. If $D(P(\alpha)) = t^2 d$, then

$$pAX^2 + BX + \frac{C}{p} = 0$$

would be the primitive equation of $P(\alpha)$ and C would be divisible by p. Further, B would also be divisible by p, because of $B^2 - 4AC = t^2 d$. This implies that $p^2 \nmid C$ and $p \nmid A$ because of $\gcd(A, B, C) = 1$. For $p \neq 2$, we conclude that $B^2 - 4AC \not\equiv 0 \bmod p^2$, which is a contradiction to $B^2 - 4AC = t^2 d \equiv 0 \bmod p^2$. For $p = 2$, it follows that $B^2 - 4AC \equiv 8, 12 \bmod 16$, which is a contradiction to $t^2 d \equiv 0, 4 \bmod 16$.

(ii) (a) We may again assume that $P = \left(\begin{smallmatrix} 1 & 0 \\ 0 & p \end{smallmatrix}\right)$. Then $[P\underline{\alpha}] \in \mathfrak{J}_t$ would imply $p | C$ as in (i). and then $t^2 d = D(\alpha) \equiv B^2 \bmod p$. But p is inert by assumption and $p \nmid t$. Therefore, $t^2 d$ cannot be a square modulo p. Hence $[P\underline{\alpha}]$ is not in $\mathfrak{J}_t$ but in $\mathfrak{J}_{pt}$.

To prove (b) we need:

Lemma 3.1.15 *Let $\mathfrak{a}$ be an integral regular ideal of $\mathfrak{O}_t$. Then for every $\mathfrak{b} \in \mathfrak{J}_t$ we have:*

$$[\mathfrak{b} : \mathfrak{b}\mathfrak{a}] = N(\mathfrak{a}).$$

Proof of the lemma If $\mathfrak{b}$ is an integral regular ideal of $\mathfrak{O}_t$, the assertion follows from Theorem 3.1.6 and the fact that the norm of ideals of $\mathfrak{O}_1$ is multiplicative. The general case can be reduced to this special case: by Theorem 3.1.10 every ideal in $\mathfrak{J}_t$ is of the form $\gamma\mathfrak{c}$ with an element

$\gamma \in K\backslash\{0\}$ and an integral regular ideal $\mathfrak{c}$ of $\mathfrak{O}_t$. Now we have $\gamma\mathfrak{c}/\gamma\mathfrak{c}\mathfrak{a} \cong \mathfrak{c}/\mathfrak{c}\mathfrak{a}$, because multiplication by γ is an isomorphism. □

Proof of (ii) (b) Let $\mathfrak{p}$ be the prime ideal of $\mathfrak{O}_1$ above p and $\mathfrak{p}_t = \mathfrak{O} \cap \mathfrak{p}$ the corresponding regular ideal of $\mathfrak{O}_t$. By Lemma 3.1.15 we then have $[[\underline{\alpha}] : [\underline{\alpha}]\mathfrak{p}_t] = N(\mathfrak{p}_t) = p$. Hence, there exists a primitive matrix $P_\mathfrak{p}$ with $[P_\mathfrak{p}\underline{\alpha}] = [\underline{\alpha}]\mathfrak{p}_t$. We contend that, up to equivalence, $P_\mathfrak{p}$ is the only primitive matrix P with $[P\underline{\alpha}] \in \mathfrak{I}_t$. Therefore, we observe that $[P\underline{\alpha}] \in \mathfrak{I}_t$ implies that $[[\underline{\alpha}] : [P\underline{\alpha}]] = p$. Thus, the integral ideal $\mathfrak{a} := [P\underline{\alpha}][\underline{\alpha}]^{-1}$ contains p, so it is regular. By Theorem 3.1.6 $\tilde{\mathfrak{a}} := \mathfrak{O}_1\mathfrak{a}$ has norm p and hence is equal to $\mathfrak{p}$. Thus, $\mathfrak{a} = \mathfrak{p}_t$ and $[P\underline{\alpha}] = [\underline{\alpha}]\mathfrak{p}_t$. Since this determines P up to equivalence we can conclude that $P \sim P_\mathfrak{p}$.

This proves part (b) of (ii), and (c) is obtained in the same way. □

To finish, we use Theorems 3.1.13 and 3.1.14 to prove the next theorem which will be needed later. Given a natural number t and a prime p, we consider the epimorphisms

$$\kappa : \mathfrak{I}_{tp} \to \mathfrak{I}_t, \quad \mathfrak{a} \mapsto \mathfrak{a}\mathfrak{O}_t,$$
$$\hat{\kappa} : \mathfrak{R}_{tp} \to \mathfrak{R}_t, \quad \mathfrak{a}\mathfrak{H}_{tp} \mapsto \kappa(\mathfrak{a})\mathfrak{H}_t.$$

By the above theorem we will be able to determine explicitly $\kappa^{-1}(\{\mathfrak{b}\})$ resp. $\hat{\kappa}^{-1}(\{\mathfrak{b}\mathfrak{H}_t\})$ for a given $\mathfrak{b} \in \mathfrak{I}_t$.

Theorem 3.1.16 *Let $[\underline{\alpha}]$ be in $\mathfrak{I}_t$, and let P_μ, $\mu = 1, \ldots, p+1$, run through a system of representatives of primitive matrices of determinant p. We assume that $P_1, \ldots, P_m$ are exactly all P_μ having the property $[P_\mu\underline{\alpha}] \in \mathfrak{I}_{tp}$. Then:*

$$m = \begin{cases} p-1 & \text{if} \quad p = \mathfrak{p}\overline{\mathfrak{p}}, \mathfrak{p} \neq \overline{\mathfrak{p}}, p \nmid t, \\ p & \text{if} \quad p = \mathfrak{p}^2, p \nmid t, \\ p+1 & \text{if} \quad p = \mathfrak{p}, p \nmid t, \\ p & \text{if} \quad p \mid t. \end{cases}$$

Furthermore, $[P_\mu\underline{\alpha}], \mu = 1, \ldots, m$, are exactly all ideals in $\mathfrak{I}_t$ that are mapped to $[\underline{\alpha}]$ by κ, and we have the equivalence

$$[P_\mu\underline{\alpha}]\mathfrak{H}_{tp} = [P_{\mu'}\underline{\alpha}]\mathfrak{H}_{tp} \iff ([P_\mu\underline{\alpha}] = \epsilon[P_{\mu'}\underline{\alpha}] \text{ with a unit } \epsilon \in \mathfrak{O}_t).$$

Clearly, by the classes of $[P_\mu\underline{\alpha}], \mu = 1, \ldots, m$, in $\mathfrak{R}_{tp}$ we are given all classes that are mapped to $[\underline{\alpha}]\mathfrak{H}_t$ under $\hat{\kappa}$. The number of different such classes is $\frac{m}{e}$, where e denotes the index of the unit groups of $\mathfrak{O}_{tp}$ in the unit group of $\mathfrak{O}_t$. Therefore, we know that $e = 1$ except for $t^2d = -3, -4$.

For the remaining $P_{m+1}, \ldots, P_{p+1}$ *we have:*

$P_p\underline{\alpha}$ *and* $P_{p+1}\underline{\alpha}$ *are bases of* $[\underline{\alpha}]\mathfrak{p}_{tp}$ *and* $[\underline{\alpha}]\overline{\mathfrak{p}}_{tp}$ *if* $p = \mathfrak{p}\overline{\mathfrak{p}}, \mathfrak{p} \neq \overline{\mathfrak{p}}, p \nmid t$,

$P_{p+1}\underline{\alpha}$ *is a basis of* $[\underline{\alpha}]\mathfrak{p}_t$ *if* $p = \mathfrak{p}^2$, $p \nmid t$,

$P_{p+1}\underline{\alpha}$ *is a basis of an ideal* $\mathfrak{a}_{tp^{-1}} \in \mathfrak{J}_{tp^{-1}}$ *if* $p|t$.

Moreover, for $p|t$ *the ideal class of* $[\underline{\alpha}]$ *is mapped to the ideal class of* $\mathfrak{a}_{tp^{-1}}\mathfrak{H}_{tp^{-1}}$ *by* $\hat{\kappa} : \mathfrak{R}_t \to \mathfrak{R}_{tp^{-1}}$.

Proof The assertions follow immediately from Theorems 3.1.13 and 3.1.14. We have only to show that the pre-images of $[\underline{\alpha}]$ under κ are of the form $[P\underline{\alpha}]$ with some primitive matrix P of determinant p. To see this, we observe that for a pre-image $\mathfrak{a} \in \mathfrak{J}_{tp}$ we have the inclusions

$$p[\underline{\alpha}] = p\mathfrak{O}_t\mathfrak{a} \subset \mathfrak{O}_{tp}\mathfrak{a} = \mathfrak{a} \subset \mathfrak{O}_t\mathfrak{a} = [\underline{\alpha}],$$

which are strict because the ideals on the left and the right are in $\mathfrak{J}_t$ and that in the middle is in $\mathfrak{J}_{tp^{-1}}$. Therefore, from $[[\underline{\alpha}] : p[\underline{\alpha}]] = p^2$ we obtain $[[\underline{\alpha}] : \mathfrak{a}] = p$. Hence, there exists a primitive matrix P of determinant p, such that $P\underline{\alpha}$ is a basis of $\mathfrak{a}$. □

3.1.4 Integral ideals that are not regular

As already mentioned, an integral ideal $\mathfrak{a}$ of an order $\mathfrak{O}_t$ is a product of primary ideals. If $\mathfrak{a}$ is regular, the primary ideals are powers of prime ideals and the factorisation is unique. If $\mathfrak{a}$ is not regular we have uniqueness only for the corresponding radicals. By the next two theorems all non-regular prime ideals and primary ideals are determined.

Theorem 3.1.17 *[Bettner] Non-regular prime ideals of* $\mathfrak{O}_t$ *are given by*

$$\mathfrak{p}_p = p\mathfrak{O}_{\frac{t}{p}},$$

where p *is a divisor of* t.

Proof First, given a prime p dividing t, the set defined by $\mathfrak{p}_p := p\mathfrak{O}_{\frac{t}{p}}$ is an integral ideal of $\mathfrak{O}_t$ that is not regular because $p \mid t$. Further, we find that

$$[\mathfrak{O}_t : \mathfrak{p}_p] = \frac{[\mathfrak{O}_{\frac{t}{p}} : \mathfrak{p}_p]}{[\mathfrak{O}_{\frac{t}{p}} : \mathfrak{O}_t]} = \frac{p^2}{p} = p.$$

Hence $\mathfrak{p}_p$ is a prime ideal. Conversely, let $\mathfrak{p}$ be a non-regular prime ideal in $\mathfrak{O}_t$. Then

$$\mathfrak{p} + \mathfrak{t} \neq \mathfrak{O}_t,$$

and we have the inclusion

$$\mathfrak{p} \subseteq \mathfrak{p} + \mathfrak{t} \subsetneq \mathfrak{O}_t.$$

Moreover, $\mathfrak{O}_t/\mathfrak{p}$ being finite, $\mathfrak{p}$ is even maximal. Therefore, in view of the above inclusion we may conclude that $\mathfrak{p} = \mathfrak{p} + \mathfrak{t}$, hence

$$t \in \mathfrak{p}.$$

This implies that $\mathfrak{p}$ contains a prime p dividing t and we have the inclusion

$$\mathfrak{p}_p^2 = p^2\mathfrak{O}_{\frac{t}{p}} \subset p\mathfrak{O}_t \subseteq \mathfrak{p},$$

which proves that $\mathfrak{p}_p \subseteq \mathfrak{p}$, because $\mathfrak{p}$ is a prime ideal by assumption. The maximality of $\mathfrak{p}_p$ now implies that $\mathfrak{p}_p = \mathfrak{p}$. □

The primary ideals belonging to a given non-regular prime ideal $\mathfrak{p}_p$ of $\mathfrak{O}_t$ are explicitly described in the next theorem.

Theorem 3.1.18 *Let p be a prime dividing t and $t = p^r t_0$, $p \nmid t_0$. Then all $\mathfrak{p}_p$-primary ideals of $\mathfrak{O}_t$ are given by*

$$\mathfrak{q} = p^\nu \left[Q\underline{\alpha}\right]$$

with some primitive matrix Q and a basis $\underline{\alpha}$ of $\mathfrak{O}_{t_0}$. The determinant of Q is a p-power and the discriminant $D(Q(\alpha))$ satisfies

$$D(Q(\alpha)) \mid t^2 d_K.$$

The exponent ν, $\nu \in \mathbb{N}_0$, must be large enough to ensure the inclusion $\mathfrak{q} \subseteq \mathfrak{O}_t$.

Proof By Theorems 3.1.2 and 3.1.17 we can immediately see that the ideals $\mathfrak{q}$ defined in the theorem are $\mathfrak{p}_p$-primary. Conversely, let $\mathfrak{q}$ be an arbitrary $\mathfrak{p}_p$-primary ideal of $\mathfrak{O}_t$. Then, $\mathfrak{O}_t$ being noetherian, a power of $\mathfrak{p}_p$ is contained in $\mathfrak{q}$:

$$p^n\mathfrak{O}_{\frac{t}{p}} = \mathfrak{p}_p^N \subseteq \mathfrak{q} \subseteq \mathfrak{O}_t.$$

Therefore

$$[\mathfrak{O}_{t_0} : \mathfrak{q}] = [\mathfrak{O}_{t_0} : \mathfrak{O}_t][\mathfrak{O}_t : \mathfrak{Q}] \mid p^{\nu+n},$$

and it follows that $\mathfrak{q}$ is of the above form. The condition on $D(Q(\alpha))$ is implied by Theorem 3.1.2 because $\mathfrak{q}$ is an ideal of $\mathfrak{O}_t$. $\square$

3.2 Density theorems

Let K be an algebraic number field and

$$\zeta_K(s) = \sum_{\mathfrak{g}} \frac{1}{N(\mathfrak{g})^s}, \quad s > 1,$$

its ζ function, where the summation is over all integral ideals $\mathfrak{g}$ of K (meaning ideals of the maximal order of K). From its product expansion

$$\zeta_K(s) = \prod_{\mathfrak{p}} \frac{1}{1 - \frac{1}{N(\mathfrak{p})^s}},$$

where $\mathfrak{p}$ runs through all prime ideals of K, one deduces that

$$\log(\zeta_K(s)) = \sum_{\mathfrak{p}} \frac{1}{N(\mathfrak{p})^s} + O(1) \text{ for } s \to 1^+.$$

Together with the well-known fact that

$$\lim_{s \to 1^+} \frac{\zeta_K(s)}{s-1} \in \mathbb{R}_{>0},$$

we thus obtain the relation

$$\lim_{s \to 1^+} \frac{-1}{\log(s-1)} \sum_{\mathfrak{p}} \frac{1}{N(\mathfrak{p})^s} = 1, \tag{3.3}$$

which motivates the following definition.

Definition 3.2.1 (Dirichlet density) Let A be a set of prime ideals of K. Then the limit

$$\delta(A) := \lim_{s \to 1^+} \frac{-1}{\log(s-1)} \sum_{\mathfrak{p} \in A} \frac{1}{N(\mathfrak{p})^s},$$

provided that it exists, is called the Dirichlet density of A.

The Dirichlet density of a set A of prime ideals is determined by the prime ideals of degree 1 in A. To be precise, we have:

Remark 3.2.2 Let A and B be two sets of prime ideals in K differing by

- finitely many prime ideals of degree 1 and
- an arbitrary set of prime ideals of degree ≥ 2.

Then the existence of $\delta(A)$ implies the existence of $\delta(B)$ and both are equal.

Proof The proof is immediate by the inequality

$$\sum_{\substack{\mathfrak{p} \\ \deg(\mathfrak{p})\geq 2}} \frac{1}{N(\mathfrak{p}))} \leq [K:\mathbb{Q}] \sum_{p} \frac{1}{p^2} < \infty,$$

where p runs through all rational primes. To derive the inequality, one has to observe that given a prime p, there are at most $[K : \mathbb{Q}]$ many prime ideals in K whose norm is a p-power. □

For a given extension L/K of number fields, we now consider the set $\mathbb{P}_{L/K}$ of prime ideals of K splitting completely in L.

Theorem 3.2.3 *The Dirichlet density of $\mathbb{P}_{L/K}$ exists, and we have*

$$\delta(\mathbb{P}_{L/K}) = \frac{1}{[\tilde{L}:K]},$$

where $\tilde{L}$ denotes the Galois closure of L/K.

Proof First, we assume L/K to be Galois, so $L = \tilde{L}$. Let $\mathbb{P}_L^{(1)}$ be the set of prime ideals of degree 1 in L, and let $\mathbb{P}_{L/K}^{(1)}$ be the set of prime ideals of degree 1 in K that split completely in L. Then, by (3.3) and Remark 3.2.2 we have

$$\delta(\mathbb{P}_L^{(1)}) = 1 \text{ and } \delta(\mathbb{P}_{L/K}^{(1)}) = \delta(\mathbb{P}_{L/K}).$$

Therefore, to prove Theorem 3.2.3, it is sufficient to derive the relation

$$1 = n\delta(\mathbb{P}_{L/K}^{(1)}) \quad \text{with} \quad n = [L:K]. \tag{3.4}$$

The proof of this equation follows from the fact that, except for a finite number, every prime ideal $\mathfrak{p} \in \mathbb{P}_{L/K}^{(1)}$ splits completely in L,

$$\mathfrak{p}\mathfrak{O}_L = \mathfrak{P}_1 \cdot \ldots \cdot \mathfrak{P}_n,$$

with prime ideals $\mathfrak{P}_i$ of norm

$$N(\mathfrak{P}_i) = N(\mathfrak{p}).$$

Conversely, every prime ideal $\mathfrak{P}$ of degree 1 in L occurs in such a decomposition for exactly one $\mathfrak{p} \in \mathbb{P}_{L/K}^{(1)}$. The relation (3.4) now follows directly from the definition of the Dirichlet density. This completes the proof of Theorem 3.2.3 for a Galois extension L/K. The general case can be reduced to this special case by showing that

$$\delta(\mathbb{P}_{L/K}) = \delta(\mathbb{P}_{\tilde{L}/K}).$$

To prove this equation, observe that

$$\tilde{L} = L_1 \cdot \ldots \cdot L_n$$

is a product of fields that are conjugate to L over K. First, every prime ideal $\mathfrak{p}$ of K that splits completely in $\tilde{L}$ also splits completely in L. Conversely, every prime ideal that splits completely in L splits completely in all conjugate fields L_i and thus also in $\tilde{L}$ up to a finite number of exceptions, as will be proved by the next lemma. □

Lemma 3.2.4 *Let L_1, L_2 be finite extensions of a number field K, and let $\mathfrak{p}$ be a prime ideal in K that splits completely in L_1 and L_2. Then $\mathfrak{p}$ also splits completely in L_1L_2.*

Proof In view of the above application it will suffice to give a proof with finitely many $\mathfrak{p}$'s excluded. We choose an integral generator α for L_2 over L_1. Then we have the inclusion

$$\mathfrak{O}_{L_1L_2} \subseteq \frac{1}{d}\mathfrak{O}_{L_1}[\alpha]$$

with the discriminant d of the minimal polynomial of α over L_1. Now, let $\mathfrak{p}$ be a prime ideal in K splitting completely in L_1 and in L_2. In order to show that $\mathfrak{p}$ also splits completely in L_1L_2, we have to prove the congruence

$$\xi^{N(\mathfrak{p})} \equiv \xi \bmod \mathfrak{P} \text{ for all } \xi \in \mathfrak{O}_{L_1L_2}$$

for every prime ideal $\mathfrak{P}$ of L_1L_2 above $\mathfrak{p}$. Therefore, let $\mathfrak{P}_i = \mathfrak{P} \cap L_i$ be the prime ideals of L_i under $\mathfrak{P}$. Then

$$\xi^{N(\mathfrak{p})} \equiv \xi \bmod \mathfrak{P}_i \text{ for all } \xi \in \mathfrak{O}_{L_i}, \quad i = 1, 2,$$

because, by assumption, $\mathfrak{p}$ splits completely in both L_i. This implies that

$$\xi^{N(\mathfrak{p})} \equiv \xi \bmod \mathfrak{P} \text{ for all } \xi \in \mathfrak{O}_{L_1}[\alpha],$$

and by the above inclusion we see that this congruence also holds for all $\xi \in \mathfrak{O}_{L_1L_2}$ if $\mathfrak{p}$ is prime to d. Therefore, every prime ideal $\mathfrak{P}$ of L_1L_2 over $\mathfrak{p}$ has relative degree 1, hence $\mathfrak{p}$ splits completely in L_1L_2. This completes the proof for all prime ideals of K except for the finitely many $\mathfrak{p}$'s dividing d. □

For later applications of the above facts we will need the following theorem:

Theorem 3.2.5 *Let N/K and L/K be finite extensions of number fields, and let N/K be Galois. Further, we assume that with a finite number of exceptions the prime ideals of degree 1 in K splitting completely in N also split completely in L. Then we have $L \subseteq N$.*

Proof By Lemma 3.2.4, up to a finite number of exceptions, all prime ideals of degree 1 in K split completely in N if and only if they split completely in NL, so by Theorem 3.2.3

$$[\tilde{NL} : K] = [N : K],$$

which implies the asserted inclusion $L \subseteq N$. □

3.3 Class field theory

Let L/K be an extension of number fields with Galois group $G = G(L/K)$. We fix two prime ideals $\mathfrak{P}$ above $\mathfrak{p}$ of L resp. K with residue fields

$$\overline{L}_{\mathfrak{P}} = \mathfrak{O}_L/\mathfrak{P} \text{ and } \overline{K}_{\mathfrak{p}} = \mathfrak{O}_K/\mathfrak{p} \cong (\mathfrak{O}_K + \mathfrak{P})/\mathfrak{P}.$$

The Galois group $G(\overline{L}_{\mathfrak{P}}/\overline{K}_{\mathfrak{p}})$ of $\overline{L}_{\mathfrak{P}}/\overline{K}_{\mathfrak{p}}$ is generated by the Frobenius map

$$\varphi : \overline{L}_{\mathfrak{P}} \to \overline{L}_{\mathfrak{P}}, \quad \varphi(\xi + \mathfrak{P}) := \xi^{N(\mathfrak{p})} + \mathfrak{P}.$$

For $\mathfrak{P}$ unramified over K we have an isomorphism between $G(\overline{L}_{\mathfrak{P}}/\overline{K}_{\mathfrak{p}})$ and the decomposition group

$$G_z(\mathfrak{P}/K) := \{\psi \in G \mid \mathfrak{P}^{\psi} = \mathfrak{P}\}$$

by

$$\psi \mapsto \bar{\psi} \text{ with } \bar{\psi}(\xi + \mathfrak{P}) := \psi(\xi) + \mathfrak{P} \text{ for all } \xi \in \mathfrak{O}_L.$$

Hence, there exists a unique automorphism $\sigma(\mathfrak{P}) \in G_z(\mathfrak{P})$ with

$$\xi^{\sigma(\mathfrak{P})} \equiv \xi^{N(\mathfrak{p})} \bmod \mathfrak{P} \text{ for all } \xi \in \mathfrak{O}_L.$$

In the following we assume G to be **abelian**. Then $\sigma(\mathfrak{P})$ only depends on $\mathfrak{p}$. Therefore, we can define the **Frobenius automorphism**

$$\sigma(\mathfrak{p}) := \sigma(\mathfrak{P})$$

with some prime ideal $\mathfrak{P}$ of L lying over $\mathfrak{p}$. Let $\mathfrak{m}$ be an integral ideal of K that is divisible by all prime ideals ramified in L, and let $\mathfrak{A}^{\mathfrak{m}}$ be the group fractional ideals of K that are prime to $\mathfrak{m}$. Then by multiplicative extension of

$$\sigma : \mathfrak{p} \mapsto \sigma(\mathfrak{p}),$$

we obtain an epimorphism

$$\sigma : \mathfrak{A}^{\mathfrak{m}} \to G$$

called an **Artin map**. In particular, we have the isomorphism

$$\mathfrak{A}^{\mathfrak{m}}/\mathfrak{U} \cong G \text{ by } \mathfrak{a}\mathfrak{U} \mapsto \sigma(\mathfrak{a}), \tag{3.5}$$

where $\mathfrak{U}$ denotes the kernel of σ. By abuse of notation we denote this map again by the symbol σ:

$$\sigma(\mathfrak{a}\mathfrak{U}) := \sigma(\mathfrak{a}).$$

As an application we can determine the factorisation in L of a prime ideal $\mathfrak{p} \in \mathfrak{A}^{\mathfrak{m}}$:

$$\mathfrak{p}\mathfrak{O}_L = \mathfrak{P}_1 \cdot \ldots \cdot \mathfrak{P}_r,$$

with different prime ideals $\mathfrak{P}_i$ of L having the same relative degree f over K and f being the order of $\mathfrak{p}\mathfrak{U}$ in $\mathfrak{A}^{\mathfrak{m}}/\mathfrak{U}$:

$$f = \left[\overline{L}_{\mathfrak{P}_i}/\overline{K}_{\mathfrak{p}}\right] = o(\mathfrak{p}\mathfrak{U}).$$

In particular, the prime ideals in $\mathfrak{U}$ are, except for a finite number, exactly the prime ideals of K splitting completely in L, and this also implies that $\mathfrak{U}$ contains infinitely many prime ideals of degree 1. By class field theory we even know this to be true for every coset modulo $\mathfrak{U}$.

By Theorem 3.2.5, L is uniquely determined by $\mathfrak{U}$. Therefore, L is called the **class field** of K corresponding to $\mathfrak{U}$. As an application of this fact we have the equivalence

$$L \subseteq L' \iff \mathfrak{U} \supseteq \mathfrak{U}'$$

for two extensions L, L' of K unramified outside $\mathfrak{m}$ and the corresponding subgroups $\mathfrak{U}, \mathfrak{U}'$ of $\mathfrak{A}^{\mathfrak{m}}$. Further, for $\mathfrak{U} \supseteq \mathfrak{U}'$ we have

$$\sigma(\mathfrak{k})|L' = \sigma(\mathfrak{k}') \quad \text{if} \quad \mathfrak{k} \supseteq \mathfrak{k}', \tag{3.6}$$

for two ideal classes $\mathfrak{k} \in \mathfrak{A}^{\mathfrak{m}}/\mathfrak{U}$, $\mathfrak{k}' \in \mathfrak{A}^{\mathfrak{m}}/\mathfrak{U}'$, which follows directly from the definition of the Artin map.

As an application of this relation we obtain:

Theorem 3.3.1 *Let $L \subset L'$ be abelian extensions of K, unramified outside $\mathfrak{m}$, and let $\mathfrak{U} \supset \mathfrak{U}'$ be the corresponding subgroups of $\mathfrak{A}^{\mathfrak{m}}$. Let s be in $\mathbb{Z} \setminus \mathbb{Z}^2$ and $\sqrt{s} \in L'$. Then for every integral ideal $\mathfrak{c}$ of $\mathfrak{O}_K$, prime to $2s\mathfrak{m}$, we have*

$$\sqrt{s}^{\sigma(\mathfrak{c})} = \left(\tfrac{s}{N(\mathfrak{c})}\right)\sqrt{s},$$

where $\left(\tfrac{s}{N(\mathfrak{c})}\right)$ denotes the Legendre symbol.

As $\sigma(\mathfrak{c})$ only depends on the ideal class $\mathfrak{k}'$ modulo $\mathfrak{U}'$ of $\mathfrak{c}$, we can define a character of $\mathfrak{A}^{\mathfrak{m}}/\mathfrak{U}'$ by

$$\chi_s(\mathfrak{k}') := \left(\tfrac{s}{N(\mathfrak{c})}\right) \text{ with } \mathfrak{c} \in \mathfrak{k}',\ \mathfrak{c} \text{ integral and } \mathfrak{c} \nmid 2s.$$

Clearly, χ_s has order 2 if $\sqrt{s}$ is not in K and is equal to the principal character for $\sqrt{s}$ in K.

In the case $[L' : K] > 2$ there always exists an ideal class $\mathfrak{k}' \in \mathfrak{A}^{\mathfrak{m}}/\mathfrak{U}'$ with

$$\mathfrak{k}' \neq \mathfrak{U}' \text{ and } \left(\tfrac{s}{N(\mathfrak{c})}\right) = 1 \text{ for all integral ideals } \mathfrak{c} \in \mathfrak{k}' \text{ with } \mathfrak{c} \nmid 2s.$$

If $\sqrt{s} \in L' \setminus L$, then every ideal class $\mathfrak{k} \in \mathfrak{A}^{\mathfrak{m}}/\mathfrak{U}$ contains two ideal classes $\mathfrak{k}'_{\pm} \in \mathfrak{A}^{\mathfrak{m}}/\mathfrak{U}'$ with

$$\mathfrak{k}'_{\pm} \subset \mathfrak{k} \text{ and } \chi_s(\mathfrak{k}'_{\pm}) = \pm 1.$$

Proof For the proof we may assume $\mathfrak{c}$ to be a prime ideal $\mathfrak{p}$. Then, by definition of $\sigma(\mathfrak{p})$, we have

$$\sqrt{s}^{\sigma(\mathfrak{p})} \equiv \sqrt{s}^{N(\mathfrak{p})} \bmod \mathfrak{p}$$

and this implies that

$$\sqrt{s}^{\sigma(\mathfrak{p})-1} \equiv s^{\frac{N(\mathfrak{p})-1}{2}} \bmod \mathfrak{p}.$$

The Euler criterion now tells us that

$$s^{\frac{N(\mathfrak{p})-1}{2}} \equiv \left(\frac{s}{N(\mathfrak{p})}\right) \bmod N(\mathfrak{p}),$$

and this proves the first assertion of our theorem because $\mathfrak{p}$ is prime to 2. The next two assertions are now immediate. To prove the last assertion, observe that the Frobenius map $\sigma(\mathfrak{k})$ of L/K has two extensions $\psi_\pm$ to L' with

$$\sqrt{s}^{\psi_\pm} = \pm\sqrt{s}$$

because $\sqrt{s} \in L' \setminus L$ and by (3.6) we know that $\psi_\pm = \sigma(\mathfrak{k}'_\pm)$ with two ideal classes $\mathfrak{k}'_\pm$ modulo $\mathfrak{U}'$ contained in $\mathfrak{k}$. This proves the last assertion of our theorem. □

In the same way we prove for later purposes:

Theorem 3.3.2 *Let L be an abelian extension of K, unramified outside $\mathfrak{m}$, and let $\mathfrak{U}$ be the corresponding subgroup of $\mathfrak{A}^{\mathfrak{m}}$. Let ζ be an n-th root of unity contained in L. Then*

$$\zeta^{\sigma(\mathfrak{c})} = \zeta^{N(\mathfrak{c})}$$

for every integral ideal $\mathfrak{c}$ of $\mathfrak{O}_K$ prime to $n\mathfrak{m}$.

Proof It is again sufficient to give the proof for a prime ideal $\mathfrak{c} = \mathfrak{p} \nmid n\mathfrak{m}$. First, $\zeta^{\sigma(\mathfrak{p})}$ is again an n-th root of unity

$$\zeta^{\sigma(\mathfrak{p})} = \zeta^{\nu}, \quad \nu \in \mathbb{Z},$$

and by definition of σ, we have

$$\zeta^{\nu} \equiv \zeta^{N(\mathfrak{p})} \bmod \mathfrak{p}.$$

If $\zeta^{\nu} - \zeta^{N(\mathfrak{p})}$ was not equal to zero, it must be a divisor of n, which contradicts the above congruence. Hence, $\zeta^{\nu} = \zeta^{N(\mathfrak{p})}$, as asserted in the theorem. □

The main theorem of class field theory determines all possible subgroups of $\mathfrak{A}^{\mathfrak{m}}$ corresponding to abelian extensions of K. To be more precise, we fix an integral ideal $\mathfrak{f}$ and define the **ray modulo** $\mathfrak{f}$:

$$\mathfrak{S}_{\mathfrak{f}} := \left\{ \left(\frac{\alpha_1}{\alpha_2}\right) \;\middle|\; \alpha_i \in \mathfrak{O}_K \text{ prime to } \mathfrak{f},\ \alpha_1 \equiv \alpha_2 \bmod \mathfrak{f} \right\}.$$

If $\mathfrak{f}$ is divisible by all prime ideals dividing $\mathfrak{m}$, then $\mathfrak{S}_{\mathfrak{f}}$ is a subgroup of $\mathfrak{A}^{\mathfrak{m}}$ and, according to class field theory, there exists a unique abelian extension $K_{\mathfrak{f}}$ of K associated with $\mathfrak{S}_{\mathfrak{f}}$. It is called the **ray class field modulo** $\mathfrak{f}$ over K. For $\mathfrak{f} = (N), N \in \mathbb{N}$, we sometimes write K_N instead of $K_{(N)}$ for simplicity.

For K totally imaginary, in particular for K imaginary quadratic, the ray class fields including their subextensions of K represent all abelian extensions of K, which means that, given an abelian extension L of K, then L is a subfield of some ray class field $K_{\mathfrak{f}}$. In this situation L/K is unramified outside $\mathfrak{m} = \mathfrak{f}$, and for the corresponding subgroup $\mathfrak{U}$ of $\mathfrak{A}^{\mathfrak{f}}$ we have

$$\mathfrak{A}^{\mathfrak{f}} \supseteq \mathfrak{U} = \sigma^{-1}(G(K_{\mathfrak{f}}/L)) \supseteq \mathfrak{S}_{\mathfrak{f}},$$

where $G(K_{\mathfrak{f}}/L)$ denotes the Galois group of $K_{\mathfrak{f}}/L$.

In complex multiplication the "natural" extensions are connected to the modular invariant j and to Weber's τ function. The corresponding subgroups are given by $\mathfrak{U}_{\mathfrak{t}}$ and $\mathfrak{U}_{\mathfrak{t},\mathfrak{f}}$, defined in Theorems 3.1.7 and 3.1.8. Let Ω_t be the class field corresponding to $\mathfrak{U}_{\mathfrak{t}}$ ($t\mathfrak{O} = \mathfrak{t}$). Then, using the Artin map, the Galois group of Ω_t/K is isomorphic to $\mathfrak{A}^{\mathfrak{t}}/\mathfrak{U}_{\mathfrak{t}}$. In view of Theorem 3.1.7 we then also have an isomorphism between the ring ideal class group of $\mathfrak{O}_t$ and the Galois group of Ω_t/K. We describe this isomorphism in the following way because this will be suitable later for the action on singular values of modular functions:

$$\sigma : \mathfrak{J}_t \to G(\Omega_t/K) \quad \text{with} \quad ker(\sigma) = \mathfrak{H}_t.$$

Again, by abuse of notation, we have used the same symbol σ as for the Artin map. σ is characterised by the following property: given an integral ideal $\mathfrak{c}$, prime to t, with corresponding ring ideal $\mathfrak{c}_t := \mathfrak{c} \cap \mathfrak{O}_t$, then the image $\sigma(\mathfrak{c}_t)$ is by definition the Frobenius automorphism of Ω_t/K associated with $\mathfrak{c}$:

$$\sigma(\mathfrak{c}_t) := \sigma(\mathfrak{c}).$$

Due to this description, Ω_t is called the **ring class field modulo** t of K. In particular,

$$\Omega := \Omega_1$$

is the maximal unramified abelian extension of K, which is called the **Hilbert class field** of K. In this case $\mathfrak{A}^{(1)}$ is the ideal group of K and the subgroup $\mathfrak{U}_1$ corresponding to Ω is the group of principal ideals of K. Hence, the Galois group of Ω/K is isomorphic to the ideal group of K.

The class field corresponding to the subgroup $\mathfrak{U}_{\mathfrak{t},\mathfrak{f}}$ is called the **ray class field modulo $\mathfrak{f}$ of $\mathfrak{O}_t$** and is denoted by $K_{\mathfrak{t},\mathfrak{f}}$. Its Galois group over K is isomorphic to the ray class group modulo $\mathfrak{f}$ of $\mathfrak{O}_t$. In analogy to the ring class fields, we have an epimorphism

$$\sigma : \mathfrak{J}_{t,\mathfrak{f}} \to G(K_{\mathfrak{t},\mathfrak{f}}/K) \quad \text{with} \quad ker(\sigma) = \mathfrak{S}_{t,\mathfrak{f}},$$

defined by

$$\sigma(\mathfrak{c}_t) := \sigma(\mathfrak{c})$$

for every integral of $\mathfrak{O}_1$ prime to $\mathfrak{f}\mathfrak{t}$ with the Frobenius automorphism $\sigma(\mathfrak{c})$ of $K_{t,\mathfrak{f}}$.

For $t = 1$ the group $\mathfrak{R}_{1,\mathfrak{f}}$ is equal to the ray class group modulo $\mathfrak{f}$ defined above and $K_{\mathfrak{f}} = K_{(1),\mathfrak{f}}$ is the ray class field modulo $\mathfrak{f}$.

The inclusion $\mathfrak{U}_{\mathfrak{t}} \supseteq \mathfrak{U}_{\mathfrak{t},\mathfrak{f}}$ implies that $\Omega_t \subseteq K_{\mathfrak{t},\mathfrak{f}}$, and we have the following diagram for class fields of K and corresponding subgroups of $\mathfrak{J}_{t,\mathfrak{f}}$:

$$\begin{array}{ccc} K_{\mathfrak{t},\mathfrak{f}} & \qquad & \mathfrak{S}_{t,\mathfrak{f}} \\ | & & | \\ \Omega_t & & \mathfrak{H}_t \cap \mathfrak{J}_{t,\mathfrak{f}} \\ | & & | \\ K & & \mathfrak{J}_{t,\mathfrak{f}} \end{array}$$

For an explicit description of the Galois groups of Ω_t/K and $K_{\mathfrak{t},\mathfrak{f}}/\Omega_t$ let

$$\mathfrak{a}_1, \ldots, \mathfrak{a}_{h_t} \in \mathfrak{J}_{t,\mathfrak{f}}$$

be a system of coset representatives modulo $\mathfrak{H}_t \cap \mathfrak{J}_{t,\mathfrak{f}}$ which is also a system of coset representatives for $\mathfrak{J}_t$ modulo $\mathfrak{H}_t$, because in every class of $\mathfrak{J}_t$ modulo $\mathfrak{H}_t$ there exist integral ideals coprime to $\mathfrak{f}\mathfrak{t}$. Therefore, the restrictions of the corresponding Frobenius automorphisms $\sigma(\mathfrak{a}_i)$ to Ω_t all yield different automorphisms of Ω_t/K:

$$G(\Omega_t/K) = \{\sigma(\mathfrak{a}_1)|\Omega_t, \ldots, \sigma(\mathfrak{a}_{h_t})|\Omega_t\}.$$

Now, choose $\lambda_1, \ldots, \lambda_m \in \mathfrak{O}_t$, prime to $\mathfrak{f}$, such that the residue classes $\lambda_i + \mathfrak{f}$ form a system of coset representatives modulo the subgroup of prime residue classes generated by roots of unity in $\mathfrak{O}_t$. Then, by $\sigma(\lambda_i), i = 1, \ldots m$, we obtain the different automorphisms of $K_{\mathfrak{f}}/\Omega$:

$$G(K_{\mathfrak{f}}/\Omega) = \{\sigma(\lambda_1), \ldots, \sigma(\lambda_m)\}.$$

Further, by Theorem 3.1.9 we obtain explicit formulae for the degrees:

$$[\Omega_t : K] = h_K \frac{w_t(\mathfrak{O}_t)\Phi_1(\mathfrak{t})}{w_1(\mathfrak{O}_1)\Phi_t(\mathfrak{t})},$$

$$[K_{t,\mathfrak{f}} : \Omega_t] = \frac{w_t(\mathfrak{f})}{w_t(\mathfrak{O}_t)}\Phi_t(\mathfrak{f}).$$

To compute the discriminant of $K_\mathfrak{f}/K$, we consider the characters of $\mathfrak{A}^\mathfrak{f}/\mathfrak{S}_\mathfrak{f}$, which in short we call "characters modulo $\mathfrak{f}$". Given an integral ideal $\mathfrak{b}$ of K, then via the epimorphism

$$\kappa : \mathfrak{A}^{\mathfrak{b}\mathfrak{f}}/\mathfrak{S}_{\mathfrak{b}\mathfrak{f}} \to \mathfrak{A}^\mathfrak{f}/\mathfrak{S}_\mathfrak{f}, \quad \mathfrak{a}\mathfrak{S}_{\mathfrak{b}\mathfrak{f}} \mapsto \mathfrak{a}\mathfrak{S}_\mathfrak{f},$$

every character χ modulo $\mathfrak{f}$ induces a character $\tilde{\chi} := \chi \circ \kappa$ modulo $\mathfrak{b}\mathfrak{f}$. Conversely, one may ask whether a given character modulo $\mathfrak{f}$ is induced by a character modulo a proper divisor $\mathfrak{f}'$ of $\mathfrak{f}$. If not, then we call χ a **primitive character**.

Theorem 3.3.3 (conductor of a character) *For every character χ modulo $\mathfrak{f}$ there exists a unique divisor $\mathfrak{f}'$ of $\mathfrak{f}$ and a unique primitive character χ' modulo $\mathfrak{f}'$, so that χ is induced by χ'. $\mathfrak{f}_\chi := \mathfrak{f}'$ is called the conductor of χ.*

Let L/K be a subextension of $K_\mathfrak{f}/K$ and let $\mathfrak{U}$ denote the corresponding subgroup of $\mathfrak{A}^\mathfrak{f}$. Then, we have the inclusion

$$\mathfrak{A}^\mathfrak{f} \supseteq \mathfrak{U} \supseteq \mathfrak{S}_\mathfrak{f}.$$

We define a character of L/K as a character χ modulo $\mathfrak{f}$ with $\ker \chi \supseteq \mathfrak{U}$.

Theorem 3.3.4 (conductor-discriminant-formula) *Let $d_{L/K}$ denote the relative discriminant of L/K. Then*

$$d_{L/K} = \prod_\chi \mathfrak{f}_\chi,$$

where χ runs through all characters of L/K.

Applying this formula in the case of a quadratic imaginary base field K, we obtain:

Theorem 3.3.5 *Let K be a quadratic imaginary number field, $\mathfrak{f}$ an integral ideal in K and $K_{\mathfrak{f}}$ the ray class field modulo $\mathfrak{f}$ over K. Let $\mathfrak{p}$ be a prime ideal dividing $\mathfrak{f}$. We write $\mathfrak{f} = \mathfrak{p}^r\mathfrak{b}$ with an integral ideal $\mathfrak{b}$, $\mathfrak{p} \nmid \mathfrak{b}$. Then the $\mathfrak{p}$-exponent of $d_{K_{\mathfrak{f}}/K}$ is given by*

$$v_{\mathfrak{p}}(d_{K_{\mathfrak{f}}/K}) = [K_{\mathfrak{b}} : K]\frac{1}{w_0}\left(w_r(r+1)\Phi(\mathfrak{p}^r) - c_r\right),$$

where

$$w_i := w(\mathfrak{p}^i\mathfrak{b})$$

is the number of roots of unity in K congruent to 1 modulo $\mathfrak{p}^i\mathfrak{b}$, and c_r is defined by

$$c_r := \sum_{j=0}^{i} w_j\Phi(\mathfrak{p}^j)$$

with Euler's function Φ.

Proof For $1 \le i \le r$ we have $v_{\mathfrak{p}}(\mathfrak{f}_\chi) \le i$ if and only if χ is a character of $K_{\mathfrak{b}\mathfrak{p}^i}/K$. Hence, the number of χ with $v_{\mathfrak{p}}(\mathfrak{f}_\chi) = i$ is given by

$$[K_{\mathfrak{b}\mathfrak{p}^i} : K] - [K_{\mathfrak{b}\mathfrak{p}^{i-1}} : K] = [K_{\mathfrak{b}} : K]\left(\frac{w_i}{w_0}\Phi(\mathfrak{p}^i) - \frac{w_{i-1}}{w_0}\Phi(\mathfrak{p}^{i-1})\right).$$

Therefore

$$\begin{aligned} v_{\mathfrak{p}}(K_{\mathfrak{f}}/K) &= [K_{\mathfrak{b}} : K]\frac{1}{w_0}\sum_{i=1}^{r}\left(w_i\Phi(\mathfrak{p}^i) - w_{i-1}\Phi(\mathfrak{p}^{i-1})\right) \\ &= [K_{\mathfrak{b}} : K]\frac{1}{w_0}\left(w_r\Phi(\mathfrak{p}^r) - \sum_{i=1}^{r-1} w_i\Phi(\mathfrak{p}^i)\right) \\ &= [K_{\mathfrak{b}} : K]\frac{1}{w_0}\left(w_r(r+1)\Phi(\mathfrak{p}^r) - c_r\right), \end{aligned}$$

and the formula is proved. □

4
Factorisation of singular values

Modular functions from F_N have the nice property that their values at a quadratic imaginary number α are algebraic and generate an abelian extension of $K = \mathbb{Q}(\alpha)$. Therefore these values are called **singular values.** Some of these values can also be factorised explicitly, which is crucial for determining the generated fields.

4.1 Singular values

In this section we show that singular values are algebraic, and we will also provide some factorisation properties that will be needed later. First, we consider the modular invariant j. Let $\mathfrak{a}$ be a lattice and α_1, α_2 a $\mathbb{Z}$-basis of $\mathfrak{a}$ with $\alpha := \frac{\alpha_1}{\alpha_2} \in \mathbb{H}$. We set

$$j(\mathfrak{a}) := j(\alpha),$$

whereby $j(\mathfrak{a})$ is well defined, because for an arbitrary basis α_1', α_2' the ratio $\alpha' = \frac{\alpha_1'}{\alpha_2'} \in \mathbb{H}$ is related to α by a unimodular transformation which leaves j invariant. Further, $j(\mathfrak{a})$ only depends on the class of the lattice, i.e. we have

$$j(\mathfrak{a}) = j(\mathfrak{a}\lambda) \text{ for } \lambda \in \mathbb{C}^*.$$

Theorem 4.1.1 *Let $\alpha \in \mathbb{H}$ be a quadratic imaginary number. Then $j(\alpha)$ is an algebraic integer.*

Proof Let $\mathfrak{a} = [\alpha, 1]$ be the $\mathbb{Z}$-module generated by α and 1, and let t be the conductor of the order of $\mathfrak{a}$ in $K = \mathbb{Q}(\alpha)$. We choose a prime ideal $\mathfrak{p}$ of degree 1 not dividing t with the property that the corresponding ring ideal $\mathfrak{p}_t$ is principal. Then

$$j(\alpha) = j(\mathfrak{a}) = j(\mathfrak{a}\mathfrak{p}_t)$$

because $\mathfrak{a}$ and $\mathfrak{a}\mathfrak{p}_t$ are in the same class. In view of $N(\mathfrak{p}_t) = p$ with a prime p there exists a primitive matrix P of determinant p, so that

$$P\begin{pmatrix}\alpha\\1\end{pmatrix}$$

is a basis of $\mathfrak{a}\mathfrak{p}_t$. We then have

$$j(P(\alpha)) = j(\mathfrak{a}\mathfrak{p}_t) = j(\mathfrak{a}) = j(\alpha),$$

and it follows that

$$I_p(j(\alpha), j(\alpha)) = I_p(j(P(\alpha), j(\alpha)) = 0$$

with the main-polynomial $I_p(X, j)$. Now, by Theorem 2.7.4, we know that $I_p(X, X)$ is a normalised polynomial in $\mathbb{Z}[X]$. Hence $j(\alpha)$ is an algebraic integer and Theorem 4.1.1 is proved. □

Theorem 4.1.1 implies:

Theorem 4.1.2 *Let f be in F_N, $N \in \mathbb{N}$, and let $\alpha \in \mathbb{H}$ be a quadratic imaginary with $f(\alpha) \neq \infty$. Then:*

(i) *$f(\alpha)$ is algebraic,*
(ii) *$f(\alpha)$ is an algebraic integer if f is holomorphic in $\mathbb{H}$ and the q-coefficients in all cusps are algebraic integers.*

Proof For f holomorphic in $\mathbb{H}$ the assertions of our theorem follow from the extended q-expansion principle of Theorem 2.6.2 together with Theorem 4.1.1.

To prove the first assertion for f having poles in $\mathbb{H}$, the following lemma provides a representation of f as a rational function with algebraic coefficients of j and some division values $\tau_{\underline{x}}, \underline{x} \in \frac{1}{N}(\mathbb{Z} \times \mathbb{Z}) \setminus \mathbb{Z} \times \mathbb{Z}$ such that the representation is also valid for the values at α unless α is equivalent to $i = \sqrt{-1}$ or $\rho = \frac{-1+\sqrt{-3}}{2}$. This proves the first assertion for f and $\alpha \not\sim i, \rho$, because j and $\tau_{\underline{x}}$ are holomorphic in $\mathbb{H}$. To prove the assertion for α equivalent to i or ρ we write

$$f(\alpha) = \tilde{f}(N_1\alpha) \text{ with } \tilde{f}(\omega) = f\left(\frac{\omega}{N_1}\right)$$

where N_1 is a natural number. Then $\tilde{f} \in F_{NN_1}$, and $N_1\alpha$ is not equivalent to i or ρ for a suitable N_1. □

Lemma 4.1.3 *Let Λ be a subfield of $\mathbb{C}$, and let f be in ΛF_N having no pole at $\alpha \in \mathbb{H}$. Let $\alpha_1, \dots, \alpha_s$ modulo $\Gamma(N)$ be all points in $\mathbb{H}$ equivalent to α modulo Γ, but different from α, where f has a pole. Let*

$$\tau_i := \tau_{\underline{x}_i}, i = 1, \dots, m,$$

denote all different functions in the set $\left\{\tau_{\underline{x}} \mid \underline{x} \in \frac{1}{N}(\mathbb{Z}\times\mathbb{Z}) \setminus \mathbb{Z}\times\mathbb{Z}\right\}$. Then there exist $i_1, \dots, i_s \in \{1, \dots, m\}$ such that

$$\tau_{i_k}(\alpha) \neq \tau_{i_k}(\alpha_k), \;\; k = 1, \dots, s,$$

and we have a representation of f as

$$f = \frac{\sum\limits_{\mu_1,\dots,\mu_m=0}^{[\Gamma:\Gamma(N)]} a_\mu(j)\tau_1^{\mu_1} \cdot \dots \cdot \tau_m^{\mu_m}}{(\prod\limits_{k=1}^{s}(\tau_{i_k} - \tau_{i_k}(\alpha_k)) \prod\limits_{1\le i<j\le m}(\tau_i - \tau_j))^n}$$

with a natural exponent n and rational functions

$$a_\mu(j) \in \Lambda(\tau_{i_1}(\alpha_1), \dots, \tau_{i_s}(\alpha_s))(j),$$

having no pole at α.

Proof We consider the tower of fields

$$L_0 = \Lambda\mathbb{Q}_N(j), L_1 = L_0(\tau_1), \dots, L_m = L_{m-1}(\tau_m) = \Lambda F_N.$$

Applying Lemma 2.7.7 to the extension L_m/L_{m-1}, shows that f can be written as

$$f = \frac{\sum\limits_{\mu=0}^{[\Gamma:\Gamma(N)]} g_{m-1,\mu}\tau_m^\mu}{\prod\limits_i(\tau_m - \tau_i)},$$

where the τ_i in the product are all conjugates of τ_m over L_{m-1} different from τ_m. The coefficients $g_{m-1,\mu}$ are in L_{m-1}. First, we assume f to have no pole at $\alpha_1, .., \alpha_s$. Then the explicit formula in Lemma 2.7.7 shows that the same is true for the $g_{m-1,\mu}$. Now, applying the same arguments to the $g_{m-1,\mu}$ and the extension L_{m-1}/L_{m-2} and so on, we end up with a representation of the form

$$f = \frac{\sum\limits_{\mu_1,\dots,\mu_m=0}^{[\Gamma:\Gamma(N)]} g_{0,\mu}(j)\tau_1^{\mu_1} \cdot \dots \cdot \tau_m^{\mu_m}}{(\prod\limits_{1\le i<j\le m}(\tau_i - \tau_j))^n},$$

where the coefficients $a_\mu(j) := g_{0,\mu} \in L_0 = \Lambda(j)$ have no poles at α, and n is some natural number.

For f having poles at $\alpha_1, \dots, \alpha_s, s \geq 1$, we conclude as follows: the α_k can be written as $M_k\alpha$ with $M_k \in \Gamma \setminus \Gamma(N)$. This implies the existence of suitable $\underline{x}_k \in \frac{1}{N}(\mathbb{Z} \times \mathbb{Z})$ with $\pm\underline{x}_k \not\equiv \underline{x}_k M_k \bmod \mathbb{Z} \times \mathbb{Z}$. Hence

$$\tau_{\underline{x}_k}(\alpha) - \tau_{\underline{x}_k}(\alpha_k) = \tau_{\underline{x}_k}(\alpha) - \tau_{\underline{x}_k M_k}(\alpha) \neq 0.$$

Therefore, with an exponent n high enough, the function

$$\hat{f} := f \prod_{k=1}^{s} (\tau_{i_k} - \tau_{i_k}(\alpha_k))^n$$

is without poles at $\alpha_1, \dots, \alpha_s$. Then, applying the above construction to $\hat{f}$, we obtain the asserted representation for f. □

In the next two sections we consider the singular values of φ_A and $\varphi_{\underline{x}}$, which are algebraic integers according to Theorem 4.1.2. Moreover, they have no zero in $\mathbb{H}$, which implies that the lowest coefficient of their main-polynomial must be a constant. Using this fact, we will be able to factorise the singular values of φ_A and $\varphi_{\underline{x}}$.

4.2 Factorisation of $\varphi_A(\alpha)$

Theorem 4.2.1 *Let K be an imaginary quadratic number field, and let $\underline{\alpha} = \binom{\alpha_1}{\alpha_2}$ be a basis of a proper ideal $\mathfrak{a}$ of $\mathfrak{O}_t$. We assume that $\alpha = \frac{\alpha_1}{\alpha_2} \in \mathbb{H}$. Let p be a prime that splits in K,*

$$p = \mathfrak{p}\overline{\mathfrak{p}},$$

and does not divide t. Then there exist two primitive matrices $P_{\mathfrak{p}}, P_{\overline{\mathfrak{p}}}$, uniquely determined up to equivalence, such that

$$P_{\mathfrak{p}}\underline{\alpha} \text{ resp. } P_{\overline{\mathfrak{p}}}\underline{\alpha}$$

are bases of $\mathfrak{a}\mathfrak{p}_t$ resp. $\mathfrak{a}\overline{\mathfrak{p}}_t$. We then have

$$\varphi_{P_{\mathfrak{p}}}(\alpha) \sim \overline{\mathfrak{p}}^{12} \text{ and } \varphi_{P_{\overline{\mathfrak{p}}}}(\alpha) \sim \mathfrak{p}^{12}.$$

Further, $\varphi_P(\alpha)$ is a unit for all primitive matrices P of determinant p that are not equivalent to $P_{\mathfrak{p}}$ and $P_{\overline{\mathfrak{p}}}$.

Proof By Theorem 4.1.2, we know that $\varphi_P(\alpha)$ is an algebraic integer for every primitive matrix P of determinant p. Further, according to Theorem 2.7.5, we have the formula

$$\prod_P \varphi_P(\alpha) = p^{12},$$

where P runs through a system of inequivalent matrices of determinant p, so $\varphi_P(\alpha)$ must divide $p^{12} = \mathfrak{p}^{12}\overline{\mathfrak{p}}^{12}$. Now we choose an exponent $m \in \mathbb{N}$, so that

$$\mathfrak{p}_t^m = (\xi)$$

is principal. Then we have the identity

$$p^{12}\frac{\Delta(\mathfrak{a}\mathfrak{p}_t^m)}{\Delta(\mathfrak{a}\mathfrak{p}_t^{m-1})}\cdot\ldots\cdot p^{12}\frac{\Delta(\mathfrak{a}\mathfrak{p}_t)}{\Delta(\mathfrak{a})} = p^{12m}\frac{\Delta(\mathfrak{a}\xi)}{\Delta(\mathfrak{a})} = \frac{p^{12m}}{\xi^{12}} \sim \overline{\mathfrak{p}}^{12m}.$$

Writing

$$\varphi_{P_\mathfrak{p}}(\alpha) = p^{12}\frac{\Delta(\mathfrak{a}\mathfrak{p}_t)}{\Delta(\mathfrak{a})},$$

and recalling that these values are divisors of $p^{12} = \mathfrak{p}^{12}\overline{\mathfrak{p}}^{12}$, it follows that each factor in the product must be associated with $\overline{\mathfrak{p}}^{12}$. This proves the first assertion of our theorem, which implies the second because the product of all $\varphi_P(\alpha)$'s is equal to p^{12}. □

By Theorem 4.2.1 we obtain:

Theorem 4.2.2 *For $\mathfrak{a}, \mathfrak{b} \in \mathfrak{J}_t$ we have*

$$\frac{\Delta(\mathfrak{a})}{\Delta(\mathfrak{a}\mathfrak{b})} \sim \mathfrak{b}\mathfrak{O}_1.$$

Proof We choose a prime ideal $\mathfrak{p}$ of degree 1 not dividing t, so that $\mathfrak{p}_t$ is in the class of $\mathfrak{b}$:

$$\mathfrak{b} = \xi\mathfrak{p}_t, \quad \xi \in K.$$

Then, by homogeneity of Δ

$$\frac{\Delta(\mathfrak{a})}{\Delta(\mathfrak{a}\mathfrak{b})} = \xi^{12}\frac{\Delta(\mathfrak{a})}{\Delta(\mathfrak{a}\mathfrak{p}_t)}.$$

Theorem 4.2.1 now implies that

$$\frac{\Delta(\mathfrak{a})}{\Delta(\mathfrak{a}\mathfrak{b})} \sim \xi^{12}p^{12}\overline{\mathfrak{p}}^{-12} = (\xi\mathfrak{p})^{12} = \mathfrak{b}^{12}\mathfrak{O}_1.$$

□

In particular, by Theorem 4.2.2 we find the following theorem used in the sequel:

Theorem 4.2.3 *Let* $\mathfrak{a}_1, \mathfrak{a}_2$ *be in* $\mathfrak{I}_{ts}$, $s \in \mathbb{N}$, *with* $\mathfrak{O}_t\mathfrak{a}_1 = \mathfrak{O}_t\mathfrak{a}_2$. *Then:*

$$\frac{\Delta(\mathfrak{a}_1)}{\Delta(\mathfrak{a}_2)} \sim 1.$$

To factorise the singular values of φ_A for an arbitrary rational matrix A of positive determinant, observe that

$$A = a \cdot Q_1 \cdot \ldots \cdot Q_n$$

is a product of primitive matrices of prime determinant with a rational factor a, and we can write

$$\varphi_A(\omega) = \varphi_{Q_1}(\omega_2) \cdot \ldots \cdot \varphi_{Q_n}(\omega_n)$$

with $\omega_n = \omega$ and $\omega_i = Q_i \cdot \ldots \cdot Q_n(\omega), i \geq 2$. Therefore, it suffices to treat the case when $\det(A)$ is a prime. Following the exposition of Deuring (1958) we prove:

Theorem 4.2.4 *Let* p *be a prime,* P *a primitive matrix of determinant* p *and* $\underline{\alpha} = \binom{\alpha_1}{\alpha_2}$ *a basis of a proper ideal* $\mathfrak{a} \in \mathfrak{I}_t$ *with* $\alpha = \frac{\alpha_1}{\alpha_2} \in \mathbb{H}$. *Let* p^e *be the* p*-part of* t. *Then:*

(i) *If* p *splits in* K, $p = \mathfrak{p}\bar{\mathfrak{p}}$:

$$\varphi_P(\alpha) \sim \begin{cases} 1, & \text{if} \quad [P\underline{\alpha}] \in \mathfrak{I}_{tp}, \\ p^{12} & \text{if} \quad [P\underline{\alpha}] \in \mathfrak{I}_{tp^{-1}}, \\ \bar{\mathfrak{p}}^{12}, & \text{if} \quad [P\underline{\alpha}] = \mathfrak{a}\mathfrak{p}_t. \end{cases}$$

(ii) *If* p *is ramified in* K, $p = \mathfrak{p}^2$:

$$\varphi_P(\alpha) \sim \begin{cases} p^{\frac{6}{p^e+1}}, & \text{if} \quad [P\underline{\alpha}] \in \mathfrak{I}_{tp}, \\ p^{12-\frac{6}{p^e}} & \text{if} \quad [P\underline{\alpha}] \in \mathfrak{I}_{tp^{-1}}, \\ \mathfrak{p}^{12}, & \text{if} \quad [P\underline{\alpha}] = \mathfrak{a}\mathfrak{p}_t. \end{cases}$$

(iii) *If* p *is inert in* K, $p = \mathfrak{p}$:

$$\varphi_P(\alpha) \sim \begin{cases} p^{\frac{12}{p^e(p+1)}}, & \text{if} \quad [P\underline{\alpha}] \in \mathfrak{I}_{tp}, \\ p^{12-\frac{12}{p^{e-1}(p+1)}} & \text{if} \quad [P\underline{\alpha}] \in \mathfrak{I}_{tp^{-1}}. \end{cases}$$

Proof As φ_P only depends on the equivalence class ΓP, it suffices to prove the assertions for a system of representatives $P_1, \dots, P_{p+1}$ of primitive matrices of determinant p. We proceed by induction on e starting with $e = 0$.

(i) For p split in K, $p = \mathfrak{p}\overline{\mathfrak{p}}$, we may assume $P_i\underline{\alpha}, i = 1, 2$, to be bases of $\mathfrak{a}\mathfrak{p}_t$ and $\mathfrak{a}\overline{\mathfrak{p}}_t$. Then for $i = 2, \dots, p+1$ the $P_i\underline{\alpha}$ must be the basis of ideals from $\mathfrak{J}_{tp}$. Keeping in mind that the product of all $\varphi_{P_i}(\alpha)$ equals p^{12}, the assertion follows from Theorem 4.2.1.

(ii) For p ramified in K, $p = \mathfrak{p}^2$, we may assume $P_1\underline{\alpha}$ to be a basis of $\mathfrak{a}\overline{\mathfrak{p}}_t$, and the other $P_i\underline{\alpha}$ must be the basis of ideals in $\mathfrak{J}_{tp}$. Moreover, for the latter, the $\varphi_{P_i}(\alpha)$ are associated with each other by Theorem 4.2.3. The assertion now follows as in (i).

(iii) For p inert in K, $p = \mathfrak{p}$, all $P_i\underline{\alpha}$ are bases of ideals in $\mathfrak{J}_{tp}$, and again by Theorem 4.2.3 the $\varphi_{P_i}(\alpha)$ are associated with each other, which proves the asserted factorisation.

Now we treat the case $e > 0$. Recalling Theorem 3.1.14, we may assume that $[P_1\underline{\alpha}] \in \mathfrak{J}_{tp}$ and $[P_i\underline{\alpha}] \in \mathfrak{J}_{tp^{-1}}$ for $i = 2, \dots, p+1$. Theorem 4.2.3 and Theorem 3.1.16 imply again that $\varphi_{P_i}(\alpha)$, $i = 2, \dots, p+1$, are associate to each other, hence

$$\varphi_{P_1}(\alpha)\varphi_{P_2}(\alpha)^p \sim p^{12}.$$

In this product the factorisation of the first factor is obtained from the case $e - 1$. To see this we write the first factor as

$$\left(p^{12}\frac{\Delta(P_1\underline{\alpha})}{\Delta(\underline{\alpha})}\right)\left(p^{12}\frac{\Delta(pP_1^{-1}P_1\underline{\alpha})}{\Delta(P_1\underline{\alpha})}\right) = p^{12}.$$

Herein, pP_1^{-1}maps the basis $P_1\underline{\alpha}$ of an ideal in $\mathfrak{J}_{tp^{-1}}$ on a basis of an ideal from $\mathfrak{J}_t$. Therefore, from the case $e - 1$ we know that the second factor in the product is associated with

$$p^{n_{e-1}} \text{ with } n_{e-1} = \begin{cases} 0, \text{ if } p = \mathfrak{p}\overline{\mathfrak{p}}, \\ \frac{6}{p^e} \text{ if } p = \mathfrak{p}^2, \\ \frac{12}{p^{e-1}(p+1)} \text{ if } p = \mathfrak{p}. \end{cases}$$

This implies that

$$\varphi_{P_1}(\alpha) \sim p^{12-n_{e-1}} \text{ and } \varphi_{P_2}(\alpha) \sim p^{\frac{n_{e-1}}{p}} = p^{n_e},$$

which is the asserted factorisation. □

4.3 Factorisation of $\varphi(\xi \mid \mathfrak{L})$

The factorisation of the singular values of Klein's normalization of the σ function will be derived from Theorem 4.2.4 using the following formulae of Theorem 1.9.3:

$$\prod_{\alpha \in \mathfrak{a}/\mathfrak{b}} \varphi(\xi + \alpha|\mathfrak{b})^{12m} = \varphi(\xi|\mathfrak{a})^{12m},$$

$$\prod_{\substack{\alpha \in \mathfrak{a}/\mathfrak{b} \\ \alpha \notin \mathfrak{b}}} \varphi(\alpha|\mathfrak{b})^{12m} = \left(\frac{\Delta(\mathfrak{a})}{\Delta(\mathfrak{b})}\right)^m .$$

Herein $\mathfrak{a} \supset \mathfrak{b}$ are proper ideals of some orders in K, $\xi \in K \setminus \mathfrak{a}$ and m a natural number depending on ξ. The following theorem essentially goes back to Ramachandra (1969) and Schertz (1989) for $t = 1$ and was then generalised by Bley (1994) for an arbitrary conductor $t \in \mathbb{N}$:

Theorem 4.3.1 *For $\mathfrak{a}_t \in \mathfrak{J}_t^0$ and $\xi \in S_t \setminus \mathfrak{a}_t$ the set*

$$o(\xi, \mathfrak{a}_t) := \{\lambda \in K \mid \lambda\xi \in \mathfrak{a}_t\}$$

is an integral regular ideal in $\mathfrak{J}_t$ and has therefore a unique factorisation with prime ideal factors. We have

$$\varphi(\xi|\mathfrak{a}_t) \sim \begin{cases} 1 \quad , & \text{if} \quad o(\xi, \mathfrak{a}_t) \text{ is composite,} \\ \mathfrak{p}^{\frac{1}{\Phi(\mathfrak{p}^s)}} & \text{if} \quad o(\xi, \mathfrak{a}_t) = \mathfrak{p}_t^s \text{ is the power of} \\ & \qquad \text{a prime ideal } \mathfrak{p}_t. \end{cases}$$

$\mathfrak{p}$ denotes the prime ideal $\mathfrak{p} = \mathfrak{p}_t\mathfrak{O}_1$ of $\mathfrak{O}_1$.

Before proving the theorem we will first make some remarks and some preparations. We use the notation

$$\varphi(\xi|\mathfrak{a}_t) := \varphi(\xi|{}^{\alpha_1}_{\alpha_2})$$

with a $\mathbb{Z}$-basis α_1, α_2 of $\mathfrak{a}_t$. This is an abuse of notation for the value still depends on the basis, but since for different bases the values only differ by a factor which is a root of unity, the factorisation is the same for every choice of basis.

The results of Theorem 4.3.1 do not contain the factorisation of the singular values $\varphi(\xi|\mathfrak{a})$ for $t \neq 1$ when the ideal $\mathfrak{a}$ is not regular. However, as Bley (1994) remarks, these cases can be derived from Theorem 4.3.1 by using the formula

$$\varphi(\xi|\mathfrak{a}) = \zeta \prod_{\lambda \in \mathfrak{a}/t\mathfrak{a}\mathfrak{O}_1} \varphi(\xi - \lambda|\, t\mathfrak{a}\mathfrak{O}_1), \quad \zeta^{12} = 1, \tag{4.1}$$

where $\mathfrak{a}$ is an integral ideal of $\mathfrak{O}_t$ and $\xi \in \mathfrak{O}_t \backslash \mathfrak{a}$. The formula shows that the algebraic integer $\varphi(\xi|\mathfrak{a})$ is a divisor of $t\mathfrak{a}\mathfrak{O}_1$. To be more precise, one can easily derive from Theorem 4.3.1 the following result:

Theorem 4.3.2 *[Bley] Let $\mathfrak{a}$ be an integral ideal of $\mathfrak{O}_t$ and $\xi \in \mathfrak{O}_t \backslash \mathfrak{a}$. Then the algebraic integer $\varphi(\xi|\mathfrak{a})$ is a divisor of $t\mathfrak{a}\mathfrak{O}_1$. Given a prime divisor $\mathfrak{p}$ of $t\mathfrak{a}\mathfrak{O}_1$, we write*

$$t\mathfrak{a}\mathfrak{O}_1 = \mathfrak{p}^n \mathfrak{b} \quad \textit{with an integral ideal } \mathfrak{b} \textit{ of } \mathfrak{O}_1 \textit{ prime to } \mathfrak{p}\ .$$

Then the ideal $\mathfrak{p}$ occurs in the factorisation of $\varphi(\xi|\mathfrak{a})$ with the following exponent:

(i) *In the case $\xi \notin \mathfrak{a} + (\mathfrak{b} \cap \mathfrak{O}_t)$:*

$$v_{\mathfrak{p}}(\varphi(\xi|\mathfrak{a})) = 0.$$

(ii) *In the case $\xi \in \mathfrak{a} + (\mathfrak{b} \cap \mathfrak{O}_t)$ let $i_0 \in \{0, \dots, n-1\}$ be minimal with*

$$\xi \in \mathfrak{a} + (\mathfrak{p}^{i_0}\mathfrak{b} \cap \mathfrak{O}_t), \qquad \xi \notin \mathfrak{a} + (\mathfrak{p}^{i_0+1}\mathfrak{b} \cap \mathfrak{O}_t).$$

Then

$$\begin{aligned} v_{\mathfrak{p}}(\varphi(\xi|\mathfrak{a})) = \quad & \frac{1}{\Phi(\mathfrak{p}^n)}\left(\frac{|\mathfrak{a}\cap\mathfrak{b}|}{|\mathfrak{p}^n\mathfrak{b}|} - \frac{|\mathfrak{a}\cap\mathfrak{p}\mathfrak{b}|}{|\mathfrak{p}^n\mathfrak{b}|}\right) \\ & + \quad \cdots \quad + \\ & + \frac{1}{\Phi(\mathfrak{p}^{n-i_0+1})}\left(\frac{|\mathfrak{a}\cap\mathfrak{p}^{i_0-1}\mathfrak{b}|}{|\mathfrak{p}^n\mathfrak{b}|} - \frac{|\mathfrak{a}\cap\mathfrak{p}^{i_0}\mathfrak{b}|}{|\mathfrak{p}^n\mathfrak{b}|}\right) \\ & + \frac{1}{\Phi(\mathfrak{p}^{n-i_0})}\,\frac{|\mathfrak{a}\cap\mathfrak{p}^{i_0}\mathfrak{b}|}{|\mathfrak{p}^n\mathfrak{b}|}. \end{aligned}$$

Recalling homogeneity, $\varphi(\xi|{}^{\alpha_1}_{\alpha_2}) = \varphi(\gamma\xi|{}^{\gamma\alpha_1}_{\gamma\alpha_2})$, $\gamma \in \mathbb{K}^*$, it is easy to see that it is sufficient to prove Theorem 4.3.1 for regular ideals $\mathfrak{a}_t$ and $\xi \in \mathfrak{O}_t \setminus \mathfrak{a}_t$. We need:

Theorem 4.3.3 *Let $\mathfrak{a}_t$ be an integral regular ideal of $\mathfrak{O}_t$, $\xi \in \mathfrak{O}_t \setminus \mathfrak{a}_t$, and let $\gamma \in \mathfrak{O}_1$ have the property $\gamma\xi \notin \mathfrak{a}_t$. Then*

$$\varphi(\xi|\mathfrak{a}_t)|\varphi(\gamma\xi|\mathfrak{a}_t).$$

Proof Observing $\mathfrak{a}_t \subseteq \frac{\mathfrak{a}_t}{\gamma}$, the assertion follows from the product formula of Theorem 1.9.3:

$$\varphi(\gamma\xi|\mathfrak{a}_t) = \varphi\left(\xi \,\middle|\, \frac{\mathfrak{a}_t}{\gamma}\right) \sim \prod_{\alpha \in \frac{\mathfrak{a}_t}{\gamma}/\mathfrak{a}_t} \varphi(\xi + \alpha|\mathfrak{a}_t),$$

where $\varphi(\xi|\mathfrak{a}_t)$ is a factor of the right-hand side. □

Proof of Theorem 4.3.1 We conclude as in Bley (1994). Let

$$\mathfrak{a}_t = (\mathfrak{p}_1)_t^{e_1} \cdot \ldots \cdot (\mathfrak{p}_s)_t^{e_s}$$

be the factorisation of $\mathfrak{a}_t$. For $\xi \in \mathfrak{O}_t \setminus \mathfrak{a}_t$ and $\nu = 1, \ldots, e_i$ we then have

$$o(\xi, \mathfrak{a}_t) = (\mathfrak{p}_i)_t^{\nu} \iff \xi \in \frac{\mathfrak{a}_t}{(\mathfrak{p}_i)_t^{\nu}} \setminus \frac{\mathfrak{a}_t}{(\mathfrak{p}_i)_t^{\nu-1}}.$$

Now by Theorem 1.9.3 we obtain

$$\prod_{\substack{\xi \in \mathfrak{O}_t/\mathfrak{a}_t \\ o(\xi,\mathfrak{a}_t)|(\mathfrak{p}_i)_t^{\nu}}} \varphi(\xi|\mathfrak{a}_t) = \prod_{\xi \in \frac{\mathfrak{a}_t}{(\mathfrak{p}_i)_t^{\nu}}/\mathfrak{a}_t} \varphi(\xi|\mathfrak{a}_t) = \sqrt[12]{\frac{\Delta\left(\frac{\mathfrak{a}_t}{(\mathfrak{p}_i)_t^{\nu}}\right)}{\Delta(\mathfrak{a}_t)}} \sim \mathfrak{p}_i^{\nu},$$

and it follows that

$$\prod_{\substack{\xi \in \mathfrak{O}_t/\mathfrak{a}_t \\ o(\xi,\mathfrak{a}_t)=(\mathfrak{p}_i)_t^{\nu}}} \varphi(\xi|\mathfrak{a}_t) \sim \mathfrak{p}_i.$$

In the last product all factors are associated with each other. This follows from Theorem 4.3.3 because for $\xi, \xi' \in \mathfrak{O}_t$ with $o(\xi, \mathfrak{a}_t) = o(\xi', \mathfrak{a}_t)$ there are elements $\gamma, \gamma' \in \mathfrak{O}_t$ such that

$$\gamma\xi \equiv \xi' \bmod \mathfrak{a}_t \text{ and } \gamma'\xi' \equiv \xi \bmod \mathfrak{a}_t.$$

Further, the number of factors in the product is equal to

$$\left|\mathfrak{O}_t/(\mathfrak{p}_i)_t^{\nu}\right| = \left|\mathfrak{O}_1/\mathfrak{p}_i^{\nu}\right| = \Phi(\mathfrak{p}_i^{\nu}).$$

This proves our assertion for $o(\xi, \mathfrak{a}_t)$ being the power of a prime ideal. Otherwise $\varphi(\xi|\mathfrak{a}_t)$ must be a unit. This follows from

$$\prod_{\substack{\xi \in \mathfrak{O}_t/\mathfrak{a}_t \\ \xi \notin \mathfrak{a}_t}} \varphi(\xi|\mathfrak{a}_t) \sim \sqrt[12]{\frac{\Delta(\mathfrak{O}_t)}{\Delta(\mathfrak{a}_t)}} \sim \mathfrak{a}_t\mathfrak{O}_1,$$

recalling that the product of factors in which $o(\xi, \mathfrak{a}_t)$ is the power of a prime ideal, is also associated with $\mathfrak{a}_t\mathfrak{O}_1$. □

4.4 A result of Dorman, Gross and Zagier

To factorise the singular values in the preceding sections it was essential that the modular functions had no zero in the upper-half plane, so the method does not apply to functions like j or $\tau_{\underline{x}}$. However, in a special case the above-named authors obtained the following result on differences of singular values of j.

Let d_1, d_2 be relatively prime fundamental discriminants of quadratic imaginary number fields K_1, K_2, we define the quantity

$$J(d_1, d_2) := \prod_{\mathfrak{a}_1, \mathfrak{a}_2} (j(\mathfrak{a}_1) - j(\mathfrak{a}_2))^{\frac{4}{w_1 w_2}},$$

where $\mathfrak{a}_1, \mathfrak{a}_2$ run through a system of representatives for the ideal classes in K_1 resp. K_2 and w_1, w_2 denote the number of roots of unity in K_1 resp. K_2. We quote the following result from Gross and Zagier (1985):

Theorem 4.4.1

$$J(d_1, d_2)^2 = \pm \prod_{\substack{x, n, n' \in \mathbb{Z} \\ n, n' > 0 \\ x^2 + 4nn' = d_1 d_2}} n^{\epsilon(n')}.$$

To define the exponent $\epsilon(n)$ note that for a prime l dividing $\frac{d_1 d_2 - x^2}{4}$ we have $\left(\frac{d_1 d_2}{l}\right) \neq -1$. For such a prime, define

$$\epsilon(l) := \begin{cases} \left(\frac{d_1}{l}\right) & \text{if } \gcd(l, d_1) = 1, \\ \left(\frac{d_2}{l}\right) & \text{if } \gcd(l, d_2) = 1 \end{cases}$$

and then

$$\epsilon(n) := \prod_{l \mid n} \epsilon(l)^{\nu_l} \quad \text{for } n = \prod_{l \mid n} l^{\nu_l}.$$

In the same paper Gross and Zagier also derive the following result:

Theorem 4.4.2 *Let $d < -4$ be the discriminant of an order $\mathfrak{O}_t$ in a quadratic imaginary number field K and l a prime satisfying $\left(\frac{d}{l}\right) = 1$. Then, for any proper ideal $\mathfrak{a}$ of $\mathfrak{O}_t$*

$$N(j(\mathfrak{a}))N(j(\mathfrak{a}) - 12^3) \not\equiv 0 \ mod\ l,$$

where N denotes the norm of $K(j(\mathfrak{a}))/\mathbb{Q}$.

5
The Reciprocity Law

In Chapter 4 we have shown that singular values of functions from F_N generate algebraic number fields. Our aim now is to determine the action of automorphisms on these values. The main result is the assertion of Theorem 5.1.2, which tells us that the generated fields are abelian extensions of imaginary quadratic number fields. In the following section we will discuss some applications which we will need later.

5.1 The Reciprocity Law of Weber, Hasse, Söhngen, Shimura

The "source" of the Reciprocity Law 5.1.2 is the following theorem:

Theorem 5.1.1 *Let K be a quadratic imaginary number field and p a prime that splits in K: $p = \mathfrak{p}\overline{\mathfrak{p}}$. Let $\mathfrak{a}$ be a proper ideal of $\mathfrak{O}_t$, α_1, α_2 a basis of $\mathfrak{a}$ and $\alpha := \frac{\alpha_1}{\alpha_2} \in \mathbb{H}$. We assume $p \nmid t$. Let $P_{\overline{\mathfrak{p}}}$ be a primitive matrix of determinant p, so that $P_{\overline{\mathfrak{p}}}\underline{\alpha}$ is a basis of $\mathfrak{a}\overline{\mathfrak{p}}_t$.*

Let Λ be an algebraic number field, P a primitive matrix of determinant p and $f = f_P$ a function from $\Lambda\mathbb{Q}_{\Gamma_P}$, holomorphic in $\mathbb{H}$. We assume f to have integral q-coefficients in all cusps and $f_{\left(\begin{smallmatrix} p & 0 \\ 0 & 1 \end{smallmatrix}\right)}$ to have q-coefficients divisible by p. Then the algebraic integer $f_{P_{\overline{\mathfrak{p}}}}(\alpha)$ is divisible by $\mathfrak{p}$.

Proof According to Theorem 2.7.6 we can write

$$f_{P_{\overline{\mathfrak{p}}}}(\alpha) \prod_{P \not\sim P_{\overline{\mathfrak{p}}}} (\varphi_{P_{\overline{\mathfrak{p}}}}(\alpha) - \varphi_P(\alpha))$$
$$= a_0(j(\alpha)) + a_1(j(\alpha))\varphi_{P_{\overline{\mathfrak{p}}}}(\alpha) + \cdots + a_p(j(\alpha))\varphi_{P_{\overline{\mathfrak{p}}}}(\alpha)^p$$

with polynomials $a_\mu(j)$ satisfying

$$a_0(j) \in p\mathfrak{O}_\Lambda[j] \text{ and } a_\mu(j) \in \mathfrak{O}_\Lambda[j], \mu = 1, \ldots, p.$$

Herein $j(\alpha)$ is an algebraic integer by Theorem 4.1.1. $\varphi_{P_{\overline{\mathfrak{p}}}}(\alpha)$ is even divisible by $\mathfrak{p}$ according to Theorem 4.2.1 and the other values $\varphi_P(\alpha)$ in the product are units for $\mathfrak{p}$. The assertion now follows from the above representation of $f_{P_{\overline{\mathfrak{p}}}}(\alpha)$. □

To state the Reciprocity Law, we set up the following notation. Let $A \in \mathbb{Z}^{2\times 2}$ with determinant $a > 0$ coprime to $N \in \mathbb{N}$ and $f \in F_N$. Then we define a function f^A by

$$f^A(\omega) := [f \circ aA^{-1}](A(\omega)).$$

The following properties of f^A are immediate:

(i) $f^M = f$ for all $M \in \Gamma$.
(ii) $f^{MA} = f^A$ for all $M \in \Gamma$, i.e. f^A only depends on ΓA.
(iii) $f^A \circ M = (f \circ M)^A$ for all $M \in \Gamma_A = \Gamma \cap A^{-1}\Gamma A$.

Theorem 5.1.2 (Weber, Hasse, Söhngen, Shimura) *Let K be a quadratic imaginary number field and $\mathfrak{a} = [\underline{\alpha}] \in \mathfrak{I}_t$ with basis $\underline{\alpha} = \binom{\alpha_1}{\alpha_2}$ and $\alpha = \frac{\alpha_1}{\alpha_2} \in \mathbb{H}$. Let f be in F_N with $f(\alpha) \neq \infty$. Then*

$$f(\alpha) \in K_{(tN)}.$$

Further, let $\mathfrak{c}$ be an integral ideal in K of norm $c > 0$, coprime to tN, and let $\mathfrak{c}_t$ denote the corresponding ideal in $\mathfrak{I}_t$. There exists $C \in \mathbb{Z}^{2\times 2}$ of determinant c such that

$$C\underline{\alpha} \text{ is a basis of } \mathfrak{a}\overline{\mathfrak{c}}_t.$$

Let $\sigma(\mathfrak{c})$ denote the Frobenius automorphism of $K_{(Nt)}/K$ belonging to $\mathfrak{c}$. Then

$$f(\alpha)^{\sigma(\mathfrak{c})} = f^C(\alpha).$$

Proof First let $\mathfrak{c} = \mathfrak{p}$ be a prime ideal of degree 1 and norm p in K with $\overline{\mathfrak{p}} \neq \mathfrak{p}$, and let $P\underline{\alpha}$ be a basis of $\mathfrak{a}\mathfrak{p}_t$ with a primitive matrix P of determinant p. The above properties of f^P then imply the polynomial

$$S_P(X,\omega) := \prod_{M \in \Gamma/\Gamma(N)} (X - ((f \circ M)^P(\omega) - (f \circ M)^p(\omega)))$$

to have coefficients in $(\mathbb{Q}_N)_{\Gamma_P}$:

$$S_P(X,\omega) \in (\mathbb{Q}_N)_{\Gamma_P}[X].$$

We first treat the case of f being holomorphic in $\mathbb{H}$ with integral q-coefficients for all $f \circ M$, $M \in \Gamma$:

$$f \circ M = \sum_{n=0}^{\infty} a_n q^{\frac{n}{N}}, \quad a_n \in \mathbb{Z}[\zeta_N].$$

Then for $P_0 = \left(\begin{smallmatrix} p & 0 \\ 0 & 1 \end{smallmatrix}\right)$ we have

$$(f \circ M)^{P_0} - (f \circ M)^p = \sum_{n=0}^{\infty} a_n^{\sigma_p} q^{\frac{n}{N}p} - \left(\sum_{n=0}^{\infty} a_n q^{\frac{n}{N}} \right)^p,$$

where σ_p denotes the automorphism $(\zeta_N \mapsto \zeta_N^p)$ of $\mathbb{Q}_N$. Using $a_n^{\sigma_p} \equiv a_n^p \bmod p$ we obtain the following congruence that is to be understood as a congruence for the q-coefficients:

$$(f \circ M)^{P_0} - (f \circ M)^p \equiv 0 \bmod p,$$

and thus

$$S_{P_0}(X, \omega) \equiv X^{[\Gamma:\Gamma(N)]} \bmod p.$$

Since $P(\alpha)$ is the ratio of a basis of $\mathfrak{a}\overline{\mathfrak{p}}_t$ by assumption, it follows from Theorem 5.1.1 that the X-coefficients of $S_P(X, \alpha)$ are contained in an algebraic extension of K satisfying the congruence

$$S_P(X, \alpha) \equiv X^{[\Gamma:\Gamma(N)]} \bmod \mathfrak{p}.$$

For every prime ideal $\mathfrak{P}$ in $L := K(\{f \circ M(\alpha) \mid M \bmod \Gamma(N)\})$, lying above $\mathfrak{p}$, this implies that

$$f^P(\alpha) - f(\alpha)^p \equiv 0 \bmod \mathfrak{P}.$$

Now the Frobenius automorphism $\sigma(\mathfrak{P})$ of $\mathfrak{P}$ over K by definition satisfies

$$f(\alpha)^{\sigma(\mathfrak{P})} \equiv f(\alpha)^p \bmod \mathfrak{P},$$

and we can conclude that

$$f(\alpha)^{\sigma(\mathfrak{P})} \equiv f^P(\alpha) \bmod \mathfrak{P}. \tag{5.1}$$

Following Söhngen (1935), we will use this congruence to show that $f(\alpha)$ is contained in the ray class field $K_{(Nt)}$. Therefore, we contend that $f^P(\alpha) = f(\alpha)$ for $\mathfrak{p}$ in the principal ray class modulo Nt. So let $\mathfrak{p} = (\pi)$ with $\pi = 1 + tN\beta$, $\beta \in \mathfrak{O}_K$. Since

$$P \begin{pmatrix} \alpha_1 \\ \alpha_2 \end{pmatrix} \text{ and } \begin{pmatrix} \pi\alpha_1 \\ \pi\alpha_2 \end{pmatrix}$$

are bases of $\mathfrak{a}\overline{\mathfrak{p}}_t$, we have

$$P\begin{pmatrix}\alpha_1\\ \alpha_2\end{pmatrix} = M\begin{pmatrix}\pi\alpha_1\\ \pi\alpha_2\end{pmatrix}$$

with a unimodular matrix M. This implies that

$$P\begin{pmatrix}\alpha_1\\ \alpha_2\end{pmatrix} = M(E + NtD_\beta)\begin{pmatrix}\alpha_1\\ \alpha_2\end{pmatrix}$$

with the matrix D_β of β with respect to the basis α_1, α_2 and the unit matrix E. The last equality shows that $P = M(E + tND_\beta)$, hence

$$P \equiv M \bmod tN.$$

Keeping in mind $p = N(\pi) \equiv 1 \bmod tN$, it follows that

$$pP^{-1} \equiv M^{-1} \bmod tN,$$

and then $f^P(\alpha) = f^M(\alpha) = f(\alpha)$, as was to be shown. Therefore, we have

$$f(\alpha) \in K_{(tN)}.$$

Now we contend that the congruence in (5.1) becomes equality

$$f(\alpha)^{\sigma(\mathfrak{P})} = f^P(\alpha)$$

for $\mathfrak{c} = \mathfrak{p}$ satisfying the assumption of our theorem. To show this, it suffices to prove the congruence

$$f(\alpha)^{\sigma(\mathfrak{P})} \equiv f^P(\alpha) \bmod \mathfrak{Q}$$

for infinitely many prime ideals $\mathfrak{Q}$ of $K_{(tN)}$. Therefore, in the ray class modulo tN of $\mathfrak{p}$ we choose infinitely many prime ideals $\mathfrak{q}$ of degree 1 with $\overline{\mathfrak{q}} \neq \mathfrak{q}$ and for each a prime ideal $\mathfrak{Q}$ of $K_{(tN)}$ above $\mathfrak{q}$. Then we have

$$\mathfrak{a}\overline{\mathfrak{q}}_t = \mathfrak{a}\overline{\mathfrak{p}}_t\beta \text{ with } \beta \equiv 1 \bmod tN.$$

Let P resp. Q be the primitive matrix such that

$$P\begin{pmatrix}\alpha_1\\ \alpha_2\end{pmatrix} \text{ resp. } Q\begin{pmatrix}\alpha_1\\ \alpha_2\end{pmatrix}$$

is a basis of $\mathfrak{a}\overline{\mathfrak{p}}_t$ resp. $\mathfrak{a}\overline{\mathfrak{q}}_t$. Then we have

$$P \equiv MQ \bmod tN$$

with a unimodular matrix M. This implies that $f^P(\alpha) = f^Q(\alpha)$, and since the Frobenius automorphisms of $\mathfrak{P}$ and $\mathfrak{Q}$ are equal, we obtain

$$f(\alpha)^{\sigma(\mathfrak{P})} - f^P(\alpha) = f(\alpha)^{\sigma(\mathfrak{Q})} - f^Q(\alpha) \equiv 0 \bmod \mathfrak{Q}.$$

This congruence holding for infinitely many prime ideals $\mathfrak{Q}$, we can conclude that $f(\alpha)^{\sigma(\mathfrak{P}\mathfrak{p})} = f^P(\alpha)$. This implies the second assertion for $\mathfrak{c} = \mathfrak{p}$ a prime ideal of degree 1, $\mathfrak{p} \neq \overline{\mathfrak{p}}$, because $\sigma(\mathfrak{p})|K(f(\alpha)) = \sigma(\mathfrak{P})|K(f(\alpha))$ by definition of $\sigma(\mathfrak{p})$.

Now let f be an arbitrary function in F_N with $f(\alpha) \neq \infty$. To prove the assertions in this general case, we proceed as in the proof of Theorem 4.1.2 and write

$$f = R(j, \tau_1, \ldots, \tau_m), \tag{5.2}$$

with a rational function $R \in \mathbb{Q}^a(X_0, \ldots, X_m))$. Since j and τ_i satisfy the hypothesis made at the beginning of the proof, we can derive that the coefficients of R are in the N-th cyclotomic field, hence in $K_{(tN)}$. For α not equivalent to $i = \sqrt{-1}$ or ρ modulo Γ, the representation is also valid for the value at α, and it follows that

$$f(\alpha) \in K_{(tN)}.$$

Now let $\mathfrak{P}$ be a prime ideal in $K_{(tN)}$ above $\mathfrak{p}$ and let $\sigma(\mathfrak{P})$ be the corresponding Frobenius automorphism of $K_{(tN)}/K$. To compute the action of $\sigma(\mathfrak{P})$ on $f(\alpha)$ using the representation, we extend the action of $g \mapsto g \circ \left(\begin{smallmatrix} 1 & 0 \\ 0 & p \end{smallmatrix}\right)$ to $K_{(tN)}F_N$ by

$$\left(\sum_{n=0}^{\infty} a_n q^{\frac{n}{N}}\right) \circ \sigma(\mathfrak{P}) := \sum_{n=0}^{\infty} a_n^{\sigma(\mathfrak{P}\mathfrak{p})} q^{\frac{n}{N}}.$$

To be precise, this is even an automorphism of $K_{(tN)}F_N/\mathbb{Q}(j)$, because the generating functions $\tau_{\underline{u}}, \underline{u} \in \frac{1}{N}(\mathbb{Z} \times \mathbb{Z}) \setminus \mathbb{Z} \times \mathbb{Z}$, over $K_{(tN)}$ are permuted by this extension. Let h be one of function involved in the above representation (5.2) of f or a coefficient of the rational function R, then with the matrices $M_1, M_2 \in \Gamma$ in a decomposition $pP^{-1} = M_1\left(\begin{smallmatrix} 1 & 0 \\ 0 & p \end{smallmatrix}\right)M_2$ the action of $\sigma(\mathfrak{P})$ on $h(\alpha)$ is given by the formula

$$h(\alpha)^{\sigma(\mathfrak{P})} = [[[h \circ M_1] \circ \sigma(\mathfrak{P})] \circ M_2](P(\alpha)).$$

From the above representation of f we obtain

$$\begin{aligned} f(\alpha)^{\sigma(\mathfrak{P})} &= [[[f \circ M_1] \circ \sigma(\mathfrak{P})] \circ M_2](P(\alpha)) \\ &= [[[f \circ M_1] \circ \left(\begin{smallmatrix} 1 & 0 \\ 0 & p \end{smallmatrix}\right)] \circ M_2](P(\alpha)) = f^P(\alpha). \end{aligned}$$

In the remaining cases, when α is equivalent to $i = \sqrt{-1}$ or ρ modulo Γ we adopt the method applied in the proof of Theorem 4.1.2

$$f(\alpha) = \tilde{f}(N^l \alpha), \qquad \tilde{f}(\omega) := f\left(\frac{\omega}{N^l}\right)$$

with a suitable power N^l of N, such that $N^l\alpha$ is not equivalent to $i = \sqrt{-1}$ or ρ modulo Γ. Since $\tilde{f}$ is in $F_{N^{l+1}}$, we have a representation

$$\tilde{f} = R(j, \tau_1, \ldots, \tau_m) \text{ mit } R \in K_{(tN^{l+1})}(X_0, \ldots, X_m),$$

where now $\tau_1, \ldots, \tau_m$ are the different functions of $\tau_{\underline{x}}$, $\underline{x} \in \frac{1}{N^{l+1}}(\mathbb{Z}\times\mathbb{Z}) \setminus \mathbb{Z}\times\mathbb{Z}$. Via the identity

$$f = R(j \circ \kappa, \tau_1 \circ \kappa, \ldots, \tau_m \circ \kappa), \;\; \kappa(\omega) = N^{l+1}\omega,$$

gives

$$f(\alpha) \in K_{(tN^{l+1})}$$

and in the same way we find the asserted formula for the action of the Frobenius automorphisms of $K_{(tN^{l+1})}/K$. To show that $f(\alpha)$ is invariant under the action of the automorphisms of $K_{(tN^{l+1})}/K_{(tN)}$, we observe that these automorphisms are given by $\sigma(\mathfrak{P})$, with $\mathfrak{P}$ above a prime ideal $\mathfrak{p}$ of degree 1 in K, generated by an element $\pi \equiv 1 \bmod tN$. An integral matrix P, transforming a basis of $\mathfrak{a}$ into a basis of $\mathfrak{a}\overline{\mathfrak{p}}_t$ then satisfies the congruence

$$P \equiv M \bmod tN$$

with a unimodular matrix M, and it follows that

$$f(\alpha)^{\sigma(\mathfrak{P})} = f^P(\alpha) = f^M(\alpha) = f(\alpha).$$

Hence $f(\alpha) \in K_{(tN)}$.

It still remains to prove that the asserted formula for the Galois action holds for an arbitrary integral ideal $\mathfrak{c}$ coprime to tN. To show this, we pick a prime ideal $\mathfrak{p}$ of degree 1 and norm p in the ray class modulo Nt of $\mathfrak{c}$. Then we have

$$\lambda_1\mathfrak{p} = \lambda_2\mathfrak{c}$$

with integral numbers $\lambda_i \in K$, coprime to Nt, satisfying $\lambda_1 \equiv \lambda_2 \bmod tN$. Here, by multiplying with a suitable power of λ_1 we can achieve that

$$\lambda_1 \equiv \lambda_2 \equiv 1 \bmod tN.$$

Two matrices $P, C \in \mathbb{Z}^{2\times 2}$ of positive determinant transforming a basis of $\mathfrak{a}_t$ into a basis of $\overline{\mathfrak{p}}_t$ resp. $\overline{\mathfrak{c}}_t$ are related by

$$PD_1 = MCD_2$$

with a unimodular matrix M and the matrices D_i of λ_i with respect to the basis $\underline{\alpha}$ von $\mathfrak{a}_t$. In view of $\lambda_i \equiv 1 \bmod tN$ we here have

$$D_i \equiv \begin{pmatrix} 1\,0 \\ 0\,1 \end{pmatrix} \mod N,$$

and it follows that

$$\begin{aligned} f \circ pP^{-1} &= f \circ p\det(D_1)(PD_1)^{-1} = f \circ \det(CD_2)(MCD_2)^{-1} \\ &= f \circ c(MC)^{-1} = f \circ cC^{-1}M^{-1}. \end{aligned}$$

Recalling $P(\alpha) = MC(\alpha)$ and $\sigma(\mathfrak{p}) = \sigma(\mathfrak{c})$, we finally obtain

$$\begin{aligned} f(\alpha)^{\sigma(\mathfrak{c})} &= f(\alpha)^{\sigma(\mathfrak{p})} = [f \circ pP^{-1}](P(\alpha)) = f \circ cC^{-1}M^{-1}(MC(\alpha)) \\ &= [f \circ cC^{-1}](C(\alpha)) = f^C(\alpha). \end{aligned}$$

□

5.2 Applications of the Reciprocity Law

For many functions the first assertion of Theorem 5.1.2 is rather imprecise, because the singular values turn out to be in a proper subfield of K_{tN}. Using the following theorem, the ray class field K_{tN} can be replaced by the smaller ring class field Ω_{tN} for certain functions.

Theorem 5.2.1 *Let f be in F_N with $f\left(\frac{-1}{z}\right)$ having rational q-coefficients, and we assume that $f \circ M = f$ for all unimodular matrices M with*

$$M \equiv \begin{pmatrix} *\,0 \\ 0\,* \end{pmatrix} \mod N.$$

Then for $\alpha \in \mathbb{H}$ quadratic imaginary with conductor t we have

$$f(\alpha) \in \Omega_{tN}$$

if $f(\alpha) \neq \infty$.
Further, if $\mathfrak{O}_t = [\alpha, 1]$ and α coprime to tN, then

$$f(\alpha)^{\sigma(\overline{\alpha})} = f\left(\frac{1}{\overline{\alpha}}\right)$$

with the Frobenius automorphism $\sigma(\overline{\alpha})$ of Ω_{tN}/K belonging to the ideal $\overline{\alpha}\mathfrak{O}_1$.

Proof Since $f(\alpha) \in K_{tN}$ by Theorem 5.1.2, it remains to show that $f(\alpha)$ is fixed under the automorphisms of K_{tN}/Ω_{tN}. These automorphisms are given by $\sigma(r) = \sigma(r\mathfrak{O}_1)$, $r \in \mathbb{Z}$, r coprime to tN. The matrix C in Theorem 5.1.2 is then of the form

$$C = \begin{pmatrix} r & 0 \\ 0 & r \end{pmatrix}.$$

Since r is coprime to N, we can find a unimodular matrix M with

$$r^2C^{-1} \equiv \begin{pmatrix} 1 & 0 \\ 0 & r^2 \end{pmatrix} M \bmod N,$$

$$\text{and } M \equiv \begin{pmatrix} r & 0 \\ 0 & r' \end{pmatrix} \bmod N,$$

where r' is a natural number satisfying $rr' \equiv 1 \bmod N$. Now we find

$$f \circ r^2C^{-1} = f \circ \begin{pmatrix} 1 & 0 \\ 0 & r^2 \end{pmatrix} = f,$$

because f has rational q-coefficients, and by Theorem 5.1.2 it follows that $f(\alpha)^{\sigma(r)} = f(\alpha)$. Hence $f(\alpha) \in \Omega_{tN}$.

To prove the second assertion, observe $\mathfrak{O}_t = [\alpha, 1] = [\overline{\alpha}, 1]$, which implies that $\alpha\mathfrak{O}_t = [\alpha, a]$ with $a = \alpha\overline{\alpha}$. By Theorem 5.1.2 and the fact that $\left[f \circ \left(\begin{smallmatrix} 0 & -1 \\ 1 & 0 \end{smallmatrix}\right)\right](z) = f\left(\frac{-1}{z}\right)$ has rational q-coefficients, we then obtain

$$\begin{aligned} f(\alpha)^{\sigma(\overline{\alpha})} &= \left[f \circ \left(\begin{smallmatrix} a & 0 \\ 0 & 1 \end{smallmatrix}\right)\right]\left(\tfrac{\alpha}{a}\right) = \left[f \circ \left(\begin{smallmatrix} 0 & -1 \\ 1 & 0 \end{smallmatrix}\right)\left(\begin{smallmatrix} 1 & 0 \\ 0 & a \end{smallmatrix}\right)\left(\begin{smallmatrix} 0 & 1 \\ -1 & 0 \end{smallmatrix}\right)\right]\left(\tfrac{\alpha}{a}\right) \\ &= f\left(\tfrac{\alpha}{a}\right) = f\left(\tfrac{1}{\overline{\alpha}}\right), \end{aligned}$$

which is our assertion. □

Theorem 5.2.2 *Let f be in $\mathbb{Q}_{\Gamma\left(\begin{smallmatrix} 1 & 0 \\ 0 & r \end{smallmatrix}\right)}$, $r \in \mathbb{N}$, with $f\left(\frac{-1}{z}\right)$ having rational q-coefficients. Let $\underline{\alpha}$ be a basis of an ideal in $\mathfrak{I}_t$ with ratio $\alpha \in \mathbb{H}$. Let $\left(\begin{smallmatrix} 1 & 0 \\ 0 & r \end{smallmatrix}\right)\underline{\alpha}$ be a basis of an ideal in $\mathfrak{I}_{t'}$. Then, if $f(\alpha) \neq \infty$, we have:*

$$f(\alpha) \in \Omega_{t''} \text{ with } t'' = \mathrm{lcm}(t, t').$$

Proof First, we have $f(\alpha) \in K_{tr}$ according to Theorem 5.1.2, and by Theorem 3.1.13 we know $t'' \mid tr$, hence $\Omega_{t''} \subseteq K_{tr}$. Therefore, we have to show that $f(\alpha)$ is left fixed by all automorphisms $\sigma(\mathfrak{c})$, where $\mathfrak{c} = \lambda\mathfrak{O}_1$, $\lambda \in \mathfrak{O}_{t''}$, with norm c coprime to tr. To compute $f(\alpha)^{\sigma(\mathfrak{c})}$, observe that

$$\mathfrak{a} := [A\alpha, A] \in \mathfrak{I}_t \quad \text{and} \quad \mathfrak{O}_t = [A\alpha, 1],$$

where A is the leading coefficient of the primitive equation of α. Substituting α by $\alpha + \mu$ with some $\mu \in r\mathbb{Z}$ we can further achieve

$$\overline{\mathfrak{c}}_t = \lambda\mathfrak{O}_t = [A\alpha, Ac] \text{ and } \mathfrak{a}\overline{\mathfrak{c}}_t = [A\alpha, Ac].$$

Since f has period r, this substitution does not change $f(\alpha)$, and the assumptions about $[\underline{\alpha}]$ and $[\left(\begin{smallmatrix}1&0\\0&r\end{smallmatrix}\right)\underline{\alpha}]$ remain valid. Now, by Theorem 5.1.2 it follows that

$$f(\alpha)^{\sigma(\mathfrak{c})} = f(\tfrac{\alpha}{c}),$$

and herein we have $\frac{\alpha}{c} = M(\alpha)$ with a unimodular matrix M because the ideals $\mathfrak{a}$ and $\mathfrak{a}\mathfrak{c}_{t'}$ are in the same class. To determine M we need the primitive equation $AX^2 + BX + C = 0$ of α. We find that

$$\overline{\lambda}\begin{pmatrix} A\alpha \\ Ac \end{pmatrix} = \begin{pmatrix} u & -vC \\ -v & u - vB \end{pmatrix}\begin{pmatrix} A\alpha \\ A \end{pmatrix},$$

where $u, v \in \mathbb{Z}$ are defined by the representation $\overline{\lambda} = u + vA\alpha \in \mathfrak{O}_t = [A\alpha, 1]$. Since

$$[A\alpha, Ac] = \mathfrak{a}\overline{\mathfrak{c}}_t = \overline{\lambda}[A\alpha, A],$$

there exists a unimodular matrix M with

$$\begin{pmatrix} A\alpha \\ Ac \end{pmatrix} = M\left(\overline{\lambda}\begin{pmatrix} A\alpha \\ A \end{pmatrix}\right) = M\begin{pmatrix} u & -vC \\ -v & u - vB \end{pmatrix}\begin{pmatrix} A\alpha \\ A \end{pmatrix}, \tag{5.3}$$

and by comparing coefficients we obtain

$$\begin{pmatrix} 1 & 0 \\ 0 & c \end{pmatrix} = M\begin{pmatrix} u & -vC \\ -v & u - vB \end{pmatrix}.$$

Now we contend that r divides vC. This implies that

$$M \equiv \left(\begin{smallmatrix} * & 0 \\ * & * \end{smallmatrix}\right) \bmod r,$$

hence $M \in \Gamma_{\left(\begin{smallmatrix}1&0\\0&r\end{smallmatrix}\right)}$, and therefore

$$f(\alpha)^{\sigma(\mathfrak{c})} = f(\tfrac{\alpha}{c}) = f(M(\alpha)) = f(\alpha).$$

To prove $r|vC$, we write $t'' = ts$, where s is a divisor of r, according to Theorem 3.1.13. Then we first have $s|v$, because $\mathfrak{O}_{t''} = [sA\alpha, 1]$. Further, $\frac{\alpha}{r}$ is a root of the equation $r^2AX^2 + rBX + C = 0$, and the discriminant of the primitive equation of $\frac{\alpha}{r}$ divides t^2s^2d. Hence C must be divisible by $\frac{r}{s}$, and we have proved that r divides vC. This concludes the proof of Theorem 5.2.1. □

Now we apply Theorem 5.1.2 to the singular values of j and $\tau^{(e)}$.

Theorem 5.2.3 *Let $\mathfrak{a}$ be a proper ideal of $\mathfrak{O}_t$. Let $\mathfrak{c}$ be an integral ideal of $\mathfrak{O}_1$ coprime to t and $\mathfrak{c}_t = \mathfrak{c} \cap \mathfrak{O}_t$ the corresponding ideal of $\mathfrak{O}_t$.*

Then

$$j(\mathfrak{a}) \in \Omega_t \text{ and } j(\mathfrak{a})^{\sigma(\mathfrak{C})} = j(\mathfrak{a}\mathfrak{c}_t^{-1}).$$

Further, let N be a natural number and $\xi \in \frac{1}{N}\mathfrak{a} \setminus \mathfrak{a}$. Then

$$\tau^{(e)}(\xi|\mathfrak{a}) \in K_{tN} \text{ and } \tau^{(e)}(\xi|\mathfrak{a})^{\sigma(\mathfrak{C})} = \tau^{(e)}(\xi|\mathfrak{a}\mathfrak{c}_t^{-1}),$$

where $\tau = \tau^{(e)}$ denotes the τ function. (2e is the number of roots of unity in $\mathfrak{O}_t$).

Proof Let $\underline{\alpha}$ be a basis of $\mathfrak{a}$ with ratio $\alpha \in \mathbb{H}$. Then by definition $j(\mathfrak{a}) = j(\alpha)$, hence $j(\mathfrak{a}) \in K_t$, because j is in F_1 and has rational q-coefficients. Further, $\tau(\xi|\mathfrak{a}) = \tau_{\underline{x}}(\alpha)$ with some $\underline{x} \in \frac{1}{N}(\mathbb{Z}\times\mathbb{Z}) \setminus (\mathbb{Z}\times\mathbb{Z})$, and it follows that $\tau(\xi|\mathfrak{a}) \in K_{tN}$ because $\tau_{\underline{x}}$ is in F_N.

To compute the Galois action of $\sigma(\mathfrak{c})$ let C be an integer matrix such that $C\underline{\alpha}$ is a basis of $\mathfrak{a}\overline{\mathfrak{c}}_t$. Then, using Theorem 5.1.2, we find that

$$j(\mathfrak{a})^{\sigma(\mathfrak{C})} = [j \circ cC^{-1}](C(\alpha)) \text{ and } \tau(\xi|\mathfrak{a})^{\sigma(\mathfrak{C})} = [\tau_{\underline{x}} \circ cC^{-1}](C(\alpha)).$$

Since j has rational q-coefficients, we have $j \circ cC^{-1} = j$, hence

$$j(\mathfrak{a})^{\sigma(\mathfrak{C})} = j(C(\alpha)).$$

Now $\mathfrak{a}\overline{\mathfrak{c}_t} = c\mathfrak{a}\mathfrak{c}_t^{-1}$, so $C(\alpha)$ is the ratio of a basis of $\mathfrak{a}\mathfrak{c}_t^{-1}$ too, and the asserted Galois action on $j(\mathfrak{a})$ follows.

Further, using the rule $\tau_{\underline{x}} \circ cC^{-1} = \tau_{\underline{x}cC^{-1}}$, we find that

$$\begin{aligned} \tau(\xi|\mathfrak{a})^{\sigma(\mathfrak{C})} &= \tau\left(\underline{x}cC^{-1}\begin{pmatrix} C(\alpha) \\ 1 \end{pmatrix}\middle| \begin{matrix} C(\alpha) \\ 1 \end{matrix}\right) \\ &= \tau(\underline{x}\,\underline{\alpha}|\tfrac{1}{c}C\underline{\alpha}) \\ &= \tau(\xi|\mathfrak{a}\mathfrak{c}_t^{-1}), \end{aligned}$$

keeping in mind that $cC^{-1}\underline{\alpha}$ is a basis of $\mathfrak{a}\mathfrak{c}_t^{-1}$. □

For most of the modular functions $f \in F_N$ considered in the sequel the value $f(\frac{\alpha_1}{\alpha_2})$, unlike the singular values of j and τ, depends on the choice of basis α_1, α_2 for the lattice $[\alpha_1, \alpha_2]$. To compute the conjugates of such singular values, the following two theorems are useful.

Theorem 5.2.4 *Let $\alpha_1, .., \alpha_{h_t} \in \mathbb{H}$ be a $2N$-system, i.e. the ratios of basis of a system of coset representatives for the ring ideal classes of conductor t with primitive equations $A_iX^2 + B_iX + C_i = 0$ normalised by*

$$\gcd(A_i, N) = 1, A_i > 0 \text{ and } B_i \equiv B_1 \bmod 2N$$

with a natural number N.
For $f \in F_N$ *we set*

$$f_i := f \circ \begin{pmatrix} A_i & 0 \\ 0 & 1 \end{pmatrix}.$$

Then, if $f(A_1\alpha_1) \neq \infty$, *we have*

$$f_i(\alpha_i) \in K_{tN}, \quad i = 1, \ldots, h_t,$$

and there exist automorphisms $\sigma_1, \ldots, \sigma_{h_t}$ *of* K_{tN} *with*

$$f_1(\alpha_1)^{\sigma_i} = f_i(\alpha_i), \quad i = 1, \ldots, h_t,$$

where the restrictions of σ_i *to* Ω_t *are the different automorphisms of* Ω_t/K.
If, in particular, $f_1(\alpha_1) \in \Omega_t$, *then*

$$\prod_{i=1}^{h_t}(X - f_i(\alpha_i)) \in K[X].$$

Proof By Theorem 3.1.10 we know that there exist unimodular transformations $M_i \in \Gamma(N)$, so that the $\alpha_i' := M_i(\alpha_i)$ have primitive equations of the form

$$A_i'X^2 + B_i' + C_i' = 0$$

with

$$\gcd(A_i', tN) = 1, A_i' > 0 \text{ and } B_i' \equiv B_i \bmod N.$$

Since $f(\alpha_i) = f(\alpha_i')$, we may assume the A_i to be coprime to Nt. Then

$$\mathfrak{a}_i := \overline{[A_i\alpha_i, A_i]}$$

are integral regular ideals of $\mathfrak{O}_t$ and the corresponding ideals of $\mathfrak{O}_1$,

$$\mathfrak{c}_i := \mathfrak{O}_1\mathfrak{a}_i$$

satify

$$\mathfrak{a}_i = \mathfrak{c}_i \cap \mathfrak{O}_t.$$

Now we set $\sigma_i := \sigma(\mathfrak{c}_i)$. Then the restrictions to Ω_t are the different automorphisms of Ω_t/K. We define

$$\alpha_0 := A_1\alpha_1 = \frac{-B_1 + t\sqrt{d}}{2},$$

which is the ratio of a basis of $\mathfrak{O}_t$ and, since $f(\alpha_0) \neq \infty$ by assumption, we have

$$f(\alpha_0) \in K_{tN}$$

by Theorem 5.1.2. Herein

$$f(\alpha_0) = f\left(\tfrac{-B_i+t\sqrt{d}}{2}\right), \quad i = 1, \dots, h_t,$$

because $B_1 \equiv B_i \bmod 2N$. By Theorem 5.1.2 we further obtain

$$f(\alpha_0)^{\sigma_i} = \left[f \circ \left(\begin{smallmatrix} A_i & 0 \\ 0 & 1 \end{smallmatrix}\right)\right]\left(\tfrac{-B_i+t\sqrt{d}}{2A_i}\right) = f_i(\alpha_i),$$

which implies that

$$f_1(\alpha_1)^{\sigma_1^{-1}\sigma_i} = f_i(\alpha_i), \quad i = 1, \dots, h_t.$$

The restrictions of $\sigma_1^{-1}\sigma_i$ to Ω_t constituting again all different automorphisms of Ω_t/K, this completes the proof of Theorem 5.2.4. □

Later we will apply Theorem 5.2.4 to compute the conjugates of the singular values of $\gamma_2 = \sqrt[3]{j}$ and $\gamma_3 = \sqrt[2]{j - 12^3}$. In particular, Theorem 5.2.4 will be useful, when we are dealing with functions of the type

$$f(\omega) = \prod_{i=1}^{n} \eta(N_i\omega)^{e_i} \text{ with } N_i \in \mathbb{N},\ e_i \in \mathbb{Z} \text{ and } \sum_{i=1}^{n} e_i = 0.$$

All these functions have the poperty of $f\left(\frac{-1}{z}\right)$ having rational q-coefficients, which implies that f is invariant under the action of $\left(\begin{smallmatrix} A_i & 0 \\ 0 & 1 \end{smallmatrix}\right)$. Hence the conjugates of $f(\alpha_1)$ in Theorem 5.2.4 are given by

$$f(\alpha_1)^{\sigma_i} = f(\alpha_i).$$

In what follows, let f be one of the functions

$$p(z|{\textstyle\frac{\omega_1}{\omega_2}}) = \frac{\wp(z|\frac{\omega_1}{\omega_2})}{\sqrt[6]{\Delta(\frac{\omega_1}{\omega_2})}}, \quad \varphi(z|{\textstyle\frac{\omega_1}{\omega_2}}) = \sigma^*(z|{\textstyle\frac{\omega_1}{\omega_2}}) \sqrt[12]{\Delta({\textstyle\frac{\omega_1}{\omega_2}})}.$$

defined already in section 2.4. Unlike the τ function they depend on the choice of basis ω_1, ω_2. As in section 2.4 we use the following notation for the values of f at torsion points of $\mathbb{C}/[\omega, 1]$:

$$f_{\underline{x}}(\omega) := f(\underline{x}({\textstyle\frac{\omega}{1}})|{\textstyle\frac{\omega}{1}})$$

with $\underline{x} \in \frac{1}{N}(\mathbb{Z} \times \mathbb{Z}) \setminus (\mathbb{Z} \times \mathbb{Z})$. For a unimodular matrix M we then have the transformation formulae

$$[f_{\underline{x}} \circ M] = \kappa(M) f_{\underline{x}M}$$

with

$$\kappa(M) = \begin{cases} \epsilon(M)^{-4}, & \text{for } f = p, \\ \epsilon(M)^2, & \text{for } f = \varphi, \end{cases}$$

where $\epsilon(M)$ denotes the root of unity from the η-transformation formula. To determine the conjugates of the singular values of $f_{\underline{x}}$ we need, apart from Theorem 5.2.4:

Theorem 5.2.5 *Let $\alpha \in \mathbb{H}$ be the ratio of a basis of an ideal $[\underline{\alpha}] \in \mathfrak{J}_t$, and let $\xi \in \frac{1}{N}[\underline{\alpha}] \setminus [\underline{\alpha}]$ with some $N \in \mathbb{N}$. Then*

$$f(\xi|\underline{\alpha}) \in \begin{cases} K_{6Nt}, & \textit{for } f = p, \\ K_{12N^2t}, & \textit{for } f = \varphi. \end{cases}$$

Further, let $\mathfrak{c}$ be an integral ideal of $\mathfrak{O}_1$, coprime to $6Nt$ and $\mathfrak{c}_t$ the corresponding ideal of $\mathfrak{O}_t$. Let C denote a rational matrix of determinant $c > 0$, such that $C\underline{\alpha}$ is a basis of $[\underline{\alpha}]\bar{\mathfrak{c}}_t$. With unimodular matrices M_1, M_2, defined by the decomposition

$$cC^{-1} \equiv M_1\left(\begin{smallmatrix}1 & 0\\ 0 & c\end{smallmatrix}\right)M_2 \ \mathit{mod}\ 6N$$

resp.

$$cC^{-1} \equiv M_1\left(\begin{smallmatrix}1 & 0\\ 0 & c\end{smallmatrix}\right)M_2 \ \mathit{mod}\ 12N^2$$

we then have

$$f(\xi|\underline{\alpha})^{\sigma(\mathfrak{c})} = \kappa(M_1)^c\kappa(M_2)f(\xi|\tfrac{1}{c}C\underline{\alpha}).$$

In particular, for $\mathfrak{c}_t = \lambda\mathfrak{O}_t$ with $\lambda \in \mathfrak{O}_t$ and

$$C\underline{\alpha} = \bar{\lambda}\underline{\alpha},$$

we have $\frac{1}{c}C\underline{\alpha} = \frac{1}{\lambda}\underline{\alpha}$ and then

$$f(\xi|\underline{\alpha})^{\sigma(\lambda)} = \kappa(M_1)^c\kappa(M_2)f(\xi\lambda|\underline{\alpha}).$$

If the leading coefficient A in the primitive equation of α is coprime to $6N$, the unimodular matrices M_1, M_2 can be chosen such that the congruences

$$M_1 \equiv \left(\begin{smallmatrix}a(us+v)+bun & -1\\ 1 & 0\end{smallmatrix}\right) \ \mathit{mod}\ 12, \quad M_2 \equiv \left(\begin{smallmatrix}uA & v\\ b & a\end{smallmatrix}\right) \ \mathit{mod}\ 12$$

are satisfied. Herein $n = A\alpha\bar{\alpha}$, $s = A(\alpha+\bar{\alpha})$, u and v are defined by the representation $\lambda = uA\alpha + v$, and a and b are solutions of the congruence

$$uAa - vb \equiv 1 \ \mathit{mod}\ 12.$$

Proof According to our notation we first have

$$f(\xi|\underline{\alpha}) = f_{\underline{x}}(\alpha) \quad \text{with some} \quad \underline{x} \in \frac{1}{N}(\mathbb{Z}\times\mathbb{Z}) \setminus (\mathbb{Z}\times\mathbb{Z}).$$

Since $f_{\underline{x}}$ is in F_{6N} resp. in F_{12N^2}, we can conclude by Theorem 5.1.2 that

$$f_{\underline{x}}(\alpha) \in K_{6Nt} \text{ resp. } f_{\underline{x}}(\alpha) \in K_{12N^2t}$$

and

$$f_{\underline{x}}(\alpha)^{\sigma(\mathfrak{C})} = [f_{\underline{x}} \circ cC^{-1}](C(\alpha)).$$

Herein $f_{\underline{x}} \circ cC^{-1} = f_{\underline{x}} \circ M_1 \left(\begin{smallmatrix} 1 & 0 \\ 0 & c \end{smallmatrix}\right) M_2$, because the matrices cC^{-1} and $M_1 \left(\begin{smallmatrix} 1 & 0 \\ 0 & c \end{smallmatrix}\right) M_2$ are congruent modulo $6N$ resp. modulo $12N^2$. The computation shows that

$$f_{\underline{x}} \circ M_1 \left(\begin{smallmatrix} 1 & 0 \\ 0 & c \end{smallmatrix}\right) M_2 = [[f_{\underline{x}} \circ M_1] \circ \left(\begin{smallmatrix} 1 & 0 \\ 0 & c \end{smallmatrix}\right)] \circ M_2 = [[\kappa(M_1) f_{\underline{x}M_1}] \circ \left(\begin{smallmatrix} 1 & 0 \\ 0 & c \end{smallmatrix}\right)] \circ M_2.$$

Looking at the q-expansion of $f_{\underline{x}M_1}$, we find that the action of $\left(\begin{smallmatrix} 1 & 0 \\ 0 & c \end{smallmatrix}\right)$ is given by

$$f_{\underline{x}M_1} \circ \left(\begin{smallmatrix} 1 & 0 \\ 0 & c \end{smallmatrix}\right) = f_{\underline{x}M_1 \left(\begin{smallmatrix} 1 & 0 \\ 0 & c \end{smallmatrix}\right)}.$$

Further, we have $\kappa(M_1) \circ \left(\begin{smallmatrix} 1 & 0 \\ 0 & c \end{smallmatrix}\right) = \kappa(M_1)^c$ for the constant $\kappa(M_1)$, hence

$$[\kappa(M_1) f_{\underline{x}M_1}] \circ \left(\begin{smallmatrix} 1 & 0 \\ 0 & c \end{smallmatrix}\right) = \kappa(M_1)^c f_{\underline{x}M_1 \left(\begin{smallmatrix} 1 & 0 \\ 0 & c \end{smallmatrix}\right)}.$$

Now, applying M_2, it follows that

$$f_{\underline{x}} \circ cC^{-1} = \kappa(M_1)^c \kappa(M_2) f_{\underline{x}M_1 \left(\begin{smallmatrix} 1 & 0 \\ 0 & c \end{smallmatrix}\right) M_2} = \kappa(M_1)^c \kappa(M_2) f_{\underline{x}cC^{-1}}$$

and then, by homogeneity of f,

$$f(\xi|\underline{\alpha})^{\sigma(\mathfrak{C})} = \kappa(M_1)^c \kappa(M_2) f(\underline{x}cC^{-1}\left(\begin{smallmatrix} C(\alpha) \\ 1 \end{smallmatrix}\right) | \left(\begin{smallmatrix} C(\alpha) \\ 1 \end{smallmatrix}\right))$$
$$= \kappa(M_1)^c \kappa(M_2) f(\underline{x}\left(\begin{smallmatrix} \alpha \\ 1 \end{smallmatrix}\right) | \tfrac{1}{c}\left(\begin{smallmatrix} C(\alpha) \\ 1 \end{smallmatrix}\right)) = \kappa(M_1)^c \kappa(M_2) f(\xi | \tfrac{1}{c} C\underline{\alpha}).$$

In the case $\mathfrak{c}_t = \lambda \mathfrak{O}_t$ and $C\underline{\alpha} = \overline{\lambda}\underline{\alpha}$ we further have by homogeneity of f

$$f(\xi | \tfrac{1}{c} C\underline{\alpha}) = f(\xi | \tfrac{1}{\lambda}\underline{\alpha}) = f(\xi\lambda | \underline{\alpha}).$$

Since $\sigma(\mathfrak{c})$ is depending on λ only modulo $6N$ resp. modulo $12N^2$, we can change u and v modulo $6N$ resp. modulo $12N^2$ such that $\gcd(u, v) = 1$. If, in addition, the leading coefficient A of the primitive equation of α is coprime to $6N$, we can further achieve that $\gcd(Au, v) = 1$. Hence there exist $a, b \in \mathbb{Z}$ with

$$uAa - vb = 1.$$

Therefore, we can define unimodular matrices by

$$M_1 := \left(\begin{smallmatrix} a(us+v)+bun & -1 \\ 1 & 0 \end{smallmatrix}\right), \quad M_2 := \left(\begin{smallmatrix} uA & v \\ b & a \end{smallmatrix}\right).$$

The computation now shows that

$$cC^{-1} = \left(\begin{smallmatrix} us+v & -un \\ uA & v \end{smallmatrix}\right).$$

Finally, verifying the identity

$$M_1\left(\begin{smallmatrix} 1 & 0 \\ 0 & c \end{smallmatrix}\right)M_2 = cC^{-1}$$

the last assertion of Theorem 5.2.5 is proved. □

We now apply Theorem 5.2.5 to the function $h_\gamma(z|L)$ defined in (1.14):

Theorem 5.2.6 *Let $\mathfrak{b} \in \mathfrak{J}_t$, $N \in \mathbb{N}$, and $\lambda, \gamma \in \frac{1}{N}\mathfrak{b}, \lambda \notin \mathfrak{b}$. Then*

$$h_\gamma(\lambda|\mathfrak{b}) \in K_{12tN^2},$$

and for every $\nu \in \mathfrak{O}_t$ with

$$\nu \equiv 1 \textit{ mod } \frac{\mathfrak{b}}{\gamma}, \quad \nu \textit{ coprime to } 12tN^2,$$

we have

$$h_\gamma(\lambda|\mathfrak{b})^{\sigma(\nu)} = \epsilon(\nu)h_\gamma(\lambda\nu|\mathfrak{b})$$

with $\epsilon(\nu)$ defined in (1.17).

Proof Let $\underline{\beta}$ be a basis of $\mathfrak{b}$ with ratio $\beta \in \mathbb{H}$. Then

$$\tfrac{1}{2}l_{\mathfrak{b}}(\lambda,\gamma) \in \tfrac{2\pi i}{2N^2}\mathbb{Z}$$

and so the exponential factor in the definition of h_γ is a $2N^2$-th root of unity. Further,

$$\varphi(\lambda+\gamma|\mathfrak{b}) = \varphi_{\underline{x}}(\beta) \text{ and } \varphi(\lambda|\mathfrak{b}) = \varphi_{\underline{x}'}(\beta) \text{ with } \underline{x}, \underline{x}' \in \frac{1}{N}(\mathbb{Z}\times\mathbb{Z}).$$

Therefore, by Theorem 5.2.5 we have

$$h_\gamma(\lambda|\mathfrak{b}) \in K_{12tN^2}.$$

Now, for $\nu \in \mathfrak{O}_t$ coprime to $12tN^2$ it follows by Theorem 5.2.5

$$h_\gamma(\lambda|\mathfrak{b})^{\sigma(\nu)} = e^{-\frac{1}{2}l_{\mathfrak{b}}(\lambda,\nu)N(\nu)}\frac{\varphi(\lambda+\gamma|\mathfrak{b}\nu^{-1})}{\varphi(\lambda|\mathfrak{b}\nu^{-1})}.$$

Here, according to the rules for $l_{\mathfrak{L}}$, we have

$$l_{\mathfrak{b}}(\lambda,\nu)N(\nu) = l_{\mathfrak{b}\nu^{-1}}(\lambda,\nu)$$

and hence by (1.17)

$$h_\gamma(\lambda|\mathfrak{b})^{\sigma(\nu)} = \epsilon(\nu)h_\gamma(\lambda\nu|\mathfrak{b})$$

if $\nu \equiv 1 \bmod \frac{\mathfrak{b}}{\gamma}$. □

6
Generation of ring class fields and ray class fields

The first two sections of this chapter are devoted to the classical constructions of ring class fields and ray class fields by singular values of j and the τ function. This is the analogue of the construction of cyclotomic fields by roots of unity. However, the algebraic equation satisfied by the singular values of j and the τ function are far from being as simple as the equations for roots of unity. Therefore, in the next sections we consider elliptic and modular functions other than j and τ, whose singular values have simpler equations. As an application we include Heegner's solution of the class number one problem for imaginary quadratic number fields using Schläfli's functions discussed in section 6.4. Section 6.6 also contains the Principal Ideal Theorem relying on properties of the η function and in section 6.9 we prove a generalization using the σ function.

6.1 Generation of ring class fields by singular values of j

In this section we will show that the ring class field Ω_t modulo t of a quadratic imaginary number field K is generated over K by a singular value of the modular invariant j. Therefore, we associate with every ring ideal class $\mathfrak{k}$ modulo t a singular value of j by

$$j(\mathfrak{k}) := j(\mathfrak{a}) := j(\tfrac{\alpha_1}{\alpha_2}) \quad \text{with} \quad \mathfrak{a} = [\alpha_1, \alpha_2] \in \mathfrak{k}, \tfrac{\alpha_1}{\alpha_2} \in \mathbb{H},$$

and call it the **modular invariant of $\mathfrak{k}$**. Clearly, this value only depends on $\mathfrak{k}$, because j is invariant by all unimodular transformations. $j(\mathfrak{k})$ is an algebraic integer according to Theorem 4.1.1. More precisely, we have:

Theorem 6.1.1 *For every ideal class $\mathfrak{k} \in \mathfrak{R}_t$ we have*

$$\Omega_t = K(j(\mathfrak{k})).$$

There exists an isomorphism

$$\sigma : \mathfrak{R}_t \to Gal(\Omega_t/K),$$

with the property that, given a prime ideal $\mathfrak{p}$ *of* K *coprime to* t, *then*

$$\sigma((\mathfrak{p} \cap \mathfrak{O}_t)\mathfrak{H}_t)$$

is the Frobenius automorphism associated with $\mathfrak{p}$. *Further,*

$$j(\mathfrak{k})^{\sigma(\mathfrak{h})} = j(\mathfrak{k}\mathfrak{h}^{-1})$$

for all $\mathfrak{h} \in \mathfrak{R}_t$.

Proof By Theorem 5.2.3 we know $j(\mathfrak{k}) \in \Omega_t$. To prove that $\mathfrak{R}_t$ is isomorphic to the Galois group of Ω_t/K, observe that by Theorem 3.1.7 we have the relations

$$\mathfrak{A}^{\mathfrak{t}}/\mathfrak{U}_{\mathfrak{t}} \cong \mathfrak{R}_t \text{ by } \mathfrak{a}\mathfrak{U}_{\mathfrak{t}} \mapsto (\mathfrak{a} \cap S_t)\mathfrak{H}_t$$

and

$$\mathfrak{A}^{\mathfrak{t}}/\mathfrak{U}_{\mathfrak{t}} \cong Gal(K_{\mathfrak{t}}/\Omega_t) \quad \text{by} \quad \mathfrak{a}\mathfrak{U}_{\mathfrak{t}} \mapsto \sigma(\mathfrak{a}),$$

where $\sigma(\mathfrak{a})$ denotes the Frobenius automorphism of Ω_t/K associated with $\mathfrak{a}$. The isomorphism $\mathfrak{R}_t \to G(\Omega/K)$, obtained by combination of the two isomorphisms above, will also be denoted by σ (there will be no danger of confusion).

We still have to show that $j(\mathfrak{k})$ is a generator for the extension Ω_t/K. To do this we observe that the singular values $j(\mathfrak{k}), \mathfrak{k} \in \mathfrak{R}_t$, are all conjugate to each other. Hence, it suffices to prove that they are all different, so let $\mathfrak{k}, \mathfrak{h} \in \mathfrak{R}_t$ with representatives $\mathfrak{a} \in \mathfrak{k},\ \mathfrak{b} \in \mathfrak{h}$ and ratios of basis $\alpha = \frac{\alpha_1}{\alpha_2}, \beta = \frac{\beta_1}{\beta_2} \in \mathbb{H}$. We assume that $j(\mathfrak{k}) = j(\mathfrak{h})$, which means that $j(\alpha) = j(\beta)$. Since j is a modular function of order 1, this implies that $\alpha = M\beta$ with a unimodular matrix M, so we have

$$\binom{\alpha_1}{\alpha_2} = \xi M \binom{\beta_1}{\beta_2}$$

with some $\xi \in K$, and it follows that $\mathfrak{a} = \xi\mathfrak{b}$, hence $\mathfrak{k} = \mathfrak{h}$. This completes the proof of Theorem 6.1.1. □

The following theorem gives a characterisation of the subfields $\mathbb{Q}(j(\mathfrak{k})),\ \mathfrak{k} \in \mathfrak{R}_t$.

Theorem 6.1.2 *The class polynomial*

$$m_t(X) := \prod_{\mathfrak{k}\in\mathfrak{R}_t} (X - j(\mathfrak{k}))$$

has coefficients in $\mathbb{Z}$. *Hence* Ω_t *is Galois over* $\mathbb{Q}$. *We have*

$$Gal(\Omega_t/\mathbb{Q}) = Gal(\Omega_t/K) \cup Gal(\Omega_t/K)\tau,$$

where τ *denotes complex conjugation in* Ω_t. *The action of* τ *on the invariants* $j(\mathfrak{k})$ *is given by*

$$j(\mathfrak{k})^\tau = j(\mathfrak{k}^{-1})$$

which implies the relation

$$\sigma(\mathfrak{h}) \circ \tau = \tau\sigma(\mathfrak{h}^{-1}).$$

For the principal class $\mathfrak{e}$ *in* $\mathfrak{R}_t$ *we have*

$$\mathbb{Q}(j(\mathfrak{e})) = \Omega_t \cap \mathbb{R}$$

and

$$[\mathbb{Q}(j(\mathfrak{e})) : \mathbb{Q}] = [\Omega_t : K].$$

The fields $\mathbb{Q}(j(\mathfrak{k})), \mathfrak{k} \in \mathfrak{R}_t$, *are conjugate to each other over* $\mathbb{Q}$.

Proof The assertions of our theorem are immediate by Theorem 6.1.1. The only thing to prove is the action of τ on the invariants $j(\mathfrak{k})$. To see this, we pick an ideal $\mathfrak{a} = [\alpha_1, \alpha_2]$ in $\mathfrak{k}$ with $\alpha = \frac{\alpha_1}{\alpha_2} \in \mathbb{H}$. Now, j having rational q-coefficients, it follows that

$$j(\mathfrak{k})^\tau = j(\alpha)^\tau = j(-\alpha^\tau).$$

Herein $-\alpha^\tau$ is the ratio of a basis of the ideal $\mathfrak{a}^\tau$, which is in $\mathfrak{k}^{-1}$. This completes the proof of Theorem 6.1.2. □

Unfortunately it transpires that the generation of Ω_t by the modular invariants $j(\mathfrak{k})$, $\mathfrak{k} \in \mathfrak{R}_t$, is rather useless for numerical purposes, because the coefficients of the class polynomial are excessively high:

Example 6.1.3 Let $K = \mathbb{Q}(\sqrt{-47})$ and $t = 1$. Then:

$$\begin{aligned} m_1(X) = X^5 \; &+ \; 2257834125X^4 - 9987963828125X^3 \\ &+ \; 5115161850595703125X^2 \\ &- \; 14982472850828613281250X \\ &+ \; 6042929600623870849609375 \end{aligned}$$

Therefore sections 6.3 to 6.6 of this chapter will provide much simpler generators for ring class fields.

Using the relation $\tau \circ \sigma(\mathfrak{k}) = \sigma(\mathfrak{k}^{-1}) \circ \tau$ of Theorem 6.1.2, the commutator group of $G = Gal(\Omega_t/\mathbb{Q})$ turns out to be

$$G' = \{\sigma(\mathfrak{k}^2) \mid \mathfrak{k} \in \mathfrak{R}_t\}.$$

Hence the maximal subfield of Ω_t that is abelian over $\mathbb{Q}$ is a product of quadratic fields. We quote the following result of Halter-Koch (1971):

Theorem 6.1.4 (Halter-Koch) *Let d be the discriminant of K and $t \in \mathbb{N}$. Let $p_1, \ldots, p_n$ be the odd primes dividing td. Further, let 2^s resp. 2^{s_t} be the powers of 2 dividing d resp. t. We set $p^* := (-1)^{\frac{p-1}{2}} p$ for an odd prime p and*

$$L_t^{(0)} := \mathbb{Q}(\sqrt{p_1^*}, \ldots, \sqrt{p_n^*}).$$

Then the maximal subfield L_t of Ω_t that is abelian over $\mathbb{Q}$ is given by

$$L_t = \begin{cases} L_t^{(0)} & \text{if} \quad s = 0, s_t \le 1, \\ L_t^{(0)}(\sqrt{-1}) & \text{if} \quad s = 0, s_t = 2 \\ & \quad\; \text{or } s = 2, s_t \le 1, \\ L_t^{(0)}(\sqrt{(-1)^e 2}) & \text{if} \quad s = 3, s_t = 0 \\ & \quad\; \text{and } \frac{d}{8} \equiv (-1)^e \bmod 4, \\ L_t^{(0)}(\sqrt{-1}, \sqrt{2}) & \text{otherwise.} \end{cases}$$

6.2 Generation of ray class fields by τ and j

For the following we fix an order $\mathfrak{O}_t$ and a proper ideal $\mathfrak{a}$ of $\mathfrak{O}_t$. Let $\tau = \tau^{(e)}$ be the Weber τ function for the lattice $\mathfrak{a}$. Then by Theorem 5.2.5 for $\xi \in K \setminus \mathfrak{a}$ the singular value $\tau(\xi|\mathfrak{a})$ is contained in a ray class field of K. Our aim is to characterise the field generated over Ω_t by $\tau(\xi|\mathfrak{a})$ as a class field over K and to find explicit formulae for the conjugates over K. In view of homogeneity of the τ function it will be sufficient to treat the case $\xi = 1$.

Proposition 6.2.1 *Let* $\mathfrak{f} = \mathfrak{a} \cap \mathfrak{O}_t$*, and let* $\mathfrak{c}, \mathfrak{c}'$ *integral ideals in* $\mathfrak{I}_{t,\mathfrak{f}}$*. Then*

$$\mathfrak{c}\mathfrak{S}_{t,\mathfrak{f}} = \mathfrak{c}'\mathfrak{S}_{t,\mathfrak{f}} \implies \tau(1|\mathfrak{a}\mathfrak{c}^{-1}) = \tau(1|\mathfrak{a}\mathfrak{c}'^{-1}).$$

In particular, if $\mathfrak{c} \in \mathfrak{S}_{t,\mathfrak{f}}$*, then*

$$\tau(1|\mathfrak{a}\mathfrak{c}^{-1}) = \tau(1|\mathfrak{a}).$$

Proof For $\mathfrak{c}$ and $\mathfrak{c}'$ in the same ray class modulo $\mathfrak{f}$ of $\mathfrak{O}_t$ we have

$$\lambda\mathfrak{c} = \lambda'\mathfrak{c}'$$

with $\lambda, \lambda' \in \mathfrak{O}_t$, coprime to $\mathfrak{f}$ and

$$\lambda \equiv \lambda' \bmod \mathfrak{f}.$$

Multiplying the above equation by a power of λ, we can achieve

$$\lambda \equiv \lambda' \equiv 1 \bmod \mathfrak{f}.$$

The last congruence then implies that

$$\lambda \equiv 1 \bmod \mathfrak{a}\mathfrak{c}^{-1} \text{ and } \lambda' \equiv 1 \bmod \mathfrak{a}\mathfrak{c}'^{-1}$$

because $\mathfrak{c}$ and $\mathfrak{c}'$ are integral ideals and thus

$$\mathfrak{f} = (\mathfrak{a} \cap \mathfrak{O}_t) \subseteq \mathfrak{a} \subseteq \mathfrak{a}\mathfrak{c}^{-1} \text{ and } \mathfrak{f} = (\mathfrak{a} \cap \mathfrak{O}_t) \subseteq \mathfrak{a} \subseteq \mathfrak{a}\mathfrak{c}'^{-1}.$$

Now, because of the periodicity of the τ function we obtain

$$\begin{aligned} \tau(1|\mathfrak{a}\mathfrak{c}^{-1}) &= \tau(\lambda|\mathfrak{a}\mathfrak{c}^{-1}) = \tau(1|\mathfrak{a}(\lambda\mathfrak{c})^{-1}) \\ &= \tau(1|\mathfrak{a}(\lambda'\mathfrak{c}')^{-1}) = \tau(\lambda'|\mathfrak{a}\mathfrak{c}^{-1}) = \tau(1|\mathfrak{a}\mathfrak{c}'^{-1}). \end{aligned}$$

□

By proposition 6.2.1 the singular value $\tau(1|\mathfrak{f}\mathfrak{c}^{-1})$ only depends on the ray class $\mathfrak{k}$ modulo $\mathfrak{f}$ of $\mathfrak{c}^{-1}$ in $\mathfrak{O}_t$. Therefore, we define:

Definition 6.2.2 Let $\mathfrak{c}$ be an integral ideal in the ray class $\mathfrak{k}$ modulo $\mathfrak{f} = \mathfrak{a} \cap \mathfrak{O}_t$ in $\mathfrak{O}_t$. Then we set

$$\tau_{\mathfrak{a}}(\mathfrak{k}) := \tau(1|\mathfrak{a}\mathfrak{c}^{-1}).$$

With this notation we have:

Theorem 6.2.3 *Let $\mathfrak{a}$ be a proper ideal of $\mathfrak{O}_t$ and $\mathfrak{f} = \mathfrak{a} \cap \mathfrak{O}_t \neq \mathfrak{O}_t$. Then for every ray class $\mathfrak{k} \in \mathfrak{R}_{t,\mathfrak{f}}$ we have*

$$\Omega_t(\tau_{\mathfrak{a}}(\mathfrak{k})) = K_{t,\mathfrak{f}},$$

where $K_{t,\mathfrak{f}}$ denotes the class field of K associated with the subgroup $\mathfrak{U}_{t,\mathfrak{f}}$. There exists an isomorphism

$$\sigma : \mathfrak{R}_{t,\mathfrak{f}} \to G(K_{t,\mathfrak{f}}/K),$$

with the property that, given a prime ideal $\mathfrak{p}$ of K coprime to $t\mathfrak{f}$, then

$$\sigma((\mathfrak{p} \cap \mathfrak{O}_t)\mathfrak{S}_{t,\mathfrak{f}})$$

is the Frobenius automorphism associated with $\mathfrak{p}$. Further,

$$\tau_{\mathfrak{a}}(\mathfrak{k})^{\sigma(\mathfrak{h})} = \tau_{\mathfrak{a}}(\mathfrak{k}\mathfrak{h}^{-1})$$

for every $\mathfrak{h} \in \mathfrak{R}_{t,\mathfrak{f}}$.

Proof Let f be the smallest natural number in $\mathfrak{f}$. Then $f \cdot 1 \in \mathfrak{f} \subseteq \mathfrak{a} \subseteq \mathfrak{a}\mathfrak{c}^{-1}$ and hence by Theorem 5.2.3

$$\tau_{\mathfrak{a}}(\mathfrak{k}) = \tau(1|\mathfrak{a}\mathfrak{c}^{-1}) \in K_{tf}.$$

To prove that $\tau_{\mathfrak{a}}(\mathfrak{k})$ is in the subfield $K_{t,\mathfrak{f}}$ of K_{tf}, we have to show that $\tau_{\mathfrak{a}}(\mathfrak{k})$ is left fixed under the relative automorphisms of $K_{tf}/K_{t,\mathfrak{f}}$. The latter are given by the Frobenius automorphisms $\sigma(\mathfrak{c})$ of integral ideals $\mathfrak{c} \in \mathfrak{U}_{t,\mathfrak{f}} \cap \mathfrak{A}^{(f)}$. Let $\mathfrak{c}_t = \mathfrak{c} \cap \mathfrak{O}_t$ be the ring ideal associated with such an ideal $\mathfrak{c}$, then by Theorem 5.2.5 the action of $\sigma(\mathfrak{c})$ on $\tau_{\mathfrak{a}}(\mathfrak{k})$ is given by

$$\tau_{\mathfrak{a}}(\mathfrak{k})^{\sigma(\mathfrak{c})} = \tau(1|\mathfrak{a})^{\sigma(\mathfrak{c})} = \tau(1|\mathfrak{a}\mathfrak{c}_t^{-1}).$$

Recalling $\mathfrak{c}_t \in \mathfrak{S}_{t,\mathfrak{f}}$, it follows that $\tau(1|\mathfrak{a}\mathfrak{c}_t^{-1}) = \tau(1|\mathfrak{a})$, hence $\tau_{\mathfrak{a}}(\mathfrak{k})$ is left fixed by $\sigma(\mathfrak{c})$, so $\tau_{\mathfrak{a}}(\mathfrak{k})$ must be in $K_{t,\mathfrak{f}}$.

To show that $\tau(\mathfrak{k})$ is a generator for $K_{t,\mathfrak{f}}$ over Ω_t, we observe that the Galois group of $K_{t,\mathfrak{f}}/\Omega_t$ is the set of all Frobenius automorphisms

$$\sigma(\mathfrak{c}) \text{ of principal ideals } \mathfrak{c} = (\lambda), \lambda\mathfrak{O}_t + \mathfrak{f} = \mathfrak{O}_t.$$

Given such an automorphism, we know by Theorem 5.2.3 that

$$\tau(1|\mathfrak{a})^{\sigma(\mathfrak{c})} = \tau(\lambda|\mathfrak{a})$$

and, using the periodicity of the τ function, we obtain

$$\tau(1|\mathfrak{a})^{\sigma(\mathfrak{c})} = \tau(1|\mathfrak{a}) \iff \lambda \equiv \epsilon \bmod \mathfrak{f}$$

with a root of unity $\epsilon \in \mathfrak{O}_t$. This is, in turn, equivalent to $\mathfrak{c} \in \mathfrak{S}_{t,\mathfrak{f}}$. Therfore, $\tau(1|\mathfrak{a})$ is a generator for $K_{t,\mathfrak{f}}$ over Ω_t. The same holds for the other values $\tau_{\mathfrak{a}}(\mathfrak{k}), \mathfrak{k} \in \mathfrak{R}_{t,\mathfrak{f}}$, because they are conjugate to $\tau(1|\mathfrak{a})$ over K. □

6.3 The singular values of γ_2 and γ_3

As we have seen, the ring class field modulo t over an imaginary quadratic number field K can be generated by a singular value $j(\alpha)$. However, numerically this generation is not very suitable, because the minimal polynomial of $j(\alpha)$ has astronomically high coefficients, as we have seen above. One way to simpler generators is to take singular values of

$$\gamma_2 := \sqrt[3]{j} \text{ and } \gamma_3 = \sqrt{j - 12^3},$$

defined in section 2.4.2. Namely, it transpires that in many cases the singular values $\gamma_2(\alpha)$ and $\gamma_3(\alpha)$ are already in $K(j(\alpha))$. Moreover, the singular values of γ_2 and γ_3 play a crucial role in the normalisation of the Weierstrass $\wp$ function in section 7.1 that leads to the construction of integral bases in ray class fields.

Theorem 6.3.1 *Let $\alpha \in \mathbb{H}$ be a root of the primitive quadratic equation*

$$X^2 + BX + C = 0 \quad \textit{with} \quad 3 \nmid A, \ B \equiv 0 \ \textit{mod} \ 3$$

and discriminant $D(\alpha) = B^2 - 4AC = t^2 d$.
Then

$$\mathbb{Q}(\gamma_2(\alpha)) = \begin{cases} \mathbb{Q}(j(\alpha)), & \textit{if } 3 \nmid D(\alpha), \\ \mathbb{Q}(j(3\alpha)) & \textit{if } 3|D(\alpha). \end{cases}$$

Herein $\mathbb{Q}(j(3\alpha))$ is conjugate to the maximal real subfield of Ω_{3t} and has degree 3 over $\mathbb{Q}(j(\alpha))$ if $3|D(\alpha)$ and $D(\alpha) \neq -3$. In the case $D(\alpha) = -3$ we have $\mathbb{Q}(j(3\alpha)) = \mathbb{Q}$.

In the case $3 \nmid D(\alpha)$ we further have: let $\alpha \in \mathbb{H}$ run through a 3-system of ratios of basis of representatives for the ring ideal classes modulo t, where the α's are roots of primitive equations normalised as above. Then the $\gamma_2(\alpha)$'s are a complete system of conjugate numbers over $\mathbb{Q}$.

For the singular values of γ_3 we have:

Theorem 6.3.2 *Let $\alpha \in \mathbb{H}$ be a root of the primitive quadratic equation*

$$AX^2 + BX + C = 0 \quad \textit{with} \quad 2 \nmid A$$

and discriminant $D(\alpha) = B^2 - 4AC = t^2 d$, and we assume that

$$B \equiv \begin{cases} 0 \ \textit{mod}\ 4 \ \textit{if}\ 2|D(\alpha), \\ 1 \ \textit{mod}\ 4 \ \textit{if}\ 2 \nmid D(\alpha). \end{cases}$$

Then

$$\begin{array}{r} \mathbb{Q}(\sqrt{d}\gamma_3(\alpha)) = \mathbb{Q}(j(\alpha)), \ \textit{if}\ 2 \nmid D(\alpha), \\ \mathbb{Q}(\gamma_3(\alpha)) = \mathbb{Q}(j(2\alpha)) \ \textit{if}\ 2|D(\alpha). \end{array}.$$

Herein $\mathbb{Q}(j(2\alpha))$ is conjugate to the maximal real subfield of Ω_{2t} and has degree 2 over $\mathbb{Q}(j(\alpha))$ if $2|D(\alpha)$ and $D(\alpha) \neq -4$. In the case $D(\alpha) = -4$ we have $\mathbb{Q}(j(2\alpha)) = \mathbb{Q}$.

In the case $2 \nmid D(\alpha)$ we further have: let $\alpha \in \mathbb{H}$ run through a 4-system of ratios of basis of representatives for the ring ideal classes modulo t, where the α's are roots of primitive equations normalised as above. Then the $\gamma_3(\alpha)$'s are a complete system of conjugate numbers over $\mathbb{Q}$.

Proof of *Theorem 6.3.1* Let $M = \left(\begin{smallmatrix} a & b \\ c & d \end{smallmatrix}\right)$ be a unimodular matrix. Then by Theorem 1.10.1 we obtain

$$\gamma_2(M(\omega)) = \zeta_3^{-ba-c(d(1-a^2)-a)}\gamma_2(\omega)$$

with $\zeta_3 = e^{\frac{2\pi i}{3}}$. Keeping in mind that γ_2 has rational q-coefficients, this implies that γ_2 satisfies the hypothesis od Theorem 5.2.1. First, we assume that

$$\alpha = \begin{cases} \sqrt{-m} & \text{or} \\ \frac{3m+\sqrt{-m}}{2}, & -m \equiv 1 \bmod 4, \end{cases}$$

with a natural number m. Then, by Theorem 5.2.1 it follows that

$$\gamma_2(\alpha) \in \Omega_{3t}.$$

For $\alpha = \frac{9+\sqrt{-3}}{2}$ the assertion of our theorem is immediate, because $\gamma(\alpha) = 0$. In the remaining cases we have

$$[\Omega_{3t} : \Omega_t] = \begin{cases} 3 & \text{if } m \equiv \ \ 0 \bmod 3, \\ 2 & \text{if } m \equiv -1 \bmod 3, \\ 4 & \text{if } m \equiv \ \ 1 \bmod 3 \end{cases}$$

and

$$Gal(\Omega_{3t}/\Omega_t) = \begin{cases} <\sigma(\overline{\alpha+1})> & \text{if } m \equiv \ \ 0 \bmod 3, \\ <\sigma(\overline{\alpha+3})> & \text{if } m \equiv -1 \bmod 3, \\ <\sigma(\overline{\alpha \pm 1})> & \text{if } m \equiv \ \ 1 \bmod 3, \end{cases}$$

where $\sigma(\overline{\alpha+\mu}), \mu = \pm 1, 3$ denotes the Frobenius automorphism corresponding to the ideal $\mathfrak{O}_1(\overline{\alpha+\mu})$. Observe that $\overline{\alpha+\mu}$ is coprime to $3t$ because $t|m$. Further, the automorphisms listed above are different from the identity and are all distinct for $m \equiv 1 \bmod 3$ because then the congruences

$$\alpha + 1 \equiv r \bmod 3t, \;\; \alpha + 3 \equiv r \bmod 3t \text{ resp. } \alpha \pm 1 \equiv r \bmod 3t$$

as well as

$$\alpha + 1 \equiv r(\alpha - 1) \bmod 3t$$

with a rational number r coprime to $3t$ are not solvable. Finally, they are relative automorphisms of Ω_{3t}/Ω_t because the numbers $\alpha + \mu$ are congruent to a rational number modulo t. Now, the Galois action can be computed by Theorem 5.2.1:

$$\gamma_2(\alpha)^{\sigma(\overline{\alpha+\mu})} = (\gamma_2(\alpha+\mu)\zeta_3^{\mu})^{\sigma(\overline{\alpha+\mu})} = \gamma_2\left(\tfrac{1}{\overline{\alpha+\mu}}\right)\zeta_3^{\mu n_\mu}.$$

Herein n_μ is the norm of $\overline{\alpha+\mu}$, and n_μ is congruent to $m+\mu^2$ modulo 3. Using the transformation formula $\gamma_2(\frac{-1}{z}) = \gamma_2(z)$, this becomes

$$\gamma_2(\alpha)^{\sigma(\overline{\alpha+\mu})} = \gamma_2(\alpha)\zeta_3^{\mu(m+\mu^2+1)}.$$

Further, recalling that $\gamma_2(\alpha)$ is real, it follows that

$$\mathbb{Q}(\gamma_2(\alpha)) = \begin{cases} \Omega_t \cap \mathbb{R} = \mathbb{Q}(j(\alpha)) \text{ if } 3 \nmid m, \\ \Omega_{3t} \cap \mathbb{R} = \mathbb{Q}(j(3\alpha)) \text{ if } 3|m. \end{cases}$$

This completes the proof of Theorem 6.3.1 for the above special arguments. Now let α' be the ratio of a basis of an arbitrary proper ideal of $\mathfrak{O}_t$, with a primitive equation $A'X^2 + B'X + C' = 0$ normalised by $3 \nmid A'$ and $B' \equiv 0 \bmod 3$. Then by Theorem 5.2.4 there exists an automorphism σ' of Ω_{3t} such that

$$\gamma_2(\alpha') = \gamma_2(\alpha)^{\sigma'} \text{ and } j(3\alpha') = j(3\alpha)^{\sigma'}.$$

This implies the remaining assertions of Theorem 6.3.1. □

Proof of Theorem 6.3.2 The proof of this theorem is completely analogous to the proof of 6.3.1. Again it suffices to consider arguments of the form

$$\alpha = \begin{cases} \frac{-m+\sqrt{-m}}{2}, & -m \equiv 1 \bmod 4, \text{ or} \\ \sqrt{-m} & \end{cases}$$

with a natural number m. For a unimodular matrix $M = \left(\begin{smallmatrix} a & b \\ c & d \end{smallmatrix}\right)$ we here have the transformation formula

$$\gamma_3(M(\omega)) = (-1)^{-ba-c(d(1-a^2)-a)-3(a-1)}\gamma_3(\omega).$$

This implies that γ_3 satisfies the hypothesis of Theorem 5.2.1 with $N = 2$. Hence

$$\gamma_3(\alpha) \in \Omega_{2t}.$$

Here for $D(\alpha) \neq -4$ we have

$$[\Omega_{2t} : \Omega_t] = \begin{cases} 1 \text{ or } 3, & \text{if } 2 \nmid D(\alpha), \\ \quad 2 \text{ if } 2|D(\alpha). \end{cases}$$

When $2 \nmid D(\alpha)$ we immediately have $\gamma_3(\alpha) \in \Omega_t$, because the square of $\gamma_3(\alpha)$ lies in Ω_t. Further, in this case $\sqrt{d}\gamma_3(\alpha)$ is real and thus

$$\mathbb{Q}(\sqrt{d}\gamma_3(\alpha)) = \mathbb{Q}(j(\alpha)) \text{ if } 2 \nmid D(\alpha).$$

When $2|D(\alpha)$ the Galois group of Ω_{2t}/Ω_t is generated by $\sigma(\overline{\alpha+1})$ and, using the identities $\gamma_3(z+1) = -\gamma(z)$ and $\gamma_3(\frac{-1}{z}) = -\gamma_3(z)$, we obtain $\gamma_3(\alpha)^{\sigma(\overline{\alpha+1})} = -\gamma_3(\alpha)$. This implies that $\mathbb{Q}(\gamma_3(\alpha)) = \Omega_{2t} \cap \mathbb{R} = \mathbb{Q}(j(2\alpha))$ because in this case $\gamma_3(\alpha)$ is real. The remaining part of the proof is now analogous to the proof of Theorem 6.3.1. □

Examples 6.3.3 Let $K = \mathbb{Q}(\sqrt{-47})$. Then the Hilbert class field Ω is of degree 5 over K and the above theorems tell us that

$$\Omega = K(\gamma_3(\alpha)), \quad \Omega = K(\gamma_2(\alpha)) \quad \text{with} \quad \alpha = \frac{-3+\sqrt{-47}}{2}$$

with smaller minimal equations than for $j(\alpha)$ in Example 6.1.3:

$$\begin{aligned} m_{\gamma_3(\alpha),K} &= X^5 - 6893\sqrt{-47}X^4 + 12355331\,X^3 + 227234667\sqrt{-47}X^2 \\ &\quad +99447994734\,X - 338821592087\sqrt{-47} \end{aligned}$$

$$\begin{aligned} m_{\gamma_2(\alpha),K} &= X^5 + 1320\,X^4 + 12100\,X^3 \\ &\quad +1927375\,X^2 + 13571250\,X + 252209375 \end{aligned}$$

Smaller generating equations for Ω will be provided by Schläfli's function f in Example 6.4.3 and by the double eta-quotients in Example 6.6.8.

6.4 The singular values of Schläfli's functions

Particularly simple generators for the ring class fields are obtained by Schläfli's functions first used by Weber (1908) for this purpose. We follow the exposition in Schertz (2002). Schläfli's functions are defined by

$$f(z) = e^{-\frac{\pi i}{24}} \frac{\eta\left(\frac{z+1}{2}\right)}{\eta(z)}, \quad f_1(z) = \frac{\eta\left(\frac{z}{2}\right)}{\eta(z)}, \quad f_2(z) = \sqrt{2}\frac{\eta(2z)}{\eta(z)}, \quad \Im(z) > 0.$$

They are obviously related by modular transformations and satisfy

$$f f_1 f_2 = \sqrt{2}.$$

The modular equations computed according to Theorem 2.10.1 can be written as

$$\gamma_2 = \frac{f^{24} - 16}{f^8} = \frac{f_1^{24} + 16}{f_1^8} = \frac{f_2^{24} + 16}{f_2^8}.$$

Therefore, in view of Theorem 6.3.1, it will be sufficient to consider the singular values of f^3. By Theorem 1.10.1 we find the transformation formula

$$f(M(\omega))^3 = f(\omega)^3 \text{ for } M = \left(\begin{smallmatrix} a & b \\ c & d \end{smallmatrix}\right) \in \Gamma, M \equiv \left(\begin{smallmatrix} * & 0 \\ 0 & * \end{smallmatrix}\right) \bmod 16. \tag{6.1}$$

Further, f having rational q-coefficients, we can apply Theorem 5.2.1 with $N = 16$, and we obtain:

Theorem 6.4.1 *Let $\alpha \in \mathbb{H}$ be a root of the primitive quadratic equation*

$$AX^2 + 2BX + C = 0 \text{ with } 2 \nmid A, \; B \equiv 0 \bmod 16$$

and discriminant $D(\alpha) = B^2 - 4AC = -4m = t^2 d$.
Then the following singular values $g(\alpha)$ are generators of $\mathbb{Q}(j(\alpha))$ over $\mathbb{Q}$:

$$\begin{array}{lll}
\left(\left(\frac{2}{A}\right)\frac{1}{\sqrt{2}} f(\alpha)^2\right)^3 & \text{if} & m \equiv 1 \bmod 8, \\
f(\alpha)^3 & \text{if} & m \equiv 3 \bmod 8, \\
\left(\frac{1}{2} f(\alpha)^4\right)^3 & \text{if} & m \equiv 5 \bmod 8, \\
\left(\left(\frac{2}{A}\right)\frac{1}{\sqrt{2}} f(\alpha)\right)^3 & \text{if} & m \equiv 7 \bmod 8, \\
 & & \\
\left(\left(\frac{2}{A}\right)\frac{1}{\sqrt{2}} f_1(\alpha)^2\right)^3 & \text{if} & m \equiv 2 \bmod 4, \\
\left(\left(\frac{2}{A}\right)\frac{1}{2\sqrt{2}} f_1(\alpha)^4\right)^3 & \text{if} & m \equiv 4 \bmod 8.
\end{array}$$

Further, if α runs through a 16-system modulo t with $B \equiv 0 \bmod 16$, then the $g(\alpha)$ constitute a complete system of conjugate numbers over $\mathbb{Q}$. Hence the minimal polynomial over $\mathbb{Q}$ is given by

$$\prod_{\{A,B,C\}} (X - g(\alpha))$$

and has coefficients in $\mathbb{Z}$, because $g(\alpha)$ is an algebraic integer.

Remark 6.4.2 For $D(\alpha)$ not divisible by 3 the assertions of Theorem 6.4.1 hold without the exponent 3 if the primitive equation of α satisfies the additional condition $3 \nmid A$ and $3|B$. This is immediate by the relations between γ_2 and f resp. f_1.

Proof of Theorem 6.4.1 With $m \in \mathbb{N}$, as defined in the theorem, we set

$$\beta_\mu = \begin{cases} \sqrt{-m} + 2\mu \text{ if } m \equiv 1 \bmod 2, \\ \sqrt{-m} + 1 + 2\mu \text{ if } m \equiv 0 \bmod 2. \end{cases}$$

By (6.1) and Theorem 5.2.1 we first have

$$f^3(\beta_\mu) \in \Omega_{16t}.$$

Here $[\Omega_{16t} : \Omega_t] = 16$ and, if we choose $\mu \in \mathbb{Z}$, $\mu \bmod 8$, then β_μ is coprime to t and

$$Gal(\Omega_{16t}/\Omega_t) =< \{\sigma(\overline{\beta_\mu}) \;:\; \mu \bmod 8\} > .$$

Further, using $f(\omega+2\mu) = \zeta_{24}^{-\mu} f(\omega)$, it follows from Theorem 5.2.1 that:

$$f^3(\beta_0)^{\sigma(\overline{\beta_\mu})} = \left(f^3(\beta_\mu)\zeta_8^\mu\right)^{\sigma(\overline{\beta_\mu})} = f^3\left(\frac{-1}{-\overline{\beta_\mu}}\right)\zeta_8^{\mu\sigma(\overline{\beta_\mu})}.$$

Theorem 3.3.2 tells us that $\zeta_8^{\sigma(\overline{\beta_\mu})} = \zeta_8^{n_\mu}$ with the complex norm n_μ of $\overline{\beta_\mu}$. Using the identity $f(\frac{-1}{\bar{z}}) = f(z)$ we then obtain

$$f^3(\sqrt{-m})^{\sigma(\overline{\beta_\mu})-1} = \zeta_8^{\mu(m+1+4\mu^2)} \qquad \text{if } 2 \nmid m,$$

and, using $f_1^3(z) = \zeta_{16} f^3(z+1)$, we further obtain

$$f_1^6(\sqrt{-m})^{\sigma(\overline{\beta_\mu})-1} = \zeta_8^{(m+2)(2\mu+1)} \qquad \text{if } 2|m.$$

Finally, by Theorem 3.3.1 we find that

$$\sqrt{2}^{\sigma(\overline{\beta_\mu})-1} = \left(\frac{2}{n_\mu}\right).$$

Using these formulae, we can easily verify that the singular values $g(\alpha))$ listed in Theorem 6.4.1 lie in $K(j(\alpha))$ for $\alpha = \sqrt{-m}$ and, since they are real, we can conclude that they are even in $\mathbb{Q}(j(\alpha))$. Moreover, $g(\alpha)$ is a generator of $\mathbb{Q}(j(\alpha))$, because in all cases $j(\alpha)$ is a rational function of $g(\alpha)$ with rational coefficients. This completes the proof for $\alpha = \sqrt{-m}$.

The general case is settled as in the proof of Theorem 6.3.1 using Theorem 5.2.4. Observe that for a root α of a primitive equation $AX^2 + 2BX + C = 0$ normalised as in Theorem 6.4.1 we have

$$\sqrt{2}^{\sigma(\mathfrak{a})} = \left(\frac{2}{A}\right)\sqrt{2}, \qquad \mathfrak{a} = [\alpha, 1],$$

according to Theorem 3.3.1. □

In contrast to the singular values of j the singular values from Theorem 6.4.1 have minimal equations with remarkably small coefficients, particularly for m not divisible by 3 because then the assertions of Theorem 6.4.1 hold without the exponent 3.

Examples 6.4.3

(i) $m = 17, \Theta = \frac{1}{\sqrt{2}} f(\sqrt{-m})^2$

$$m_{\Theta,K}(X) = X^4 - X^3 - 2X^2 - X + 1$$

(ii) $m = 11, \Theta = f(\sqrt{-m})$

$$m_{\Theta,K}(X) = X^3 - 2X^2 + 2X - 2$$

(iii) $m = 13, \Theta = \frac{1}{2} f(\sqrt{-m})^4$

$$m_{\Theta,K}(X) = X^2 - 3X - 1$$

(iv) m=23 , $\Theta = \frac{1}{\sqrt{2}} f(\sqrt{-m})$

$$m_{\Theta,K}(X) = X^3 - X - 1$$

(v) m=22, $\Theta = \frac{1}{\sqrt{2}} f_1(\sqrt{-m})^2$

$$m_{\Theta,K}(X) = X^2 - 2X - 1$$

(vi) m=28, $\Theta = \frac{1}{2\sqrt{2}} f_1(\sqrt{-m})^4$

$$m_{\Theta,K}(X)(X) = X^2 - 6X + 2$$

(vii) m=47, $\Theta = \frac{1}{\sqrt{2}} f(\sqrt{-m})$

$$m_{\Theta,K}(X) = X^5 - X^3 - 2X^2 - 2X - 1$$

(viii) m=407, $\Theta = \frac{1}{\sqrt{2}} f(\sqrt{-m})$

$$m_{\Theta,K}(X) = X^{16} - 10X^{15} + 4X^{14} - 33X^{13} + 17X^{12} - 19X^{11} - 4X^{10} - 12X^8 - 15X^6 + 2X^5 - 9X^4 - X^3 - 2X^2 - X - 1$$

The following section deals with one of the nicest applications of singular values of Schläfli's functions.

6.5 Heegner's solution of the class number one problem

In his *Disquisitiones Arithmeticae* 1801 Gauss conjectured that there are exactly nine quadratic imaginary number fields of class number 1, namely

$$K = \mathbb{Q}(\sqrt{-m}), \quad m = 1, 2, 3, 7, 11, 19, 43, 67, 163.$$

The first complete solution of the problem is due to Heegner (1952), who crucially used Weber's results on Schläfli's functions.

We follow the the exposition of H.M. Stark, who explained Heegner's proof to the author in July 1969 during a boat trip on the River Rhein near the Cathedral of Cologne. We start our exposition by some remarks that reduce the problem to the non-trivial cases. Let d be the discriminant of a quadratic imaginary number field K of class number one. Then by Theorem 6.1.4 we conclude that d must be a prime discriminant, i.e.

$$d = -4, -8 \quad \text{or} \quad d = -p, \quad p \text{ a prime}, \ p \equiv 3 \bmod 4.$$

Further, we know that 2 splits in K for $p \equiv 7 \bmod 8$, hence 2 is the norm of a non-rational integer in K. Therefore, in this case the class number can only be 1 for $p = 7$. Therefore, to prove Gauss's conjecture, it suffices to show that for a prime $p \equiv 3 \bmod 8$:

$$h_{\mathbb{Q}(\sqrt{-p})} = 1 \Longrightarrow p = 3, 11, 19, 43, 67, 163.$$

Heegner considers the function

$$F(z) := 2^{12} \frac{\Delta(2z)}{\Delta(z)}.$$

F is related to Schläfli's function f by

$$F\left(\frac{z+3}{2}\right) = -\left(\frac{\sqrt{2}}{f(z)}\right)^{24}. \tag{6.2}$$

and its modular equation is given by

$$(F + 16)^3 = jF. \tag{6.3}$$

Now, for the following let p be a prime, $p \equiv 3 \bmod 8$, with $h_{\mathbb{Q}(\sqrt{-p})} = 1$. We set

$$\alpha := \frac{3 + \sqrt{-p}}{2},$$

which is the ratio of a basis of the maximal order of $K = \mathbb{Q}(\sqrt{-p})$, and we study the 12-th root of $-F(\alpha)$:

$$h(\alpha) := \frac{2}{f(\sqrt{-p})^2},$$

defined by (6.2). Then, according to Theorems 6.3.1 and 6.4.1, the powers $h(\alpha), h(\alpha)^2$ and $h(\alpha)^4$ as generators of the maximal real subfield of Ω_2 are cubic irrationalities. The minimal equation of $h(\alpha)^4$ is obtained by (6.3):

$$X^3 + \gamma_2(\alpha)X - 16 \tag{6.4}$$

because we know that $\gamma_2(\alpha)$ is in $\mathbb{Z}$. On the other hand the minimal equation of $h(\alpha)^4$ can be derived from the minimal equation of $h(\alpha)$. Let

$$m_l(X) = X^3 + A_l X^2 + B_l X - 2^{2^l}, \quad l = 0, 1, 2,$$

be the minimal equations for $h(\alpha)^{2^l}$ over $\mathbb{Q}$. They have coefficients in $\mathbb{Z}$, since $h(\alpha)$ is integral. The sign in -2^{2^l} is determined by observing that $h(\alpha)$ is real and positive and has two other non-real conjugates over $\mathbb{Q}$ that are complex conjugates of each other. The coefficients of $m_l(X)$ and $m_{l+1}(X)$ are related by

$$\left.\begin{array}{rcl} A_{l+1} & = & 2B_l - A_l^2 \\ B_{l+1} & = & B_l^2 + 2^{(2^l+1)} A_l \end{array}\right\}, \quad l = 1, 2. \tag{6.5}$$

To see this, we write $m_l(X) = 0$ as

$$X^3 + B_l X = -(A_l X^2 - 2^{2^l}).$$

Taking squares on both sides, this leads us to

$$X^6 + (2B_l - A_l^2)X^4 + (B^2 + 2^{2^l+1}A_l)X^2 - 2^{2^{l+1}}.$$

Hence $h(\alpha)^{2^{l+1}} = \left(h(\alpha)^{2^l}\right)^2$ is a root of

$$X^3 + (2B_l - A_l)X^2 + (B^2 + 2^{2^l+1}A_l)X - 2^{2^{l+1}}.$$

Since this must be the minimal equation for the cubic irrationality $h(\alpha)^{2^{l+1}}$ it follows (6.5). Using (6.5) and (6.4), we then find a Diophantine equation for A_0 and B_0, by which all possible primes $p \equiv 3 \bmod 8$ with $h_{\mathbb{Q}(\sqrt{-p})} = 1$ can be determined: first, we have $A_2 = 0$ by (6.4), and by (6.5) we then obtain:

$$0 = A_2 = 2B_1 - A_1^2 = 2(B_0^2 + 4A_0) - (2B_0 - A_0^2)^2, \tag{6.6}$$

$$\gamma_2(\alpha) = B_2 = B_1^2 + 8A_1 = (B_0^2 + 4A_0)^2 + 8(2B_0 - A_0^2). \tag{6.7}$$

From (6.6) it follows that

$$B_0 = 4b, \quad A_0 = 2a \quad \text{with } a, b \in \mathbb{Z},$$

and, setting

$$y := 2(b - a^2) \quad \text{and} \quad x := -a,$$

(6.6) becomes the **Heegner equation**

$$y^2 = 2x(x^3 - 1). \tag{6.8}$$

According to Heegner all solutions $(x, y) \in \mathbb{Z} \times \mathbb{Z}$ of this equation are given by

$$(x, y) = (0, 0),\ (1, 0),\ (-1, 2),\ (-1, -2),\ (-2, 6),\ (-2, -6). \tag{6.9}$$

By (6.7) this leads to six possible values for $\gamma_2(\alpha)$ and, because $\gamma_2(\alpha)$ uniquely determines $\mathbb{Q}(\sqrt{-p})$, it follows that there exist at most six quadratic imaginary number fields of discriminant $d \equiv -3 \bmod 8$ and class number one. Since on the other hand we know that six such fields exist, the problem is solved. To be more precise, the solutions in (6.9) are in bijection to the following values of γ_2:

$$0, \quad -2^5 \cdot 3, \quad -2^5 \cdot 3 \cdot 11, \quad -2^5, \quad -2^6 \cdot 3 \cdot 5 \cdot 23 \cdot 29, \quad -2^6 \cdot 3 \cdot 5$$

and the latter to the discriminants $d = -p$,

$$p = 3, \quad 19, \quad 67, \quad 11, \quad 163, \quad 43.$$

Remark 6.5.1 The formulae (6.5) have already been used in 1908 in Weber's Algebra III (1908) on page 475 to compute the cubic equations for $f(\sqrt{-m})$, $m = 11, 19, 43, 163$. However, it took 44 years before Heegner had the idea to use the relation to derive a Diophantine equation leading to the solution of Gauss's problem.

At first Heegner's proof was believed to be invalid because it seemed that he had used the result of Theorem 6.4.1 for $m \equiv 3 \bmod 8$, which at that time had not yet been proved. But in fact, as the above exposition shows, he had only made use of the weaker statement about f^6, which had been proved by Weber (1908). A gap in this proof was rectified later by Stark (1969) and, finally Birch (1969) and Meyer (1970) gave a proof for special cases of Weber's assertion about f^3, including when $h_{\mathbb{Q}(\sqrt{-m})} = 1$.

Before complete evidence was provided for the requisites of Heegner's solution, Stark (1967) published an independent proof of Gauss's conjecture. Remarkably, the heart of this proof was again the singular value of a modular function, and this is also true for the proof of Siegel (1968), by which he intended to interpret Stark's solution.

6.6 Generation of ring class fields by η-quotients

In view of the results in Theorem 6.4.1, one may ask whether more general η-quotients can supply simple generators of ring class fields. To obtain such results we first prove:

Theorem 6.6.1 *Let $\mathfrak{a}, \mathfrak{b}$ be proper ideals of the orders $\mathfrak{O}_{t'}$ resp. $\mathfrak{O}_{t''}$ in K. Then*

$$\frac{\Delta(\mathfrak{a})}{\Delta(\mathfrak{b})} \in \Omega_t \text{ with } t = lcm(t', t'').$$

The action of the Frobenius automorphism corresponding to a proper ideal $\mathfrak{c}$ of $\mathfrak{O}_t$ is given by

$$\left(\frac{\Delta(\mathfrak{a})}{\Delta(\mathfrak{b})}\right)^{\sigma(\mathfrak{c})} = \frac{\Delta(\mathfrak{a}\mathfrak{c}^{-1})}{\Delta(\mathfrak{b}\mathfrak{c}^{-1})}.$$

Proof First, we assume $\mathfrak{a}$, $\mathfrak{b}$ and $\mathfrak{c}$ to be of the form

$$\mathfrak{a} = \mathfrak{p}_{t'}, \quad \mathfrak{b} = \mathfrak{q}_{t''}, \quad \mathfrak{c} = \mathfrak{r}_t$$

with prime ideals $\mathfrak{p}, \mathfrak{q}$, $\mathfrak{r}$ of degree 1 and norm p, q resp. r not dividing $t^2 d$. Then, for some $\alpha \in \mathbb{H}$

$$\mathfrak{O}_t = [\alpha, 1], \quad \mathfrak{a} = [s'\alpha, p], \quad \mathfrak{b} = [s''\alpha, q],$$

$$\mathfrak{a}\bar{\mathfrak{c}} = [s'\alpha, rp], \quad \mathfrak{b}\bar{\mathfrak{c}} = [s''\alpha, rq], \quad s' = \frac{t}{t'}, s'' = \frac{t}{t''},$$

and we have

$$\frac{\Delta(\mathfrak{a})}{\Delta(\mathfrak{b})} = g(\alpha) \quad \text{with} \quad g(\omega) = \frac{\Delta\left(\frac{s'\omega}{p}\right)}{\Delta\left(\frac{s''\omega}{q}\right)}.$$

Since g satisfies the hypothesis of Theorem 5.2.1 with $N = tpq$, it follows that

$$\frac{\Delta(\mathfrak{a})}{\Delta(\mathfrak{b})} \in \Omega_{t^2pq}$$

and then by Theorem 5.1.2

$$\left(\frac{\Delta(\mathfrak{a})}{\Delta(\mathfrak{b})}\right)^{\sigma(\mathfrak{r})} = g(\alpha)^{\sigma(\mathfrak{c})} = g\left(\frac{\alpha}{r}\right) = \frac{\Delta(\mathfrak{a}\bar{\mathfrak{c}})}{\Delta(\mathfrak{b}\bar{\mathfrak{c}})} = \frac{\Delta(\mathfrak{a}\mathfrak{c}^{-1})}{\Delta(\mathfrak{b}\mathfrak{c}^{-1})}. \tag{6.10}$$

For the last equation note that $\bar{\mathfrak{c}} = N(\mathfrak{c})\mathfrak{c}^{-1}$ together with the homogeneity of Δ. By the last formula it is evident that the image $\left(\frac{\Delta(\mathfrak{a})}{\Delta(\mathfrak{b})}\right)^{\sigma(\mathfrak{r})}$

only depends on the ring ideal class modulo t of $\mathfrak{r}_t$. Hence

$$\left(\frac{\Delta(\mathfrak{a})}{\Delta(\mathfrak{b})}\right)^{\sigma(\mathfrak{r})} = \frac{\Delta(\mathfrak{a})}{\Delta(\mathfrak{b})}, \quad \text{for } \mathfrak{r}_t \in \mathfrak{H}_t,$$

and therefore $\frac{\Delta(\mathfrak{a})}{\Delta(\mathfrak{b})} \in \Omega_t$, because all automorphisms of Ω_{t^2pq}/Ω_t are given by $\sigma(\mathfrak{r})$ with some prime ideal $\mathfrak{r}$ of degree 1 not dividing t^2dpq. The same argument applies to the Galois action because in every ring ideal class there exists an integral regular ideal coprime to t^2dpq. This completes the proof of our theorem for $\mathfrak{a}$ and $\mathfrak{b}$ of the special form above. In general, we have

$$\mathfrak{a} = \xi'\mathfrak{p}_{t'}, \quad \mathfrak{b} = \xi''\mathfrak{q}_{t''}, \quad \xi', \xi'' \in K \setminus \{0\},$$

with prime ideals $\mathfrak{p}, \mathfrak{q}$ of degree 1 not dividing t because there exists some prime ideal of degree 1 in every ideal class modulo $\mathfrak{U}_t$. We then have

$$\frac{\Delta(\mathfrak{a})}{\Delta(\mathfrak{b})} = \left(\frac{\xi''}{\xi'}\right)^{12} \frac{\Delta(\mathfrak{p}_{t'})}{\Delta(\mathfrak{q}_{t''})},$$

whereby the assertion of our theorem is reduced to the special cases considered earlier. □

Our aim now is to prove that ring class fields can be generated by the Δ-quotients of Theorem 6.6.1. By Theorem 6.4.1 we already know that Schläfli's functions have this property. However, the proof does not extend to most cases because, unlike Schläfli's functions, the modular invariant is in general not a rational function of the function g in the proof of Theorem 6.6.1. Therefore, following Schertz (1978), we will use the results from Chapter 11, which imply that the L-function s of the extension Ω_t/K does not vanish at $s = 1$.

Theorem 6.6.2 *Let $\mathfrak{a}, \mathfrak{b} \in \mathfrak{J}_t$ and $\mathfrak{O} = \mathfrak{O}_t$. Then for every intermediate field $K \subseteq L \subseteq \Omega_t$ and every natural exponent e we have*

$$L = K\left(N_{\Omega_t/L}\left(\left(\frac{\Delta(\mathfrak{a})}{\Delta(\mathfrak{b})}\right)^{e}\right)\right) \quad \textit{if} \quad \mathfrak{a}\mathfrak{U} \neq \mathfrak{b}\mathfrak{U},$$

$$L = K\left(N_{\Omega_t/L}\left(\left(\frac{\Delta(\mathfrak{a})\Delta(\mathfrak{b})}{\Delta(\mathfrak{a}\mathfrak{b})\Delta(\mathfrak{O})}\right)^{e}\right)\right) \quad \textit{if} \quad \mathfrak{a} \notin \mathfrak{U} \textit{ and } \mathfrak{b}^2 \notin \mathfrak{U},$$

where $\mathfrak{U}$ is the subgroup of $\mathfrak{R}_t$ corresponding to L.

For the proof of Theorem 6.6.2 we need:

Lemma 6.6.3 *Let G be a finite abelian group. Then:*

(i) *Given two elements $a, b \in G \setminus \{1\}$ there exists a character χ of G with $\chi(a), \chi(b) \neq 1$.*

(ii) *Given three elements $a, b, c \in G \setminus \{1\}$ with $c^2 \neq 1$ then there exists a character χ with $\chi(a), \chi(b), \chi(c) \neq 1$.*

Proof For the first assertion we have to distinguish the cases "$b \in < a >$" and "$b \notin < a >$". In the first case a generating character χ_0 of $< a >$ has the property $\chi_0(a), \chi_0(b) \neq 1$, and χ is obtained by extension of χ_0 to G. In the second case there exists an extension χ_1 of χ_0 to $< a, b >$ with $\chi_1(b) \neq 1$. Any extension of χ_1 to G then has the desired property.

To prove the second assertion we may assume the orders of a and b to be primes p_a, p_b and the order of c to be the power of a prime p_c. If not all of these primes are equal, $p_c \neq p_a, p_b$ say, then $c \notin < a, b >$ and the character we want is obtained as in the first part by extension of a character χ_0 of $< a, b >$ with $\chi_0(a), \chi_0(b) \neq 1$, so for the following we assume all three primes to be equal to p. Then for $< a > = < b >$ we can conclude as in the first part. Otherwise we have

$$< a, b > = < a > \times < b > .$$

For c not in $< a, b >$ we proceed again by extending a character of $< a, b >$. In the remaining case we have

$$c = a^{\mu} b^{\nu}, \quad (\mu, \nu) \not\equiv (0, 0) \bmod p,$$

the order of c is equal to p and p must be different from 2 because by assumption $c^2 \neq 1$. Now, because of $< a, b > = < a > \times < b >$ we can find characters χ_a, χ_b of $< a, b >$ with

$$\chi_a(a) = \chi_b(b) = \zeta \text{ and } \chi_a(b) = \chi_b(a) = 1,$$

where ζ is a primitive p-th root of unity. We set

$$\chi := \chi_a^m \chi_b^n, \quad m, n \in \mathbb{Z}.$$

Then we have

$$\chi(a) = \zeta^m, \chi(b) = \zeta^n, \chi(c) = \zeta^{m\mu + n\nu},$$

and, noting that $p \neq 2$, we see that the exponents m, n can be chosen such that $\chi(a), \chi(b) \neq 1$. Now any extension of χ to G has the desired property. □

Proof of Theorem 6.6.2 From homogeneity of Δ it follows that for $\mathfrak{c} \in \mathfrak{J}_t$

$$\delta(\mathfrak{c}) := N(\mathfrak{c})^{12}|\Delta(\mathfrak{c})|^2, \quad \mathfrak{c} \in \mathfrak{J}_t,$$

only depends on the ideal class of $\mathfrak{c}$, and for any character $\chi \neq 1$ of $\mathfrak{R}_t$ we know by Theorem 11.1.1

$$\sum_{\mathfrak{c} \bmod \mathfrak{H}_t} \overline{\chi}(\mathfrak{c}) \log(\delta(\mathfrak{c})) \neq 0,$$

where for simplicity we have written $\chi(\mathfrak{c})$ instead of $\chi(\mathfrak{c}\mathfrak{H}_t)$.

Now, assume the first assertion of our theorem to be false. Then

$$\Theta := N_{\Omega_t/L}\left(\left(\frac{\Delta(\mathfrak{a})}{\Delta(\mathfrak{b})}\right)^e\right)$$

is contained in a proper subfield L', $K \subseteq L' \subset L$. For the corresponding subgroup $\mathfrak{U}'$ of $\mathfrak{R}_t$ we then have

$$\mathfrak{U}' \supset \mathfrak{U}, \mathfrak{U}' \neq \mathfrak{U}.$$

Therefore, by Lemma 6.6.3 there exists a character χ of $\mathfrak{R}_t/\mathfrak{U}$ with

$$\chi|\mathfrak{U}' \neq 1 \quad \text{and} \quad \chi(\mathfrak{a}\mathfrak{b}^{-1}) \neq 1.$$

Since Θ lies in L' by assumption, it follows that

$$\sum_{\mathfrak{c}_1 \bmod \mathfrak{U}'} \left(\sum_{\substack{\mathfrak{c}_2 \in \mathfrak{U}' \\ \mathfrak{c}_2 \bmod \mathfrak{U}}} \overline{\chi}(\mathfrak{c}_2) \right) \log|\Theta^{\sigma(\mathfrak{c}_1)}| = 0 \tag{6.11}$$

using the relation

$$\sum_{\substack{\mathfrak{c}_2 \in \mathfrak{U}' \\ \mathfrak{c}_2 \bmod \mathfrak{U}}} \overline{\chi}(\mathfrak{c}_2) = 0.$$

On the other hand by Theorem 6.6.1 we have

$$2 \sum_{\mathfrak{c} \bmod \mathfrak{U}} \overline{\chi}(\mathfrak{c}) \log|\Theta^{\sigma(\mathfrak{c})}| =$$

$$= e \sum_{\mathfrak{c} \bmod \mathfrak{U}} \overline{\chi}(\mathfrak{c}) \log \left| \prod_{\substack{\mathfrak{d}\mathfrak{s} \in \mathfrak{U} \\ \mathfrak{d}\mathfrak{s} \bmod \mathfrak{U}_t}} \left(\frac{N(\mathfrak{b})}{N(\mathfrak{a})}\right)^{12} \frac{\delta(\mathfrak{a}\mathfrak{d}^{-1}\mathfrak{c}^{-1})}{\delta(\mathfrak{b}\mathfrak{d}^{-1}\mathfrak{c}^{-1})} \right|.$$

Using the above character relation and the rule $\delta(\mathfrak{a}) = \delta(\overline{\mathfrak{a}})$, we further obtain

$$2 \sum_{\mathfrak{c} \bmod \mathfrak{U}} \overline{\chi}(\mathfrak{c}) \log |\Theta^{\sigma(\mathfrak{c})}| = 2e(\chi(\mathfrak{a}) - \chi(\mathfrak{b})) \sum_{\mathfrak{c} \bmod \mathfrak{h}_t} \overline{\chi}(\mathfrak{c}) \log(\delta(\mathfrak{c})).$$

Herein the right-hand side is different from zero because $\chi(\mathfrak{a}\mathfrak{b}^{-1}) \neq 1$. This contradiction to (6.11) then proves the first assertion of Theorem 6.6.2.

To prove the second assertion, we set

$$\Theta := N_{\Omega_t/L}\left(\left(\frac{\Delta(\mathfrak{a})\Delta(\mathfrak{b})}{\Delta(\mathfrak{a}\mathfrak{b})\Delta(\mathfrak{O})}\right)^e\right).$$

The assumption of Θ lying in a proper subfield L' of L then leads to the equation

$$\begin{aligned} 0 &= 2 \sum_{\mathfrak{c} \bmod \mathfrak{U}} \overline{\chi}(\mathfrak{c}) \log |\Theta^{\sigma(\mathfrak{c})}| \\ &= 2e(\chi(\mathfrak{a}) - 1)(\chi(\mathfrak{b}) - 1) \sum_{\mathfrak{c} \bmod \mathfrak{h}_t} \overline{\chi}(\mathfrak{c}) \log(\delta(\mathfrak{c})), \end{aligned}$$

which is contradictory if, according to Lemma 6.6.3, the character χ is chosen such that

$$\chi|\mathfrak{U} = 1, \quad \chi|\mathfrak{U}' \neq 1, \quad \chi(\mathfrak{a}) \neq 1, \quad \chi(\mathfrak{b}) \neq 1.$$

□

For many constructions in complex multiplication it is necessary to prove similar results for certain roots of the singular values in Theorem 6.6.1. This is in fact possible since Δ is the 24-th root of a modular form of "weight $\frac{1}{2}$":

$$\Delta\binom{\omega}{1} = (2\pi)^{12}\eta(\omega)^{24}.$$

As in section 2.4.3 we consider the following functions g:

$$\begin{array}{ll}
\left(\frac{\eta(\frac{\omega}{n})}{\eta(\omega)}\right)^8 \gamma_2(\omega)^{n-1} & \text{for } 3 \nmid n, \\
\left(\frac{\eta(\frac{\omega}{n})}{\eta(\omega)}\right)^6 \gamma_3(\omega)^{\frac{n-1}{2}} & \text{for } 2 \nmid n, \\
\left(\frac{\eta(\frac{\omega}{n})}{\eta(\omega)}\right)^m, \ m = \gcd(3, n) & \text{for } n = t^2, \ t \in \mathbb{N} \text{ and } 2 \nmid n, \\
\frac{\eta(\frac{\omega}{p})\eta(\frac{\omega}{q})}{\eta(\frac{\omega}{pq})\eta(\omega)} (\gamma_2(\omega)\gamma_3(\omega))^{\frac{p-1}{2}\frac{q-1}{2}} & \text{for } n = pq, \ p, q \in \mathbb{N}, \\
 & \gcd(6, n) = 1.
\end{array} \tag{6.12}$$

From section 2.4.3 we know that these functions are in $\mathbb{Q}_{\Gamma_{\left(\begin{smallmatrix} 1 & 0 \\ 0 & n \end{smallmatrix}\right)}}$. The next theorem tells us that their singular values lie in ring class fields and in combination with Theorems 6.3.1 and 6.3.2 we will obtain information about roots of the numbers in Theorem 6.6.1.

Theorem 6.6.4 *Let g be one of the functions in (6.12). Let $\alpha \in \mathbb{H}$ be the ratio of a basis of a proper ideal of $\mathfrak{O}_t$, and $\frac{\alpha}{n}$ the ratio of a basis of a proper ideal of $\mathfrak{O}_{t'}$ with a divisor t' of t. Then*

$$g(\alpha) \in \Omega_t.$$

Further, let $\alpha = \alpha_1, \ldots, \alpha_g$ be an n-system modulo t, then the numbers $g(\alpha_i)$ are the images of $g(\alpha)$ under the distinct automorphisms of Ω_t/K.

The proof of Theorem 6.6.4 follows immediately from Theorem 5.2.2 and the theorems about singular values of γ_2 and γ_3. In particular, under the hypothesis of Theorem 6.6.4 it follows that

$$\prod_{i=1}^{h_t} (X - g(\alpha_i)) \in K[X],$$

where h_t denotes the ring class number modulo t.

In particular, by Theorem 5.2.4 we obtain the following results for the functions

$$g_{p,q}(\omega) := \frac{\eta(\frac{\omega}{p})\eta(\frac{\omega}{q})}{\eta(\frac{\omega}{pq})\eta(\omega)},$$

defined already in (2.4).

Theorem 6.6.5 *Let $p, q \in \mathbb{N}$ be prime to 6 and let $\alpha = \alpha_1, \ldots, \alpha_{h_t}$ as in Theorem 6.6.4 have the property that $\frac{\alpha}{pq}$ is the ratio of a basis of a proper ideal of $\mathfrak{O}_{t'}$ with $t' \mid t$. Further, we assume that*

$$(\gamma_2(\alpha)\gamma_3(\alpha))^{\frac{p-1}{2}\frac{q-1}{2}} \in \Omega_t.$$

Then

$$m_{p,q}(X) := \prod_{i=1}^{h_t} (X - g_{p,q}(\alpha_i)) \in \mathbb{Q}[X]$$

if we assume (a little stronger than in Theorem 6.6.4) that the α_i are a $12pq$-system modulo t. Of course, for $g_{p,q}(\alpha)$ integral, it follows that the coefficients of $m_{p,q}(X)$ are in $\mathbb{Z}$.

Proof By Theorems 6.3.1, 6.3.2 and 6.6.4 we know that for two elements β, β' of a $12pq$-system the singular values $g_{p,q}(\beta)$ and $g_{p,q}(\beta')$ are conjugate over K. We choose an element $\beta \in \mathbb{H} \cap K$ satisfying a primitive equation

$AX^2 + BX + pq$ with $\gcd(6pq, A) = 1, A > 0$ and $B \equiv B_1 \bmod 12pq$.

It follows easily from the hypothesis on $\frac{\alpha}{pq}$ that such a choice is possible. From the theorems quoted above, we can then conclude that $g_{p,q}(\beta)$ is conjugate over K to any of the numbers $g_{p,q}(\alpha_i)$, hence is equal to one of them. The same holds for $g_{p,q}(A\beta)$, and we show that

$$g_{p,q}(\beta)^\tau = g_{p,q}(A\beta) \tag{6.13}$$

with τ denoting complex conjugation. By (6.13) it follows that the conjugates of $g_{p,q}(\alpha)$ are permuted under τ. Hence the polynomial $m_{p,q}(X)$ must have coefficients in $K \cap \mathbb{R} = \mathbb{Q}$. To prove (6.13), we use the identity

$$g_{p,q}\left(-\frac{pq}{\omega}\right) = g_{p,q}(\omega)$$

and we find that

$$g_{p,q}(\beta)^\tau = g_{p,q}(-\beta^\tau) = g_{p,q}\left(-\frac{\beta\beta^\tau}{\beta}\right) = g_{p,q}\left(-\frac{pq}{A\beta}\right) = g_{p,q}(A\beta).$$

This completes the proof. □

By Theorem 6.6.4 we can further prove:

Theorem 6.6.6 *We assume that $\mathfrak{R}_t \not\cong \mathbb{Z}/2\mathbb{Z}$ resp. $\not\cong \mathbb{Z}/2\mathbb{Z} \times \mathbb{Z}/2\mathbb{Z}$ depending on whether one resp. two of the roots $\sqrt{-1}, \sqrt{-3}$ are in Ω_t. Then there exists a primitive ideal $\mathfrak{c}$ of $\mathfrak{O}_1$ prime to t so that the 24-th root defined in the proof below,*

$$\sqrt[24]{\frac{\Delta(\mathfrak{c}_t)\Delta(\mathfrak{c}_t)}{\Delta(\mathfrak{c}_t^2)\Delta(\mathfrak{O}_t)}}$$

is a generator of Ω_t over K.

Proof The case $\Omega_t = K$ being trivial, we may assume that $[\mathfrak{A}^{\mathfrak{t}} : \mathfrak{U}_{\mathfrak{t}}] = |\mathfrak{R}_t| \neq 1$. Let $\mathfrak{c}$ be an integral ideal of $\mathfrak{O}_1$ of norm c prime to $6t$, satisfying the congruence

$$\frac{c-1}{2}\frac{c-1}{2} \equiv 0 \bmod 6.$$

In $\mathfrak{O}_t$ we choose a basis $\alpha, 1$ with $\alpha \in \mathbb{H}$, such that α, c and α, c^2 are bases of $\mathfrak{c}_t$ and $\mathfrak{c}_t^2$. Then by Theorem 5.2.2

$$\frac{\eta\left(\frac{\alpha}{c}\right)\eta\left(\frac{\alpha}{c}\right)}{\eta\left(\frac{\alpha}{c^2}\right)\eta\left(\alpha\right)}\left(\gamma_2(\alpha)\gamma_3(\alpha)\right)^{\frac{c-1}{2}\frac{c-1}{2}} \in \Omega_t,$$

and since the exponent $\frac{c-1}{2}\frac{c-1}{2}$ is divisible by 6 and $\gamma_2(\alpha)\gamma_3(\alpha)$ being non-zero because of $\Omega_t \neq K$, it follows that

$$\frac{\eta\left(\frac{\alpha}{c}\right)\eta\left(\frac{\alpha}{c}\right)}{\eta\left(\frac{\alpha}{c^2}\right)\eta\left(\alpha\right)} \in \Omega_t.$$

Now

$$\frac{\Delta(\mathfrak{c}_t)\Delta(\mathfrak{c}_t)}{\Delta(\mathfrak{c}_t^2)\Delta(\mathfrak{O}_t)} = \left(\frac{\eta\left(\frac{\alpha}{c}\right)\eta\left(\frac{\alpha}{c}\right)}{\eta\left(\frac{\alpha}{c^2}\right)\eta\left(\alpha\right)}\right)^{24},$$

and, according to Theorem 6.6.2, the left-hand side is a generator for Ω_t/K if $\mathfrak{c} \notin \mathfrak{U}_t$. Note that for $\mathfrak{c}_t^2 \sim \lambda \in \mathfrak{O}_t$ we have

$$\frac{\Delta(\mathfrak{c}_t)\Delta(\mathfrak{c}_t)}{\Delta(\mathfrak{c}_t^2)\Delta(\mathfrak{O}_t)} = \lambda^{24}\left(\frac{\Delta(\mathfrak{c}_t)}{\Delta(\mathfrak{O}_t)}\right)^2$$

so that in this case we can apply the first assertion of Theorem 6.6.2. Now we are going to prove the existence of an ideal $\mathfrak{c}$ with the above properties that is not in $\mathfrak{U}_t$. We distinguish the following cases:

(i) "$\sqrt{-1}, \sqrt{-3} \notin \Omega_t$",
(ii) "$\sqrt{-1} \in \Omega_t, \sqrt{-3} \notin \Omega_t$",
(iii) "$\sqrt{-1} \notin \Omega_t, \sqrt{-3} \in \Omega_t$",
(iv) "$\sqrt{-1}, \sqrt{-3} \in \Omega_t$".

(i) We chose $\mathfrak{k} \in \mathfrak{A}^{\mathfrak{t}}/\mathfrak{U}_{\mathfrak{t}}$ different from $\mathfrak{U}_{\mathfrak{t}}$. Then, according to Theorem 3.3.1 we can find an ideal $\mathfrak{c}$ of norm c in $\mathfrak{k}$ such that

$$\left(\tfrac{-1}{c}\right) = \left(\tfrac{-3}{c}\right) = 1.$$

Further, we may assume $\mathfrak{c}$ to be primitive, because an integer divisor of $\mathfrak{c}$ lies in $\mathfrak{U}_t$ and its norm is a square. Then, for c we have the congruence

$$c \equiv 1 \bmod 4 \text{ and } c \equiv 1 \bmod 3.$$

Hence, the above congruence modulo 6 is satisfied.
(ii) In this case we have $\mathfrak{R}_t \not\cong \mathbb{Z}/2\mathbb{Z}$ by assumption, so by Lemma 6.6.3

there exists $\mathfrak{k} \in \mathfrak{A}^{\mathfrak{t}}/\mathfrak{U}_{\mathfrak{t}}$, $\mathfrak{k} \neq \mathfrak{U}_{\mathfrak{t}}$ with $\chi_{-1}(\mathfrak{k}) = 1$. Now, let $\mathfrak{U}'$ be the subgroup of $\mathfrak{U}^{3\mathfrak{t}}$ associated with $\Omega_t(\sqrt{-3})$. Then by Lemma 6.6.3 there exists a class $\mathfrak{k}'_+$ modulo $\mathfrak{U}'$ in $\mathfrak{k}$ with $\chi_{-3}(\mathfrak{k}'_+) = 1$. In $\mathfrak{k}'_+$ we choose an integral primitive ideal $\mathfrak{c}$ prime to 6. Its norm c must satisfy the congruence $\frac{c-1}{2}\frac{c-1}{2} \equiv 0 \bmod 6$, and then we proceed as in case (i).
(iii) is settled analogously to (ii).
(iv) Here, by assumption $\mathfrak{R}_t \not\cong \mathbb{Z}/2\mathbb{Z}, \mathbb{Z}/2\mathbb{Z} \times \mathbb{Z}/2\mathbb{Z}$. Therefore, by Lemma 6.6.3 there exists $\mathfrak{k} \neq \mathfrak{U}_{\mathfrak{t}}$ in $\mathfrak{A}^{\mathfrak{t}}/\mathfrak{U}_{\mathfrak{t}}$ with $\chi_{-1}(\mathfrak{k}) = \chi_{-3}(\mathfrak{k}) = 1$. Any primitive ideal $\mathfrak{c} \in \mathfrak{k}$ prime to 6 then has the desired property. □

Remark 6.6.7 In the case $\gcd(6, t^2d) = 1$, one may pick any primitive ideal $\mathfrak{c} \notin \mathfrak{U}_{\mathfrak{t}}$ prime to $6t$ for the construction of the 24-th root in Theorem 6.6.6 because then $\gamma_2(\alpha)$ and $\gamma_3(\alpha)$ lie in Ω_t if the trace of α is divisible by 3.

Examples 6.6.8

(i) Let $d_K = -47$. Then

$$\frac{\eta\left(\frac{\alpha}{7}\right)^2}{\eta\left(\frac{\alpha}{49}\right)\eta\left(\alpha\right)}, \qquad \alpha = \frac{549 + \sqrt{-47}}{2}$$

is a generator of Ω/K. Its minimal equation over K,

$$X^5 + 2\,X^4 + 2\,X^3 + X^2 - 1,$$

has quite small coefficients, compared to Example 6.1.3.

(ii) Let $d_K = -4447$. Then

$$\frac{\eta\left(\frac{\alpha}{13}\right)^2}{\eta\left(\frac{\alpha}{169}\right)\eta\left(\alpha\right)}, \quad \alpha = \frac{2397 + \sqrt{-4447}}{2}$$

is a generator of Ω/K. Its minimal equation over K is

$$\begin{aligned} X^{17} &+ 4356\,X^{16} + 4469408\,X^{15} - 577415049\,X^{14} \\ &+ 63837613820\,X^{13} + 422739382392\,X^{12} \\ &+ 73026311344680\,X^{11} + 657942992235294\,X^{10} \\ &+ 2422818074079732\,X^{9} + 3631848712112989\,X^{8} \\ &+ 2571388650262763\,X^{7} + 887568818022539\,X^{6} \\ &+ 608523191355758\,X^{5} + 51489792551504\,X^{4} \\ &+ 2122601876446\,X^{3} + 66143066\,X^{2} + 2765955\,X + 1 \end{aligned}$$

The corresponding equation for $j(\alpha)$ would fill too much space, so we only note its constant term:

$$\begin{aligned}&2027256793176619858847293796948033762652840532920 6\\&0255831619247753493290896027594804483259330540838 3\\&8915802206964211370576146346255990861330008601481 6\\&5426140268063854623280563133058625470027242430694 7\\&3008614937134623135608402218369231603217258915554 6\\&6470127233602482874630368314683437347412109375\end{aligned}$$

Using the same arguments as in the proof of Theorem 6.6.6, we obtain the following result that will be needed later for a suitable normalisation of the $\wp$ function:

Theorem 6.6.9 *For $d < -4$ and $t \in \mathbb{N}$ there exist primitive ideals $\mathfrak{p}, \mathfrak{q}$ of $\mathfrak{O}_1$ of norm p and q prime to $6t$ satisfying*

$$\begin{array}{lll} p \equiv q \equiv -1 \bmod 4 & \text{if} & 2|td \text{ and } 3 \nmid td, \\ p \equiv q \equiv -1 \bmod 3 & \text{if} & 2 \nmid td \text{ and } 3|td, \\ p \equiv q \equiv -1 \bmod 12 & \text{if} & 2|td \text{ and } 3|td. \end{array}$$

Given two such ideals $\mathfrak{p}, \mathfrak{q}$ with $\mathfrak{pq}$ being primitive too and any ideal $\mathfrak{a} \in \mathfrak{J}_t$, we can choose the ratio $\alpha \in \mathbb{H}$ of a basis of $\mathfrak{a}$ so that

$$\tfrac{\alpha}{p}, \tfrac{\alpha}{q} \text{ resp. } \tfrac{\alpha}{pq} \text{ are ratios of basis of } \mathfrak{ap}_t, \mathfrak{aq}_t \text{ resp. } \mathfrak{ap}_t\mathfrak{q}_t.$$

Further, we can achieve the coefficients A, B of the primitive equation $X^2 + BX + C = 0$ of α to satisfy

$$\gcd(A, 6) = 1 \quad \text{and} \quad \begin{cases} B \equiv 0 \bmod 3 & \text{for} \quad 3 \nmid td, \\ B \equiv 1 \bmod 4 & \text{for} \quad 2 \nmid td. \end{cases}$$

Then we have

$$\frac{\eta(\frac{\alpha}{p})\eta(\frac{\alpha}{q})}{\eta(\frac{\alpha}{pq})\eta(\alpha)}\gamma_2(\alpha)\gamma_3(\alpha) \in \Omega_t.$$

Proof The existence of primitive ideals $\mathfrak{p}, \mathfrak{q}$ with norms satisfying the above congruences follows from Theorem 3.3.1, bearing in mind that for $d \neq -3, -4$ the roots $\sqrt{-3}, \sqrt{-4}$ are not in K. Now by Theorem 6.6.4 it follows that

$$\frac{\eta(\frac{\alpha}{p})\eta(\frac{\alpha}{q})}{\eta(\frac{\alpha}{pq})\eta(\alpha)}(\gamma_2(\alpha)\gamma_3(\alpha))^{\frac{p-1}{2}\frac{q-1}{2}} \in \Omega_t.$$

Since $t^2d \neq -3, -4$, we know that $\gamma_2(\alpha)\gamma_3(\alpha) \neq 0$. Further, in view of the congruences satisfied by A, B, p, q, Theorems 6.3.1 and 6.3.2 tell us that

$$\frac{(\gamma_2(\alpha)\gamma_3(\alpha))^{\frac{p-1}{2}\frac{q-1}{2}}}{\gamma_2(\alpha)\gamma_3(\alpha)} \in \Omega_t,$$

thereby proving the assertion of our theorem. □

A consequence of Theorem 6.6.9 is the following result first obtained by Fueter in a different way.

Theorem 6.6.10 *We assume that $d < -4$ and $t \in \mathbb{N}$ or $d = -3, -4$ and $t = 1$. Let $\alpha \in \mathbb{H}$ be the ratio of a basis of $\mathfrak{a} \in \mathfrak{J}_t$. Then there exist algebraic units $\epsilon_2(\alpha), \epsilon_3(\alpha)$ such that*

$$\frac{\gamma_2(\alpha)}{\epsilon_2(\alpha)}, \frac{\gamma_3(\alpha)}{\epsilon_3(\alpha)} \in \Omega_t.$$

Proof For $d \neq -3, -4$ the assertion is immediate by Theorem 6.6.9, because the η-product there is a unit according to Theorem 4.2.1. For $d = -3$ we have $\gamma_2(\alpha) = 0$ and $\gamma_3(\alpha) \in \Omega_1$ by Theorem 6.3.2, hence the assertion is trivial. For $d = -4$ we have $\gamma_3(\alpha) = 0$ and $\zeta\gamma_2(\alpha) \in \Omega_1$ with a third root of unity ζ by Theorem 6.3.1. Since in this case $\zeta \in K = \Omega_1$ Theorem 6.6.10 again implies our assertion. □

For some later applications we also need the following result obtained, like Theorem 6.6.9, by using the fact that

$$\left(\frac{\eta(\frac{\omega}{n})}{\eta(\omega)}\right)^6 \gamma_3(\omega)^{\frac{n-1}{2}}, \text{ for } 2 \nmid n,$$

known from (6.12):

Theorem 6.6.11 *For $d \neq -4$ and $t \in \mathbb{N}$ there exists a primitive ideal $\mathfrak{p}$ of $\mathfrak{O}_1$ of norm p prime to $2t$ satisfying*

$$p \equiv -1 \ mod\ 4.$$

Given such an ideal $\mathfrak{p}$ and any ideal $\mathfrak{a} \in \mathfrak{J}_t$, we can choose the ratio $\alpha \in \mathbb{H}$ of a basis of $\mathfrak{a}$ so that

$$\tfrac{\alpha}{p} \text{ is the ratio of a basis of } \mathfrak{a}\mathfrak{p}_t.$$

Then we have

$$\left(\frac{\eta(\frac{\alpha}{p})}{\eta(\alpha)}\right)^6 \gamma_3(\alpha) \in \Omega_t.$$

A further application of Theorem 6.6.4 is:

Theorem 6.6.12 (principal ideal theorem) *Let* $\mathfrak{a}$ *be any ideal of* $\mathfrak{O}_1$. *Then in the Hilbert class field* Ω *the ideal* $\mathfrak{O}_\Omega \mathfrak{a}$ *is principal.*

Proof First let $\mathfrak{a} = \mathfrak{p}$ be a primitive ideal of norm $p \nmid 6$ with $\mathfrak{p}^2$ being primitive, too. Then there exists a basis $\alpha, 1$ of $\mathfrak{O}_1$ with $\alpha \in \mathbb{H}$, such that α, p resp. α, p^2 are basis of $\overline{\mathfrak{p}}$ resp. $\overline{\mathfrak{p}}^2$. Therefore, according to Theorem 6.6.4,

$$\theta := \frac{\eta\left(\frac{\alpha}{p^2}\right)}{\eta(\alpha)}$$

is in Ω, and is associated with $\mathfrak{p}$ by Theorem 4.2.2. This proves the assertion of the principal ideal theorem for primitive ideals $\mathfrak{p} \neq \overline{\mathfrak{p}}$ of norm $p \nmid 6$. To prove the assertion for an arbitrary ideal, observe that in every ideal class there exists a prime ideal $\mathfrak{p} \neq \overline{\mathfrak{p}}$ of norm $p \nmid 6$. Our assertion then becomes immediate. □

6.7 Double η-quotients in the ramified case

The content of this section forms part of work done jointly with Andreas Enge. Let K be a quadratic imaginary number field of discriminant d. Then

$$D = t^2 d$$

is the discriminant of $\mathfrak{O}_t$. Further, let p, q be two primes not dividing $2t$ that are ramified in K. Let $\mathfrak{p}, \mathfrak{q}$ be the prime ideals in K above p resp. q and $\mathfrak{p}_t, \mathfrak{q}_t$ the corresponding proper ideals of $\mathfrak{O}_t$.

We consider again the function

$$g_{p,q}(\omega) = \frac{\eta\left(\frac{\omega}{p}\right)\eta\left(\frac{\omega}{q}\right)}{\eta\left(\frac{\omega}{pq}\right)\eta\left(\omega\right)}.$$

From Theorem 6.6.4 we know that the singular values of $g_{p,q}$ at ratios of basis of proper ideals of $\mathfrak{O}_t$ sometimes lie in Ω_t. In what follows we will show that in many cases these values even lie in a proper subfield of Ω_t. As we will see later in section 10.2.3, this fact is very useful for the determination of cryptographically relevant elliptic curves over finite fields.

Theorem 6.7.1 *We assume the above two primes* p, q *to satisfy one of the conditions (a)–(c) and further, to satisfy (d):*

(a) $p = 3,\ q \equiv 1 \bmod 3$,
(b) $p, q \neq 3$ *and* $3 \nmid D$,

(c) $(p-1)(q-1) \equiv 0 \bmod 3$ *and* $3|D$,
(d) $(p-1)(q-1) \equiv 0 \bmod 8$ *if* $2|D$.

We set $\alpha := \frac{3D+\sqrt{D}}{2}$. *Then*

(i) $g_{p,q}(\alpha) \in \Omega_t$,

(ii) $g_{p,q}(\alpha)^{\sigma(\mathfrak{p})} = \left(\frac{p}{q}\right)\frac{1}{g_{p,q}(\alpha)}$, $\quad g_{p,q}(\alpha)^{\sigma(\mathfrak{q})} = \left(\frac{q}{p}\right)\frac{1}{g_{p,q}(\alpha)}$,

(iii) $g_{p,q}(\alpha)^{\sigma(\mathfrak{pq})} = (-1)^{\frac{p-1}{2}\frac{q-1}{2}} g_{p,q}(\alpha)$.

According to Theorem 3.1.10 we can choose a $6pq$-system $\alpha = \alpha_1, \ldots, \alpha_h \bmod t$. Then by Theorem 6.6.5

$$m(X) := \prod_{i=1}^{h} (X - g_{p,q}(\alpha_i))$$

has coefficients in $\mathbb{Z}$:

$$m(X) \in \mathbb{Z}[X].$$

Further, by the above theorem $m(X)$ has the following properties: for $(\mathfrak{pq})_t$ not principal $\sigma(\mathfrak{pq})$ is of order 2, hence Ω_t is of degree 2 over the fixed field L of $\sigma(\mathfrak{pq})$. For the following, we assume that

$$\sigma([\alpha_i, 1]), i = 1, \ldots, \tfrac{h}{2},$$

to induce the different automorphisms of L/K.

<u>Case 1:</u> $\frac{p-1}{2}\frac{q-1}{2} \equiv 0 \bmod 2$:
Then

$$m(X) = \tilde{m}(X)^2 \text{ with } \tilde{m}(X) := \prod_{i=1}^{h/2} (X - g_{p,q}(\alpha_i)) \in \mathbb{Z}[X],$$

and $\tilde{m}(X)$ is a product of factors

$$X^2 + uX + \left(\tfrac{p}{q}\right) \text{ with integral numbers } u \in L.$$

<u>Case 2:</u> $\frac{p-1}{2}\frac{q-1}{2} \equiv 1 \bmod 2$:
Then

$$m(X) = \tilde{m}(X^2) \text{ with } \tilde{m}(Y) := \prod_{i=1}^{h/2} (Y - g_{p,q}(\alpha_i)^2) \in \mathbb{Z}[Y],$$

and $\tilde{m}(Y)$ is a product of factors

$$Y^2 + uY + 1 \text{ with integral numbers } u \in L.$$

The determination of the "half-system" $\alpha_1, \ldots, \alpha_{h/2}$ is easy for Ω_t/L being generated by the square root of a rational number. According to Theorem 6.1.4 such numbers are given by the divisors

$$s|D \text{ with } s \equiv 1 \bmod 4$$

satisfying

$$\sqrt{s}^{\sigma(\mathfrak{pq})} = -\sqrt{s}. \tag{6.14}$$

Clearly, in such a case we can find s with $|s|$ being a prime or equal to 1. Then the "half-system" is obtained by taking all α_i having a primitive equation

$$A_i X^2 + B_i X + C_i = 0$$

with

$$\left(\frac{s}{A_i}\right) = 1 \text{ if } s \nmid A_i,$$
$$\left(\frac{s}{C_i}\right) = 1 \text{ if } s \nmid C_i.$$

Note that for such a divisor s of D we always have $s \nmid A_i$ or $s \nmid C_i$.

Verification of (6.14) is done by the formula

$$\sqrt{s}^{\sigma(\mathfrak{p})} = \left(\frac{s}{p}\right)\sqrt{s}, \text{ for } \gcd(s,p) = 1$$

and the analogous formula for q instead of p, following from the Frobenius congruence. For $s = \pm p$, $\pm q$ we can prove the relation

$$\sqrt{p^*}^{\sigma(\mathfrak{p})} = \left(\frac{p^*}{qm}\right)\sqrt{p^*}, \quad p^* := (-1)^{\frac{p-1}{2}} p,$$

and analogously for $\sqrt{q^*}^{\sigma(\mathfrak{q})}$, where m is the natural number defined by $D = -pqm$.

Proof of Theorem 6.7.1 To prove the first assertion, we observe that by 2.4.3

$$\frac{\eta(\frac{\omega}{p})\eta(\frac{\omega}{q})}{\eta(\frac{\omega}{pq})\eta(\omega)}(\gamma_2(\omega)\gamma_3(\omega))^{\frac{p-1}{2}\frac{q-1}{2}} \in \mathbb{Q}_{\Gamma^0(pq)}.$$

Further, by the results on the singular values of γ_2 and γ_3, we know

$$(\gamma_2(\alpha)\gamma_3(\alpha))^{\frac{p-1}{2}\frac{q-1}{2}}$$

to be in Ω_t. For $(p-1)(q-1) \equiv 0 \bmod 24$ we even have

$$\frac{\eta(\frac{\omega}{p})\eta(\frac{\omega}{q})}{\eta(\frac{\omega}{pq})\eta(\omega)} \in \mathbb{Q}_{\Gamma^0(pq)}.$$

Therefore, in this case assumption (i) is immediate by Theorem 6.6.4, and the above consequence holds for any pq-system modulo t. Further, in this case the factor 3 in the definition of α can be omitted.

To prove the remaining assertions, we consider the function

$$g_n(\omega) = \frac{\eta(\frac{\omega}{n})}{\eta(\omega)},$$

defined already in (2.3), where n is an **odd** natural number. The transformation formula of the η function then yields

$$g_n \circ M = \left(\frac{a}{n}\right) \zeta_{24}^{(1-n)\{\frac{b}{n}a - c(d(1-a^2)-a)-3(a-1)c_1\}} g_n$$
$$\text{for } M = \left(\begin{smallmatrix} a & b \\ c & d \end{smallmatrix}\right) \in \Gamma^0(n) \text{ with } c \geq 0 \text{ and } d > 0 \text{ if } c = 0,$$

where c_1 denotes the odd part of $c = 2^\lambda c_1$ for $c \neq 0$ and is set to be $c_1 = 1$ for $c = 0$. The transformation formula now shows that

$$g_n \in F_N \;\text{ with }\; N = \begin{cases} 24n^2 & if \quad n \equiv 1 \bmod 2, \\ 8n^2 & if \quad n \equiv 1 \bmod 6. \end{cases} \tag{6.15}$$

Using the Reciprocity Law, we obtain

$$g_{p,q}(\alpha) = \left(\frac{p}{q}\right) g_p(\alpha)^{1-\sigma(\mathfrak{q})} = \left(\frac{q}{p}\right) g_q(\alpha)^{1-\sigma(\mathfrak{p})}$$

and further, in view of $\sigma(\mathfrak{p})^2 = \sigma(p)$,

$$g_{p,q}(\alpha)^{\sigma(\mathfrak{p})} = \left(\frac{p}{q}\right) g_q(\alpha)^{(1-\sigma(\mathfrak{p}))\sigma(\mathfrak{p})} = \frac{1}{g_{p,q}(\alpha)} g_q(\alpha)^{1-\sigma(p)}.$$

Again by the Reciprocity Law we have

$$g_q(\alpha)^{\sigma(p)} = \left[g_q \circ cC^{-1}\right](C(\alpha)), \quad c = \det(C), \text{ with } C = \left(\begin{smallmatrix} p & 0 \\ 0 & p \end{smallmatrix}\right).$$

To determine $g_q \circ cC^{-1}$ we decompose

$$C \equiv \left(\begin{smallmatrix} 1 & 0 \\ 0 & p^2 \end{smallmatrix}\right) M \bmod N$$

with a unimodular matrix

$$M \equiv \left(\begin{smallmatrix} p & 0 \\ 0 & p' \end{smallmatrix}\right) \bmod N,$$

where N is defined in (6.15). The η transformation formula then yields

$$g_q \circ cC^{-1} = g_q \circ \left(\begin{smallmatrix} 1 & 0 \\ 0 & p^2 \end{smallmatrix}\right) M = g_q \circ M = \left(\frac{p}{q}\right) g_q.$$

This proves the first assertion in (ii). The second is obtained analogously.

Finally (iii) follows from (ii). □

Examples 6.7.2 Suitable p, q-pairs are, for instance,
$(3, q)$, $q = 13, 37, 61$,
$(5, q)$, $q = 7, 11, 13, 17, 19$,
$(7, q)$, $q = 13, 17, 29, 37$,
$(11, q)$, $q = 13, 29, 37$,

$(13, q)$, $q = 17, 19, 23$,
on condition that $3 \nmid D$ for $(p-1)(q-1) \not\equiv 0 \bmod 3$ and $2 \nmid D$ for $(p-1)(q-1) \not\equiv 0 \bmod 8$.

Remark 6.7.3 If the discriminant d is divisible by a further prime $r \neq 2, p, q$, coprime to $2t$, the construction of the theorem can of course be continued by

$$g_{p,q,r}(\alpha) := g_{p,q}(\alpha)^{(1-\sigma(\mathfrak{r}))} = g_{p,r}(\alpha)^{(1-\sigma(\mathfrak{q}))} = g_{p,q}(\alpha) g_{p,q}(\tfrac{\alpha}{r})^{-1}.$$

The number obtained in this way is fixed under $\sigma(\mathfrak{pq})$ and $\sigma(\mathfrak{pr})$ if (a)–(c) are satisfied for two of the primes p, q, r. A sufficient condition is, for example, the congruence

$$(p-1)(q-1)(r-1) \equiv 0 \bmod 3.$$

Furthermore, if the ring ideals $(\mathfrak{pq})_t$, $(\mathfrak{pr})_t$ and $(\mathfrak{pqr})_t$ are not principal, this implies that $g_{p,q,r}(\alpha)$ lies in a subfield L of Ω_t with $[\Omega_t : L] = 4$. The generalisation of this construction using more primes now becomes obvious.

6.8 Generation of ray class fields by $\varphi(z|{\omega_1 \atop \omega_2})$

The results in this section are mainly based on Bettner and Schertz (2001) and Schertz (2005). Given a lattice $\mathfrak{L} = \mathbb{Z}\omega_1 + \mathbb{Z}\omega_2$, the relation

$$\varphi(z|{\omega_1 \atop \omega_2})^{12} = \sigma^*(z|\mathfrak{L})^{12} \Delta(\mathfrak{L})$$

shows that the 12-th power $\varphi(z|{\omega_1 \atop \omega_2})$ is independent of the choice of basis. We therefore write

$$\varphi(z|\mathfrak{L})^e := \varphi(z|{\omega_1 \atop \omega_2})^e \text{ if } e \in 12\mathbb{N}.$$

For $\mathfrak{L} = \mathfrak{a}$ a proper ideal of the order of conductor t in a quadratic imaginary number field K and $\xi \in K \backslash \mathfrak{a}$ the number $\varphi(\xi|\mathfrak{a})^{12}$ is algebraic according to Theorem 4.1.2. In view of homogeneity we may assume that $\xi \in \mathfrak{O}_t$. For these values we have:

Proposition 6.8.1 *Let $\mathfrak{a} \in \mathfrak{J}_t$, $\mathfrak{f} := \mathfrak{a} \cap \mathfrak{O}_t$ and $f := \min(\mathfrak{f} \cap \mathbb{N})$. We set $e_f := lcm(12, f)$. Then*

$$\varphi(\xi|\mathfrak{a})^{e_f} \in K_{t,\mathfrak{f}} \quad \text{if} \quad \xi \in \mathfrak{O}_t,$$

and for $\xi, \xi' \in \mathfrak{O}_t$ we have

$$\varphi(\xi'|\mathfrak{a})^{e_f} = \varphi(\xi|\mathfrak{a})^{e_f} \quad \text{if} \quad \xi' \equiv \epsilon\xi \bmod \mathfrak{f},$$

with a unit $\epsilon \in \mathfrak{O}_t$.

Let, further, $\mathfrak{c}$ be an integral ideal in $\mathfrak{k} \in \mathfrak{R}_{t,\mathfrak{f}}$. Then the action of the Frobenius automorphism $\sigma(\mathfrak{k})$ is given by

$$\varphi(\xi|\mathfrak{a})^{e_f \sigma(\mathfrak{k})} = \varphi(\xi|\mathfrak{a}\mathfrak{c}^{-1})^{e_f}.$$

Proof First we prove the second assertion. Let $\xi \in \mathfrak{O}_t$ and

$$\xi' = \epsilon\xi + \alpha \text{ with } \alpha \in \mathfrak{f}.$$

Then, using the transformation formula from Theorem 1.2.4 for σ^*, we find that

$$\varphi(\xi'|\mathfrak{a})^{e_f} = \psi(\alpha)^{e_f} e^{l_{\mathfrak{a}}(\epsilon\xi,\alpha)e_f} \varphi(\epsilon\xi|\mathfrak{a})^{e_f}.$$

Herein we have

$$l_{\mathfrak{a}}(\epsilon\xi,\alpha)e_f = l_{\mathfrak{a}}(e_f\epsilon\xi,\alpha) \in 2\pi i\mathbb{Z}$$

because $e_f\epsilon\xi$ and α are in $\mathfrak{a}$. Further, e_f is divisible by 2, hence $\psi(\alpha)^{e_f} = 1$. It follows that

$$\varphi(\xi'|\mathfrak{a})^{e_f} = \varphi(\epsilon\xi|\mathfrak{a})^{e_f} = \varphi(\xi|\epsilon^{-1}\mathfrak{a})^{e_f} = \varphi(\xi|\mathfrak{a})^{e_f}.$$

This proves the second assertion.

For $\xi \in \mathfrak{a}$ we know $\varphi(\xi|\mathfrak{a})^{e_f} = 0$, so the first and third assertions are obvious. For $\xi \in \mathfrak{O}_t \setminus \mathfrak{a}$ Theorem 5.2.5 tells us that

$$\varphi(\xi|\mathfrak{a})^{e_f} \in K_{12tf^2}$$

because $\xi \in \frac{1}{f}\mathfrak{a}$. Now let $\mathfrak{b}$ be an integral ideal of $\mathfrak{O}_1$, coprime to $12tf$ and $\sigma(\mathfrak{b})$ the corresponding Frobenius automorphism of K_{12tf^2}/K. Then, again by Theorem 5.2.5

$$\varphi(\xi|\mathfrak{a})^{e_f\sigma(\mathfrak{b})} = \varphi(\xi|\mathfrak{a}\mathfrak{b}_t^{-1})^{e_f} \text{ with } \mathfrak{b}_t = \mathfrak{b} \cap \mathfrak{O}_t.$$

The right side of this equation can be rewritten as

$$\varphi(\xi|\mathfrak{a}\mathfrak{b}_t^{-1})^{e_f} = \varphi(\xi|\mathfrak{a}\mathfrak{c}^{-1})^{e_f},$$

where $\mathfrak{c}$ denotes an arbitrary integral ideal from $\mathfrak{b}_t\mathfrak{S}_{t,\mathfrak{f}}$. This implies the first and last assertions of our proposition because in every class of $\mathfrak{R}_{t,\mathfrak{f}}$, there exists an integral regular ideal. Therefore, let $\mathfrak{c}$ be an integral ideal in $\mathfrak{b}_t\mathfrak{S}_{t,\mathfrak{f}}$. Then there are integers $\lambda, \lambda' \in \mathfrak{O}_t$, coprime to $\mathfrak{f}$, such that

$$\lambda\mathfrak{c} = \lambda'\mathfrak{b}_t \text{ and } \lambda \equiv \epsilon\lambda' \bmod \mathfrak{f}$$

with a unit $\epsilon \in \mathfrak{O}_t$. Multiplying this equation by a suitable power of λ' we can further achieve that

$$\lambda \equiv \epsilon \bmod \mathfrak{f} \text{ and } \lambda' \equiv \epsilon' \bmod \mathfrak{f}$$

with two units ϵ, ϵ' from $\mathfrak{O}_t$. Then

$$\xi\lambda \equiv \xi\epsilon \bmod \mathfrak{f} \text{ and } \xi\lambda' \equiv \xi\epsilon' \bmod \mathfrak{f},$$

hence, in particular,

$$\xi\lambda \equiv \xi\epsilon \bmod \mathfrak{a}\mathfrak{c}^{-1} \cap \mathfrak{O}_t \text{ and } \xi\lambda' \equiv \xi\epsilon' \bmod \mathfrak{a}\mathfrak{b}_t^{-1} \cap \mathfrak{O}_t.$$

Applying the second assertion of our proposition for the ideals $\mathfrak{a}\mathfrak{c}^{-1}$ and $\mathfrak{a}\mathfrak{b}_t^{-1}$, we obtain

$$\varphi(\xi|\mathfrak{a}\mathfrak{c}^{-1})^{e_f} = \varphi(\xi\lambda|\mathfrak{a}\mathfrak{c}^{-1})^{e_f}$$

$$= \varphi(\xi|\mathfrak{a}(\lambda\mathfrak{c})^{-1})^{e_f} = \varphi(\xi|\mathfrak{a}(\lambda'\mathfrak{b}_t)^{-1})^{e_f}$$

$$= \varphi(\xi\lambda'|\mathfrak{a}\mathfrak{b}_t^{-1})^{e_f} = \varphi(\xi|\mathfrak{a}\mathfrak{b}_t^{-1})^{e_f},$$

and this finishes the proof. □

By the third assertion of proposition 6.8.1 we see that $\varphi(1|\mathfrak{a}\mathfrak{c}^{-1})^{e_f}$ only depends on the ray class $\mathfrak{k} = \mathfrak{c}\mathfrak{S}_{t,\mathfrak{f}}$. Therefore, similarly to the singular values of the τ function, we set

$$\Phi_{\mathfrak{a}}(\mathfrak{k}) := \varphi(1|\mathfrak{a}\mathfrak{c}^{-1})^{e_f}$$

with an integral ideal $\mathfrak{c} \in \mathfrak{k}^{-1}$. Immediately from Proposition 6.8.1 we then obtain:

Theorem 6.8.2 *Let $\mathfrak{a} \in \mathfrak{J}_t$ and $\mathfrak{f} := \mathfrak{a} \cap \mathfrak{O}_t$. Then for all $\mathfrak{k}, \mathfrak{h} \in \mathfrak{R}_{t,\mathfrak{f}}$ we have*

$$\Phi_{\mathfrak{a}}(\mathfrak{k}) \in K_{t,\mathfrak{f}} \quad \text{and} \quad \Phi_{\mathfrak{a}}(\mathfrak{k})^{\sigma(\mathfrak{h})} = \Phi_{\mathfrak{a}}(\mathfrak{k}\mathfrak{h}^{-1}).$$

As we will explain later in Chapter 11, Meyer (1957) has shown that for $t = 1$ the above numbers act similarly in the values of certain L-function s at $s = 1$ as the Δ quotients considered in 6.6. In the following, we let $\mathfrak{a}$ be an integral ideal of $\mathfrak{O}_1$, different from $\mathfrak{O}_1$. Then in Theorem 6.8.2 $\mathfrak{f} = \mathfrak{a} \cap \mathfrak{O}_1 = \mathfrak{a}$. In view of the analogy to Δ quotients, we may now conjecture as follows:

Conjecture 6.8.3 *Let $\mathfrak{f}$ be an integral ideal of $\mathfrak{O}_1$, different from $\mathfrak{O}_1$, and let $\mathfrak{k}, \mathfrak{h} \in \mathfrak{K}_\mathfrak{f}$. Then for every intermediate field $K \subseteq L \subseteq K_\mathfrak{f}$ and for every natural exponent e*

$$L = K\left(N_{K_\mathfrak{f}/L}\left(\Phi_\mathfrak{f}(\mathfrak{k})^e\right)\right),$$

$$L = K\left(N_{K_\mathfrak{f}/L}\left(\left(\frac{\Phi_\mathfrak{f}(\mathfrak{k})}{\Phi_\mathfrak{f}(\mathfrak{h})}\right)^e\right)\right) \quad \text{if } \mathfrak{k}\mathfrak{U} \neq \mathfrak{h}\mathfrak{U},$$

where $\mathfrak{U}$ denotes the subgroup of $\mathfrak{K}_\mathfrak{f}$ corresponding to L.

Following the exposition in Bettner and Schertz (2001), we will now prove this conjecture for certain cases – roughly speaking for $\mathfrak{f}$ not divisible by "too small" prime ideal factors. Given a character χ of $\mathfrak{K}_\mathfrak{f}$, we consider the expression

$$A_\mathfrak{f}(\chi) := \sum_{\mathfrak{k} \in \mathfrak{K}_\mathfrak{f}} \overline{\chi}(\mathfrak{k}) \log |\Phi_\mathfrak{f}(\mathfrak{k})|.$$

Unlike the sum in the proof of Theorem 6.6.2, this sum may be zero for some character $\chi \neq 1$, because Theorem 11.2.1 only tells us that

$$A_{\mathfrak{f}_\chi}(\chi) \neq 0$$

for any character $\chi \neq 1$ of $\mathfrak{K}_\mathfrak{f}$ with $\mathfrak{f}_\chi \neq (1)$. The relation between $A_\mathfrak{f}(\chi)$ and $A_{\mathfrak{f}_\chi}(\chi)$ is given by

$$A_\mathfrak{f}(\chi) = \frac{w_{\mathfrak{f}_\chi}}{w_\mathfrak{f}} \left(\prod_{\mathfrak{p} \mid \frac{\mathfrak{f}}{\mathfrak{f}_\chi}} (1 - \chi'(\mathfrak{p}) \right) A_{\mathfrak{f}_\chi}(\chi),$$

where $w_\mathfrak{f}$ resp. $w_{\mathfrak{f}_\chi}$ denote the number of roots of unity in K that are congruent to 1 modulo $\mathfrak{f}$ resp. modulo $\mathfrak{f}_\chi$. χ' denotes the primitive character modulo $\mathfrak{f}_\chi$ belonging to χ. This explains why the proof of Theorem 6.6.2 does not carry over to this case. However, under certain conditions on $\mathfrak{f}$ it will be possible to construct sufficient characters χ of $\mathfrak{K}_\mathfrak{f}$ satisfying $A_\mathfrak{f}(\chi) \neq 0$. Let

$$\mathfrak{f} = \mathfrak{p}_1^{m_1} \cdot \ldots \cdot \mathfrak{p}_s^{m_s}$$

therefore, be the decomposition of $\mathfrak{f}$ into a product of powers of prime ideals. We let e_i be the exponent of the prime residue class, $(\mathfrak{O}_1/\mathfrak{p}_i^{m_i})^*$, and we note the following two conditions:

(i) Let $\mathfrak{f} = \mathfrak{p}^m \neq (1)$ be a power of a prime ideal and also the conductor of $K_\mathfrak{f}/K$.

(ii) With the above notation let

$$e_i = 2,\ i = 1, \dots, s_0,$$
$$e_i \nmid 2,\ i = s_0 + 1, \dots, s,$$
$$e_s \nmid 4,$$

$$\mathfrak{p}_s^{m_s} \nmid \gcd(6, w),$$

where w denotes the number of roots of unity in K. Further, we assume the existence of classes $\mathfrak{k}_i \in \mathfrak{K}_{\mathfrak{f}\mathfrak{p}_i^{-m_i}}$, $i = 1, \dots, s_0$, with $\mathfrak{p}_i \in \mathfrak{k}_i$, such that the orders of $\mathfrak{k}_i$ in $\mathfrak{K}_{\mathfrak{f}\mathfrak{p}_i^{-m_i}}$ are divisible by a prime $q_i > s_0 + 2$.

We can now prove the following special case of the above conjecture:

Theorem 6.8.4 *Let $\mathfrak{f} \neq (1)$ be an integral ideal satisfying (i) or (ii).*

Then for $\mathfrak{k} \in \mathfrak{K}_\mathfrak{f}$ and $e \in \mathbb{N}$

$$K_\mathfrak{f} = K(\Phi_\mathfrak{f}(\mathfrak{k})^e).$$

Further, for $\mathfrak{h} \in \mathfrak{K}_\mathfrak{f}$ not having order 1 or 3 and any $e \in \mathbb{N}$

$$K_\mathfrak{f} = K\left(\left(\frac{\Phi_\mathfrak{f}(\mathfrak{k}\mathfrak{h})}{\Phi_\mathfrak{f}(\mathfrak{k})}\right)^e\right)$$

if in addition to (i) $[K_\mathfrak{f} : \Omega] \neq 2$ for $\mathfrak{h}$ having order 2.

Proof We conclude as in the proof of Theorem 6.6.2. The assertion of Theorem 6.8.4 is then immediate by the following lemma. □

Lemma 6.8.5 *We assume the conditions of Theorem 6.8.4 to be satisfied. Then, for every $\mathfrak{k} \in \mathfrak{K}_\mathfrak{f}$ different from the main class, there exists a character χ of $\mathfrak{K}_\mathfrak{f}$ with*

$$A_\mathfrak{f}(\chi) \neq 0 \text{ und } \chi(\mathfrak{k}) \neq 1.$$

Further, for $\mathfrak{k}, \mathfrak{h} \in \mathfrak{K}_\mathfrak{f}$ different from the main class and $\mathfrak{h}$ having an order other than 3 there exists a character χ of $\mathfrak{K}_\mathfrak{f}$ with

$$A_\mathfrak{f}(\chi) \neq 0, \quad \chi(\mathfrak{k}) \neq 1 \text{ and } \chi(\mathfrak{h}) \neq 1.$$

For the proof of Lemma 6.8.5 we need the following analogon of Lemma 6.6.3.

Lemma 6.8.6 *Let $g_1, .., g_n$ be elements of a finite abelian group G of order $o(g_i) \neq 1$. Further, for every prime number p we assume that*

$$\#\{g_i \mid p | o(g_i)\} \leq p.$$

Then, there exists a character χ *of* G *with*

$$\chi(g_i) \neq 1 \text{ for } i = 1, \ldots, n.$$

Proof of Lemma 6.8.6 In view of the main theorem on finite abelian groups, it is sufficient to prove the lemma for all elements of $G \setminus \{1\}$ having order p with a fixed prime p and when $n = p$. Therefore, we may assume that

$$G = < g_1 > \times \ldots \times < g_k >;$$

where $k \geq 2$ because of $n = p$. The remaining $g_{k+1}, .., g_p$ are products of powers of $g_1, \ldots, g_k$:

$$g_j = g_1^{\mu_{j,1}} \cdot \ldots \cdot g_k^{\mu_{j,k}}$$

with at least one exponent not divisible by p. For $i = 1, \ldots, k$ we denote by χ_i the character of G uniquely defined by

$$\chi_i(g_j) = \zeta_p \delta_{ij}$$

with a fixed primitive p-th root of unity ζ_p and the Kronecker symbol δ_{ij}. Then, by

$$\chi = \chi_1^{\nu_1} \cdot \ldots \cdot \chi_k^{\nu_k}, \quad \nu_i = 1, \ldots, p-1,$$

$(p-1)^k$ characters of G with $\chi(g_i) \neq 1, i = 1, \ldots, k$, are defined. The condition $\chi(g_j) \neq 1$ for some $j \geq k+1$ is satisfied if exactly one exponent $\mu_{j,i}$ is not divisible by p. Otherwise, we have to exclude a certain number of χ's. Therefore, noting that

$$\chi(g_j) = \zeta_p^{\mu_{j,1}\nu_1 + \cdots + \mu_{j,k}\nu_k},$$

we consider the matrix

$$M := \begin{pmatrix} \mu_{k+1,1} & \cdots & \mu_{k+1,k} \\ \mu_{p,1} & \cdots & \mu_{p,k} \end{pmatrix}.$$

If M has rank 1, all $\chi(g_j)$, $j \geq k+1$, are powers of $\chi(g_{k+1})$, and to satisfy the condition $\chi(g_j) \neq 1$ for all $j \geq k+1$, it suffices to exclude $p^{k-1} - 1$ characters χ. The number of characters left is then

$$(p-1)^k - (p^{k-1} - 1) = p^k \left[\left(1 - \frac{1}{p}\right)^k - \frac{1}{p} \right] + 1$$

$$\geq p^k \left[\left(1 - \frac{1}{p}\right)^{p-1} - \frac{1}{p} \right] + 1 \geq p^k \left[1 - \frac{p-1}{p} - \frac{1}{p} \right] + 1 \geq 1.$$

If M has rank ≥ 2, then for each $\chi(g_j) \neq 1$, $j \geq k+1$, we have to exclude at most $p^{k-2}-1$. Therefore, the remaining number of characters is

$$(p-1)^k - (p-k)(p^{k-2}-1) = p^k \left[\left(1-\frac{1}{p}\right)^k - \frac{1}{p}\right] + kp^{k-2} + (p-k)$$

$$\geq p^k \left[\left(1-\frac{1}{p}\right)^{p-1} - \frac{1}{p}\right] + kp^{k-2} + (p-k)$$

$$\geq p^k \left[1 - \frac{p-1}{p} - \frac{1}{p}\right] + kp^{k-2} + (p-k) \geq 1.$$

This proves Lemma 6.8.6. □

Proof of Lemma 6.8.5 It is sufficient, to prove Lemma 6.8.5 for $\mathfrak{k}$ and $\mathfrak{h}$ having prime order:

$$o(\mathfrak{k}) = p, \quad o(\mathfrak{h}) = q \neq 3.$$

According to the remark following Theorem 6.8.4, the number s_0 is at most equal to 4 and $\mathfrak{k}_i, i = 1, \ldots, s_0$, have an order divisible by a prime $q_i \geq 7$. We chose classes $\mathfrak{k}_i' \in \mathfrak{K}_\mathfrak{f}$ with $\mathfrak{k}_i' \subset \mathfrak{k}_i$. Their orders are then also divisible by q_i. Applying Lemma 6.8.4 to the elements $\mathfrak{k}, \mathfrak{h}$ and $\mathfrak{k}_i', i = 1, \ldots, s_0$, of $G = \mathfrak{K}_\mathfrak{f}$ and noting that $s_0 \leq 4$ and $q_i \geq 7$, we can find a character χ of $\mathfrak{K}_\mathfrak{f}$ with

$$\chi(\mathfrak{k}) \neq 1,\ \chi(\mathfrak{h}) \neq 1 \text{ and } \chi(\mathfrak{k}_i') \neq 1, i = 1, \ldots, s_0. \tag{6.16}$$

The properties in (6.16) imply that

$$\chi'(\mathfrak{p}_i) \neq 1 \text{ if } \mathfrak{p}_i \nmid \mathfrak{f}_\chi \text{ for } i = 1, \ldots, s_0,$$

where χ' denotes the primitive character belonging to χ. We will now modify χ without changing (6.16) such that $\mathfrak{f}_\chi$ is divisible by all $\mathfrak{p}_i, i = s_0+1, \ldots, s$. The modified χ then has the properties of Lemma 6.8.5. Assuming $\mathfrak{p}_i$ to be no divisor of $\mathfrak{f}_\chi$ for some $i = s_0+1, \ldots, s-1$, we have to construct a character ψ of $\mathfrak{K}_\mathfrak{f}$ with $\mathfrak{p}_i|\mathfrak{f}_\psi|\mathfrak{p}_i^{m_i}\mathfrak{p}_s^{m_s}$. Then $\mathfrak{p}_i\mathfrak{f}_\chi\mathfrak{p}_s^{-m_s}|\mathfrak{f}_{\chi\psi}$. Further, ψ will be constructed in such a way that $\tilde{\chi} := \chi\psi$ satisfies (6.16).

To find ψ we consider the homomorphism

$$\kappa : (\mathfrak{O}/\mathfrak{f})^* \to \mathfrak{K}_\mathfrak{f}, \quad \alpha + \mathfrak{f} \mapsto (\alpha)\mathfrak{e},$$

where $\mathfrak{e}$ denotes the main class of $\mathfrak{K}_\mathfrak{f}$. We have

$$ker\kappa = (W + \mathfrak{f})/\mathfrak{f},$$

where W denotes the group of roots of unity in K. By restriction to the image of κ, every character χ of $\mathfrak{K}_\mathfrak{f}$ defines a character $\hat{\chi}$ of $(\mathfrak{O}/\mathfrak{f})^*$ that is trivial on $(W+\mathfrak{f})/\mathfrak{f}$. To determine the prime ideals dividing the conductor of χ we have to write $\hat{\chi}$ according to the isomorphism

$$(\mathfrak{O}/\mathfrak{f})^* \cong (\mathfrak{O}/\mathfrak{p}_1^{m_1})^* \times \ldots \times (\mathfrak{O}/\mathfrak{p}_s^{m_s})^*$$

as a product

$$\hat{\chi} = \chi_1 \cdot \ldots \cdot \chi_s$$

of characters χ_i of $(\mathfrak{O}/\mathfrak{p}_i^{m_i})^*$. Then

$$\chi_i \neq 1 \iff \mathfrak{p}_i | \mathfrak{f}_\chi.$$

To construct ψ we now choose a character ψ_i of $(\mathfrak{O}/\mathfrak{p}_i^{m_i})^*$ with $o(\psi_i) = e_i$ and a character ψ_s of $(\mathfrak{O}/\mathfrak{p}_s^{m_s})^*$ with $o(\psi_s) = e_s$ that, in addition, satisfies

$$\psi((W+\mathfrak{p}_s^{m_s})/\mathfrak{p}_s^{m_s})^*) = W.$$

Such a choice of ψ_s is possible because of the assumption $\mathfrak{p}_s^{m_s} \nmid \gcd(6, w)$. Due to the last condition we can now find $k \in \mathbb{N}$ such that

$$\psi_i(\zeta + \mathfrak{p}_i^{m_i})\psi_s(\zeta + \mathfrak{p}_s^{m_s})^k = 1$$

for all $\zeta \in W$. Hence

$$\psi := \psi_i \psi_s^k,$$

the product being understood in the sence of the above isomorphism as a character of $(\mathfrak{O}/\mathfrak{f})^*$, is trivial on $(W+\mathfrak{f})/\mathfrak{f}$. Therefore, ψ defines a character on the image of κ that can be continued to a character of $\mathfrak{K}_\mathfrak{f}$. For this continuation that we also denote by ψ, we then have

$$\mathfrak{p}_i | \mathfrak{f}_{\psi^{2\mu}} | \mathfrak{p}_i^{m_i} \mathfrak{p}_s^{m_s} \text{ for } \mu \in \mathbb{Z} \text{ with } e_i \nmid 2\mu.$$

We now define

$$\tilde{\chi} := \chi\psi^{2\mu}, \; e_i \nmid 2\mu,$$

and we will show that μ can be chosen such that $\tilde{\chi}$ also satisfies (6.16), i.e.

$$\psi^{2\mu}(\mathfrak{x}) \neq \chi^{-1}(\mathfrak{x}), \;\; \mathfrak{x} = \mathfrak{k}, \mathfrak{h}, \; \mathfrak{k}_j', j = 1, \ldots s_0.$$

If a class $\mathfrak{x}$ has order 2, then $\psi^{2\mu}(\mathfrak{x}) = 1$ and condition $\psi^{2\mu}(\mathfrak{x}) \neq \chi^{-1}(\mathfrak{x})$ is satisfied. But order 2 can only occur for $\mathfrak{x} = \mathfrak{k}, \mathfrak{h}$ because of the above assuptions on the $\mathfrak{k}_i'$. For the following, we may assume that all $\mathfrak{x}$ have an order divisible by a prime ≥ 3. More precisely, by hypothesis of Lemma 6.8.6, the $\mathfrak{k}'_i$ have orders divisible by a prime $q_j > s_0 + 3$. We set

$$q_{\mathfrak{x}} := \begin{cases} p & \text{if} \quad \mathfrak{x} = \mathfrak{k}, \\ q & \text{if} \quad \mathfrak{x} = \mathfrak{h}, \\ q_j & \text{if} \quad \mathfrak{x} = \mathfrak{k}_j, \ j = 1, \ldots, s_0. \end{cases}$$

Now, $\mu \in \mathbb{Z}$ satisfies the above conditions if it is a solution of the following congruences

$$\begin{aligned} \mu &\not\equiv 0 \bmod l, \\ \mu &\not\equiv \mu_{\mathfrak{x}} \bmod l \text{ if } q_{\mathfrak{x}} = l, \end{aligned}$$

for all odd primes l with

$$l | e_i p q q_1 \cdot \ldots \cdot q_{s_0}$$

and $\mathfrak{x} = \mathfrak{k}, \mathfrak{h}, \mathfrak{k}'_j, j = 1, \ldots, s_0$. For $l > s_0 + 3$ the existence of such a solution is immediate. For $l = 3$ we have $q_{\mathfrak{x}} = l$ at most for one class $\mathfrak{x}$ and in the case $5 \leq l \leq s_0 + 3$ there are at most two classes $\mathfrak{x}$ with $q_{\mathfrak{x}} = l$, hence the congruences can also be solved in these cases.

In this way we obtain a modification of χ such that $\mathfrak{p}_{s_0+1}, \ldots, \mathfrak{p}_{s-1}$ become divisors of $\mathfrak{f}_\chi$, thereby maintaining property (6.16). If, in addition, $\mathfrak{p}_s$ is a divisor of $\mathfrak{f}_\chi$, we have done. Otherwise we multiply χ by a character whose conductor is a power of $\mathfrak{p}_s$. We can find such a character as follows: first, ψ_s^w is obviously trivial on $(W + \mathfrak{f})/\mathfrak{f}$, hence defines a character of the image of κ that we continue to a character ψ of $\mathfrak{K}_\mathfrak{f}$. Then the conductor of ψ^2 is a $\mathfrak{p}_s$-power, which is different from (1) because our assumption $e_s \nmid 4$ implies that the conductor of ψ_s^{2w} cannot be (1). Then, as for $s_0 < i < s$, we end up with a character χ satisfying (6.16) and whose conductor is divisible by $\mathfrak{p}_{s_0+1}, \ldots, \mathfrak{p}_s$. This implies that

$$A_\mathfrak{f}(\chi) = \frac{w_{\mathfrak{f}_\chi}}{w_\mathfrak{f}} \left(\prod_{j=1}^{s_0} (1 - \chi'(\mathfrak{p}_j)) \right) A_{\mathfrak{f}_\chi}(\chi) \neq 0,$$

thereby proving the second assertion of Lemma 6.8.5 under the second condition in Theorem 6.8.4. The proof of the first assertion is analogous.

Now we prove the second assertion of Lemma 6.8.5 under the assumption (i). If $\mathfrak{f} = \mathfrak{p}_s^m \neq (1)$ is the power of a prime ideal and also the conductor of $K_\mathfrak{f}/K$, then $K_\mathfrak{f} \neq \Omega$. Let $\mathfrak{U}$ be the subgroup of $\mathfrak{K}_\mathfrak{f}$ corresponding to Ω. Then the characters of $\mathfrak{K}_\mathfrak{f}$ with conductor divisible by $\mathfrak{p}$ are exactly those who are non-trivial on $\mathfrak{U}$. We chose a class $\mathfrak{u} \in \mathfrak{U} \setminus \{\mathfrak{e}\}$. Then, by Lemma 6.8.6, there exists a character χ of $\mathfrak{K}_\mathfrak{f}$ with

$$\chi(\mathfrak{k}) \neq 1, \ \chi(\mathfrak{h}) \neq 1 \text{ and } \chi(\mathfrak{u}) \neq 1$$

if not all three elements $\mathfrak{k}, \mathfrak{h}, \mathfrak{u}$ have order 2. If one can find an element $u \in \mathfrak{U}$ of order $o(\mathfrak{u}) > 2$, then Lemma 6.8.5 is proved. Otherwise $\mathfrak{U}$ is the direct product of cyclic groups of order 2 and not cyclic because in this case $[K_\mathfrak{f} : \Omega] \neq 2$ by assumption. Hence, there are two classes $u_1, u_2 \in \mathfrak{U}$ of order 2. If, then, $\mathfrak{h} \notin < \mathfrak{k}, \mathfrak{u}_1 >$, we chose a character χ of $< \mathfrak{k}, \mathfrak{u}_1 >$ with $\chi(\mathfrak{k}) \neq 1$ and $\chi(\mathfrak{u}_1) \neq 1$, which we continue to a character of $\mathfrak{K}_\mathfrak{f}$ with $\chi(\mathfrak{h}) \neq 1$. However, if $\mathfrak{h}$ is in both $< \mathfrak{k}, \mathfrak{u}_1 >$ and $< \mathfrak{k}, \mathfrak{u}_2 >$, then $\mathfrak{h}$ must be equal to $\mathfrak{k}, \mathfrak{u}_1$ or to $\mathfrak{u}_2$. For $\mathfrak{h} = \mathfrak{k}$ or $\mathfrak{h} = \mathfrak{u}_1$, the character χ from Lemma 6.8.6 with $\chi(\mathfrak{k}) \neq 1$ and $\chi(\mathfrak{u}_1) \neq 1$ has the desired properties.

This proves the second assertion of Lemma 6.8.5. The first assertion is obtained analogously. □

The class invariants $\Phi_\mathfrak{a}(\mathfrak{k})$, $\mathfrak{k} \in \mathfrak{K}_{t,\mathfrak{f}}$, are by definition high powers of singular values of the Klein functions $\varphi_{\underline{x}}$:

$$\Phi_\mathfrak{a}(\mathfrak{k}) = \varphi(1|\mathfrak{f}\mathfrak{c}^{-1})^{e_f}.$$

In particular, for the numerical construction of class fields it is interesting to know whether lower powers of $\varphi(1|\mathfrak{f}\mathfrak{c}^{-1})$ are already in $K_{t,\mathfrak{f}}$. To answer this question one has to apply Theorem 5.2.5 to obtain the Galois action of $K_{t12f^2}/K_{t,\mathfrak{f}}$ on these singular values. It then transpires that mostly $\varphi(1|\mathfrak{f}\mathfrak{c}^{-1})$ generates an extension of $K_{t,\mathfrak{f}}$ that under certain conditions is even of degree e_f. However, we will show that simple products of different values $\varphi(1|\mathfrak{f}\mathfrak{c}^{-1})$ are in $K_{t,\mathfrak{f}}$, thus providing simple generators for $K_{t,\mathfrak{f}}/K$, similar to those given by Theorem 6.6.6 for ring class fields. Such products are given by the following theorem, where for simplicity we consider only the case of $t = 1$.

Theorem 6.8.7 *Let $\mathfrak{f} \neq (1)$ be an integral ideal of $\mathfrak{O}_1$ and $\xi_1, \ldots, \xi_s \in \frac{1}{\mathfrak{f}} \setminus \mathfrak{O}_1$. Then*

$$(\xi_i) = \frac{\mathfrak{c}_i}{\mathfrak{f}} = \frac{\tilde{\mathfrak{c}}_i}{\mathfrak{t}_i}, \quad \gcd(\tilde{\mathfrak{c}}_i, \mathfrak{t}_i) = 1,$$

with integral ideals $\mathfrak{c}_i, \tilde{\mathfrak{c}}_i$ and divisors $\mathfrak{t}_i \neq (1)$ of $\mathfrak{f}$. We choose a basis $\alpha, 1$ for $\mathfrak{O}_1$ with $\alpha \in \mathbb{H}$ and

$$tr(\alpha) \equiv 0 \ mod\ 3, \qquad tr(\alpha) \equiv \begin{cases} 0 \ mod\ 4 \ if\ 2|d, \\ 1 \ mod\ 4 \ if\ 2 \nmid d, \end{cases}$$

which is always possible. By $\mathfrak{k}_{\tilde{\mathfrak{c}}_i}$ we denote the ray class modulo $\mathfrak{t}_i$ that contains $\tilde{\mathfrak{c}}_i^{-1}$. Then

$$\Phi_{\mathfrak{t}_i}(\mathfrak{k}_{\tilde{\mathfrak{c}}_i}) = \varphi(\xi_i|{}^{\alpha}_{1})^{e_{t_i}}, \quad i = 1, \ldots, s.$$

We write $\mathfrak{f} = f_1\mathfrak{f}_2$ *with a natural number* f_1 *and a primitive ideal* $\mathfrak{f}_2$*. In* $\mathfrak{f}_2$ *we choose a canonical basis*

$$\mathfrak{f}_2 = [\tilde{\alpha}, f_2] \text{ with } f_2 = N(\mathfrak{f}_2) \text{ and } \tilde{\alpha} \in \mathbb{H}.$$

$f := f_1 f_2$ *is the smallest natural number in* $\mathfrak{f}$*. Further, we decompose* $f_2 = f_2^* f_2^{**}$*, where* f_2^* *denotes the ramified and* f_2^{**} *the split part of* f_2*. According to Theorem 5.2.5 we then have*

$$\varphi(\xi_i|{\alpha \atop 1}) \in K_{12f^2}, \quad i = 1, \dots, s.$$

Further, let

$$\gamma_i = x_{i,1}\sigma(\mathfrak{b}_{i,1}) + \dots + x_{i,s_i}\sigma(\mathfrak{b}_{i,s_i}) \in \mathbb{Z}[Gal(K_{12f^2}/K)], \quad i = 1, \dots, s,$$

with integral ideals $\mathfrak{b}_{i,j}$ *prime to* $6f$ *and corresponding automorphisms* $\sigma(\mathfrak{b}_{i,j}) \in Gal(K_{12f^2}/K)$*. We set*

$$N(\gamma_i) := x_{i,1}N(\mathfrak{b}_{i,1}) + \dots + x_{i,s_i}N(\mathfrak{b}_{i,s_i}),$$

and we assume that the following conditions are satisfied:

$$\begin{aligned} &N(\gamma_i) \equiv 0 \bmod 2, \ i = 1, \dots, s, \\ &N(\gamma_1) + \dots + N(\gamma_s) \equiv 0 \bmod 4 \text{ if } 2|d \text{ and } 2 \nmid \mathfrak{f}, \\ &N(\gamma_1) + \dots + N(\gamma_s) \equiv 0 \bmod 3 \text{ if } 3|d \text{ and } 3 \nmid \mathfrak{f}, \\ &N(\mathfrak{c}_1)N(\gamma_1) + \dots + N(\mathfrak{c}_s)N(\gamma_s) \equiv 0 \bmod \tfrac{2f}{f_2^{**}\gcd(2, f_2^{**})}. \end{aligned}$$

As the trace of $\tilde{\alpha}$ *is coprime to* f_2^{**}*, there exists a solution* ν *of the congruence*

$$\mu\nu f_1 tr(\tilde{\alpha}) - (N(\gamma_1) + \dots + N(\gamma_s)) \equiv 0 \bmod 2f,$$

where $\mu = 1$ *resp.* $\mu = 2$ *for* f_2^{**} *even resp. odd. Then, with the root of unity* $\zeta = exp(\frac{2\pi i\mu\nu}{2f})$*, we have*

$$\zeta \prod_{i=1}^{s} \varphi(\xi_i|{\alpha \atop 1})^{\gamma_i} \in K_{\mathfrak{f}}.$$

The action of the Galois group of $K_{\mathfrak{f}}/K$ *on these values can be computed by Theorem 5.2.5.*

In view of Theorems 6.8.3 and 6.8.4 we note two special cases that are useful for numerical constructions of $K_{\mathfrak{f}}$.

Theorem 6.8.8 *With the notation of Theorem 6.8.7 we have*

(i) $\zeta\varphi(\xi|{\alpha \atop 1})^{2\gcd(d,12)} \in K_{\mathfrak{f}}$ *if* $\gcd(\mathfrak{f}, \bar{\mathfrak{f}}) = 1$.

(ii) $\zeta\dfrac{\varphi(\xi|{\alpha \atop 1})^{\sigma(\mathfrak{b})}}{\varphi(\xi|{\alpha \atop 1})} \in K_{\mathfrak{f}}$ *if* $N(\mathfrak{b}) \equiv 1 \bmod \frac{12f}{f_2^{**}}$,

for any integral ideal $\mathfrak{b}$ *of* $\mathfrak{O}_1$ *coprime to* $6f$*.*

Proof of Theorem 6.8.7 By definition of $\Phi_{\mathfrak{t}_i}(\mathfrak{k}_{\tilde{\mathfrak{c}}_i})$ and by homogeneity of φ we obtain, using Theorem 5.2.5,

$$\Phi_{\mathfrak{t}_i}(\mathfrak{k}_{\tilde{\mathfrak{c}}_i}) = \varphi(\xi_i|{}^{\alpha}_{1})^{e_{t_i}}$$

and then

$$\varphi(\xi_i|{}^{\alpha}_{1}) \in K_{12f^2}.$$

To prove the remaining assertions, we fix i, and we write $\xi = \xi_i$, $\mathfrak{c} = \mathfrak{c}_i$, $\gamma = \gamma_i$. To compute the action of the automorphisms of $K_{12f^2}/K_{\mathfrak{f}}$ on $\varphi(\xi|{}^{\alpha}_{1})^{\gamma}$ we note that the relative automorphisms are of the form

$$\sigma(\lambda), \lambda \text{ coprime to } 12f \text{ with } \lambda \equiv 1 \bmod \mathfrak{f}.$$

Using Theorem 5.2.5, we find that

$$\varphi(\xi|{}^{\alpha}_{1})^{\sigma(\lambda)} = \epsilon(\lambda, \alpha)\varphi(\xi\lambda|{}^{\alpha}_{1})$$

with

$$\epsilon(\lambda, \alpha) = \epsilon(M_1)^{2l}\epsilon(M_2)^2, \quad l = N(\lambda),$$

where $\epsilon(M_i)$ denotes the 24-th root of unity in the transformation formula of the η function with unimodular matrices

$$M_1 \equiv \left({}^{a(us+v)+bun}_{1} \;\; {}^{-1}_{0}\right) \bmod 12, \quad M_2 \equiv \left({}^{u}_{b} \; {}^{v}_{a}\right) \bmod 12.$$

Herein, s and n denote trace and norm of α, and $u, v \in \mathbb{Z}$ are the coordinates in the representation

$$\lambda = u\alpha + v.$$

The numbers $a, b \in \mathbb{Z}$ are chosen in such a way that

$$det\left({}^{u}_{b} \; {}^{v}_{a}\right) \equiv 1 \bmod 12.$$

Modifying λ modulo $\mathfrak{f}$, one can further achieve $u \neq 0$ and $b > 0$. Then, observing $l = u^2n + uvs + v^2$, we obtain by Theorem 1.10.1:

$$\epsilon(\lambda, \alpha)^3 = \begin{cases} \zeta_4^{(-uvn+1)l-uv-v} & \text{if } 2 \nmid v, \\ \zeta_4^{sl+l-u+1} & \text{if } 2|v, \end{cases}$$

$$\epsilon(\lambda, \alpha)^4 = \begin{cases} \zeta_3^{-uv(nl+1)} & \text{if } 3 \nmid v, \\ \zeta_3^{sl} & \text{if } 3|v. \end{cases}$$

Keeping in mind the congruence conditions on s and n, these formulae lead us to

$$\begin{aligned} \epsilon(\lambda, \alpha)^6 &= 1 \;\text{ if }\; 2 \nmid d \;\text{ or }\; 2|(\lambda - 1), \\ \epsilon(\lambda, \alpha)^4 &= 1 \;\text{ if }\; 3 \nmid d \;\text{ or }\; 3|(\lambda - 1). \end{aligned}$$

Writing

$$\lambda = 1 + \omega \text{ mit } \omega \in \mathfrak{f},$$

the transformation formula of the σ^* function yields

$$\varphi(\xi\lambda|{}^{\alpha}_{1}) = \psi(\xi\omega)e^{\frac{1}{2}l_{\mathfrak{O}_1}(\xi,\xi(\lambda-1))}\varphi(\xi|{}^{\alpha}_{1}) = \psi(\xi\omega)e^{\frac{1}{2}N(\xi)l_{\mathfrak{O}_1}(1,\omega)}\varphi(\xi|{}^{\alpha}_{1}).$$

To evaluate $l_{\mathfrak{O}_1}(1,\omega)$, we note that $\mathfrak{f} = f_1[\tilde{\alpha}, f_2]$. Hence, ω can be written as

$$\omega = wf + yf_1\tilde{\alpha} \text{ with } w, y \in \mathbb{Z}.$$

From $\tilde{\alpha} \equiv \alpha \bmod \mathbb{Z}$ it follows that

$$l_{\mathfrak{O}_1}(1,\omega) = -2\pi i f_1 y.$$

Further, by definition of ξ,

$$N(\xi) = \frac{N(\mathfrak{c})}{N(\mathfrak{f})} = \frac{N(\mathfrak{c})}{f_1^2 f_2}.$$

Finally, we have

$$\varphi(\xi|{}^{\alpha}_{1})^{\sigma(\lambda)-1} = \epsilon(\lambda,\alpha)\psi(\xi\omega)e^{-\frac{2\pi i}{2f}N(\mathfrak{C})y},$$

and this implies the Galois action as asserted:

$$\left(\varphi(\xi|{}^{\alpha}_{1})^{\gamma}\right)^{\sigma(\lambda)-1} = \epsilon(\lambda,\alpha)^{N(\gamma)}\psi(\xi\omega)^{N(\gamma)}e^{-\frac{2\pi i}{2f}N(\mathfrak{C})N(\gamma)y}.$$

Further, we find that

$$\zeta_{2f}^{\mu(\sigma(\lambda)-1)} = \zeta_{2f}^{\mu(N(\lambda)-1)} = \zeta_{2f}^{\mu f_1 tr(\tilde{\alpha})y}$$

with $\zeta_{2f} = exp(\frac{2\pi i}{2f})$. The hypothesis on $N(\gamma_i)$, together with the properties of $\epsilon(\lambda,\alpha)$ that we have deduced, now show us that $\zeta\prod_{i=1}^{s}\varphi(\xi_i|{}^{\alpha}_{1})^{\gamma_i}$ is fixed by all automorphisms of $K_{12f^2}/K_{\mathfrak{f}}$, thereby finishing the proof of Theorem 6.8.7. □

Remark 6.8.9 From the last part of the proof we see that the condition "$N(\gamma_i) \equiv 0 \bmod 2$" can be weakened according to the individual choice of ξ_i. Namely, it suffices to have

$$\prod_{i=1}^{s}\psi(\xi_i\omega)^{N(\gamma_i)} = 1$$

for all $\omega \in \mathfrak{f}$.

Examples 6.8.10 By the following examples it is shown on the one hand, that the minimal polynomials of the numbers constructed in Theorem 6.8.7 mostly have rather small coefficients. On the other hand all numbers considered in the following are generators for $K_{\mathfrak{f}}/K$.

In the following, $m_{\theta,K}$ denotes the minimal polynomial of θ over K. Further, we denote by $\mathfrak{p}_2, \mathfrak{p}_3, \mathfrak{p}_5, \mathfrak{p}_7$ prime ideals dividing $2, 3, 5, 7$ and the number α is assumed to be of the form $\alpha = \frac{u+\sqrt{d}}{2}$ with the discriminant d of K and $u = tr(\alpha) \in \mathbb{Z}$.

(i) Let $d = -7$, $\mathfrak{f} = (9)$ and

$$\Theta = \varphi\left(\xi \,\middle|\, \begin{matrix} \frac{\alpha}{11} \\ 1 \end{matrix}\right) \varphi\left(\xi \,\middle|\, \begin{matrix} \frac{\alpha}{7} \\ 1 \end{matrix}\right)$$

where $tr(\alpha) = 189$, $\xi = \frac{1}{9}$. Then $\zeta = 1$ and

$$\begin{aligned} m_{\Theta,K} = \; & X^{36} - 6X^{33} + 15X^{32} + 27X^{30} - 45X^{29} + 63X^{28} \\ & -81X^{27} + 54X^{26} - 54X^{25} + 414X^{24} + 117X^{23} \\ & +135X^{22} - 747X^{21} + 162X^{20} + 351X^{19} \\ & +1269X^{18} + 108X^{17} - 387X^{16} - 576X^{15} \\ & +450X^{14} + 1638X^{13} + 1683X^{12} + 891X^{11} \\ & -171X^{10} - 1002X^{9} - 1044X^{8} - 351X^{7} \\ & +531X^{6} + 729X^{5} + 531X^{4} + 297X^{3} + 117X^{2} \\ & +27X + 3. \end{aligned}$$

(ii) Let $d = -24$, $\mathfrak{f} = (2)\mathfrak{p}_3\mathfrak{p}_5$ and

$$\Theta = \varphi\left(\xi \,\middle|\, \begin{matrix} \alpha \\ 1 \end{matrix}\right) \varphi\left(\xi \,\middle|\, \begin{matrix} \frac{\alpha}{59} \\ 1 \end{matrix}\right)$$

where $tr(\alpha) = 624$, $\xi = \frac{3+16\sqrt{-24}}{30}$. Then $\zeta = 1$ and

$$\begin{aligned} m_{\Theta,K} = \; & X^{16} + (2 - 2\sqrt{-6})X^{15} + (-17 - \sqrt{-6})X^{14} \\ & +(-6 + 22\sqrt{-6})X^{13} + (98 + 4\sqrt{-6})X^{12} \\ & +(-16 - 57\sqrt{-6})X^{11} + (-159 + 24\sqrt{-6})X^{10} \\ & +(62 + 56\sqrt{-6})X^{9} + (41 - 37\sqrt{-6})X^{8} \\ & +(-112 - 26\sqrt{-6})X^{7} + (-87 + 15\sqrt{-6})X^{6} \\ & +(2 + 27\sqrt{-6})X^{5} + (35 - 5\sqrt{-6})X^{4} \\ & +(-24 - 17\sqrt{-6})X^{3} + (-26 + 2\sqrt{-6})X^{2} \\ & +(2 + 3\sqrt{-6})X + 1. \end{aligned}$$

(iii) Let $d = -11$, $\mathfrak{f} = (9)$ and

$$\Theta = \frac{\varphi\left(\xi \,\middle|\, \begin{matrix} \frac{\alpha}{47} \\ 1 \end{matrix}\right) \varphi\left(2\xi \,\middle|\, \begin{matrix} \alpha \\ 1 \end{matrix}\right)^2}{\varphi\left(\xi \,\middle|\, \begin{matrix} \alpha \\ 1 \end{matrix}\right)}$$

where $tr(\alpha) = 417$, $\xi = \frac{1}{9}$. Then $\zeta = 1$ and

$$\begin{aligned} m_{\Theta,K} = \ & X^{18} + 9X^{17} + 36X^{16} + (95 - 4\sqrt{-11})X^{15} \\ & +(189 - 39\sqrt{-11})X^{14} + (192 - 147\sqrt{-11})X^{13} \\ & +(-233 - 246\sqrt{-11})X^{12} + (-960 - 60\sqrt{-11})X^{11} \\ & +(-735 + 408\sqrt{-11})X^{10} + (935 + 534\sqrt{-11})X^{9} \\ & +(1716 - 9\sqrt{-11})X^{8} + (84 - 441\sqrt{-11})X^{7} \\ & +(-1379 - 144\sqrt{-11})X^{6} + (-603 + 258\sqrt{-11})X^{5} \\ & +(345 + 195\sqrt{-11})X^{4} + (218 + \sqrt{-11})X^{3} \\ & +(-12 - 3\sqrt{-11})X^{2} + (12 + 3\sqrt{-11})X + 1. \end{aligned}$$

(iv) Let $d = -31$, $\mathfrak{f} = \mathfrak{p}_2^3$ and

$$\Theta = \frac{\varphi\left(\xi \,\middle|\, \begin{matrix} \frac{\alpha}{31} \\ 1 \end{matrix}\right)}{\varphi\left(\xi \,\middle|\, \begin{matrix} \alpha \\ 1 \end{matrix}\right)^3}$$

where $tr(\alpha) = 93$, $\xi = \frac{15+\sqrt{-31}}{16}$. Then $\zeta = 1$ and

$$\begin{aligned} m_{2\Theta,K} = \ & X^6 + 8X^5 + (18 - 2\sqrt{-31})X^4 + 24X^3 \\ & +(66 - 2\sqrt{-31})X^2 + (40 - 8\sqrt{-31})X \\ & +(4 - 4\sqrt{-31}). \end{aligned}$$

The denominator of Θ can be obtained by the known factorisation of the singular values of φ.

6.9 Generalised principal ideal theorem

In this section we will essentially follow the exposition presented in Schertz (2006). Let $\mathfrak{p}^n$ be a power of a prime ideal in K. Then, by the generalised principal ideal theorem, Schertz(1990, 1999)† we obtain

† In Schertz (1999) the following has to be corrected:

(i) The prime ideal $\mathfrak{q}$ in the definition of $H_{\mathfrak{q}}(z)$ has to satisfy the additional condition $\gcd(\mathfrak{q}, \overline{\mathfrak{q}}) = 1$,
(ii) $H_{\mathfrak{q}}(1)$ has to be replaced by $H_{\mathfrak{q}}(\omega)$ with $\omega \equiv 1 \bmod \mathfrak{q}$, $\omega \equiv 0 \bmod \overline{\mathfrak{q}}$.

an expicit construction of an element $\pi_n \in K_{\mathfrak{p}^n}$ with

$$\pi_n \sim \mathfrak{p}^{\frac{1}{[K_{\mathfrak{p}^n}:\Omega]}}. \tag{6.17}$$

π_n can be viewed as an analogue of the cyclotomic unit

$$\omega_n = e^{\frac{2\pi i}{p^n}} - 1$$

for the power p^n of a prime number p. As an element of the p^n-th cyclotomic field $\mathbb{Q}_{p^n}$ it has the factorisation

$$\omega_n \sim (p)^{\frac{1}{[\mathbb{Q}_{p^n}:\mathbb{Q}]}}.$$

Furthermore, ω_n is endowed with the following nice properties, that can easily be verified:

- $\omega_n = e_n(1)$ with p^n-periodic function $e_n(z) = e^{\frac{2\pi i}{p^n}z} - 1$.
- Let $\mathbb{C}_{p^n\mathbb{Z}}$ denote the field of p^n-periodic meromorphic functions on $\mathbb{C}$. Then we have the norm relation
$$e_n(z) = N_{\mathbb{C}_{p^{n+1}\mathbb{Z}}/\mathbb{C}_{p^n\mathbb{Z}}}(e_{n+1}(z)) = \prod_{\xi \in p^n\mathbb{Z} \bmod p^{n+1}\mathbb{Z}} e_{n+1}(z+\xi).$$
- For $z = 1$ and $n \geq 1$ the last relation becomes a norm relation between number fields:
$$\omega_n = N_{\mathbb{Q}_{p^{n+1}}/\mathbb{Q}_{p^n}}(\omega_{n+1}) = \prod_{\xi \in p^n\mathbb{Z} \bmod p^{n+1}\mathbb{Z}} e_{n+1}(1+\xi),$$
- and for $n = 0$ we have
$$\left.\frac{e_0(z)}{e_1(z-1)}\right|_{z=1} = p.$$

Using the normalised Weierstrass σ function

$$\varphi(z|\mathfrak{L}) = e^{-\frac{zz^*}{2}}\sigma(z|\mathfrak{L})\sqrt[12]{\Delta(\mathfrak{L})},$$

we will now construct an element π_n in the ray class field $K_{\mathfrak{p}^n}$ having a similar factorisation and satisfying an analogous norm relation.

In what follows, let $\mathfrak{p}$ be a fixed prime ideal and $\mathfrak{q} \nmid 2$ an auxiliary ideal coprime to $\mathfrak{p}$ satisfying the condition

$$\gcd(\mathfrak{q}, \overline{\mathfrak{q}}) = 1.$$

For $n \in \mathbb{N}$ we define

$$E_n(z) := \frac{\varphi(z - \gamma_n|\,\mathfrak{q}\mathfrak{p}^n)\,\varphi(z + \gamma_n|\,\mathfrak{q}\mathfrak{p}^n)}{\varphi^2(z|\,\mathfrak{q}\mathfrak{p}^n)}$$

with a solution γ_n of the simultaneous congruences

$$\begin{aligned}\gamma_n &\equiv 0 \bmod \mathfrak{p}^n,\\ \gamma_n &\equiv 1 \bmod \mathfrak{q},\\ \gamma_n &\equiv 0 \bmod \overline{\mathfrak{q}}.\end{aligned}$$

For the definition of $E_n(z)$ it is assumed that all φ-values involved contain the same root $\sqrt[12]{\Delta(\mathfrak{L})}$. In fact, $E_n(z)$ is then independent of this root as it cancels out in the defining product. Using the identity $\wp(u) - \wp(v) = -\frac{\sigma(u-v)\sigma(u+v)}{\sigma^2(u)\sigma^2(v)}$ we can express E_n by the Weierstrass $\wp$ function:

$$E_n(z) = -\varphi^2(\gamma_n|\,\mathfrak{q}\mathfrak{p}^n)\left(\frac{\wp(z|\,\mathfrak{q}\mathfrak{p}^n)}{\sqrt[6]{\Delta(\mathfrak{q}\mathfrak{p}^n)}} - \frac{\wp(\gamma_n|\,\mathfrak{q}\mathfrak{p}^n)}{\sqrt[6]{\Delta(\mathfrak{q}\mathfrak{p}^n)}}\right).$$

Therefore, E_n is elliptic with respect to $\mathfrak{q}\mathfrak{p}^n$. Furthermore, E_n satisfies the following norm relation:

Theorem 6.9.1 *Let $\mathbb{C}_{\mathfrak{q}\mathfrak{p}^n}$ denote the field of elliptic functions with respect to $\mathfrak{q}\mathfrak{p}^n$. Then $\mathbb{C}_{\mathfrak{q}\mathfrak{p}^{n+1}}/\mathbb{C}_{\mathfrak{q}\mathfrak{p}^n}$ is a Galois extension and its Galois group consists of all substitutions*

$$g(z) \mapsto g(z+\xi), \qquad \xi \in \mathfrak{q}\mathfrak{p}^n \bmod \mathfrak{q}\mathfrak{p}^{n+1},$$

for $g \in \mathbb{C}_{\mathfrak{q}\mathfrak{p}^{n+1}}$. The function $E_n(z)$ satisfies the norm relation

$$E_n(z) = \mathbf{N}_{\mathbb{C}_{\mathfrak{q}\mathfrak{p}^{n+1}}/\mathbb{C}_{\mathfrak{q}\mathfrak{p}^n}}(E_{n+1}(z)) = \prod_{\xi\in\mathfrak{q}\mathfrak{p}^n \bmod \mathfrak{q}\mathfrak{p}^{n+1}} E_{n+1}(z+\xi).$$

For the singular values $E_n(1)$ we thereby obtain:

Theorem 6.9.2 *Let $\mathfrak{p}$ and $\mathfrak{q}$ be as above, and let Φ denote Euler's function in K. Then:*

(i) $E_n(1) \in K_{\mathfrak{q}\mathfrak{p}^n}$ *for* $n \geq 0$,

(ii) $E_n(1) \sim \mathfrak{p}^{\frac{1}{\Phi(\mathfrak{p}^n)}}$ *for* $n \geq 1$,

(iii) $E_n(1) = \mathbf{N}_{K_{\mathfrak{q}\mathfrak{p}^{n+1}}/K_{\mathfrak{q}\mathfrak{p}^n}}(E_{n+1}(1)) = \prod_{\xi\in\mathfrak{q}\mathfrak{p}^n \bmod \mathfrak{q}\mathfrak{p}^{n+1}} E_{n+1}(1+\xi)$ *for* $n \geq 1$,

$$\text{(iv)}\quad \left.\frac{E_0(z)}{E_1(z-1+\gamma_1)}\right|_{z=1} = \mathbf{N}_{K_{\mathfrak{q}\mathfrak{p}}/K_{\mathfrak{q}}}(E_1(1)) = \frac{\varphi(2|\mathfrak{q})\varphi(\gamma_1|\mathfrak{q}\mathfrak{p})^2}{\varphi(2\gamma_1|\mathfrak{q}\mathfrak{p})\varphi(1|\mathfrak{q})^2}\sqrt[12]{\frac{\Delta(\mathfrak{q})}{\Delta(\mathfrak{q}\mathfrak{p})}} \sim \mathfrak{p}.$$

To obtain an analogous result for the extensions $K_{\mathfrak{p}^{n+1}}/K_{\mathfrak{p}^n}$, we have to lose the auxiliary ideal $\mathfrak{q}$. We need the following well-known lemma:

Lemma 6.9.3 *For every integral ideal* $\mathfrak{a}$ *in* K *we have*

$$\gcd\{N(\mathfrak{q})-1 \mid \mathfrak{q} \textit{ prime ideal in } K,\ \mathfrak{q} \nmid 2\overline{\mathfrak{q}}\mathfrak{a}\} = w_K,$$

where w_K *denotes the number of roots of unity in* K.

Therefore, there are finitely many prime ideals $\mathfrak{q}_i$, $i = 1, \ldots, s$, of degree 1, not dividing $N(\mathfrak{p})$, and $x_i \in \mathbb{Z}$ with

$$x_1(N(\mathfrak{q}_1)-1) + \cdots + x_s(N(\mathfrak{q}_s)-1) = w_K.$$

As above, we define for every $\mathfrak{q}_i$ a sequence of functions $E_{n,i}(z)$ with parameters $\gamma_{n,i}$, and by taking norms we obtain:

$$\mathbf{N}_{K_{\mathfrak{q}_i\mathfrak{p}^n}/K_{\mathfrak{p}^n}}(E_{n,i}(1)) \sim \mathfrak{p}^{\frac{N(\mathfrak{q}_i)-1}{\Phi(\mathfrak{p}^n)}}.$$

Therefore,

$$\pi_n := \prod_{i=1}^{s}\left(\mathbf{N}_{K_{\mathfrak{q}_i\mathfrak{p}^n}/K_{\mathfrak{p}^n}}(E_{n,i}(1))\right)^{x_i},$$

defines an element in $K_{\mathfrak{p}^n}$ having the factorisation

$$\pi_n \sim \mathfrak{p}^{\frac{w_K}{\Phi(\mathfrak{p}^n)}}.$$

In view of the formula

$$[K_{\mathfrak{p}^n} : \Omega] = \frac{w(\mathfrak{p}^n)}{w_K}\Phi(\mathfrak{p}^n),$$

where $w(\mathfrak{p}^n)$ denotes the number of roots of unity in K that are congruent to 1 modulo $\mathfrak{p}^n$, we can write the factorisation of π_n as

$$\pi_n \sim \mathfrak{p}^{\frac{w(\mathfrak{p}^n)}{[K_{\mathfrak{p}^n}:\Omega]}}.$$

Herein

$$w(\mathfrak{p}^n) = 1$$

except where

(i) $\mathfrak{p} \mid 2$, $n \leq 2$, where $w(\mathfrak{p}^n) \in \{1, 2\}$, for $d_K \neq -4$ and $w(\mathfrak{p}^n) \in \{1, 2, 4\}$

for $d_K = -4$,
(ii) $\mathfrak{p} \mid 3, n = 1, d_K = -3$, where $w(\mathfrak{p}^n) = 2$.

Moreover, we will show now that this element can be written analogously to the cyclotomic case. We therefore note that by Reciprocity Law the conjugates of the singular values $E_{n,i}(1)$ over $K_{\mathfrak{p}^n}$ are given by

$$E_{n,i}(1)^{\sigma(\lambda)} = \frac{\varphi(\lambda - \gamma_{n,i}\lambda | \mathfrak{q}_i\mathfrak{p}^n)\, \varphi(\lambda + \gamma_{n,i}\lambda | \mathfrak{q}_i\mathfrak{p}^n)}{\varphi^2(\lambda | \mathfrak{q}_i\mathfrak{p}^n)},$$

where $\sigma(\lambda)$ denotes the Frobenius automorphism of $K_{\mathfrak{q}\mathfrak{p}^n}/K_{\mathfrak{p}^n}$ belonging to $(\lambda), \lambda \equiv 1 \bmod \mathfrak{p}^n$. Therefore, we define the function

$$E_n^*(z) :=$$

$$\prod_{i=1}^{s} \prod_{j=1}^{N(\mathfrak{q}_i)-1} \left(\frac{\varphi\left(z + (\lambda_{i,j}^{(n)} - 1) - \gamma_{n,i}\lambda_{i,j}^{(n)} \middle| \mathfrak{q}_i\mathfrak{p}^n \right) \varphi\left(z + (\lambda_{i,j}^{(n)} - 1) + \gamma_{n,i}\lambda_{i,j}^{(n)} \middle| \mathfrak{q}_i\mathfrak{p}^n \right)}{\varphi^2\left(z + (\lambda_{i,j}^{(n)} - 1) \middle| \mathfrak{q}_i\mathfrak{p}^n \right)} \right)^{x_i},$$

where for fixed i and n the numbers

$$\lambda_{i,j}^{(n)}, \quad j = 1, \dots, N(\mathfrak{q}_i) - 1,$$

run through a complete system of prime residue classes mod $\mathfrak{q}_i$ satisfying

$$\lambda_{i,j}^{(n)} \equiv 1 \bmod \mathfrak{p}^n.$$

Now we can prove the following two theorems:

Theorem 6.9.4 *Let $\mathfrak{p}$ and $\mathfrak{q} = \mathfrak{q}_1 \cdot \ldots \cdot \mathfrak{q}_s$ with $\mathfrak{q}_i$ as above. Then the functions $E_n^*(z)$ are in $\mathbb{C}_{\mathfrak{q}\mathfrak{p}^n}$ for $n \geq 0$ and satisfy the norm relation*

$$E_n^*(z) = \mathbf{N}_{\mathbb{C}_{\mathfrak{q}\mathfrak{p}^{n+1}}/\mathbb{C}_{\mathfrak{q}\mathfrak{p}^n}}\left(E_{n+1}^*(z)\right) = \prod_{\xi \in \mathfrak{q}\mathfrak{p}^n \ mod \ \mathfrak{q}\mathfrak{p}^{n+1}} E_{n+1}^*(z + \xi).$$

Theorem 6.9.5 *Let $\mathfrak{p}$ and $\mathfrak{q} = \mathfrak{q}_1 \cdot \ldots \cdot \mathfrak{q}_s$ with $\mathfrak{q}_i$ as above and let Φ denote the Euler function in K. Then*

(i) $E_n^*(1) \in K_{\mathfrak{p}^n}$ *for* $n \geq 0$,

(ii) $E_n^*(1) \sim \mathfrak{p}^{\frac{w(\mathfrak{p}^n)}{[K_{\mathfrak{p}^n}:\Omega]}}$ *for* $n \geq 1$,

(iii) $E_n^*(1) =$

$$\mathbf{N}_{K_{\mathfrak{p}^{n+1}}/K_{\mathfrak{p}^n}}\left(E_{n+1}^*(1)\right)^{\frac{w(\mathfrak{p}^n)}{w(\mathfrak{p}^{n+1})}} = \prod_{\substack{\xi \in \mathfrak{q}\mathfrak{p}^n \\ mod \ \mathfrak{q}\mathfrak{p}^{n+1}}} E_{n+1}^*(1 + \xi)$$

for $n \geq 1$,

(iv) $\mathbf{N}_{K_{\mathfrak{p}^n}/\Omega}\left(E_n^*(1)\right) \sim \mathfrak{p}^{w(\mathfrak{p}^n)}$.

Remark 6.9.6 The constructions of the above theorems can clearly be generalised to any integral ideal $\mathfrak{a}$ prime to $\mathfrak{q}$ instead of $\mathfrak{p}^n$ with obvious norm relations for two ideals $\mathfrak{a}$, $\mathfrak{b}$ with $\mathfrak{a} \mid \mathfrak{b}$. Of course, for a composite ideal $\mathfrak{a}$ the singular values will be units.

Proof of Theorems 6.9.1, 6.9.2 and 6.9.5

Proof of Theorem 6.9.1: By Theorem 1.9.3 we have the formula

$$\prod_{\xi} e^{-\frac{1}{2} l_{\mathfrak{L}}(z,\xi)} \varphi(z+\xi|\mathfrak{L}) = \zeta \varphi(z|\hat{\mathfrak{L}}) \tag{6.18}$$

for two complex lattices $\hat{\mathfrak{L}} \supset \mathfrak{L}$ and an arbitrary system of representatives $\{\xi\}$ of $\hat{\mathfrak{L}}/\mathfrak{L}$ with a $12[\hat{\mathfrak{L}} : \mathfrak{L}]$-th root of unity ζ only depending on the choice of representatives and the root $\sqrt[12]{\Delta(\hat{\mathfrak{L}})}$ resp. $\sqrt[12]{\Delta(\mathfrak{L})}$ in the definition of φ.

We choose $\hat{\mathfrak{L}} = \mathfrak{q}\mathfrak{p}^n$, $\mathfrak{L} = \mathfrak{q}\mathfrak{p}^{n+1}$. Then we obtain

$$\prod_{\xi \in \mathfrak{q}\mathfrak{p}^n \, mod \, \mathfrak{q}\mathfrak{p}^{n+1}} E_{n+1}(z+\xi) = \frac{\varphi(z-\gamma_{n+1}|\mathfrak{q}\mathfrak{p}^n)\varphi(z+\gamma_{n+1}|\mathfrak{q}\mathfrak{p}^n)}{\varphi^2(z|\mathfrak{q}\mathfrak{p}^n)}, \tag{6.19}$$

noting that $l_{\mathfrak{L}}(z, \xi)$ is a linear function of ξ, which implies that the $e^{-\frac{1}{2} l_{\mathfrak{L}}(z,\xi)}$ factors in (6.18) cancel out. Using the transformation formula for φ, we can replace γ_{n+1} on the right side in (6.19) by γ_n, thereby finding the asserted norm relation. □

Proof of Theorem 6.9.2 The Reciprocity Law tells us that

$$\varphi(\delta|\mathfrak{q}\mathfrak{p}^n) \in K_{12N(\mathfrak{q}\mathfrak{p}^n)^2} \quad \text{for } \delta \in \mathfrak{O}_K,$$

and the action of the Frobenius automorphisms $\sigma(\lambda)$ of $K_{12N(\mathfrak{q}\mathfrak{p}^n)^2}/K$, $\lambda \in \mathfrak{O}_K$ coprime to $12N(\mathfrak{q}\mathfrak{p}^n)$, is given by

$$\varphi(\delta|\mathfrak{q}\mathfrak{p}^n)^{\sigma(\lambda)} = \epsilon \, \varphi(\delta\lambda|\mathfrak{q}\mathfrak{p}^n)$$

with ϵ a root of unity that is independent of δ. This implies that

$$E_n(\delta) \in K_{12N(\mathfrak{q}\mathfrak{p}^n)^2}$$

and

$$E_n(\delta)^{\sigma(\lambda)} = \frac{\varphi(\delta\lambda - \gamma_n\lambda|\, \mathfrak{q}\mathfrak{p}^n)\, \varphi(\delta\lambda + \gamma_n\lambda|\, \mathfrak{q}\mathfrak{p}^n)}{\varphi^2(\delta\lambda|\, \mathfrak{q}\mathfrak{p}^n)} \quad \text{for } \delta \in \mathfrak{O}_K \setminus \{0\}.$$

For $\lambda = 1 + \tau$, $\tau \in \mathfrak{q}\mathfrak{p}^n$, the φ-values in the numerator can be simplified using the transformation formula of φ:

$$\begin{aligned}\varphi(\delta\lambda \pm \gamma_n\lambda | \mathfrak{q}\mathfrak{p}^n) &= \varphi(\delta\lambda \pm \gamma_n \pm \gamma_n\tau | \mathfrak{q}\mathfrak{p}^n) \\ &= \psi(\tau\gamma_n)e^{\frac{1}{2}l(\delta\lambda \pm \gamma_n, \pm\gamma_n\tau)}\varphi(\delta\lambda \pm \gamma_n | \mathfrak{q}\mathfrak{p}^n),\end{aligned}$$

writing l instead of $l_{\mathfrak{q}\mathfrak{p}^n}$. Hence,

$$E_n(\delta)^{\sigma(\lambda)} = e^{l(\gamma_n, \gamma_n\tau)}E_n(\delta\lambda).$$

Herein, the rule $l(a, bc) = l(a\bar{b}, c)$ implies that

$$l(\gamma_n, \gamma_n\tau) = l(\gamma_n\overline{\gamma}_n, \tau) \in 2\pi i\mathbb{Z}$$

because $\gamma_n\overline{\gamma}_n, \tau \in \mathfrak{q}\mathfrak{p}^n$. Therefore,

$$E_n(\delta)^{\sigma(\lambda)} = E_n(\delta\lambda).$$

E_n being elliptic with respect to $\mathfrak{q}\mathfrak{p}^n$, we further obtain

$$E_n(1)^{\sigma(\lambda)} = E_n(1) \quad \text{for } \lambda \equiv 1 \bmod \mathfrak{q}\mathfrak{p}^n,$$

and it follows that $E_n(1)$ is in $K_{\mathfrak{q}\mathfrak{p}^n}$.

The third assertion of Theorem 6.9.2 is proved similarly: we have

$$Gal(K_{\mathfrak{q}\mathfrak{p}^{n+1}}/K_{\mathfrak{q}\mathfrak{p}^n}) = \{\sigma(1+\xi) \mid \xi \in \mathfrak{q}\mathfrak{p}^n \bmod \mathfrak{q}\mathfrak{p}^{n+1}\}$$

and

$$E_{n+1}(1)^{\sigma(1+\xi)}) = e^{l(\gamma_{n+1}\overline{\gamma_{n+1}}, \xi)}E_{n+1}(1+\xi)$$

with $l = l_{\mathfrak{q}\mathfrak{p}^{n+1}}$. Herein again, $l(\gamma_{n+1}\overline{\gamma_{n+1}}, \xi)$ is in $2\pi i\mathbb{Z}$ because $\xi \in \mathfrak{q}\mathfrak{p}^n$ and $\gamma_{n+1}\overline{\gamma_{n+1}}$ is in $\mathfrak{q}\mathfrak{p}^{n+1}\overline{\mathfrak{p}}^{n+1}$. Hence

$$E_{n+1}(1)^{\sigma(1+\xi)} = E_{n+1}(1+\xi),$$

and the third assertion is proved.

Finally, the second assertion of Theorem 6.9.2 follows from the factorisation of the singular φ-values in Theorem 4.3.1. This factorisation implies that the first φ factor in the numerator of the definition of $E_n(1)$ has the factorisation $\mathfrak{p}^{\frac{1}{\Phi(\mathfrak{p}^n)}}$, whereas the other φ values are units. □

Proof of Theorem 6.9.5 The first two assertions have already been shown. The third assertion follows analogously the proof of Theorem 6.9.2 from the Reciprocity Law, and the last follows from the third. □

7
Integral basis in ray class fields

We start by some motivating remarks on cyclotomic fields. Let $f > 1$ be a natural number and ζ a primitive f-th root of unity. Then

$$k_f := \mathbb{Q}(\zeta)$$

is the ray class field modulo f over $\mathbb{Q}$, and ζ is known to generate a power basis over $\mathbb{Z}$ for the ring of integers $\mathfrak{o}_f$ in k_f:

$$\mathfrak{o}_f = \mathbb{Z}[\zeta].$$

Therefore, the ray class field modulo f over $\mathbb{Q}$ is generated by a torsion point of the unit circle and the same torsion point yields a power basis for ring of integers in k_f.

We replace $\mathbb{Q}$ by the Hilbert class field Ω of an imaginary quadratic number field K and, given an integral ideal $\mathfrak{f}$ in K, we consider the ray class field $K_{\mathfrak{f}}$ instead of k_f. According to Theorem 6.2.3, $K_{\mathfrak{f}}$ is generated over Ω by singular values of Weber's τ function :

$$K_{\mathfrak{f}} = \Omega(\tau(\delta|\mathfrak{O})),$$

where $\delta \in K \setminus \mathfrak{O}$ has order $\mathfrak{f}$, i.e. $\mathfrak{f}$ is the denominator of the ideal (δ). Geometrically, $\tau(\delta|\mathfrak{O})$ is the x-coordinate of a torsion point of an elliptic curve defined over Ω. We may pose the question whether analogous to $\mathfrak{o}_f$ the ring of integers in $K_{\mathfrak{f}}$ can explicitly be constructed by torsion points of an elliptic curve. However, studying simple examples, it transpires that the singular values $\tau(\delta|\mathfrak{O})$ are not suitable for this purpose because their discriminants have too many divisors. Fueter then (1924, 1927) introduced the normalisation of $\wp$ given by the Fueter model in section 8.7.2, to determine the divisors of the relative different of $K_{\mathfrak{f}}/\Omega$ without proving any result on an integral basis in ray class fields – probably due to the lack of the conductor–discriminant formula as in Theorem 3.3.4.

First results on an integral basis were proved 60 years later by Cassou-Noguès and Taylor (1987), again using the function from the model of Fueter. Their book was the "kick-off" for the integral complex multiplication. Then, using other models, Cougnard and Fleckinger proved similar results, for example those in their 1989 paper. However, these results did not cover all cases and most of them assumed a higher base field than Ω.

In this chapter we will follow and refine the ideas used in Schertz(1989, 1990). First, in section 7.1, we introduce a suitable normalisation of the Weierstrass $\wp$ function that is different from the normalisations used before. Then, after a calculation of the discriminant in section 7.2 that uses the first identity of Theorem 1.3.2, together with the factorisation of φ values in Theorem 4.3.1, we obtain an integral basis for all extensions $K_{\mathfrak{f}}/\Omega$ in sections 7.4 and 7.5. The main results in Theorems 7.4.1 and 7.5.3 do not provide power bases in all cases, and in fact one can prove that some extensions $K_{\mathfrak{f}}/\Omega$ do not have an integral power basis, as in the cyclotomic case. Finally, by reformulating the result for $k_f/\mathbb{Q}$, we will see in section 7.4.1 that Theorem 7.4.1 is the elliptic analogue of the cyclotomic result.

We conclude this chapter by explaining the generalisation of Bley (1994) to the extensions $K_{t,\mathfrak{f}}/\Omega_t$ with $t > 1$.

7.1 A normalisation of the Weierstrass $\wp$ function

In what follows, let $\mathfrak{O} = \mathfrak{O}_t$ be an order in a quadratic imaginary number field K. In $\mathfrak{O}$ we choose a fixed basis

$$\mathfrak{O} = [\alpha, 1], \quad \Im(\alpha) > 0.$$

Assuming that

$$t = 1 \text{ in the cases } d = -3, -4,$$

we know by Theorem 6.6.10 that there are algebraic units ϵ_2, ϵ_3 with

$$\frac{\gamma_2(\alpha)}{\epsilon_2}, \ \frac{\gamma_3(\alpha)}{\epsilon_3} \in \Omega.$$

We define the following normalisation of the $\wp$ function for the lattice $\mathfrak{O}$.

Definition 7.1.1 (normalisation of the $\wp$ function) For $\delta \in K \setminus \mathfrak{O}$ we set

$$\mathcal{P}(\delta) := \mathcal{P}(\delta|{\textstyle\binom{\alpha}{1}}) := \left(\epsilon \frac{\wp(\delta|\mathfrak{O})}{\sqrt[6]{\Delta\binom{\alpha}{1}}}\right)^{e}$$

with

$$\epsilon = \begin{cases} \epsilon_2\epsilon_3 & \text{if} \quad d \neq -3, -4, \\ \epsilon_3 & \text{if} \quad d = -3, \\ \epsilon_2 & \text{if} \quad d = -4. \end{cases}$$

e denotes the half of the number of roots of unity in $\mathfrak{O}$: e=1, 3, 2 according to whether $d \neq -3, -4$, $d = -3$ resp. $d = -4$.

We note the relation between Weber's τ function and $\mathcal{P}$:

$$\mathcal{P}(\delta) = \begin{cases} -\frac{1}{12}\frac{\epsilon_2\epsilon_3}{\gamma_2(\alpha)\gamma_3(\alpha)}\tau^{(1)}(\delta|\mathfrak{O}) & \text{if} \quad d \neq -3, -4, \\ \frac{1}{12^2}\frac{\epsilon_2^2}{\gamma_2(\alpha)^2}\tau^{(2)}(\delta|\mathfrak{O}) & \text{if} \quad d = -3, \\ -\frac{1}{12^3}\frac{\epsilon_3^3}{\gamma_3(\alpha)}\tau^{(3)}(\delta|\mathfrak{O}) & \text{if} \quad d = -4. \end{cases}$$

Using this relation, we can apply Theorem 6.2.3 to derive:

Theorem 7.1.2 *For $\delta \in K \setminus \mathfrak{O}$ and an integral ideal $\mathfrak{f}$ of $\mathfrak{O}$ with $\delta\mathfrak{f} \subseteq \mathfrak{O}$ we have*

$$\mathcal{P}(\delta) \in K_{t,\mathfrak{f}},$$

and the action of the Galois group of $K_{t,\mathfrak{f}}/\Omega_t$ on $\mathcal{P}(\delta)$ is given by

$$\mathcal{P}(\delta)^{\sigma(\nu)} = \mathcal{P}(\delta\nu)$$

for every $\nu \in \mathfrak{O}$ coprime to $t\mathfrak{f}$. $\sigma(\nu)$ denotes the Frobenius automorphism of $K_{t,\mathfrak{f}}/K$ belonging to the ideal $\nu\mathfrak{O}_1$.

For explicit calculations it is necessary to have an explicit formula for the unit ϵ and its conjugates. Such a formula is provided by the next theorem.

Theorem 7.1.3 *Let $\mathfrak{p}, \mathfrak{q}$ be two primitive ideals of $\mathfrak{O}$, such that their product is primitive as well, with norms p and q prime to $6t$. These ideals can always be chosen in such a way that their norms satisfy the conditions*

$$\begin{array}{lll} p \equiv q \equiv -1 \ mod\ 4 & if & 2|td \ and\ 3 \nmid td, \\ p \equiv q \equiv -1 \ mod\ 3 & if & 2 \nmid td \ and\ 3|td, \\ p \equiv q \equiv -1 \ mod\ 12 & if & 2|td \ and\ 3|td. \end{array}$$

We can choose $\alpha \in \mathbb{H}$ *such that*

$$\alpha, \tfrac{\alpha}{p}, \tfrac{\alpha}{q} \text{ resp. } \tfrac{\alpha}{pq} \text{ are quotients of a basis of } \mathfrak{O}, \mathfrak{p}_t, \mathfrak{q}_t \text{ resp. } \mathfrak{p}_t\mathfrak{q}_t.$$

Further, we assume the coefficient B *in the equation* $X^2 + BX + C = 0$ *of* α *to satisfy the congruences*

$$\begin{array}{lll} B \equiv 0 \bmod 3 & \text{if} & 3 \nmid td \text{ or } d = -3, \\ B \equiv 1 \bmod 4 & \text{if} & 2 \nmid td, \\ B \equiv 0 \bmod 4 & \text{if} & d = -4. \end{array}$$

Then

$$\epsilon(\alpha) := \begin{cases} \dfrac{\eta(\frac{\alpha}{pq})\eta(\alpha)}{\eta(\frac{\alpha}{p})\eta(\frac{\alpha}{q})} & \text{for} \quad d \neq -3, -4, \\ \\ 1 & \text{for} \quad d = -3, -4 \end{cases}$$

is an algebraic unit, and for δ *and* $\mathfrak{f}$ *as in Theorem 7.1.2 we have*

$$\mathcal{P}(\delta|{}^{\alpha}_{1}) := \left(\epsilon(\alpha) \frac{\wp(\delta|{}^{\alpha}_{1})}{\sqrt[6]{\Delta({}^{\alpha}_{1})}} \right)^{e} \in K_{t,\mathfrak{f}}.$$

The conjugates over K *are obtained as follows: we choose an* N*-system* $\alpha_i \in K$ *for* $N = 12pqN(\mathfrak{f})$ *and further, a system of integers* $\nu \in K$ *prime to* $t\mathfrak{f}$ *such that the corresponding Frobenius automorphisms* $\sigma(\nu)$ *are just the different elements of the Galois group of* $K_{t,\mathfrak{f}}/\Omega_t$*. Then, the conjugates are given by*

$$\mathcal{P}(\delta\nu|{}^{\alpha_i}_{1}).$$

7.2 The discriminant of $\mathcal{P}(\delta)$

In the following, we will assume that $t = 1$. Then, $\mathfrak{O}$ is the maximal order in K, and $\mathfrak{f}$ is an integral ideal of $\mathfrak{O}$. In this case $K_{t,\mathfrak{f}} = K_{\mathfrak{f}}$ is the ray class field modulo $\mathfrak{f}$ over K and $\Omega_t = \Omega$ is the Hilbert class field of K.

According to Theorem 7.1.2 the discriminant of $\mathcal{P}(\delta)$ with respect to the extension $K_{\mathfrak{f}}/\Omega$ can be written as

$$d_{K_{\mathfrak{f}}/\Omega}(\mathcal{P}(\delta)) = N_{K_{\mathfrak{f}}/\Omega}\left(\prod_{\sigma(\nu)} (\mathcal{P}(\delta\nu) - \mathcal{P}(\delta)) \right),$$

where $\sigma(\nu)$ runs through all non-trivial automorphisms of $K_{\mathfrak{f}}/\Omega$. Its factorisation is given by the next theorem:

Theorem 7.2.1 *Let $\delta \in K \setminus \mathfrak{O}$ have order $\mathfrak{f}$. Then*

$$d_{K_{\mathfrak{f}}/\Omega}(\mathcal{P}(\delta)) \sim \begin{cases} d_{K_{\mathfrak{f}}/\Omega}^{w_{\mathfrak{f}}} & \text{for } \mathfrak{f} \text{ composite,} \\ \frac{1}{\mathfrak{p}^{2w_{\mathfrak{f}}(n-1)}} d_{K_{\mathfrak{f}}/\Omega}^{w_{\mathfrak{f}}} & \text{for } \mathfrak{f} = \mathfrak{p}^r \end{cases}$$

with a prime ideal $\mathfrak{p}$. Herein $n = [K_{\mathfrak{f}} : \Omega]$, and $w_{\mathfrak{f}}$ denotes the number of roots of unity in K congruent to 1 modulo $\mathfrak{f}$.

Proof By the formula of Theorem 1.3.2 the factorisation of the discriminant is reduced to the factorisation of values of the Klein form φ. In this way, we obtain the relation

$$\mathcal{P}(\delta\nu) - \mathcal{P}(\delta) = \prod_{s=0}^{e-1} (-\epsilon) \frac{\varphi(\delta(\nu - \lambda^s))\varphi(\delta(\nu + \lambda^s))}{\varphi(\delta\nu)^2 \varphi(\lambda^s \delta)^2} \tag{7.1}$$

with a primitive $2e$-th root of unity λ, and for abbreviation we have set

$$\varphi(z) := \varphi(z|{}^{\alpha}_{1}).$$

The factorisation we are aiming at is now obtained from Theorem 4.3.1. As δ has denominator $\mathfrak{f}$, the φ-values in the denominator of the right-hand side in (7.1) are associated with 1 or to $\mathfrak{p}^{\frac{1}{\Phi(\mathfrak{f})}}$ according to whether $\mathfrak{f}$ is composite or the power of a prime ideal $\mathfrak{p}$. It follows that

$$\mathcal{P}(\delta\nu) - \mathcal{P}(\delta) \sim \prod_{\zeta \in E} \varphi(\delta(\nu - \zeta)) \quad \text{if } \mathfrak{f} \text{ is composite}$$

and

$$\mathcal{P}(\delta\nu) - \mathcal{P}(\delta) \sim \frac{1}{\mathfrak{p}^{\frac{4e}{\Phi(\mathfrak{p}^r)}}} \prod_{\zeta \in E} \varphi(\delta(\nu - \zeta)) \quad \text{if } \mathfrak{f} = \mathfrak{p}^r$$

is the power of a prime ideal. E denotes the group of roots of unity in $\mathfrak{O}$. Since $o(\delta(\nu - \zeta), \mathfrak{O})$ divides $o(\delta, \mathfrak{O}) = \mathfrak{f}$, Theorem 4.3.1 implies that

$$\varphi(\delta(\nu - \zeta)) \not\sim 1 \iff o(\delta(\nu - \zeta), \mathfrak{O}) = \mathfrak{p}^t$$

with the power $\mathfrak{p}^t$ of a prime ideal dividing $\mathfrak{f}$, and in this case we have

$$\varphi(\delta(\nu - \zeta)) \sim \mathfrak{p}^{\frac{1}{\Phi(\mathfrak{p}^t)}}.$$

Now let $\mathfrak{p}$ be a prime ideal dividing $\mathfrak{f}$ and

$$\mathfrak{f} = \mathfrak{b}\mathfrak{p}^r, \quad \mathfrak{p} \nmid \mathfrak{b},$$

with an integral ideal $\mathfrak{b}$. For $\nu \in \mathfrak{O}, \zeta \in E$ and $0 \leq i \leq r-1$ we then have the equivalence

$$o(\delta(\nu-\zeta),\mathfrak{O}) = \mathfrak{p}^{r-i} \iff \begin{cases} \delta(\nu-\zeta) \equiv 0 \bmod \mathfrak{b}\mathfrak{p}^i, \\ \delta(\nu-\zeta) \not\equiv 0 \bmod \mathfrak{b}\mathfrak{p}^{i+1}, \end{cases}$$

and, since δ has the denominator $\mathfrak{f}$, this means that

$$o(\delta(\nu-\zeta),\mathfrak{O}) = \mathfrak{p}^{r-i} \iff \begin{cases} \nu \equiv \zeta \bmod \mathfrak{b}\mathfrak{p}^i, \\ \nu \not\equiv \zeta \bmod \mathfrak{b}\mathfrak{p}^{i+1}. \end{cases}$$

The last condition implies that $\sigma(\nu)$ is in the Galois group of $K_{\mathfrak{f}}/K_{\mathfrak{b}\mathfrak{p}^i}$. Hence, only those differences $\mathcal{P}(\delta\nu) - \mathcal{P}(\delta)$ give us a "$\mathfrak{p}$-contribution" to the discriminant, for which $\sigma(\nu)$ is in the Galois group of $K_{\mathfrak{f}}/K_{\mathfrak{b}\mathfrak{p}^i}$. For every such ν, there is exactly one $i = i(\nu)$, $0 \leq i \leq r-1$, with

$$\sigma(\nu) \in Gal(K_{\mathfrak{f}}/K_{\mathfrak{b}\mathfrak{p}^i}) \setminus Gal(K_{\mathfrak{f}}/K_{\mathfrak{b}\mathfrak{p}^{i+1}}),$$

which is equivalent to

$$\begin{array}{lll} \nu \equiv \zeta & \bmod \mathfrak{b}\mathfrak{p}^i & \text{for a } \zeta \in E, \\ \nu \not\equiv \zeta & \bmod \mathfrak{b}\mathfrak{p}^{i+1} & \text{for all } \zeta \in E. \end{array}$$

Now we set

$$E_j(\nu) := \{\zeta \in E \mid \nu \equiv \zeta \bmod \mathfrak{b}\mathfrak{p}^j\}.$$

Then

$$E_0(\nu) \supseteq E_1(\nu) \supseteq \ldots \supseteq E_i(\nu) \not\supseteq E_{i+1} = \phi$$

and

$$|E_j(\nu)| = w(\mathfrak{b}\mathfrak{p}^j) \text{ f|r } j = 0, \ldots, i.$$

Writing

$$w_j := w(\mathfrak{b}\mathfrak{p}^j),$$

we have the factorisation

$$\prod_{\zeta \in E_i} \varphi(\delta(\nu-\zeta)) \sim \mathfrak{p}^{e_i}$$

with

$$e_i = w_0 \frac{1}{\Phi(\mathfrak{p}^r)} \qquad \text{for } i = 0,$$

$$e_i = (w_0 - w_1)\frac{1}{\Phi(\mathfrak{p}^r)} + \cdots + (w_{i-1} - w_i)\frac{1}{\Phi(\mathfrak{p}^{r-i+1})} + w_i \frac{1}{\Phi(\mathfrak{p}^{r-i})}$$
$$\text{for } 1 \leq i \leq r-1.$$

We contend that

$$e_i = \frac{1}{\Phi_r} c_i \text{ with } c_i := \sum_{j=0}^{i} w_j \Phi_j,$$

where for brevity we have set

$$\Phi_j := \Phi(\mathfrak{p}^j).$$

For verification, we first write

$$e_i \Phi_r := \sum_{j=0}^{i} w_i \left(\frac{\Phi_r}{\Phi_{r-j}} - \frac{\Phi_r}{\Phi_{r-j+1}} \right).$$

Keeping in mind $0 \le i \le r-1$, we have

$$\frac{\Phi_r}{\Phi_{r-j}} - \frac{\Phi_r}{\Phi_{r-j+1}} = \frac{(p-1)p^{r-1}}{(p-1)p^{r-j-1}} - \frac{(p-1)p^{r-1}}{(p-1)p^{r-j}} = p^j - p^{j-1} = \Phi_j,$$

where p denotes the norm of $\mathfrak{p}$. Now we can conclude the $\mathfrak{p}$-part of the discriminant to be

$$\mathfrak{p}^{v_{\mathfrak{p}}(d_{K_{\mathfrak{f}}/\Omega}((\mathcal{P}(\delta))} = \begin{cases} \mathfrak{p}^m \text{ for } \mathfrak{f} \text{ composite,} \\ \left(\frac{1}{\mathfrak{p}^{\frac{1}{\Phi_r}}} \right)^{4e(n-1)n} \mathfrak{p}^m \text{ for } \mathfrak{f} = \mathfrak{p}^r \end{cases}$$

with $n = [K_{\mathfrak{f}} : \Omega]$ and

$$m = [K_{\mathfrak{f}} : \Omega] \sum_{i=0}^{r-1} e_i([K_{\mathfrak{f}} : K_{\mathfrak{b}\mathfrak{p}^i}] - [K_{\mathfrak{f}} : K_{\mathfrak{b}\mathfrak{p}^{i+1}}]).$$

Using the formulae

$$[K_{\mathfrak{f}} : \Omega] = [K_{\mathfrak{b}} : \Omega] \frac{w_r}{w_0} \Phi_r \text{ and } [K_{\mathfrak{f}} : K_{\mathfrak{b}\mathfrak{p}^i}] = \frac{w_0 \Phi_r}{w_i \Phi_i},$$

we can now write m as

$$m = [K_{\mathfrak{b}} : \Omega] \frac{w_r^2}{w_0} \Phi_r \sum_{i=0}^{r-1} c_i \left(\frac{1}{w_i \Phi_i} - \frac{1}{w_{i+1} \Phi_{i+1}} \right).$$

By partial summation this becomes

$$\begin{aligned} m &= [K_{\mathfrak{b}} : \Omega] \tfrac{w_r^2}{w_0} \Phi_r \left(c_0 \tfrac{1}{w_0 \Phi_0} - c_r \tfrac{1}{\Phi_r} + \sum_{i=1}^{r} (c_i - c_{i-1}) \tfrac{1}{w_i \Phi_i} \right) \\ &= [K_{\mathfrak{b}} : \Omega] \tfrac{w_r^2}{w_0} \Phi_r \left(1 - c_r \tfrac{1}{w_r \Phi_r} + r \right) \\ &= [K_{\mathfrak{b}} : \Omega] \tfrac{w_r}{w_0} \left((r+1) w_r \Phi_r - c_r \right). \end{aligned}$$

Comparing this formula to that for the relative discriminant of $K_{\mathfrak{f}}/\Omega$ in Theorem 3.3.5, we then obtain the formula asserted in Theorem 7.2.1, bearing in mind that Ω/K is unramified and hence

$$\mathbf{N}_{\Omega/K}(d_{K_{\mathfrak{f}}/\Omega}) = d_{K_{\mathfrak{f}}/K},$$

which implies that

$$d_{K_{\mathfrak{f}}/\Omega}^{[\Omega:K]} = d_{K_{\mathfrak{f}}/K}.$$

□

7.3 The denominator of $\mathcal{P}(\delta)$

As in section 7.2 we are still assuming that $t = 1$. In view of the integral basis to be constructed and the discriminant formula of Theorem 7.2.1, it is interesting to ask whether the values $\mathcal{P}(\delta)$ are integral. Then, for composite order $\mathfrak{f}$ of δ and $w_{\mathfrak{f}} = 1$ the value $\mathcal{P}(\delta)$ would generate a relative integral power basis for $K_{\mathfrak{f}}/\Omega$. However, $\mathcal{P}(\delta)$, even for composite $\mathfrak{f}$, is in general not integral. A suitable modification is provided by the following theorem:

Theorem 7.3.1 *There exists a number $P \in \Omega$ that is independent of δ and has the following properties:*

$$\mathcal{P}(\delta) - P \textit{ is integral for } \mathfrak{f} \textit{ composite,}$$

$\mathfrak{p}^{\frac{2e}{\Phi(\mathfrak{p}^r)}}(\mathcal{P}(\delta) - P)$ *is integral for $\mathfrak{f} = \mathfrak{p}^r$ the power of a prime ideal.*

Proof First we assume 2 and 3 not to be inert in K. In this case there exist two prime ideals $\mathfrak{p}_2, \mathfrak{p}_3$ of K of norm 2 and 3. We chose $\delta_0 \in K$ of order $\mathfrak{p}_2\mathfrak{p}_3$, and we set

$$P := \mathcal{P}(\delta_0).$$

Then, by Theorem 7.1.2 $\mathcal{P}(\delta_0) \in K_{\mathfrak{p}_2\mathfrak{p}_3}$ and, since in this case $K_{\mathfrak{p}_2\mathfrak{p}_3} = \Omega$, we have $P \in \Omega$. As in section 7.2 we use the formula

$$\mathcal{P}(\delta) - \mathcal{P}(\delta_0) = \prod_{s=0}^{e-1}(-\epsilon)\frac{\varphi(\delta - \lambda^s\delta_0)\varphi(\delta + \lambda^s\delta_0)}{\varphi(\delta)^2\varphi(\lambda^s\delta_0)^2}.$$

Herein $\varphi(\lambda^s\delta_0) \sim 1$, since the order of δ_0 is composite, and we have

$$\varphi(\delta)^2 \sim \mathfrak{p}^{\frac{2}{\Phi(\mathfrak{p}^r)}}$$

if $\mathfrak{f} = \mathfrak{p}^r$ the power of a prime ideal. This implies that $P = \mathcal{P}(\delta_0)$ has the required properties.

This construction of P is always possible if there exists a composite ideal $\mathfrak{g}$ in K with $K_{\mathfrak{g}} = \Omega$. For example, this is true for $d = -3, -4$, taking $\mathfrak{g} = \mathfrak{p}_2\mathfrak{p}_3$ resp. $\mathfrak{g} = \mathfrak{p}_2\mathfrak{p}_5$ with prime ideals dividing 2, 3 and 5.

In the remaining cases, we let δ_0 be of composite order $\mathfrak{g}$. Then we first obtain by summation of the numbers $\mathcal{P}(\delta) - \mathcal{P}(\delta_0\nu)$ with the conjugates $\mathcal{P}(\delta_0\nu)$ of $\mathcal{P}(\delta_0)$ that

$$n\mathcal{P}(\delta) - tr_{K_{\mathfrak{g}}/\Omega}(P(\delta_0)) \text{ with } n = [K_{\mathfrak{g}} : \Omega],$$

satisfies the required integrality conditions. To lose the factor n, we then have to take suitable gcd's for suitable $\mathfrak{g}$'s. We may assume that $d \neq -3, -4$, since P has already been constructed in these cases. Then $K \not\subseteq \mathbb{Q}(\sqrt{-3}), \sqrt{-4})$ and, by Theorem 3.2.3 we conclude that there exists a prime ideal $\mathfrak{q}$ of degree 1 in K of norm q that is prime to 6 and inert in both $\mathbb{Q}(\sqrt{-3})$ and $\mathbb{Q}(\sqrt{-4})$. Hence, the decomposition law for $\mathbb{Q}(\sqrt{-3})$ and $\mathbb{Q}(\sqrt{-4})$ implies that

$$q \equiv -1 \bmod 12.$$

Now, choosing prime ideals $\mathfrak{p}_2, \mathfrak{p}_3$ in K dividing 2 and 3, we set

$$\mathfrak{g}_1 := \mathfrak{p}_2\mathfrak{p}_3, \quad \mathfrak{g}_2 = \mathfrak{p}_2\mathfrak{q}, \quad \mathfrak{g}_3 = \mathfrak{p}_3\mathfrak{q},$$

and we choose $\delta_i \in K \setminus \mathfrak{O}$, $i = 1, 2, 3$, of order $\mathfrak{g}_i$. Then, the gcd of the degrees $n_i = [K_{\mathfrak{g}_i} : \Omega]$ is equal to 1 because of

$$\begin{aligned} n_1 &= 1, 3, 4, 12, \\ n_2 &= m_2\tfrac{q-1}{2}, m_2 = 1, 3, \\ n_3 &= m_3\tfrac{q-1}{2}, m_3 = 2, 8, \end{aligned}$$

and

$$\gcd(q-1, 12) = 2.$$

Hence, there exist $a_1, a_2, a_3 \in \mathbb{Z}$ with

$$a_1n_1 + a_2n_2 + a_3n_3 = 1.$$

Thus, by

$$P = n_1 tr_{K_{\mathfrak{g}_1}/\Omega}(\mathcal{P}(\delta_1)) + n_2 tr_{K_{\mathfrak{g}_2}/\Omega}(\mathcal{P}(\delta_2)) + a_3 tr_{K_{\mathfrak{g}_3}/\Omega}(\mathcal{P}(\delta_3))$$

we have constructed a number $P \in K$ with the desired properties. □

The following theorem describes the denominators of the numbers P in Theorem 7.3.1, where the denominator of P is understood as the largest ideal $\mathfrak{n}$ of Ω such that $\mathfrak{n}P$ is integral.

Theorem 7.3.2 *$12^e P$ is always integral. More precisely:*
$3^e P$ has denominator $(1), (2)^e$ resp. $(4)^e$ according to whether 2 is inert, ramified resp. split in K.
$4^e P$ has denominator $(1), (3)^{\frac{e}{2}}$ resp. $(3)^e$ according to whether 3 is inert, ramified resp. split in K.

Proof First we assume that $d \neq -3, -4$. Then, for every $N \in \mathbb{N} \setminus \{1\}$ we have the formula

$$\sum_{\substack{\xi \in \frac{1}{N}\mathfrak{O} \setminus \mathfrak{O} \\ \pm\xi \bmod \mathfrak{O}}} \mathcal{P}(\xi) = 0.$$

To prove this formula, note that

$$\sum_{\substack{\pm(x,y) \bmod N \\ \gcd(x,y,N) \neq N}} \wp\left(\frac{x\omega + y}{N} \,\middle|\, \begin{matrix}\omega \\ 1\end{matrix}\right)$$

defines a modular form of weight 2, holomorphic in $\mathbb{H}^*$, which by Theorem 2.5.3 must be equal to zero.

We use this formula for $N = 2$ to prove the first assertion of our theorem. In this case the system of representatives of $\frac{1}{2}\mathfrak{O} \setminus \mathfrak{O}$ modulo $\mathfrak{O}$ is given by the elements ξ_1, ξ_2, ξ_3. We further choose $\xi \in K$ of composite order prime to 2. Then, P and $\mathcal{P}(\xi)$ have the same denominator. Using the above formula, we can write

$$3\mathcal{P}(\xi) = (\mathcal{P}(\xi) - \mathcal{P}(\xi_1)) + (\mathcal{P}(\xi) - \mathcal{P}(\xi_2)) + (\mathcal{P}(\xi) - \mathcal{P}(\xi_3)).$$

Herein, according to Theorem 1.3.2, for each summand we have

$$\mathcal{P}(\xi) - \mathcal{P}(\xi_i) \sim \frac{\varphi(\xi - \xi_i)\varphi(\xi + \xi_i)}{\varphi(\xi)^2\varphi(\xi_i)^2}.$$

By Theorem 4.3.1 it follows that their denominators $\mathfrak{n}_i$ are associated with $\varphi(\xi_i)^2$, which again by Theorem 4.3.1 implies the factorisation

$$\mathfrak{n}_i \sim \begin{cases} (1) & \text{if} \quad o(\xi_i, \mathfrak{O}) \text{ composite}, \\ \mathfrak{p}^{\frac{2}{\Phi(\mathfrak{p}^r)}} & \text{if} \quad o(\xi_i, \mathfrak{O}) = \mathfrak{p}^r \text{ a power of a prime ideal.} \end{cases}$$

For $o(\xi_i, \mathfrak{O})$ we find, according to the decomposition of 2 in K, the ideals:

$$\begin{array}{llllr} (2), & (2), & (2) & \text{if} & 2{=}\mathfrak{p}^2, \\ \mathfrak{p}, & \mathfrak{p}^2, & \mathfrak{p}^2 & \text{if} & 2{=}\mathfrak{p}^2, \\ \mathfrak{p}, & \overline{\mathfrak{p}}, & (2) & \text{if} & 2 = \mathfrak{p}\overline{\mathfrak{p}}. \end{array}$$

Hence, the denominators $\mathfrak{n}_i$ are equal to

$$\begin{array}{llllr} (2)^{\frac{2}{3}}, & (2)^{\frac{2}{3}}, & (2)^{\frac{2}{3}} & \text{if} & 2{=}\mathfrak{p}^2, \\ \mathfrak{p}^2, & \mathfrak{p}, & \mathfrak{p} & \text{if} & 2{=}\mathfrak{p}^2, \\ \mathfrak{p}^2, & \overline{\mathfrak{p}}^2, & (1) & \text{if} & 2 = \mathfrak{p}\overline{\mathfrak{p}}, \end{array}$$

For the denominator of $3\mathcal{P}(\xi)$ and hence for the denominator $\mathfrak{n}$ of $3P$ this implies that

$$\begin{array}{rlr} \mathfrak{n}|(2)^{\frac{2}{3}} & \text{if} & 2 = \mathfrak{p}, \\ \mathfrak{n} = \mathfrak{p}^2 = (2) & \text{if} & 2 = \mathfrak{p}^2, \\ \mathfrak{n} = \mathfrak{p}^2\overline{\mathfrak{p}}^2 = (4) & \text{if} & 2 = \mathfrak{p}\overline{\mathfrak{p}}, \end{array}$$

and, more precisely, for $2 = \mathfrak{p}$ we have $\mathfrak{n} = (1)$ since P is in Ω and 2 is unramified in Ω. This proves the first assertion of Theorem 7.3.2. The proof of the second assertion is obtained in the same way using the above formula for $N = 3$.

For $d = -3, -4$ an element P with the properties of Theorem 7.3.1 is given by

$$\mathcal{P}(\delta_0) \text{ with } o(\delta_0, \mathfrak{O}) = \begin{cases} (2\sqrt{-3}) & \text{if} \quad d = -3, \\ (1+\sqrt{-1})(1+2\sqrt{-1}) & \text{if} \quad d = -4, \end{cases}$$

and an explicit calculation shows that in these cases $\mathcal{P}(\delta_0)$ is associated with

$$\frac{2\sqrt{-3}}{9} \text{ resp. } \frac{1+2\sqrt{-1}}{4},$$

which implies the asssertion of our theorem. □

In some cases the construction of P can be symplified using Theorem 7.3.2, which is useful for numerical purposes in particular:

Theorem 7.3.3 *Let $\mathfrak{p}_2, \mathfrak{p}_3, \mathfrak{p}_5$ be prime ideals dividing 2, 3, 5, and let $\mathfrak{p}$ be a prime ideal of degree 1 and norm $p \geq 5$ in K. Then, a number having the property of Theorem 7.3.1 is given by*

$P = 0$	if $2 = \mathfrak{p}_2$ and $3 = \mathfrak{p}_3$,
$P = \mathcal{P}(\delta_0),\ o(\delta_0, \mathfrak{O}) = \mathfrak{p}_2\mathfrak{p}_3$	if $(2 \neq \mathfrak{p}_2$ and $3 \neq \mathfrak{p}_3)$, or $d = -3$
$P = \mathcal{P}(\delta_0),\ o(\delta_0, \mathfrak{O}) = \mathfrak{p}_2\mathfrak{p}_5$	if $d = -4$,
$P = \mathcal{P}(\delta_0),\ o(\delta_0, \mathfrak{O}) = 2$	if 2 split in K,
$P = -tr_{K_\mathfrak{p}/\Omega}(\mathcal{P}(\delta_0)),\ o(\delta_0, \mathfrak{O}) = \mathfrak{p}$	if $2 = \mathfrak{p}_2$ and $p \equiv -1 \bmod 3$,
$P = -\left(\frac{2}{p}\right) tr_{K_\mathfrak{p}/\Omega}(\mathcal{P}(\delta_0)),\ o(\delta_0, \mathfrak{O}) = \mathfrak{p}$	if $3 = \mathfrak{p}_3$ and $p \equiv -1 \bmod 4$.

Proof The first assertion of Theorem 7.3.3 follows from Theorem 7.3.2. The 2nd, 3rd and 4th construction of P is obtained as at the beginning of the proof of Theorem 7.3.1. For the last two constructions we conclude as follows: let P_0 be the number from Theorem 7.3.1. By assumption we have $d \neq -3, -4$ in this case, so $e = 1$ in the definition of $\mathcal{P}$, and similar to the proof of Theorem 7.3.1, it follows that the denominator of

$$P_0 - \mathcal{P}(\delta_0)$$

divides $\mathfrak{p}^{\frac{2}{p-1}}$. We have $\mathcal{P}(\delta_0) \in K_\mathfrak{p}$ and

$$[K_\mathfrak{p} : \Omega] = \frac{p-1}{2}.$$

Hence

$$\frac{p-1}{2} P_0 - tr_{K_\mathfrak{p}/\Omega}(\mathcal{P}(\delta_0))$$

is in Ω having a denominator dividing $\mathfrak{p}^{\frac{2}{p-1}}$, but since $\mathfrak{p}$ is unramified in Ω and $p \geq 5$, the denominator must be (1).

For $2 = \mathfrak{p}_2$, $p \equiv -1 \bmod 3$ the number $3P_0$ is integral according to Theorem 7.3.1 and $\frac{p-1}{2} \equiv -1 \bmod 3$. Hence

$$-P_0 - tr_{K_\mathfrak{p}/\Omega}(\mathcal{P}(\delta_0))$$

is integral. This proves the fifth assertion of Theorem 7.3.2. The sixth assertion is proved completely analogously. □

7.4 Construction of relative integral basis

We will use the preparations of sections 7.1 to 7.3 to prove the following theorem:

Theorem 7.4.1 *Let $\mathfrak{f}$ be an integral ideal of K, and let $\delta \in K \setminus \mathfrak{O}$ have denominator $\mathfrak{f}$. Let P be any number in Ω having the properties* *described in Theorem 7.3.1. We set*

$$\Theta := \mathcal{P}(\delta) - P.$$

Then a basis for the ring of integers $\mathfrak{O}_{K_\mathfrak{f}}$ of $K_\mathfrak{f}$ is given by

$$\begin{aligned}
\mathfrak{O}_{K_\mathfrak{f}} &= \mathfrak{O}_\Omega[\Theta] && \text{if } \mathfrak{f} \text{ is composite,}\\
\mathfrak{O}_{K_\mathfrak{f}} &= \mathfrak{O}_\Omega + \mathfrak{O}_\Omega\alpha\Theta + \cdots + \mathfrak{O}_\Omega\alpha\Theta^{n-1} && \text{if } \mathfrak{f} = \mathfrak{p}^r \nmid 2 \text{ is a prime ideal power,}\\
\mathfrak{O}_{K_\mathfrak{f}} &= \mathfrak{O}_\Omega[\alpha\Theta] && \text{if } \mathfrak{f} = \mathfrak{p}^r | 2 \text{ is a prime ideal power,}
\end{aligned}$$

where $\mathfrak{O}_\Omega$ denotes the ring of integers of Ω, $n = [K_\mathfrak{f} : \Omega]$, and α is a number in Ω with $\alpha \sim \mathfrak{p}$.

Proof The assertions are obtained by Theorems 7.2.1 and 7.3.1. For $\mathfrak{f}$ composite we have $w_\mathfrak{f} = 1$, and Θ is integral according to Theorem 7.3.1. By Theorem 7.2.1 we then have

$$d_{K_\mathfrak{f}/\Omega}(\Theta) = d_{K_\mathfrak{f}/\Omega}(\mathcal{P}(\delta)) \sim d_{K_\mathfrak{f}/\Omega},$$

thereby proving the first assertion. For $\mathfrak{f} = \mathfrak{p}^r$ a prime ideal power not dividing 2 again $w_\mathfrak{f} = 1$. With a number $\alpha \in \Omega$ associated with $\mathfrak{p}$, Theorem 7.2.1 again yields

$$d_{K_\mathfrak{f}/\Omega}(1, \alpha\Theta, \ldots, \alpha\Theta^{n-1}) = \alpha^{2(n-1)} d_{K_\mathfrak{f}/\Omega}(P(\delta_0)) \sim \frac{\alpha^{2(n-1)}}{\mathfrak{p}^{2(n-1)}} \sim d_{K_\mathfrak{f}/\Omega}.$$

This is the second assertion since, according to Theorem 7.3.1, the numbers $\alpha\Theta^i, i = 1, \ldots, n-1$, are integral. In the remaining case, when $\mathfrak{f} = \mathfrak{p}^r | 2$, we have $w_\mathfrak{f} = 2$, so the assertion also follows from Theorems 7.3.1 and 7.2.1. □

Examples 7.4.2

(i) Let $d_K = -20, \mathfrak{f} = (\sqrt{-5})(3\mathbb{Z} + (1+\sqrt{-5})\mathbb{Z})$. Then $\Omega = K(\sqrt{-1})$, $[K_\mathfrak{f} : \Omega] = 4$ and

$$O_{K_\mathfrak{f}} = \mathfrak{O}_\Omega[\Theta],$$

with

$$m_{\Theta,\Omega} = X^4 + \frac{1}{2}(1 + 3\sqrt{5} + 3\sqrt{-5} - \sqrt{-1})X^3$$

$$+(\sqrt{5}+\sqrt{-5}+7\sqrt{-1})X^2$$
$$+\frac{1}{2}(-3-3\sqrt{5}+3\sqrt{-5}+7\sqrt{-1})X$$
$$-\frac{1}{2}(3+\sqrt{5}).$$

(ii) Let $d_K=-4$, $\mathfrak{f}=(\pi)$, $\pi=1+6i$. Then $\Omega=K$, $\left[K_{\mathfrak{f}}:\Omega\right]=9$ and

$$\mathfrak{O}_{K_{\mathfrak{f}}}=\mathfrak{O}_{\Omega}+\mathfrak{O}_{\Omega}\pi\Theta+\cdots+\mathfrak{O}_{\Omega}\pi\Theta^8,$$

with

$$\begin{aligned} m_{\Theta,\Omega}=X^9 &- (3-8i)X^8-(31-4i)X^7+(43+50i)X^6\\ &+ (31-83i)X^5+(75-5i)X^4+(17+36i)X^3\\ &- (9-9i)X^2-(2+i)X+\frac{1-6i}{37}\end{aligned}$$

(iii) Let $d_K=-163$, $\mathfrak{f}=(2)$. Then $\Omega=K$, $\left[K_{\mathfrak{f}}:\Omega\right]=3$ and

$$\mathfrak{O}_{K_{\mathfrak{f}}} = \mathfrak{O}_{\Omega}[2\Theta],$$

with

$$m_{\Theta,\Omega}=X^3+2^2\cdot 5\cdot 23\cdot 29X-\frac{7\cdot 11\cdot 19\cdot 127}{4}\sqrt{-163}.$$

(iv) Let $d_K=-7$, $\mathfrak{f}=(12)$. Then $\Omega=K$, $\left[K_{\mathfrak{f}}:\Omega\right]=16$ and

$$\mathfrak{O}_{K_{\mathfrak{f}}} = \mathfrak{O}_{\Omega}[\Theta],$$

with

$$\begin{aligned} m_{\Theta,\Omega}= \quad & X^{16}+\left(4+4\sqrt{-7}\right)X^{15}+\left(-57+15\sqrt{-7}\right)X^{14}\\ &+\left(-193-83\sqrt{-7}\right)X^{13}+\left(\tfrac{1403}{2}-\tfrac{477}{2}\sqrt{-7}\right)X^{12}\\ &+\left(1533+735\sqrt{-7}\right)X^{11}+\left(-\tfrac{9123}{2}+\tfrac{2063}{2}\sqrt{-7}\right)X^{10}\\ &+\left(-3235-3257\sqrt{-7}\right)X^{9}+\left(\tfrac{25179}{2}-\tfrac{1395}{2}\sqrt{-7}\right)X^{8}\\ &+\left(-1067+5239\sqrt{-7}\right)X^{7}+\left(-\tfrac{22605}{2}-\tfrac{2483}{2}\sqrt{-7}\right)X^{6}\\ &+\left(3273-2541\sqrt{-7}\right)X^{5}+\left(2869+699\sqrt{-7}\right)X^{4}\\ &+\left(-617+325\sqrt{-7}\right)X^{3}+\left(-174-42\sqrt{-7}\right)X^{2}\\ &+\left(8-8\sqrt{-7}\right)X+1\end{aligned}$$

7.4.1 Analogy to cyclotomic fields

The result of Theorem 7.4.1 is unexpected in view of the result on cyclotomic fields mentioned at the beginning of this chapter. However, as we are going to show, there is a remarkable analogy to cyclotomic fields. For explanation, we start by considering the limit

$$\lim_{\Im(\omega)\to\infty} \frac{\wp\left(\frac{1}{N}\middle|\begin{smallmatrix}\omega\\1\end{smallmatrix}\right)}{\sqrt[6]{\Delta\left(\begin{smallmatrix}\omega\\1\end{smallmatrix}\right)}} = \frac{1}{12} + \frac{\zeta}{(1-\zeta)^2}$$

whith the primitive N-th root of unity

$$\zeta = \exp\left(\tfrac{2\pi i}{N}\right).$$

By $\mathbb{Q}_N^+$ we denote the maximal real subfield of the N-th cyclotomic field and by $\mathfrak{O}_N^+$ its ring of integers. Then, with the number

$$\theta := \frac{\zeta}{(1-\zeta)^2} = \frac{1}{\left(\zeta^{\frac{1}{2}} - \zeta^{-\frac{1}{2}}\right)^2}$$

from the above limit the following theorem holds:

Theorem 7.4.3 *For $N > 3$ we have*

$$\begin{aligned} \mathfrak{O}_N^+ &= \mathbb{Z}[\theta] && \text{if } N \text{ is composite,} \\ \mathfrak{O}_N^+ &= \mathbb{Z} + \mathbb{Z}p\theta + \cdots + \mathbb{Z}p\theta^{n-1} && \text{if } N = p^r \text{ is a prime power,} \end{aligned}$$

where n denotes the degree of $\mathbb{Q}_N^+/\mathbb{Q}$.

Proof First, θ is obviously in $\mathbb{Q}_N^+$. To prove the theorem, we apply on the one hand the conductor discriminant formula to the extension $\mathbb{Q}_N^+/\mathbb{Q}$ to calculate its discriminant. On the other hand the discriminant of θ with respect to this extension has to be computed. Both computations are formally completely analogous to the computations in section 7.2, and can be seen by the following proof sketch: first, we have

$$d_{\mathbb{Q}_N^+/\mathbb{Q}}(\theta) = N_{\mathbb{Q}_N^+/\mathbb{Q}}\left(\prod_{\sigma(\nu)}\left(\theta - \theta^{\sigma(\nu)}\right)\right),$$

where $\sigma : \overline{\nu} \mapsto \sigma(\nu),\ \nu > 0$, is the usual parametrisation by class field theory for the Galois group of $\mathbb{Q}_N^+/\mathbb{Q}$. In the product $\sigma(\nu)$ runs through all non-trivial automorphisms of $\mathbb{Q}_N^+/\mathbb{Q}$. The factors of the product can be written in the form

$$\theta - \theta^{\sigma(\nu)} = \frac{\psi(\delta(\nu-1)\psi(\delta(\nu+1))}{\psi(\delta)^2\psi(\delta\nu)^2}$$

with the function

$$\psi(x) = 1 - exp(2\pi i x) \text{ and } \delta = \frac{1}{N}.$$

From cyclotomic theory we know that $\psi(x)$, $x \in \mathbb{Q} \setminus \mathbb{Z}$, has the factorisation

$$\psi(x) \sim \begin{cases} 1 & \text{if the denominator of } x \text{ is composite,} \\ (p)^{\frac{1}{\phi(p^r)}} & \text{if the denominator of } x \text{ is a prime power } p^r, \end{cases}$$

where ϕ denotes the Euler function in $\mathbb{Z}$. Hence, the above difference has the factorisation

$$\begin{aligned} \theta - \theta^{\sigma(\nu)} &\sim \prod_{\xi \in E} \psi(\delta(\nu - \xi)) \quad \text{if } N \text{ is composite,} \\ \theta - \theta^{\sigma(\nu)} &\sim \frac{1}{(p)^{\frac{4}{\phi(p^r)}}} \prod_{\xi \in E} \psi(\delta(\nu - \xi)) \quad \text{if } N = p^r \text{ is a prime power,} \end{aligned}$$

where $E = \{\pm 1\}$ denotes the unit group of $\mathbb{Z}$. The computation of the discriminant of θ is now completely analogous to the computation in section 7.2. □

7.5 Relative integral power basis

In this section we are interested in the relative power basis for the extension $K_{\mathfrak{f}}/\Omega$. For composite $\mathfrak{f}$ such bases are given by Theorem 7.4.1, whereas for non-composite $\mathfrak{f} \nmid 2$, the bases of Theorem 7.4.1 are no power bases. By modifying our construction we will show in this section that "for enough torsion points in the base field" a relative integral power basis can always be constructed. Given an integral ideal $\mathfrak{f}$ in K, we choose a prime ideal $\mathfrak{p}^*$ not dividing $\mathfrak{f}$, and we will construct integral power bases for the extension

$$K_{\mathfrak{f}} K_{\mathfrak{p}^*} / K_{\mathfrak{p}^*}.$$

For $N(\mathfrak{p}^*)|6$ we have $K_{\mathfrak{p}^*} = \Omega$, so that the following theorem also contains an integral power basis for $K_{\mathfrak{f}}/\Omega$ in some cases when $\mathfrak{f}$ is a prime ideal power.

For the following we will again restrict ourselves to the case of $t = 1$. Instead of $\mathcal{P}(\delta)$, we will now use the function

$$\mathcal{Q}(\delta) := \frac{1}{\alpha^{2w_{\mathfrak{p}^*}} (\mathcal{P}(\delta) - \mathcal{P}(\delta^*))}$$

with

$$\delta, \delta^* \in K \setminus \mathfrak{O}, o(\delta, \mathfrak{O}) = \mathfrak{f}, \quad o(\delta^*, \mathfrak{O}) = \mathfrak{p}^*.$$

The number α is chosen according to Theorem 6.9.5 in $K_{\mathfrak{p}^*}$ having the factorisation

$$\alpha \sim \mathfrak{p}^{*\frac{1}{n^*}}, \quad n^* = [K_{\mathfrak{p}^*} : \Omega].$$

By Theorem 7.1.2 $\mathcal{Q}(\delta)$ is in $K_{\mathfrak{f}}K_{\mathfrak{p}^*}$, and to compute the relative discriminant of $\mathcal{Q}(\delta)$ we have, in analogy to section 7.2, to factorise the differences

$$\mathcal{Q}(\delta) - \mathcal{Q}(\delta\nu) = \frac{\mathcal{P}(\delta\nu) - \mathcal{P}(\delta)}{\alpha^{2w_{\mathfrak{p}^*}}(\mathcal{P}(\delta) - \mathcal{P}(\delta^*))(\mathcal{P}(\delta\nu) - \mathcal{P}(\delta^*))}.$$

As in section 7.2 we obtain

$$\mathcal{Q}(\delta) - \mathcal{Q}(\delta\nu) \sim \prod_{\zeta \in E} \varphi(\delta(\nu - \zeta)),$$

and since the Galois group of $K_{\mathfrak{f}}K_{\mathfrak{p}^*}/K_{\mathfrak{p}^*}$ is isomorphic to the Galois group of $K_{\mathfrak{f}/\Omega}$ by $\sigma(\nu) \mapsto \sigma(\nu)|K_{\mathfrak{f}}$, we find analogously to Theorem 7.2.1

$$d_{K_{\mathfrak{f}}K_{\mathfrak{p}^*}/K_{\mathfrak{p}^*}}(\mathcal{Q}(\delta)) \sim d_{K_{\mathfrak{f}}/\Omega} \quad \text{if } w_{\mathfrak{f}} = 1.$$

For the following, we assume that $w_{\mathfrak{f}} = 1$. Then, we have the diagram

$$\begin{array}{ccccc} & & LM & & \\ & / & & \backslash & \\ L & & & & M \\ & \backslash & & / & \\ & & L \cap M & = \Omega & \end{array}$$

for the extensions $L = K_{\mathfrak{f}},\ M = K_{\mathfrak{p}^*}$. By the tower discriminant formula we obtain

$$N_{M/\Omega}\left(d_{LM/M}\right) d_{M/\Omega}^{[LM:M]} = d_{L/\Omega}^{[LM:L]} N_{L/\Omega}\left(d_{LM/L}\right).$$

Since $\mathfrak{p}^* \nmid \mathfrak{f}$, the discriminants $d_{M/\Omega}$ and $d_{L/\Omega}$ must be coprime. Hence

$$d_{L/\Omega}^{[LM:L]} | N_{M/\Omega}\left(d_{LM/M}\right).$$

On the other hand the above computation of the discriminant of $\mathcal{Q}(\delta)$ yields

$$N_{M/\Omega}\left(d_{LM/M}\right) | N_{M/\Omega}\left(d_{LM/M}(\mathcal{Q}(\delta)\right) = N_{M/\Omega}\left(d_{L/\Omega}\right) = d_{L/\Omega}^{[M:\Omega]},$$

and this implies the equality in the next theorem because $[LM : M] = [L : \Omega]$.

Theorem 7.5.1 *We assume that* $w_{\mathfrak{f}} = 1$. *Then*

$$N_{K_{\mathfrak{p}^*}/\Omega}(d_{K_{\mathfrak{f}}K_{\mathfrak{p}^*}/K_{\mathfrak{p}^*}}(\mathcal{Q}(\delta))) \sim N_{K_{\mathfrak{p}^*}/\Omega}(d_{K_{\mathfrak{f}}K_{\mathfrak{p}^*}/K_{\mathfrak{p}^*}}).$$

Now we have again to deal with the problem of $\mathcal{Q}(\delta)$ not being integral in general, which is solved by:

Theorem 7.5.2 *There exist a number $Q \in K_{\mathfrak{p}^*}$, independent of $\mathfrak{f}$, such that*

$$\mathcal{Q}(\delta) - Q$$

is integral.

The condition $w_{\mathfrak{f}} = 1$ in Theorem 7.5.1 is always satisfied for $\mathfrak{f} \nmid 2$. Therefore, by Theorems 7.5.1 and 7.5.2 we obtain:

Theorem 7.5.3 *Let $\mathfrak{f} \nmid 2$ be an integral ideal and $\mathfrak{p}^*$ a prime ideal in K not dividing $\mathfrak{f}$. Then, with the above notations,*

$$\mathfrak{O}_{K_{\mathfrak{f}}K_{\mathfrak{p}^*}} = \mathfrak{O}_{K_{\mathfrak{p}^*}}\left[\mathcal{Q}(\delta) - Q\right].$$

Proof of Theorem 7.5.2 We choose a prime ideal $\mathfrak{p}_1$ different from $\mathfrak{p}^*$ and an element $\delta_1 \in K$ with $o(\delta_1, \mathfrak{O}) = \mathfrak{p}_1$. Then

$$\mathcal{Q}(\delta) - \mathcal{Q}(\delta_1) \text{ is integral,}$$

as can be seen from the factorisation

$$\begin{aligned}
\mathcal{Q}(\delta) - \mathcal{Q}(\delta_1) &= \frac{\mathcal{P}(\delta_1) - \mathcal{P}(\delta)}{\alpha^{2w_{\mathfrak{p}^*}}(\mathcal{P}(\delta) - \mathcal{P}(\delta^*))(\mathcal{P}(\delta_1) - \mathcal{P}(\delta^*))} \\
&\sim \frac{\varphi(\delta)^{2e}\varphi(\delta^*)^{2e}\varphi(\delta_1)^{2e}\varphi(\delta^*)^{2e}}{\alpha^{2w_{\mathfrak{p}^*}}\varphi(\delta)^{2e}\varphi(\delta_1)^{2e}}\varphi(\delta+\delta_1)^e\varphi(\delta-\delta_1)^e \\
&\sim \varphi(\delta+\delta_1)^e\varphi(\delta-\delta_1)^e.
\end{aligned}$$

If $\mathcal{Q}(\delta_1)$ is in $K_{\mathfrak{p}^*}$, then clearly $Q = \mathcal{Q}(\delta_1)$ has the desired properties. Otherwise, as we will show, one can find a prime ideal $\mathfrak{p}_2 \neq \mathfrak{p}^*$, such that the degrees $n_i = [K_{\mathfrak{p}_i} : \Omega]$, $i = 1, 2$, are coprime. Then, as in the proof of Theorem 7.3.1 it is clear that

$$Q = a_1 tr_{K_{\mathfrak{p}_1}K_{\mathfrak{p}^*}/K_{\mathfrak{p}^*}}(\mathcal{Q}(\delta_1)) + a_2 tr_{K_{\mathfrak{p}_2}K_{\mathfrak{p}^*}/K_{\mathfrak{p}^*}}(\mathcal{Q}(\delta_2))$$

for suitable $a_i \in \mathbb{Z}$ with $a_1n_1 + a_2n_2 = 1$ has the property we want.

To show the existence of two prime ideals $\mathfrak{p}_1, \mathfrak{p}_2$ necessary for the above construction of Q, we first assume that $d < -4$, and we take, as in the proof of Theorem 7.3.1, a prime ideal $\mathfrak{p}_1$ of degree 1 with

$$N(\mathfrak{p}_1) \equiv -1 \bmod 12$$

and a prime ideal $\mathfrak{p}_2$ dividing one of the primes 2, 5, 7, 17. The corresponding degrees n_i then satisfy

$$\gcd(n_1, 6) = 1,$$
$$n_2 = 1, 2, 3, 6, 8, 12, 24, 144.$$

If $d = -3, -4$, then there exist three prime ideals $\mathfrak{p}_i$, such that the degrees $n_i = [K_{\mathfrak{p}_i} : \Omega]$ for any two of the $\mathfrak{p}_i$ are coprime. For $d = -3$ the three ideals can be chosen as divisors of 2, 3 and 5, with degrees n_i equal to 1, 1 and 4. If $d = -4$, we take the prime ideal divisors of 2 and 5 having degrees n_i equal to 1. □

Example 7.5.4 Let $d = -7$. Then $K = \Omega$, 2 is split in K, $2 = \mathfrak{p}_2\overline{\mathfrak{p}}_2$. Therefore, we can find elements $\delta_2, \overline{\delta}_2 \in K$ such that

$$o(\delta_2, \mathfrak{O}) = \mathfrak{p}_2 \quad \text{and} \quad o(\overline{\delta}_2, \mathfrak{O}) = \overline{\mathfrak{p}}_2.$$

Then, with $\alpha = \frac{1+\sqrt{-7}}{2} \sim \mathfrak{p}_2$, we have

$$Q = \frac{1}{\alpha^4(\mathcal{P}(\overline{\delta}_2) - \mathcal{P}(\delta_2))} \in \Omega,$$

and, according to Theorem 7.5.2, we obtain a power basis for $K_{\mathfrak{p}}/\Omega$, $\mathfrak{p} = \left[23, \frac{19+\sqrt{-7}}{2}\right]$, generated by

$$\Theta := \frac{1}{\alpha^4(\mathcal{P}(\delta) - \mathcal{P}(\delta_2))} - \frac{1}{\alpha^4(\mathcal{P}(\overline{\delta}_2) - \mathcal{P}(\delta_2))},$$

where δ has order $o(\delta, \mathfrak{O}) = \mathfrak{p}$. The minimal polynomial of Θ over K is given by

$$\begin{aligned} X^{11} &+ (3 + 3\sqrt{-7})\, X^{10} + \left(\frac{-61 + 23\sqrt{-7}}{2}\right) X^9 \\ &+ \left(\frac{-239 - 57\sqrt{-7}}{2}\right) X^8 + \left(\frac{263 - 185\sqrt{-7}}{2}\right) X^7 \\ &+ \left(\frac{559 + 125\sqrt{-7}}{2}\right) X^6 + \left(\frac{-299 + 119\sqrt{-7}}{2}\right) X^5 \\ &+ \left(\frac{5 - 73\sqrt{-7}}{2}\right) X^4 + (44 + 24\sqrt{-7})\, X^3 \\ &+ (-40 + 5\sqrt{-7})\, X^2 + (-2 - 4\sqrt{-7})\, X + 1 \end{aligned}$$

By Theorems 7.4.1 and 7.5.3 we have found a relative integral basis for the extension $K_{\mathfrak{f}}/\Omega$ up to three series of exceptions:

Remark 7.5.5 (exceptional series) Let $d \neq -3, -4$ and $[K_\mathfrak{f} : \Omega] \geq 3$. Then by Theorems 7.4.1 and 7.5.3 we do not get integral relative power bases for $K_\mathfrak{f}/\Omega$ in the following cases:

(i) 2 and 3 inert in K, $\mathfrak{f} = \mathfrak{p}^r \nmid 2$,

(ii) 2 inert and 3 not inert in $K, \mathfrak{f} = \mathfrak{p}^r, \mathfrak{p}|3$,

(iii) 3 inert and 2 not inert in $K, \mathfrak{f} = \mathfrak{p}^r, \mathfrak{p}|2$.

In fact, Cougnard and Fleckinger (1989), Klebel (1995) and Verant (1990) have proved for seven extensions $K_\mathfrak{f}/\Omega$ of series 1 the non-existence of a relative integral power basis. For the series 2, so far, only one exception is known, and for the series 3 a counterexample has not yet been discovered, mainly because of the numerical difficulties due to the degree $[K_\mathfrak{f} : \mathbb{Q}]$, which is quite high in these cases. From Klebel (1995) we quote the following table. The last row contains the number of generators for a relative integral power basis modulo numbers of Ω. It is remarkable that in the cases considered, this number is at most equal to 1.

series	$\mathfrak{f}$	d	$[K_\mathfrak{f} : \Omega]$	$\#PB - Gen.$
2 and 3 inert	$\mathfrak{p}_7$	-19	$3 \cdot 2$	0
	$\mathfrak{p}_3 = (3)$	-19	$4 \cdot 2$	0
		-43	$4 \cdot 2$	0
		-67	$4 \cdot 2$	0
		-163	$4 \cdot 2$	0
	$\mathfrak{p}_{11}$	-19	$5 \cdot 2$	1
		-43	$5 \cdot 2$	1
	$\mathfrak{p}_2^2 = (4)$	-67	$6 \cdot 2$	0
		-163	$6 \cdot 2$	0
2 inert, 3 ramified	$\mathfrak{p}_3^2 = (3)$	-51	$3 \cdot 4$	1
		-123	$3 \cdot 4$	1
		-267	$3 \cdot 4$	0
2 ramified, 3 inert	$\mathfrak{p}_2^4$	-40	$4 \cdot 4$	?

7.6 Bley's generalisation for $K_{t,\mathfrak{f}}/\Omega_t$ with $t > 1$

The results of Theorem 7.4.1 have been generalised to the extensions $K_{t,\mathfrak{f}}/\Omega_t$ by Bley (1994). Similar to Theorem 7.4.1 the construction relies on the function $\mathcal{P}(\delta|\mathfrak{O})$ for a non-maximal order $\mathfrak{O}$. To explain Bley's results, we need some preparations:

First, we have to exclude the discriminants $d = -3, -4$ because the units ϵ_1, ϵ_2 needed for the definition of $\mathcal{P}$ do not exist in these cases. Further, we have to assume $\mathfrak{f}$ to be an integral proper ideal of $\mathfrak{O}_t$. There is no loss of generality here since the ideal $\mathfrak{f}$ of $\mathfrak{O}_t$ is a proper ideal of a larger order $\mathfrak{O}_{t'}$ with $t'|t$. And by class field theory in this situation we have the equality

$$K_{t,\mathfrak{f}} = K_{t',\mathfrak{f}}$$

and the inclusion

$$\Omega_{t'} \subseteq \Omega_t.$$

Hence, the generator of a relative integral power basis for $K_{t',\mathfrak{f}}/\Omega_{t'}$ must also be a generator of a relative integral power basis for $K_{t,\mathfrak{f}}/\Omega_t$. The same argument also holds for the integral bases constructed in the following Theorem 7.6.1 which are no power bases.

The generalisation of Theorem 7.4.1 relies mainly on Theorem 4.3.2, which contains the factorisation of $\varphi(\xi|\mathfrak{a}) := \varphi(\xi|\underline{\alpha})$ for $\underline{\alpha}$ being the basis of a proper ideal $\mathfrak{a}$ of a non-maximal order. In this way Bley proves the following theorem:

Theorem 7.6.1 *Let $\mathfrak{f}$ be a proper ideal of $\mathfrak{O}_t$ and $\delta \in K$ with $\delta\mathfrak{O}_t = \frac{\mathfrak{c}}{\mathfrak{f}}$, where $\mathfrak{c}$ is an integral ideal of $\mathfrak{O}_t$ coprime to $\mathfrak{f}$. Then*

$$\mathcal{P}(\delta|\mathfrak{O}_t) \in K_{t,\mathfrak{f}},$$

and we have

$$d_{K_{t,\mathfrak{f}}/\Omega_t}(\mathcal{P}(\delta|\mathfrak{O}_t)) \sim \begin{cases} d_{K_{t,\mathfrak{f}}/\Omega_t} & \text{for} \quad \mathfrak{f}\mathfrak{O}_1 \text{ composite}, \\ \frac{1}{\mathfrak{p}^{4\kappa m(m-1)}} d_{K_{t,\mathfrak{f}}/\Omega_t} & \text{for} \quad \mathfrak{f}\mathfrak{O}_1 = \mathfrak{p}^r \nmid 2 \\ & \qquad \text{a prime ideal power.} \end{cases}$$

Herein $m := [K_{t,\mathfrak{f}} : \Omega_t]$,

$$\kappa := \frac{1}{\Phi(\mathfrak{p}^n)} \left|(\mathfrak{f} \cap \mathfrak{b})/t\mathfrak{f}\right|$$

with the integral ideal $\mathfrak{b}$ of $\mathfrak{O}_1$ and the exponent n defined by the *decomposition*

$$\mathfrak{t}\mathfrak{f} = \mathfrak{b}\mathfrak{p}^n, \quad \mathfrak{p} \nmid \mathfrak{b}.$$

However, the computation of discriminants needed for the proof are much more complicated than for $t = 1$ in section 7.2, complications coming in particular from the ideal theory of the non-maximal orders $\mathfrak{O}_t$.

Next, Bley proves analogously to Theorem 7.3.1:

Theorem 7.6.2 *Let $\mathfrak{f}$, δ and t be as in Theorem 7.6.1. Then there exists a number $P \in \Omega_t$ having the following properties:*

$$\mathcal{P}(\delta|\mathfrak{O}_t) - P \text{ is integral for } \mathfrak{O}_1\mathfrak{f} \text{ composite,}$$
$$(\mathcal{P}(\delta|\mathfrak{O}_t) - P)\mathfrak{p}^{2\kappa} \text{ is integral for } \mathfrak{O}_1\mathfrak{f} = \mathfrak{p}^r$$

a prime ideal power.

The construction of P is analogous to the proof of Theorem 7.4.1. The prime ideals $\mathfrak{p}_2, \mathfrak{p}_3$ occuring there have to be replaced by prime ideals of $\mathfrak{O}_t$ above 2 resp. 3, and instead of the prime ideal $\mathfrak{q}$ in the proof of Theorem 7.3.1 one has to take the ring ideal $\mathfrak{q}_t = \mathfrak{q} \cap \mathfrak{O}_t$, where, in addition, $\mathfrak{q}$ has to be coprime to $\mathfrak{t}$.

Theorems 7.6.1 and 7.6.2 now imply that:

Theorem 7.6.3 *Let $\mathfrak{f}$, δ and t be as in Theorem 7.6.1 and, in addition, let $\mathfrak{f}\mathfrak{O}_1 \nmid 2$. We set*

$$\Theta := \mathcal{P}(\delta|\mathfrak{O}_t) - P,$$

and we choose $\alpha \in \Omega_t$ with

$$\alpha \sim \mathfrak{p}^{\Phi_t(\mathfrak{f})\kappa}$$

if $\mathfrak{f}\mathfrak{O}_1 = \mathfrak{p}^r$ is a prime ideal power. Then

$$\mathfrak{O}_{K_{t,\mathfrak{f}}} = \mathfrak{O}_{\Omega_t}[\Theta] \qquad \text{for } \mathfrak{f}\mathfrak{O}_1 \text{ composite,}$$

$$\mathfrak{O}_{K_{t,\mathfrak{f}}} = \mathfrak{O}_{\Omega_t} + \mathfrak{O}_{\Omega_t}\alpha\Theta + \cdots + \mathfrak{O}_{\Omega_t}\alpha\Theta^{m-1} \quad \text{for } \mathfrak{f}\mathfrak{O}_1 = \mathfrak{p}^r \text{ a prime ideal power.}$$

Φ_t denotes Euler's function in $\mathfrak{O}_t$.

The construction of $\alpha \in \Omega_t$ having the above factorisation is in general not as easy as in Theorem 7.4.1, except when $\mathfrak{f}$ is regular because then $\Phi_t(\mathfrak{f})\kappa = 1$ and the existence of α follows from the principal ideal theorem. Otherwise, assuming that

$$|\mathfrak{O}_t/(\mathfrak{p} \cap \mathfrak{O}_t)| > 3,$$

we obtain such a number by taking the relative norm

$$N_{K_{t,\mathfrak{f}}/\Omega_t}\left(\frac{1}{\mathcal{P}(\delta_0\nu|\mathfrak{O}_t) - \mathcal{P}(\delta_0)|\mathfrak{O}_t)}\right),$$

where $\delta \in K$ has $\mathfrak{O}_t-$ ideal denominator $\mathfrak{f}$, i.e.

$$\delta_0\mathfrak{O}_t = \frac{\mathfrak{c}}{\mathfrak{f}}, \quad \mathfrak{c} + \mathfrak{f} = \mathfrak{O}_t,$$

and ν is a number from $\mathfrak{O}_t$ not being congruent to a root of unity modulo $\mathfrak{p}$. Alternatively, a construction of α for $|\mathfrak{O}_t/(\mathfrak{p} \cap \mathfrak{O}_t)| \leq 3$ is obtained by generalising the proof construction of the generalised principal ideal theorem to ring class fields.

8
Galois module structure

We begin with some motivating and some historical remarks.

Let N/M be a Galois extension of number fields with Galois group G. Then, according to the normal basis theorem, N is a rank one module over $M[G]$, i.e.

$$N = n \circ M[G]$$

with some element $n \in N$ and the operation of $M[G]$ on N being defined by

$$n \circ \left(\sum_{\sigma \in G} a_\sigma \sigma \right) := \sum_{\sigma \in G} a_\sigma n^\sigma .$$

One may ask whether an analogous result also holds for the ring of integers $\mathfrak{O}_N$ with $M[G]$ being replaced by the associated order

$$\mathfrak{A}_{N/M} := \left\{ \gamma = \sum_{\sigma \in G} a_\sigma \sigma \in M[G] \;\middle|\; \mathfrak{O}_N \circ \gamma \subseteq \mathfrak{O}_N \right\},$$

which means that there exists an element $\vartheta \in \mathfrak{O}_N$ with

$$\mathfrak{O}_N = \vartheta \circ \mathfrak{A}_{N/M}.$$

The existence of such an element ϑ has been shown by Leopoldt (1962) for the extensions $N/M = \mathbb{Q}(\zeta)/\mathbb{Q}$ generated by a root of unity ζ, hence by a torsion point of the unit circle.

In view of this geometric background, the obvious question may be posed whether there exist natural algebraic objects associated with torsion points of elliptic curves, for which a similar result holds. Geometrically, the analogue of $\mathbb{Q}(\zeta)/\mathbb{Q}$ is given by $K_\mathfrak{f}/\Omega$, where Ω is the Hilbert class field of a quadratic imaginary number field and $K_\mathfrak{f}$ some ray class field over K. However, Leopoldt's result does not hold in this general situation, as has been shown by the counterexamples in Ayala and Schertz (1993). The first positive results in this framework were discovered and proved by Cassou-Noguès and Taylor (1987), and further results were obtained by Srivastav and Taylor (1990), Agboola (1996) and Pappas (1998). In this chapter we will essentially follow the exposition presented in Schertz (2005). Before defining in section 8.3 the "integral objects" to be studied, we start by recalling some basic facts about elliptic curves.

In the following let x, y be a pair of Weierstrass functions with respect to an ideal $\mathfrak{m}$ in the ideal class group $\mathfrak{J}_1$ of a quadratic imaginary number field K, and let

$$y^2 + a_1xy + a_3y = x^3 + a_2x^2 + a_4x + a_6 \tag{8.1}$$

be the equation satisfied by x and y. We assume the coefficients a_i to be in a finite extension L of the Hilbert class field of K. The corresponding elliptic curve is then parametrised by

$$\mathbb{C}/\mathfrak{m} \in \xi + \mathfrak{m} \ \mapsto Q(\xi) := \begin{cases} (1 : x(\xi) : y(\xi)) \text{ for } z \notin \mathfrak{m}, \\ (\frac{1}{y(\xi)} : \frac{x(\xi)}{y(\xi)} : 1) \text{ for } y(z) \neq 0 \end{cases}$$

and possesses a group structure with neutral element $O = Q(0) = (0 : 0 : 1)$. We will also use a geometric notation for an elliptic function f with respect to $\mathfrak{m}$ by setting

$$f(Q) := f(\xi), \quad \text{for } Q = Q(\xi),$$

which will be useful in the sequel.

For simplicity we will now restrict ourselves to elliptic curves belonging to proper ideals of the maximal order in K. In fact almost all results we will obtain can be generalised to proper ideals of subordes using the results of Bley (1994) as is explained in 8.12.

First, we collect some basic facts that we will need for the sequel:

8.1 Torsion points and good reduction

$\mathfrak{O}_K$ is acting naturally on $E(\mathbb{C})$ by

$$[\gamma]Q(\xi) := Q(\gamma\xi) \text{ for } \gamma \in \mathfrak{O}_K.$$

$Q \in E(\mathbb{C})$ is called a **torsion point**, if $[\nu]Q = O$ for some element $\nu \in \mathfrak{O}_K \setminus \{0\}$. Given a torsion point $Q = Q(\xi)$ we call

$$o(Q) := o(\xi) := \{\nu \in \mathfrak{O}_K \mid [\nu]Q = O\} = \{\nu \in \mathfrak{O}_K \mid \nu\xi \in \mathfrak{m}\}$$

the **order of** Q or the **order of** ξ. Note that $o(\xi)$ is an ideal not only depending on ξ but also on $\mathfrak{m}$. To be clear we sometimes write

$$o(\xi, \mathfrak{m}) = o(\xi).$$

The torsion points of E are in $E(\mathbb{Q}^c)$, where $\mathbb{Q}^c$ denotes the algebraic closure of $\mathbb{Q}$. Given an integral ideal $\mathfrak{f}$ of $\mathfrak{O}_K$ we set

$$E[\mathfrak{f}] := \{Q \in E(\mathbb{Q}^c) \mid o(Q)|\mathfrak{f}\}.$$

The above operation of $\mathfrak{O}_K$ is compatible with the Galois action in the following sense: let Gal_L be the Galois group of $\mathbb{Q}^c/L$. As we will show in Theorem 8.7.5,

$$Q(\xi)^\sigma = Q(\lambda_\sigma \xi) \text{ for all } \xi \in \mathfrak{f}^{-1}\mathfrak{m}$$

with some element $\lambda_\sigma \in \mathfrak{O}_K$ depending only on σ and $\mathfrak{f}$. Hence, we have the following compatibility

$$[\gamma](Q^\sigma) = ([\gamma]Q)^\sigma \text{ for all } Q \in E[\mathfrak{f}],\ \gamma \in \mathfrak{O}_K \text{ and } \sigma \in Gal_L,$$

which implies that $E[\mathfrak{f}]$ is stable under the action of Gal_L.

Now we choose a fixed prime ideal $\mathfrak{p}$ in K, we assume the a_i to be in L, to be integral for $\mathfrak{p}$ and, further, E to have good reduction above $\mathfrak{p}$. We consider points $Q \in E(\mathbb{Q}^c)$ of the order $o(Q) = \mathfrak{p}^e$, $e > 0$. Then, because of good reduction, Theorem 8.7.5 implies that the values of the uniformising parameter $W = -\frac{x}{y}$ have the factorisation

$$W(Q) \sim \mathfrak{p}^{\frac{1}{\Phi(\mathfrak{p}^e)}}\mathfrak{b},\ \text{for } o(Q) = \mathfrak{p}^e \tag{8.2}$$

with the Euler function Φ in K and an ideal $\mathfrak{b}$ coprime to $\mathfrak{p}$.

A further important property of W is contained in Theorem 1.6.4, whereby for two torsion points $Q_1, Q_2 \in E[\mathfrak{p}^\infty]$ we have a $\mathfrak{P}$-adic addition formula

$$W(Q_1+Q_2) = W(Q_1) + W(Q_2) + W(Q_1)W(Q_2)G(W(Q_1), W(Q_2)) \tag{8.3}$$

with a power series $G(X,Y) \in \mathbb{Z}[a_0, \ldots, a_6][[X,Y]]$. $\mathfrak{P}$ is any prime ideal over $\mathfrak{p}$ in any extension of L containing $W(Q_1+Q_2)$ and $W(Q_1), W(Q_2)$.

8.2 Kummer theory of E

For a given prime ideal $\mathfrak{p}$ with the above properties and $s \geq 0$ we set

$$G_s := E[\mathfrak{p}^s].$$

Further, we fix $r, m \in \mathbb{N}_0$ with

$$m \geq 1 \text{ and } r \geq 0$$

and a coset $P + G_m$ of a point P such that

$$[\mathfrak{p}^m]P \subseteq G_r.$$

Moreover, we assume that

$$G_r \subseteq E(L).$$

Then $P^\sigma - P$ is in G_m for $\sigma \in Gal_L$. Therefore, the coset $P + G_m$ of all "$\mathfrak{p}^m$-th roots" of a point in $E(L)$ of order dividing $\mathfrak{p}^r$ is fixed under Gal_L. Hence, the polynomial

$$h_P(X) := \prod_{R \in P + G_m} (X - W(R))$$

has coefficients in L. Using (8.2) and (8.3), we find that $h_P(X)$ has no multiple roots, so the factorisation of $h_P(X)$ over L is of the form

$$h_P(X) = g_1(X) \cdot \ldots \cdot g_t(X)$$

with different factors $g_i(X)$. We define the following L-algebra

$$\mathbb{M}_P := \mathbb{M}_P(L) := L[X]/(h_P(X)) \cong L[X]/(g_1(X)) \oplus \cdots \oplus L[X]/(g_t(X)),$$

which is obviously the direct sum of fields and which can also be written as

$$\mathbb{M}_P = \mathrm{Map}(P + G_m, \mathbb{Q}^c)^{Gal_L} :=$$

$$\{\ (\theta_R)_{R \in P + G_m} \in (\mathbb{Q}^c)^{|G_m|} \mid \theta_R^\sigma = \theta_{R^\sigma} \forall \sigma \in Gal_L\}.$$

Another useful description of $\mathbb{M}_P$ following directly from the definition is

$$\begin{aligned} \mathbb{M}_P \quad &= \quad \{\ (f(W(R))_{R \in P + G_m} \mid f(X) \in L[X]\} \\ &= \quad L[\theta], \qquad \theta = (W(R))_{R \in P + G_m}. \end{aligned}$$

This shows that $\mathbb{M}_P$ is the direct sum:

$$\mathbb{M}_P = \bigoplus_{i=0}^{p^m - 1} L\, \theta^i.$$

In particular, $\mathbb{M}_O$ defines an L-algebra in the group ring of G_m over $\mathbb{Q}^c$ by

$$\mathbb{A} := \left\{ \sum_{Q \in G_m} a_Q Q \;\middle|\; (a_Q)_{Q \in G_m} \in \mathbb{M}_O \right\}.$$

It can also be written as

$$\mathbb{A} = \bigoplus_{i=0}^{p^m - 1} L\, \alpha^{(i)}, \quad \alpha^{(i)} = \sum_{Q \in G_m} W(Q)^i Q.$$

Then for P with the above properties, $\mathbb{M}_P$ is an $\mathbb{A}$-module with respect to the action

$$(\theta_R)_{R \in P+G_m} \circ \sum_{Q \in G_m} a_Q Q := \left(\sum_{Q \in G_m} \theta_{R-Q} a_Q \right)_{R \in P+G_m}, \tag{8.4}$$

and $\theta \in \mathbb{M}_P$ is a generating element over $\mathbb{A}$,

$$\mathbb{M}_P = \theta \circ \mathbb{A}$$

if and only if for all characters χ of G_m the resolvent

$$(\theta, \chi) := \sum_{Q \in G_m} \theta_{P+Q}\, \overline{\chi}(Q)$$

is different from zero.

In the following sections 8.3 to 8.7 we will explain the central definitions, problems and results. Most of the proofs will follow in sections 8.8 to 8.9.

8.3 Integral objects

For clarity we summarise the hypothesis made so far and which will also be assumed in the following.

General hypothesis (8.5)

(a) K is a quadratic imaginary number field and $\mathfrak{O}_K$ its maximal order.
(b) E is an elliptic curve, defined by the equation (8.1) of a pair of Weierstrass functions with respect to an ideal $\mathfrak{m}$ in $\mathfrak{I}_1$ and having coefficients in a finite extension L of the Hilbert class field of K.
(c) $\mathfrak{p}$ is a fixed prime ideal in $\mathfrak{O}_K$ of norm p, we assume the coefficients of the equation (8.1) to be $\mathfrak{p}$-integral and E to have good reduction above $\mathfrak{p}$.
(d) r, m are in $\mathbb{N}_0$, $m \geq 1$ and $P \in E(\mathbb{Q}^c)$ with

$$G_r \subseteq E(L) \quad \text{and} \quad [\mathfrak{p}^m]P \in G_r.$$

To define the "Integral Objects" in $\mathbb{M}_P$, we make some preliminary remarks: let $\mathfrak{O}$ be an order in $\mathbb{M}_P$, and for a given prime ideal $\mathfrak{p}'$ in L we denote by $\mathfrak{O}_{\mathfrak{p}'}$ the local component belonging to $\mathfrak{p}'$, i.e. the completion of $\mathfrak{O}$ in $\mathbb{M}_P(L_{\mathfrak{p}'})$. Then $\mathfrak{O}$ is uniquely determined by the completions with respect to all prime ideals $\mathfrak{p}'$ of L. Conversely, if for every prime ideal $\mathfrak{p}'$ of L an order $\mathfrak{O}_{\mathfrak{p}'}$ in $\mathbb{M}_P(L_{\mathfrak{p}'})$ is defined such that $\mathfrak{O}_{\mathfrak{p}'}$ is the maximal order for almost all $\mathfrak{p}'$, then there exists a unique order $\mathfrak{O}$ in $\mathbb{M}_P$ having local components $\mathfrak{O}_{\mathfrak{p}'}$.

The following definitions are motivated by the Leopoldt approach for cyclotomic fields, which was also the starting point for Cassou-Noguès and Taylor (1987). Let $\mathfrak{O}_P$ be the maximal order in $\mathbb{M}_P$. It is the direct sum of the maximal orders of the fields $L[X]/(g_i(X))$ in the decomposition of $\mathbb{M}_P$. The integral object, we will study, is the order $\tilde{\mathfrak{O}}_P$ defined by its local components:

$$(\tilde{\mathfrak{O}}_P)_{\mathfrak{p}'} := \begin{cases} \{\ (f(W(R)))_{R \in P+G_m} \mid f(X) \in \mathfrak{O}_{L_{\mathfrak{p}'}}[[X]]\} & \text{if} \quad \mathfrak{p}'|\mathfrak{p}, \\ (\mathfrak{O}_P)_{\mathfrak{p}'} & \text{if} \quad \mathfrak{p}' \nmid \mathfrak{p}. \end{cases}$$

Herein, the local components belonging to $\mathfrak{p}'$ with $\mathfrak{p}'|\mathfrak{p}$ can be written as a direct sum:

$$(\tilde{\mathfrak{O}}_P)_{\mathfrak{p}'} = \bigoplus_{i=0}^{p^m-1} \mathfrak{O}_{L_{\mathfrak{p}'}}\,\theta^i \quad \text{mit} \quad \theta = (W(R))_{R \in P+G_m}.$$

The associated order of $\tilde{\mathfrak{O}}_P$ in $\mathbb{A}$ is defined as the subring

$$\{\gamma \in \mathbb{A} \mid \tilde{\mathfrak{O}}_P \circ \gamma \subseteq \tilde{\mathfrak{O}}_P\},$$

and under certain conditions we will show that this associated order is given by

$$\mathfrak{A} := \frac{1}{\pi^m}\left\{ \sum_{Q \in G_m} a_Q\, Q \;\middle|\; (a_Q)_{Q \in G_m} \in \tilde{\mathfrak{O}}_O \right\}.$$

Herein, π is an element of the Hilbert class field of K that is associated with $\mathfrak{p}$, which exists according to the principal ideal theorem.

For $\mathfrak{p}' \mid \mathfrak{p}$ we can then, in analogy to $(\tilde{\mathfrak{O}}_P)_{\mathfrak{p}'}$, write the local components of $\mathfrak{A}$ as a direct sum:

$$\mathfrak{A}_{\mathfrak{p}'} = \bigoplus_{i=0}^{p^m-1} \mathfrak{O}_{L_{\mathfrak{p}'}}\,\tau_i \quad \text{with} \quad \tau_i = \frac{1}{\pi^m} \sum_{Q \in G_m} W(Q)^i Q.$$

In later results Srivastav and Taylor (1990), Agboola (1996) and Pappas (1998) adopted a different point of view for the definition of $\tilde{\mathfrak{O}}_P$ and $\mathfrak{A}$. In these papers the first object to be defined is $\mathfrak{A}$ as the

Cartier dual of the affine $\mathfrak{O}_L$ group scheme belonging to the $\mathfrak{p}^m$ torsion points and then $\tilde{\mathfrak{O}}_P$ by setting

$$\tilde{\mathfrak{O}}_P := \{\theta \in \mathfrak{O}_P \mid \theta \circ \mathfrak{A} \subseteq \mathfrak{O}_P\}.$$

In fact, this leads to the same objects as defined above. The results obtained in the following constitute a slight generalisation of the result proved by Srivastav and Taylor in 1990. Furthermore, we will include a global description of $\mathfrak{A}$.

For the reader's convenience we start by formulating the results with only part of the proofs for explanation in this section and the following sections 8.4, 8.5 and 8.6. Then, after some preparations about models of elliptic curves in section 8.7, the full proofs will be given in section 8.8.

Theorem 8.3.1 *We assume (8.5) to be given. Then*

(i) $\mathfrak{A}$ *is a ring,*

(ii) $\tilde{\mathfrak{O}}_P \circ \mathfrak{A} \subseteq \tilde{\mathfrak{O}}_P$,

(iii) *the $\mathfrak{p}$-part of the discriminant of $\tilde{\mathfrak{O}}_P$ is equal to the $\mathfrak{p}$-part of the discriminant $h_P(X)$ and is given by*

$$discr(\tilde{\mathfrak{O}}_P)_{\mathfrak{p}} = discr(h_P(X))_{\mathfrak{p}} = \mathfrak{p}^{mp^m}.$$

Moreover,

- *if E has everywhere good reduction* **or**
- *if $E(L)$ contains a point of order $\mathfrak{q}$, where $\mathfrak{q}$ is coprime to $N(\mathfrak{p})$ and $\frac{\mathfrak{q}}{\gcd(w_K,\mathfrak{q})}$ is composite,*

then

$$discr(\tilde{\mathfrak{O}}_P) = \mathfrak{p}^{mp^m}.$$

In the following section we will be concerned with the order $\tilde{\mathfrak{O}}_P$, defined above by its local components, and we will give a global construction of it as a $\mathfrak{O}_L$-algebra. Next, $\tilde{\mathfrak{O}}_P$ will be considered as a $\mathfrak{A}$-module, and we will construct Galois generators, i.e. elements $\theta \in \tilde{\mathfrak{O}}_P$ with $\tilde{\mathfrak{O}}_P = \theta \circ \mathfrak{A}$.

For the constructions we are aiming for, we have to assume the elliptic curve to have some of the following properties with respect to an integral ideal $\mathfrak{q}$ of $\mathfrak{O}_K$:

$$\begin{aligned} &K_{\mathfrak{q}} \subseteq L \text{ and} \\ &L(E[\mathfrak{f}]) \subseteq LK_{\mathfrak{q}\mathfrak{f}} \text{ for all integral ideals } \mathfrak{f} \text{ of } K, \end{aligned} \tag{8.6}$$

where $K_{\mathfrak{c}}$ denotes the ray class field modulo as $\mathfrak{c}$ over K. Later in proposition 8.7.7 we will show that the next condition (8.7) is sufficient for (8.6).

$$\begin{array}{ll} E[\mathfrak{q}] \subseteq E(L) \ \mathfrak{q} \nmid 2 & \text{if } d_K \neq -3, -4, \\ E[\mathfrak{q}] \subseteq E(L) \text{ and } 12|\mathfrak{q} & \text{if } d_K = -3, \\ E[\mathfrak{q}] \subseteq E(L) \text{ and } 4|\mathfrak{q} & \text{if } d_K = -4, \end{array} \tag{8.7}$$

where d_K denotes the discriminant of K.

In particular, for the construction of a generating element θ of $\tilde{\mathfrak{H}}_P$ over $\mathfrak{A}$ we sometimes make the following hypothesis:

$$\begin{array}{l} \mathfrak{q} \text{ is composite,} \\ \mathfrak{d}^2|\mathfrak{q} \text{ for every prime ideal } \mathfrak{d} \text{ of norm } N(\mathfrak{d}) = 2 \text{ dividing } \mathfrak{q} \end{array} \tag{8.8}$$

or a bit stronger:

$$\begin{array}{l} \mathfrak{q} \text{ is composite,} \\ \mathfrak{d}^2|\mathfrak{q} \text{ for every prime ideal } \mathfrak{d} \text{ of norm } N(\mathfrak{d}) = 2, 3 \text{ dividing } \mathfrak{q}. \end{array} \tag{8.9}$$

8.4 Global construction of $\tilde{\mathfrak{H}}_P$ and $\mathfrak{A}$ as $\mathfrak{O}_L$-algebras

The discriminant formula of Theorem 8.3.1 implies the following criterion:

Theorem 8.4.1 *We assume (8.5) to be given. We let $T \in L(x, y)$ be a uniformising parameter of the origin of E without poles outside $K \setminus \mathfrak{p}^{-\infty}\mathfrak{m}$, and we assume that:*

$$T(R+Q) - T(R) \sim \mathfrak{p}^{\frac{1}{\Phi(o(Q))}} \text{ for all } R, Q \in E[\mathfrak{p}^\infty],\ Q \neq O.$$

Then $(T(R))_{R \in P+G_m}$ is a generating element of $\tilde{\mathfrak{H}}_P$ over $\mathfrak{O}_L$:

$$\tilde{\mathfrak{H}}_P = \{\ (f(T(R)))_{R \in P+G_m} \mid f \in \mathfrak{O}_L[X]\}.$$

In particular,

$$\mathfrak{A} = \frac{1}{\pi^m} \left\{ \sum_{Q \in G_m} f(T(Q))\, Q \mid f(X) \in \mathfrak{O}_L[X] \right\}.$$

More precisely, $\tilde{\mathfrak{H}}_P$ and $\mathfrak{A}$ are free modules on $\mathfrak{O}_L$:

$$\tilde{\mathfrak{H}}_P = \bigoplus_{i=0}^{p^m-1} \mathfrak{O}_L \Theta^i \qquad \textit{with} \qquad \Theta = (T(R))_{R \in P+G_m}$$

$$\mathfrak{A} = \bigoplus_{i=0}^{p^m-1} \mathfrak{O}_L \tau_i \qquad \text{with} \qquad \tau_i = \frac{1}{\pi^m} \sum_{Q \in G_m} T(Q)^i Q.$$

To construct a function satisfying the hypothesis of Theorem 8.4.1, we use the normalisation of $\wp$ given in Definition 7.1.1:

$$\mathcal{P}(z) := \left(\epsilon \frac{\wp(z|\mathfrak{m})}{\sqrt[6]{\Delta(\mathfrak{m})}} \right)^{\frac{w_K}{2}}.$$

We set

$$T(z) := \mathcal{P}(\delta + z) - \mathcal{P}(\delta) \tag{8.10}$$

with an element δ in K of order $o(\delta) = \mathfrak{q} \neq (1)$.

Theorem 8.4.2 *We assume (8.5) to be given and, further, $E(L)$ to contain a point of order $\mathfrak{q}$ coprime to $\mathfrak{p}$ and $\frac{\mathfrak{q}}{gcd(w_K, \mathfrak{q})}$ be composite. Then the function T defined in (8.10) with an element δ of order $\mathfrak{q}$ is in $L(x, y)$ and satisfies the hypothesis of Theorem 8.4.1.*

8.5 Construction of a generating element for $\tilde{\mathfrak{O}}_P$ over $\mathfrak{A}$

Using the discriminant formula of Theorem 8.3.1, we obtain the following criterion:

Theorem 8.5.1 *We assume (8.5) to be given. Then every element $\epsilon \in \tilde{\mathfrak{O}}_P$ with*

$$(\epsilon, \chi) \sim \mathfrak{p}^m \text{ for all characters } \chi \text{ of } G_m$$

is a generating element of $\tilde{\mathfrak{O}}_P$ over $\mathfrak{A}$:

$$\tilde{\mathfrak{O}}_P = \epsilon \circ \mathfrak{A}.$$

In particular, in this case $\mathfrak{A}$ is the associated order of $\tilde{\mathfrak{O}}_P$ in $\mathbb{A}$.

In the following a generating element for $\tilde{\mathfrak{O}}_P$ over $\mathfrak{A}$ will be constructed using the normalised σ-function $\varphi(z|\mathfrak{m}) = \sigma^*(z|\mathfrak{m}) \sqrt[12]{\Delta(\mathfrak{m})}$ of a complex lattice $\mathfrak{m}$. We choose an integral ideal $\mathfrak{q}$ of K coprime to $N(\mathfrak{p})$ and satisfying (8.9). Then there exist two elements $\gamma, \delta \in \frac{N(\mathfrak{p}^m)}{\mathfrak{q}}\mathfrak{m}$ such that

$$o(\delta), o(\gamma), o(\delta + \gamma) \text{ and } o\left(\delta + \frac{\gamma}{N(\mathfrak{p}^m)}\right) \text{ are composite.} \tag{8.11}$$

This choice can also be made under the weaker condition (8.8) if $N(\mathfrak{p}^m) \equiv 1 \bmod \mathfrak{q}$ because then the last two conditions in (8.11) coincide. We set

$$h_\gamma(z) = e^{-\frac{1}{2}l(z,\gamma)} \frac{\varphi(z+\gamma|\mathfrak{m})}{\varphi(z|\mathfrak{m})}, \quad l(z,\gamma) := z\gamma^* - z^*\gamma,$$

and with an element $\omega \in \mathfrak{O}_K$ satisfying the congruences

$$\begin{aligned} \omega &\equiv 0 \bmod \mathfrak{p}^m, \\ \omega &\equiv 1 \bmod N(\mathfrak{q}) \end{aligned} \tag{8.12}$$

we define

$$g(z) := \frac{h_\gamma(\delta + z)}{h_\gamma(\omega(\delta+z))}. \tag{8.13}$$

Now we can prove the following theorem:

Theorem 8.5.2 *We assume (8.5) to be given, and we choose an integral composite ideal $\mathfrak{q}$ of K coprime to $N(\mathfrak{p})$ that satisfies the conditions (8.9) and (8.6). Then the function g, defined in (8.13), is in $L(x,y)$, the element*

$$\theta := (g(R))_{R\in P+G_m}$$

is in $\tilde{\mathfrak{O}}_P$, and its resolvents are associated with $\mathfrak{p}^m$. Hence θ is a generating element of $\tilde{\mathfrak{O}}_P$ over $\mathfrak{A}$.

For $N(\mathfrak{p}^m) \equiv 1$ mod $\mathfrak{q}$ the assertion also holds under the weaker condition (8.8). Otherwise the construction of a generating element can be reduced to the case $N(\mathfrak{p}^m) \equiv 1$ mod $\mathfrak{q}$, using the map λ described below.

Further, the numbers in Theorem 8.5.2 can be used for the construction of a generating element in the more general case

$$\mathfrak{p} \nmid w_K, \tag{8.14}$$

where w_K denotes the number of roots of unity in K. Therefore, we consider the following operation: given two points $P_1, P_2 \in E(\mathbb{Q}^c)$ of order dividing $\mathfrak{p}^{r+m}$ and two elements $\theta^{(1)} \in \tilde{\mathfrak{O}}_{P_1}$, $\theta^{(2)} \in \tilde{\mathfrak{O}}_{P_2}$, we find that by

$$\begin{aligned} \theta^{(1)} \circ \theta^{(2)} &:= (\theta_{P_1+P_2+Q})_{Q\in G_m}, \\ \theta_{P_1+P_2+Q} &:= \tfrac{1}{\pi^m} \sum_{Q'\in G_m} \theta^{(1)}_{P_1+Q+Q'} \theta^{(2)}_{P_2-Q'}, \end{aligned} \tag{8.15}$$

an element in $\tilde{\mathfrak{O}}_{P_1+P_2}$ is defined. Further, we have the following relation for resolvents:

$$(\theta^{(1)} \circ \theta^{(2)}, \chi) = \frac{1}{\pi^m}(\theta^{(1)}, \chi)(\theta^{(2)}, \chi).$$

In particular, the resolvents of $\theta^{(1)} \circ \theta^{(2)}$ are associated with $\mathfrak{p}^m$ if the same is true for the resolvents of $\theta^{(1)}$ and $\theta^{(2)}$.

Now, for $\mathfrak{p}$ prime to w_K we choose an integral composite ideal $\mathfrak{q}$ prime to $N(\mathfrak{p})$ that satisfies (8.9) and

$$\gcd(\Phi(\mathfrak{q}), \mathfrak{p}) | w_K.$$

Let $\tilde{L}$ be the extension of L generated by the torsion points of order $\mathfrak{q}$ resp. of order 12 or 4 if $d_K = -3$ or $d_K = -4$. Then

$$n := [\tilde{L} : L] \mid w_K \Phi(\mathfrak{q}).$$

Hence n is prime to $N(\mathfrak{p})$, and we can find $n' \in \mathbb{N}$ with

$$n\, n' \equiv 1 \bmod \mathfrak{p}^{r+m}.$$

Let $\theta^{(1)}$ be the element in $\tilde{\mathfrak{O}}_{[n']P}$ $(= \tilde{\mathfrak{O}}_{[n']P}(\tilde{L}))$ defined in Theorem 8.5.2, and let $\theta^{(1)}, \ldots, \theta^{(n)}$ be the image of $\theta^{(1)}$ under the automorphisms of $\tilde{L}/L$. We set

$$\theta := \theta^{(1)} \circ \ldots \circ \theta^{(n)}, \tag{8.16}$$

thereby obtaining an element in $\tilde{\mathfrak{O}}_P$ $(= \tilde{\mathfrak{O}}_P(L))$ that satisfies the hypothesis of Theorem 8.5.1. This proves the following theorem:

Theorem 8.5.3 (Srivastav–Taylor) *We assume (8.5) to be given, one of the hypotheses in Theorem 8.3.1 (iii), and further, $\mathfrak{p} \nmid w_K$. Then $\tilde{\mathfrak{O}}_P$ is a free module of rank 1 on $\mathfrak{A}$.*

The relation between $\tilde{\mathfrak{O}}_P$ for different m's that we are going to consider, can also be used for the construction of generating elements. Let $\mathfrak{m}$ and $\mathfrak{n}$ be ideals of $\mathfrak{O}_K$,

$$\mathfrak{n} \subset \mathfrak{m},$$

and let $E^{\mathfrak{m}}, E^{\mathfrak{n}}$ be elliptic curves being parametrised pairs of Weierstrass functions $x(z|\mathfrak{n}), y(z|\mathfrak{n})$ resp. $x(z|\mathfrak{m}), y(z|\mathfrak{m})$ with respect to $\mathfrak{m}$ and $\mathfrak{n}$. Then the kernel of the epimorphism

$$\lambda : E^{\mathfrak{n}} \to E^{\mathfrak{m}},$$

defined by

$$E^{\mathfrak{n}} \ni Q \mapsto \xi_Q + \mathfrak{n} \mapsto \xi_Q + \mathfrak{m} \mapsto Q' =: \lambda(Q) \in E^{\mathfrak{m}},$$

consists of all $\mathfrak{n}\mathfrak{m}^{-1}$-torsion points of $E^{\mathfrak{n}}$. We consider the special case

$$\mathfrak{n} = \mathfrak{m}\mathfrak{p}^s,\ 1 \leq s < m,$$

and we assume $E^{\mathfrak{m}}$, $E^{\mathfrak{n}}$ and $\mathfrak{p}$ to satisfy (8.5a)–(8.5d) for the same base field L. Given an elliptic curve $E^{\mathfrak{m}}$ defined over L having good reduction above $\mathfrak{p}$, we can take as $E^{\mathfrak{n}}$ the Fueter model, for example, if $\mathfrak{p}$ is odd and the Deuring model if $\mathfrak{p} \nmid 3$. Further, a common field of definition for $E^{\mathfrak{m}}$ and $E^{\mathfrak{n}}$ is obtained by a finite extension $\hat{L}$ of L, and in addition we can achieve $\hat{L}(x(z|\mathfrak{n}), y(z|\mathfrak{n}))/\hat{L}(x(z|\mathfrak{m}), y(z|\mathfrak{m}))$ to be Galois with the Galois group

$$Gal\left(\hat{L}(x(z|\mathfrak{n}), y(z|\mathfrak{n})) \middle/ \hat{L}(x(z|\mathfrak{m}), y(z|\mathfrak{m}))\right) = \{\tau_\xi \mid \xi \in \mathfrak{m} \bmod \mathfrak{n}\}. \tag{8.17}$$

Herein τ_ξ denotes the substitution $f(z) \mapsto f(z+\xi)$. (For instance, such a field $\hat{L}$ is obtained by adjunction of the $\mathfrak{n}\mathfrak{m}^{-1}$-torsion points of $E^{\mathfrak{n}}$ and the coefficients of the rational functions F, F_1 involved in the representations $x(z|\mathfrak{m}) = F(x(z|\mathfrak{n}), y(z|\mathfrak{n}))$, $y(z|\mathfrak{m}) = F_1(x(z|\mathfrak{n}), y(z|\mathfrak{n}))$.) The kernel consists of all $\mathfrak{p}^s$-torsion points of $E^{\mathfrak{n}}$, and we have

$$\lambda(P + G_m^{\mathfrak{n}}) = \lambda(P) + G_{m-s}^{\mathfrak{m}}$$

with the obvious meaning of $\mathfrak{m}$ and $\mathfrak{n}$ in the exponent. We also denote by λ the induced map

$$\lambda : \mathrm{M}_P^{\mathfrak{n}}(m) \to \mathrm{M}_{\lambda(P)}^{\mathfrak{m}}(m-s),$$
$$(\lambda(\theta))_R := \sum_{R' \in \lambda^{-1}(R)} \theta_{R'},$$

and, as we will show later, we have

$$\lambda(\tilde{\mathfrak{O}}_P^{\mathfrak{n}}(m)) \subseteq \mathfrak{p}^s \tilde{\mathfrak{O}}_{\lambda(P)}^{\mathfrak{m}}(m-s). \tag{8.18}$$

Further, considering the characters of $G_{m-s}^{\mathfrak{m}} = \lambda(G_m^{\mathfrak{n}})$ as characters of $G_m^{\mathfrak{n}}$, the resolvents of $\lambda(\theta)$ are equal to the resolvents of θ. Hence, if $\theta \in \tilde{\mathfrak{O}}_P^{\mathfrak{n}}(m)$ satisfies the condition of Theorem 8.5.1, this is also valid for

$$\frac{1}{\pi^{m-s}} \lambda(\theta) \in \tilde{\mathfrak{O}}_{\lambda(P)}^{\mathfrak{m}}(m-s).$$

Therefore, by generating elements of $\tilde{\mathfrak{O}}_P^{\mathfrak{n}}(m)$ we can obtain generating elements of $\tilde{\mathfrak{O}}_{\lambda(P)}^{\mathfrak{m}}(m-s)$.

8.6 Galois module structure of ray class fields

In this section we consider Galois extensions of ray class fields over a quadratic imaginary number field K,

$$N_{\mathfrak{q}}/M_{\mathfrak{q}} = K_{\mathfrak{q}\mathfrak{p}^{r+m}}/K_{\mathfrak{q}\mathfrak{p}^r}, \qquad [N_{\mathfrak{q}} : M_{\mathfrak{q}}] = N(\mathfrak{p}^m),$$

with an integral ideal $\mathfrak{q}$, a prime ideal $\mathfrak{p}$ of K and exponents m and r satisfying

$$1 \leq m \leq r.$$

On certain conditions for $\mathfrak{q}$ and $\mathfrak{p}$ we will be able to use Theorems 8.4.2 and 8.5.2 to find explicit constructions of the associated order of N/M as well as a Galois generating element for $\mathfrak{O}_N$

The condition $1 \leq m \leq r$ implies the Galois group $G = G(N_\mathfrak{q}/M_\mathfrak{q})$ to be isomorphic to $\mathfrak{p}^{-m}/\mathfrak{O}_K$ by

$$\begin{array}{rcl} \mathfrak{p}^{-m}/\mathfrak{O}_K & \to & G, \\ \xi + \mathfrak{O}_K & \mapsto & \sigma_\xi := \sigma(1+\xi_0\xi), \end{array} \tag{8.19}$$

with some $\xi_0 \in \mathfrak{q}\mathfrak{p}^{r+m} \setminus \mathfrak{p}^{r+m+1}$ and $\sigma(1+\xi_0\xi)$ denoting the Frobenius automorphism belonging to the ideal $(1+\xi_0\xi)$. In particular, the characters of $\mathfrak{p}^{-m}/\mathfrak{O}_K$ can be viewed as characters of G. Now we first consider the

Case $\frac{\mathfrak{q}}{\gcd(w_K,\mathfrak{q})}$ composite: Let g and T be the elliptic functions defined in (8.13) and (8.10) with respect to the lattice

$$\mathfrak{m} = \mathfrak{O}_K.$$

We set

$$\vartheta := g(\xi_1) \text{ with some } \xi_1 \in K,\ o(\xi_1) = \mathfrak{p}^{m+r},\quad \xi_1\xi_0 \equiv 1 \bmod \mathfrak{p}^{r+m},$$

$$T(\sigma_\xi) := T(\xi) \text{ for } \xi \in \mathfrak{p}^{-m}.$$

With these notations we then have the following explicit result:

Theorem 8.6.1 *Let $\mathfrak{q}$ be an integral ideal satisfying (8.9), prime to $N(\mathfrak{p})$ with composite $\frac{\mathfrak{q}}{\gcd(w_K,\mathfrak{q})}$. Then*

$$\mathfrak{O}_{N_\mathfrak{q}} = \mathfrak{O}_{M_\mathfrak{q}}[T(\xi_1)],$$

$$\mathfrak{A}_{N_\mathfrak{q}/M_\mathfrak{q}} = \frac{1}{\pi^m}\left\{ \sum_{\sigma\in G} f(T(\sigma))\sigma \;\middle|\; f(X) \in \mathfrak{O}_{M_\mathfrak{q}}[X] \right\},$$

$$\mathfrak{O}_{N_\mathfrak{q}} = \vartheta \circ \mathfrak{A}_{N_\mathfrak{q}/M_\mathfrak{q}}.$$

If $\mathfrak{q}$ only satisfies (8.8) instead of (8.9), the construction of ϑ has to be changed as explained after Theorem 8.5.3.

Case $\mathfrak{q} = (1)$: Our aim is to prove the last assertion of Theorem 8.6.1 to hold for the "natural" extensions $N_1/M_1 = K_{\mathfrak{p}^{r+m}}/K_{\mathfrak{p}^r}$, too with some possibly modified generating element ϑ'. As we will see, the proof relies on the factorisation

$$(\vartheta', \chi) := \sum_{\sigma \in G} \vartheta'^{\sigma} \overline{\chi}(\sigma) \sim \mathfrak{p}^m \tag{8.20}$$

for all characters χ of G. In fact, by computing discriminants, the third assertion of Theorem 8.6.1 follows for every element $\vartheta' \in \mathfrak{O}_{N_\mathfrak{q}}$ satisfying (8.20) for every character. Therefore, if we can find such an element $\vartheta' \in K_{\mathfrak{p}^{r+m}}$, we obtain

$$\mathfrak{O}_{N_1} = \vartheta' \circ \mathfrak{A}_{N_1/M_1},$$

by intersecting both sides of the equation $\mathfrak{O}_{N_\mathfrak{q}} = \vartheta' \circ \mathfrak{A}_{N_\mathfrak{q}/M_\mathfrak{q}}$ with $K_{\mathfrak{p}^{r+m}}$. For the construction of such an element $\vartheta' \in N_1 = K_{\mathfrak{p}^{r+m}}$ we assume $\mathfrak{p} \nmid w_K$. Then, as above, we can conclude that there exists an integral composite ideal $\mathfrak{q}$ satisfying (8.9) such that

$$\mathfrak{p} \nmid n := [N_\mathfrak{q} : N_1].$$

Therefore, we can find an $n' \in \mathbb{N}$ with $nn' \equiv 1 \bmod \mathfrak{p}^{r+m}$. As in (8.16) we set

$$\vartheta' := \frac{1}{\pi^{m(n-1)}} \sum_{\omega_1 \ldots \omega_n} g(n'\xi_1 + \omega_1)^{\sigma_1} \cdot \ldots \cdot g(n'\xi_1 + \omega_n)^{\sigma_n}, \tag{8.21}$$

where the sum is taken over all

$$(\omega_1, \ldots, \omega_n) \bmod \mathfrak{p}^m \text{ with } \omega_1 + \cdots + \omega_n \equiv 0 \bmod \mathfrak{O}_K.$$

$\sigma_1, \ldots, \sigma_n$ are the different automorphisms of $N_\mathfrak{q}/N_1$.

Theorem 8.6.2 *For $\mathfrak{p} \nmid w_K$ we have*

$$\mathfrak{O}_{N_1} = \vartheta' \circ \mathfrak{A}_{N_1/M_1}$$

with the element in (8.21).

However, for the extensions considered in Theorem 8.6.2, we obtain **no** explicit description of the associated order $\mathfrak{A}_{N_1/M_1}$. As explained in Schertz (1999), there are various other possibilities of constructing a Galois generator ϑ. In special cases we can find even simpler generators than the uniformly defined generators in this chapter. The following example relies on the construction in Schertz (1991).

Example 8.6.3 Let $K = \mathbb{Q}(\sqrt{-11})$. Then 3 is split in K, $3 = \mathfrak{p}\bar{\mathfrak{p}}$ with $\mathfrak{p} = \mathbb{Z}\alpha + \mathbb{Z}$, $\alpha = \frac{-31+\sqrt{-11}}{2}$. In this case $K_{\mathfrak{p}^4}/K_{\mathfrak{p}^2}$ is of degree

$$\left[K_{\mathfrak{p}^4} : K_{\mathfrak{p}^2}\right] = 9$$

and

$$\vartheta := \zeta \frac{\varphi\left(37\xi \,|\mathfrak{O}\right)}{\varphi(\xi|\mathfrak{O})}, \quad \zeta = e^{\frac{2\pi i}{3}}, \quad \xi = \frac{\alpha}{81},$$

is a Galois generator for this extension:

$$\mathfrak{O}_{K_{\mathfrak{p}^4}} = \vartheta \circ \mathfrak{A}_{K_{\mathfrak{p}^4}/K_{\mathfrak{p}^2}}.$$

The minimal equation of ϑ over K is

$$\begin{aligned}
X^{27} \quad &+ \quad \left(\frac{-31 - 1\sqrt{-11}}{2}\right) X^{25} + \left(\frac{33 - 3\sqrt{-11}}{2}\right) X^{24} \\
&+ \quad \left(\frac{39 + 3\sqrt{-11}}{2}\right) X^{23} + \left(\frac{-71 + 43\sqrt{-11}}{2}\right) X^{22} \\
&+ \quad \left(\frac{-117 - 171\sqrt{-11}}{2}\right) X^{21} + \left(\frac{451 + 295\sqrt{-11}}{2}\right) X^{20} \\
&+ \quad (-575 - 164\sqrt{-11})X^{19} + \left(\frac{1723 + 223\sqrt{-11}}{2}\right) X^{18} \\
&+ \quad (-657 - 63\sqrt{-11})X^{17} + (156 + 120\sqrt{-11})X^{16} \\
&+ \quad \left(\frac{-457 - 517\sqrt{-11}}{2}\right) X^{15} + (450 + 261\sqrt{-11}X^{14} \\
&+ \quad (-243 - 189\sqrt{-11})X^{13} + (106 + 202\sqrt{-11})X^{12} \\
&+ \quad \left(\frac{-963 - 315\sqrt{-11}}{2}\right) X^{11} + (376 - 68\sqrt{-11})X^{10} \\
&+ \quad \left(\frac{251 + 311\sqrt{-11}}{2}\right) X^{9} + (-245 - 68\sqrt{-11})X^{8} \\
&+ \quad \left(\frac{241 + 13\sqrt{-11}}{2}\right) X^{7} + \left(\frac{-109 - 25\sqrt{-11}}{2}\right) X^{6} \\
&+ \quad (16 + 4\sqrt{-11})X^{5} + \left(\frac{5 + 35\sqrt{-11}}{2}\right) X^{4} \\
&+ \quad \left(\frac{25 - 41\sqrt{-11}}{2}\right) X^{3} + (-18 + 9\sqrt{-11})X^{2} \\
&+ \quad \left(\frac{15 - 3\sqrt{-11}}{2}\right) X - 1
\end{aligned}$$

8.7 Models of elliptic curves

In this section we consider some models of elliptic curves that will be needed for the proofs in section 8.10 and later for the applications to cryptography in Chapter 10.

8.7.1 The Weierstrass model

We define a pair of Weierstrass functions by

$$x(z) := \frac{\wp(z)}{\sqrt[6]{\Delta(\mathfrak{L})}}$$

and

$$y(z) := \frac{\wp'(z)}{2\sqrt[4]{\Delta(\mathfrak{L})}},$$

where the roots are defined by

$$\sqrt[n]{\Delta(\mathfrak{L})} := \left(\frac{2\pi}{\omega_2}\eta\left(\frac{\omega_1}{\omega_2}\right)^2\right)^{\frac{12}{n}}, \quad n = 6, 4$$

and hence depend on the choice of basis $\begin{pmatrix}\omega_1\\ \omega_2\end{pmatrix}$, $\Im\left(\frac{\omega_1}{\omega_2}\right) > 0$, for the lattice $\mathfrak{L}$. The equation for x and y has the form

$$y^2 = x^3 + a_4 x + a_6$$

with

$$a_4 = -\frac{g_2(\mathfrak{L})}{4\sqrt[3]{\Delta(\mathfrak{L})}} = -\frac{1}{4\cdot 12}\sqrt[3]{j(\mathfrak{L})},$$

$$a_6 = -\frac{g_3(\mathfrak{L})}{4\sqrt[2]{\Delta(\mathfrak{L})}} = -\frac{1}{4\cdot 6^3}\sqrt{j(\mathfrak{L}) - 12^3},$$

and discriminant

$$\Delta_{x,y} = 1.$$

Further, y has the factorisation

$$y(z) = \frac{\varphi(2z)}{2\varphi(z)^4}.$$

Specialising this model in section 8.10, we will obtain elliptic curves defined over certain ray class fields with coefficients that are integral for 6 having good reduction outside 2 and 3. To settle the cases of good reduction over 2 and 3, too, we will need the following Fueter and Deuring models.

8.7.2 The Fueter model

Let $\psi \in \mathbb{C}$ have order 4 with respect to $\mathfrak{L}$. We set

$$x(z) := \frac{\wp(\psi) - \wp(2\psi)}{\wp(z + 2\psi) - \wp(2\psi)},$$

$$y(z) := \frac{1}{2\sqrt{\wp(\psi) - \wp(2\psi)}} \frac{dx(z)}{dz},$$

where the root is defined according to the formula of Theorem 1.3.2 by

$$\sqrt{\wp(\psi) - \wp(2\psi)} = \frac{1}{\sigma^*(2\psi)}.$$

The pair x, y of Weierstrass functions satisfies the equation

$$y^2 = x^3 + a_2 x^2 + x$$

with

$$a_2 = \frac{3\wp(2\psi)}{\wp(\psi) - \wp(2\psi)}.$$

This can easily be verified by writing down the Laurent expansion at $z = -2\psi$ of both sides of the equation and comparing coefficients. Further, using the theorem of Abel-Jacobi, we find the following representations:

$$x(z) = -\frac{\varphi(3\psi)\varphi(z + 2\psi)^2}{\varphi(\psi)\varphi(z + 4\psi)\varphi(z)},$$

$$y(z) = -\frac{\varphi(2\psi)^3\varphi(3\psi)\varphi(2z)}{2\varphi(\psi)\varphi(z)^4}$$

with

$$\varphi(z) := \sigma^*(z|\mathfrak{L}) \sqrt[12]{\Delta(\mathfrak{L})}.$$

Here, a normalisation of the 12-th root is not necessary since the roots cancel out in the representation of x and y.

The discriminant $\Delta_{x,y}$ of the Weierstrass equation is by definition the discriminant of the polynomial $x^3 + a_2 x^2 + x$, hence is equal to

$$x_1^2 x_2^2 (x_1 - x_2)^2$$

with the two non-vanishing zeros of the polynomial. They are given by

$$x_i = x(\omega_i) = \frac{\wp(\psi) - \wp(2\psi)}{\wp(\omega_i + 2\psi) - \wp(2\psi)}, i = 1, 2,$$

with the two half periods ω_i different from 2ψ. Now, using the formula of Theorem 1.3.2, we find that

$$\Delta_{x,y} = \frac{(\wp(\psi) - \wp(2\psi))^6 (\wp(\omega_1 + 2\psi) - \wp(\omega_2 + 2\psi))^2}{(\wp(\omega_1 + 2\psi) - \wp(2\psi))^4 (\wp(\omega_2 + 2\psi) - \wp(2\psi))^4} = \frac{\varphi(2\psi)^{12}}{2^4}.$$

On the other hand, expressing Δ by the coefficients of the equation for x and y, we obtain

$$\Delta_{x,y} = 16(a_2^2 - 4).$$

Therefore, we can compute a_2 via $\varphi(2\psi)^{12}$. Another formula for a_2 is given by

$$a_2 = -(x_1 + x_2) = -\left(\frac{\wp(\psi) - \wp(2\psi)}{\wp(\omega_1 + 2\psi) - \wp(2\psi)} + \frac{\wp(\psi) - \wp(2\psi)}{\wp(\omega_2 + 2\psi) - \wp(2\psi)} \right).$$

Further, writing $j(\mathfrak{L})$ in terms of the coefficients of the Weierstrass equation for x and y, we obtain the relation

$$j(\mathfrak{L}) = 2^8 \frac{(a_2^2 - 3)^3}{a_2^2 - 4}. \tag{8.22}$$

Remark 8.7.1 For a variable lattice $\mathfrak{L} = [\omega, 1]$, $\omega \in \mathbb{H}$, the "division values" $a_2(\kappa|\mathfrak{L})^2$, $\kappa \in \frac{1}{4}\mathfrak{L} \setminus 2\mathfrak{L}$, define six different modular functions for $\Gamma(4)$. So by (8.22) they constitute the six roots of the equation

$$2^8 (X^2 - 3)^3 - j(X^2 - 4) = 0, \tag{8.23}$$

having discriminant

$$2^{36} j^4 (j - 12^3)^3. \tag{8.24}$$

Conclusion: given two lattices $\mathfrak{L}, \mathfrak{L}'$ and $\kappa \in \frac{1}{4}\mathfrak{L} \setminus 2\mathfrak{L}$, $\kappa' \in \frac{1}{4}\mathfrak{L}' \setminus 2\mathfrak{L}'$. Then

$$a_2(\kappa|\mathfrak{L}) = a_2(\kappa'|\mathfrak{L}') \quad \Longrightarrow \quad j(\mathfrak{L}) = j(\mathfrak{L}') \quad \Longrightarrow \quad \mathfrak{L} \sim \mathfrak{L}' \tag{8.25}$$

and further, if the lattice $\mathfrak{L}$ is inequivalent to the maximal orders in $\mathbb{Q}(\sqrt{-3})$, $\mathbb{Q}(\sqrt{-4})$,

$$\kappa \equiv \pm\kappa' \bmod \mathfrak{L}. \tag{8.26}$$

8.7.3 The Deuring model

Let $\kappa \in \mathbb{C}$ be of order 3 with respect to $\mathfrak{L}$. Then, following Fleckinger (1987–8), we define

$$x(z) := \frac{\wp(z) - \wp(\kappa)}{\wp'(\kappa)^{\frac{2}{3}}},$$

$$y(z) := \frac{1}{2}\left(\frac{\wp'(z)}{\wp'(\kappa)} - a_1 x(z) - 1\right)$$

with

$$a_1 = \frac{\wp''(\kappa)}{\wp'(\kappa)^{\frac{4}{3}}} = \frac{12\wp(\kappa)^2 - g_2(\mathfrak{L})}{2\wp'(\kappa)^{\frac{4}{3}}}.$$

The power of $\wp'$ in the denominator is naturally defined according to the formula of Theorem 1.3.2:

$$\wp'(\kappa)^{\frac{4}{3}} = \sigma^*(\kappa|\mathfrak{L})^{-4}.$$

The equation satisfied by x and y is given by

$$y^2 + a_1 xy + y = x^3.$$

As for the Fueter model, verification can again be done by studying the Laurent series of both sides at $z = 0$. In addition, one has to use the equation for $\wp(\kappa)$ given by f_3 in (1.5). Further, we have the factorisations

$$x(z) = -\frac{\varphi(z-\kappa)\varphi(z+\kappa)}{\varphi(z)^2},$$

$$y(z) = \frac{\varphi(z-\kappa)^2\varphi(z+2\kappa)}{\varphi(z)^3},$$

$$\Delta_{x,y} = \varphi(\kappa)^{12}.$$

The factorisation of $x(z)$ is immediate by definition, keeping in mind the formula of Theorem 1.3.2. To factorise $y(z)$, we use the algebraic equation for x and y, which implies that y has a pole only at zero of order 3. Further, every zero of y is a zero of x and hence congruent to $\pm\kappa$. By definition of y we have $y(-\kappa) = -1$, so y has the only zero modulo $\mathfrak{L}$ at κ of order 3. Therefore, according to the Abel-Jacobi theorem, up to a constant factor, y must have the above representation, and by taking the limit for $z \to 0$ the factor turns out to be 1. To factorise $\Delta_{x,y}$, we observe the relation in (1.12):

$$\Delta_{x,y} = u^{12}\Delta_{\tilde{x},\tilde{y}}.$$

For the functions $\tilde{x}, \tilde{y}$ from the Weierstrass model the factor u is given by

$$u = \frac{1}{\wp'(\kappa)^{\frac{1}{3}}} = -\sigma^*(\kappa|\mathfrak{L}),$$

and this implies the above formula for $\Delta_{x,y}$. Finally, similarly to the Fueter model, we obtain the relation

$$\Delta_{x,y} = a_1^3 - 27$$

and

$$j(\mathfrak{L}) = \frac{a_1^3(a_1^3 - 24)^3}{a_1^3 - 27}. \tag{8.27}$$

Remark 8.7.2 For a variable lattice $\mathfrak{L} = [\omega, 1]$, $\omega \in \mathbb{H}$, the "division values" $a_1(\kappa|\mathfrak{L})^3$, $\kappa \in \frac{1}{3}\mathfrak{L} \setminus \mathfrak{L}$, define four different modular functions for $\Gamma(3)$, so by (8.27) they constitute the four roots of the equation

$$X(X - 24)^3 - j(X - 3) = 0, \tag{8.28}$$

having discriminant

$$-27j^2(j - 12^3). \tag{8.29}$$

Conclusion: given two lattices $\mathfrak{L}, \mathfrak{L}'$ and $\kappa \in \frac{1}{3}\mathfrak{L}\setminus\mathfrak{L}$, $\kappa' \in \frac{1}{3}\mathfrak{L}'\setminus\mathfrak{L}'$. Then

$$a_1(\kappa|\mathfrak{L}) = a_1(\kappa'|\mathfrak{L}') \quad \Longrightarrow \quad j(\mathfrak{L}) = j(\mathfrak{L}') \quad \Longrightarrow \quad \mathfrak{L} \sim \mathfrak{L}' \tag{8.30}$$

and further, if the lattice $\mathfrak{L}$ is inequivalent to the maximal orders in $\mathbb{Q}(\sqrt{-3})$, $\mathbb{Q}(\sqrt{-4})$,

$$\kappa \equiv \pm\kappa' \bmod \mathfrak{L}. \tag{8.31}$$

8.7.4 Singular values of the Weierstrass, Fueter and Deuring functions

In the following let x, y be a pair of Weierstrass functions from the models of Weierstrass, Fueter or Deuring, and let $\Psi(X, Y)$ denote the corresponding algebraic equation. We set

$$n := \begin{cases} 12 & \text{for the Weierstrass model,} \\ 8 & \text{for the Fueter model,} \\ 9 & \text{for the Deuring model.} \end{cases}$$

Further, by

$$W := -\frac{x}{y}$$

we denote the uniformising parameter at the origin.

Theorem 8.7.3 *Let K be a quadratic imaginary number field, $\mathfrak{m} \in \mathfrak{I}_1$ and $\mathfrak{p}$ a prime ideal of K not dividing n. Then the Weierstrass equation $\Psi(X, Y)$ associated with $\mathfrak{m}$ has $\mathfrak{p}$-integral coefficients in K_n, and $\Psi(X, Y)$ has good reduction modulo for all prime ideals of K_n above $\mathfrak{p}$. For every $\xi \in K \setminus \mathfrak{m}$ with $o(\xi, \mathfrak{m}) = \mathfrak{f}$ we have*

$$x(\xi), y(\xi), W(\xi) \in K_{n\mathfrak{f}}.$$

The automorphisms of $K_{n\mathfrak{f}}/K_n$ are the $\sigma(\nu)$ with $\nu \equiv 1 \bmod n$, ν prime to $n\mathfrak{f}$, and the action of $\sigma(\nu)$ on the singular values of x, y and W is given by

$$x(\xi)^{\sigma(\nu)} = x(\xi\nu), \; y(\xi)^{\sigma(\nu)} = y(\xi\nu), \; W(\xi)^{\sigma(\nu)} = W(\xi\nu).$$

If $\mathfrak{f} = \mathfrak{p}^s$, $s \in \mathbb{N}$, then $W(\xi)$ has the factorisation

$$W(\xi) \sim \mathfrak{p}^{\frac{1}{\Phi(\mathfrak{p}^s)}} \mathfrak{b}$$

with an ideal $\mathfrak{b}$ prime to $\mathfrak{p}$. For the Fueter model $\mathfrak{b}$ is at most divisible by prime divisors of 2, and for Deuring's model $\mathfrak{b} = (1)$.

If the order of ξ is not a $\mathfrak{p}$-power, then $x(\xi)$ and $y(\xi)$ are integral for $\mathfrak{p}$.

Remark 8.7.4 In special cases the above models of elliptic curves are already defined over proper subfields of K_n.

In view of the results on singular values of γ_2 and γ_3 the Weierstrass model can even be defined over K_1 if the disciminant of K is prime to 6.

By Reciprocity Law the same will be shown in section 10.3.1 for the Fueter model if 2 is split in K, $2 = \mathfrak{p}\overline{\mathfrak{p}}$, and ψ chosen such that $o(\psi) = \mathfrak{p}^2$.

For the Deuring model we will show the same result in section 10.3.2 if 3 is split, $3 = \mathfrak{p}\overline{\mathfrak{p}}$, with a κ having order $\mathfrak{p}$.

Proof of Theorem 8.7.3 First, we prove the assertions for the model of Weierstrass. The theorems in section 6.3 tell us that the coefficients of $\Psi(X, Y)$ are in K_6 and integral for 6. Further, since the discriminant is equal to 1, it follows that the curve defined by $\Psi(X, Y)$ has good reduction outside 2 and 3. Theorem 5.2.5 implies that $x(\xi), y(\xi) \in K_{n\mathfrak{f}}$ with the Galois action asserted. To derive the factorisations, we conclude as follows: by Theorem 4.3.1

$$y(\xi) \sim \frac{1}{2}\mathfrak{p}^{\frac{-3}{\Phi(\mathfrak{p}^s)}},$$

and, keeping in mind that $\Psi(x(\xi), y(\xi)) = 0$, this implies that

$$x(\xi) \sim \mathfrak{b}\mathfrak{p}^{\frac{-2}{\Phi(\mathfrak{p}^s)}}$$

with an ideal $\mathfrak{b}$ prime to $\mathfrak{p}$, which is the desired factorisation for $W(\xi)$.

For the other two models the assertions concerning the singular values $x(\xi), y(\xi)$ and $W(\xi)$ are obtained by writing them as a product of φ-values. Then, by inserting these values into the equation $\Psi(X, Y) = 0$, we can deduce that the coefficients a_2 and a_1 are in $K_{n\mathfrak{p}^s}$ for every prime ideal $\mathfrak{p}$ not dividing n and for every $s \in \mathbb{N}$. Hence a_2 and a_1 must be in K_n. Further, due to the factorisations

$$x(\xi) \sim \mathfrak{b}\mathfrak{p}^{\frac{-2}{\Phi(\mathfrak{p}^s)}}, \; y(\xi) \sim \mathfrak{b}'\mathfrak{p}^{\frac{-3}{\Phi(\mathfrak{p}^s)}}, \quad s \in \mathbb{N},$$

with ideals $\mathfrak{b}, \mathfrak{b}'$ prime to $\mathfrak{p}$, the equation $\Psi(x(\xi), y(\xi)) = 0$ tells us that a_2 and a_1 must be integral for all prime ideals of K_n not dividing n. Finally, the representations of $\Delta_{x,y}$ by φ-values together with Theorem 4.3.1 shows $\Psi(X, Y)$ to have good reduction outside n.

The last assertion for ξ having composite order also follows Theorem 4.3.1. □

8.7.5 Singular values of Weierstrass functions

In the following let x, y be an arbitrary pair of Weierstrass functions associated with $\mathfrak{m} \in \mathfrak{J}_1$. We assume the coefficients of the corresponding elliptic curve E to be in a finite extension L of the Hilbert class field of K. For such an elliptic curve Theorem 8.7.3 can be generalised as follows:

Theorem 8.7.5 *Let $\mathfrak{f}$ be an integral ideal in K and $\xi \in \mathfrak{f}^{-1}\mathfrak{m} \setminus \mathfrak{m}$. Then $x(\xi), y(\xi)$ are algebraic over L, and for every automorphism σ in* *the Galois group* Gal_L *of* $\mathbb{Q}^c/L$ *there exists* $\lambda_\sigma \in \mathfrak{O}_K$ *only depending on* σ *and* $\mathfrak{f}$ *such that*

$$x(\xi)^\sigma = x(\lambda_\sigma \xi), \quad y(\xi)^\sigma = y(\lambda_\sigma \xi).$$

In particular, if $o(\xi) = \mathfrak{p}^s$ *is a prime ideal power and* E *has good reduction above* $\mathfrak{p}$*, we have the factorisation*

$$W(\xi) \sim \mathfrak{b}\mathfrak{p}^{\frac{1}{\Phi(\mathfrak{p}^s)}}$$

with an ideal $\mathfrak{b}$ *prime to* $\mathfrak{p}$.

For ξ *having composite order, the values* $x(\xi)$ *and* $y(\xi)$ *are integral for* $\mathfrak{p}$.

Proof First, we write x and y in terms of the normalised Weierstrass $\wp$ function via the following transformation, where a_i denote the coefficients of E. We set

$$(\tilde{x}, \tilde{y}) := \left(\frac{1}{48}(12x + a_1^2 + 4a_2), \frac{1}{8}(2y + a_1 x + a_3)\right). \tag{8.32}$$

This is a pair of Weierstrass functions satisfying an equation of the form

$$\tilde{y}^2 = 4\tilde{x}^3 + A\tilde{x} + B \text{ with } A, B \in L. \tag{8.33}$$

As a pair of Weierstrass functions, $\tilde{x}$ and $\tilde{y}$ are polynomials of $\wp$ and $\wp'$ of degree 1. So

$$\tilde{x} = a\mathcal{P}_0 + b, \qquad \mathcal{P}_0 := \epsilon \frac{\wp}{\sqrt[6]{\Delta(\mathfrak{m})}},$$

$$\tilde{y} = d\mathcal{P}_1 + e\mathcal{P}_0 + f, \quad \mathcal{P}_1 := \epsilon^{\frac{3}{2}} \frac{\wp'}{\sqrt[4]{\Delta(\mathfrak{m})}},$$

with a unit ϵ from the definition of $\mathcal{P}$ and constants a, b, c, d, e, f. Inserting these equations into (8.33) and comparing coefficients and pole orders, we obtain $b = e = f = 0$ and $d^2 = a^3 \neq 0$. Hence

$$\begin{aligned} \tilde{x} &= u^2 \mathcal{P}_0, \\ \tilde{y} &= u^3 \mathcal{P}_1 \end{aligned} \tag{8.34}$$

with a constant $u \neq 0$. Now we can write the coefficients in (8.33) in terms of the coefficients of the algebraic equation

$$\wp'(z|\mathfrak{m})^2 = 4\wp(z|\mathfrak{m})^3 - g_2(\mathfrak{m})\wp(z|\mathfrak{m}) - g_3(\mathfrak{m})$$

for $\wp$ and $\wp'$. We obtain

$$\epsilon^2 \frac{g_2(\mathfrak{m})}{\sqrt[3]{\Delta(\mathfrak{m})}} u^4 = -A \text{ and } \epsilon^3 \frac{g_3(\mathfrak{m})}{\sqrt{\Delta(\mathfrak{m})}} u^6 = B. \tag{8.35}$$

By Theorem 6.6.10 we know that

$$\epsilon^2 \frac{g_2(\mathfrak{m})}{\sqrt[3]{\Delta(\mathfrak{m})}} \text{ and } \epsilon^3 \frac{g_3(\mathfrak{m})}{\sqrt{\Delta(\mathfrak{m})}} \tag{8.36}$$

are in the Hilbert class field of K, hence in L. This implies that

$$u^{w_K} \in L, \tag{8.37}$$

keeping in mind that $g_2(\mathfrak{m}) \neq 0$ for $d_K \neq -3$ and $g_3(\mathfrak{m}) \neq 0$ for $d_K \neq -4$. To prove Theorem 8.7.5, we use (8.34), so we first compute the action of σ on $\mathcal{P}_0(\xi)$ and $\mathcal{P}_1(\xi)$. By Reciprocity Law these values are in $K_{12N(\mathfrak{f})^2}$. Further, since L contains the Hilbert class field of K, it follows that the restriction of σ to $K_{12N(\mathfrak{f})^2}$ is the Frobenius automorphism of some principal ideal (λ). Theorem 5.2.5 tells us that

$$\mathcal{P}_0(\xi)^\sigma = \zeta_0 \mathcal{P}_0(\lambda\xi), \quad \mathcal{P}_1(\xi)^\sigma = \zeta_1 \mathcal{P}_1(\lambda\xi)$$

with roots of unity ζ_0, ζ_1 that are independent of ξ. According to Theorem 7.1.2 we have $\mathcal{P}(\xi)^\sigma = \mathcal{P}(\lambda\xi)$, and further

$$u^{2\sigma} = \zeta_u u^2$$

with another $\frac{w_K}{2}$ -th root of unity ζ_u since u^{w_K} is in L. So $\zeta_u\zeta_0$ is a $\frac{w_K}{2}$-th root of unity. By homogeneity of $\wp$ and the fact that K contains the w_K-th roots of unity, we can now write the action of σ as

$$\tilde{x}(\xi)^\sigma = \tilde{x}(\zeta_0'\lambda\xi) = \tilde{x}(-\zeta_0'\lambda\xi)$$

with another w_K-th root of unity ζ_0', which is independent of ξ, too. In a similar way we obtain

$$\tilde{y}(\xi)^\sigma = \zeta_1'\tilde{y}(\zeta_0'\lambda\xi)$$

with a root of unity ζ_1' independent of ξ. More precisely by (8.33) we find that

$$\zeta_1' = \pm 1$$

and further, by homogeneity of $\tilde{y}$:

$$\tilde{y}(\xi)^\sigma = \tilde{y}(\zeta_1'\zeta_0'\lambda\xi).$$

This proves the first assertion of Theorem 8.7.5 with $\lambda_\sigma = \zeta_1'\zeta_0'\lambda$.

To prove the last two assertions, let $\tilde{x}, \tilde{y}$ be one of the Fueter or Deuring models having good reduction above $\mathfrak{p}$. Then, as explained in 1.6, the two models are related by

$$x = u^2\tilde{x} + r,$$
$$y = u^3\tilde{y} + u^2 v\tilde{x} + t,$$

where u, r, t, v are integral for $\mathfrak{p}$ and u is a $\mathfrak{p}$-unit. Now the two last assertions of Theorem 8.7.5 follow from the last assertion of Theorem 8.7.3. □

From the proof of Theorem 8.7.5 we can derive the following two propositions:

Proposition 8.7.6 *Let $Q = Q(\delta) \in E(L)$ have order $\mathfrak{q} \neq (1)$. Then T, as defined in (8.10), is in $L(x, y)$.*

Proof Using (8.34) and (8.37), we can write

$$T(z) = u^{-w_K}\left(\tilde{x}(z+\delta)^{\frac{w_K}{2}} - \tilde{x}(\delta)^{\frac{w_K}{2}}\right).$$

By assumption L contains the coordinates $\tilde{x}(\delta)$ and $\tilde{x}(\delta)$, so the addition formula for $\wp$ tells us that $T \in L(\tilde{x}, \tilde{y}) = L(x, y)$. □

Proposition 8.7.7 *Condition (8.7) implies condition (8.6).*

Proof First we consider the cases when $d_K \neq -3, -4$. Then u^2 is in L according to (8.37), and by Theorem 7.1.2 the values $\mathcal{P}(\xi)$, $o(\xi)|\mathfrak{f}$, $\xi \notin \mathfrak{m}$, are in $K_{\mathfrak{f}}$. Hence by (8.34) the $\tilde{x}$-coordinate of points of order dividing $\mathfrak{f}$ must be in $LK_{\mathfrak{qf}}$. We are left with proving the $\tilde{y}$-coordinate to lie in $LK_{\mathfrak{qf}}$ also. By assumption, L contains a point of order $\mathfrak{q}$. Its $\tilde{y}$-coordinate $\tilde{y}(\xi_0)$ is non-zero since $\mathfrak{q} \nmid 2$, so it suffices to show that

$$\frac{\tilde{y}(\xi)}{\tilde{y}(\xi_0)} \in K_{\mathfrak{qf}} \text{ for } o(\xi)|\mathfrak{f},\ \xi \notin \mathfrak{m}.$$

We use (8.34) and write

$$\frac{\tilde{y}(\xi)}{\tilde{y}(\xi_0)} = \frac{\wp'(\xi\,|\mathfrak{m})}{\wp'(\xi_0\,|\mathfrak{m})},$$

whereafter, according to the addition formula for $\wp$, it is easy to show that the quotient is in $K_{\mathfrak{qf}}$. Hence $\tilde{y}(\xi) \in LK_{\mathfrak{qf}}$, as asserted.

For $d_K = -3, -4$ let ξ_0 be the parameter of a point in $\mathbb{C}/\mathfrak{m}$ of order $t = 12$ resp. $t = 4$. In these cases, using the Reciprocity Law, it follows that $\mathcal{P}_0(\xi_0)$ and $\mathcal{P}_1(\xi_0)$ are in $K_{(t)}$. Then by (8.34) we first obtain $u \in LK_{(t)}$, and we contend that $LK_{(t)} = L$. According to (8.37) we have $u^6 \in L$ resp. $u^4 \in L$, and therefore (8.34) implies that $LK_{(t)}$ is generated over L by $\tilde{x}^3(\xi_0) = u^6\mathcal{P}(\xi_0)$ resp. $\tilde{x}(\xi_0)^2 = u^4\mathcal{P}(\xi_0)$. Hence, $K_{(t)} \subseteq L$ and $u \in L$. The assertion of our proposition then follows as in the cases where $d_K \neq -3, -4$. □

8.8 Proofs of Theorems 8.3.1 and 8.5.1

For the following we assume the properties of (8.5) to be given. The proofs will essentially rely on the $\mathfrak{P}$-adic addition formula (8.3) with a prime ideal $\mathfrak{P}$ above $\mathfrak{p}$ in an extension $\hat{L}$ of L so that $G_m \subseteq E(\hat{L})$. To prove $\mathfrak{A}$ to be a ring, we need the following lemma for the polynomial

$$h_O(X) := \prod_{Q \in G_m} (X - W(Q)).$$

Lemma 8.8.1 *There exists a power series*

$$u(X) = u_0 + u_1 X + \cdots \in \mathfrak{O}'_L[[X]],$$

with u_0 prime to $\mathfrak{p}$ such that

$$h'_O(W(Q)) = \pi^m u(W(Q))$$

for all $Q \in G_m$, where $\mathfrak{O}'_L$ denotes the ring of elements in L that is integral for $\mathfrak{p}$.

Proof Let $G(X, Y)$ be the power series in Theorem 1.6.6. Then we find that

$$\begin{aligned} h'_O(W(Q)) &= \prod_{Q' \neq Q} (W(Q) - W(Q')) \\ &= \prod_{Q' \neq O} (W(Q) - W(Q + Q')) \\ &= \prod_{Q' \neq O} (-W(Q') - W(Q)W(Q')G(W(Q), W(Q'))) \\ &= \pm \left(\prod_{Q' \neq O} W(Q') \right) \left(\prod_{Q' \neq O} (1 + W(Q)G(W(Q), W(Q'))) \right). \end{aligned}$$

Herein, according to Theorem 8.7.5, we have the factorisation

$$\prod_{Q' \neq O} W(Q') \sim \mathfrak{b} \prod_{Q' \neq O} \mathfrak{p}^{\frac{1}{\Phi(o(Q'))}}$$

with an ideal $\mathfrak{b}$ prime to $\mathfrak{p}$, and since for $s = 1, \ldots, m$ there are exactly $\Phi(\mathfrak{p}^s)$ elements of order $\mathfrak{p}^s$ in G_m; this implies that

$$\prod_{Q' \neq O} W(Q') \sim \mathfrak{b}\mathfrak{p}^m.$$

As the automorphisms of $\mathbb{Q}^c/L$ act as a permutation on the $W(Q')$ in the product, we further obtain

$$\prod_{Q' \neq O} W(Q') = \pi^m u_0$$

with a $\mathfrak{p}$-unit $u_0 \in L$, hence

$$h_O'(W(Q)) = \pm\pi^m u_0 \tilde{u}(W(Q)) \quad \text{with} \quad \tilde{u}(X) := \prod_{Q' \neq O} (1 + XG(X, W(Q'))),$$

showing us that the coefficients of $\tilde{u}(X)$ are integral for $\mathfrak{P}$. We are now left with the proof of $\tilde{u}(X)$ being a power series with coefficients in the completion L' of L with respect to $\mathfrak{P}$. Therefore, we observe that the coefficients of $\tilde{u}$ are power series of $W(Q')$ with $\mathfrak{P}$-integral coefficients in L and that the application of an automorphism can be interchanged with summation because the automorphisms are continous. This finishes the proof of Lemma 8.8.1. □

Lemma 8.8.2 *For $i \geq 0$ we have*

$$\sum_{Q \in G_m} \frac{W(Q)^i}{h_O'(W(Q))} \in \mathfrak{O}_L',$$

wherein $W(Q)^0$ is set to be 1 if $W(Q) = 0$.

Proof By substituting $X = \frac{1}{T}$ in the partial fraction decomposition

$$\frac{1}{h_O(X)} = \sum_{Q \in G_m} \frac{1}{h_O'(W(Q))} \frac{1}{X - W(Q)}$$

we obtain

$$\frac{T^{p^m}}{T^{p^m} h_O(\frac{1}{T})} = \sum_{Q \in G_m} \frac{1}{h_O'(W(Q))} \frac{T}{1 - W(Q)T},$$

where p^m is the norm of $\mathfrak{p}^m$. Now, writing the right-hand side as a power series in T, it becomes

$$\frac{T^{p^m}}{T^{p^m} h_O(\frac{1}{T})} = \sum_{i \geq 0} \left(\sum_{Q \in G_m} \frac{W(Q)^i}{h_O'(W(Q))} \right) T^{i+1},$$

where

$$T^{p^m} h_O\left(\frac{1}{T}\right) = 1 + a_1 T^1 + \cdots + a_{p^m} T^{p^m}$$

with $\mathfrak{p}$-integral coefficients a_i. Therefore, the inverse of this polynomial is a power series in T with $\mathfrak{p}$-integral coefficients, too, so we have an identity of the form

$$\sum_{\nu \geq 0} b_\nu T^\nu = \sum_{i \geq 0} \left(\sum_{Q \in G_m} \frac{W(Q)^i}{h_O'(W(Q))} \right) T^{i+1},$$

with $\mathfrak{p}$-integral coefficients b_ν. Comparing coefficients now yields the assertion of the lemma. More precisely, the proof shows $b_0 = \cdots = b_{p^m-1} = 0$. Hence, for $i = 0, \dots, p^m - 1$ the sum in the lemma is equal to zero. □

Lemma 8.8.3 *For $i \geq 0$ we have*

$$\sum_{Q \in G_m} W(Q)^i \equiv 0 \ mod\ \mathfrak{p}^m.$$

Proof The power series $u(X)$ in Lemma 8.8.1 is a unit in $R_\mathfrak{p}[[X]]$, where $R_\mathfrak{p}$ denotes the ring of $\mathfrak{p}$-integral algebraic numbers. Its inverse

$$\hat{u}(X) = \hat{u}_0 + \hat{u}_1 X + \cdots$$

again has $\mathfrak{p}$-integral coefficients, and in particular, the leading coefficient $\hat{u}_0$ is prime to $\mathfrak{p}$. By Lemma 8.8.1 and 8.8.2 it follows for all $i \geq 0$:

$$\hat{u}_0 S_i + \hat{u}_1 S_{i+1} + \cdots \equiv 0 \bmod \pi^m,$$

where S_i denotes the sum in Lemma 8.8.3. The factorisation of $W(Q)$ in Theorem 8.7.5 implies the existence of a natural number N, so that trivially

$$S_i \equiv 0 \bmod \pi^m \text{ for all } i \geq N.$$

By the above congruence for S_i we then recursively obtain for $i = N-1, \dots, 0$:

$$S_{i+1} \equiv 0 \bmod \pi^m \Longrightarrow S_i \equiv 0 \bmod \pi^m,$$

thereby finishing the proof. □

To show that $\mathfrak{A}$ is a ring, we use Lemma 8.8.3. This is trivial for the local components $\mathfrak{p}'$ outside $\mathfrak{p}$. Two elements of a local component above $\mathfrak{p}$ can be written as

$$\gamma = \frac{1}{\pi^m} \sum_{Q \in G_m} f(W(Q))Q, \quad \delta = \frac{1}{\pi^m} \sum_{Q \in G_m} g(W(Q))Q$$

with power series

$$f(X) = \sum_{n \geq 0} a_n X^n, \ \ g(X) = \sum_{n \geq 0} b_n X^n \ \in \mathfrak{O}_{L_{\mathfrak{p}'}}[[X]].$$

We contend that the product

$$\gamma\delta = \frac{1}{\pi^{2m}} \sum_{Q' \in G_m} c_{Q'} Q',$$

$$c_{Q'} = \sum_{Q \in G_m} f(W(Q))g(W(Q'-Q)),$$

is in $\mathfrak{A}_{\mathfrak{p}'}$ again. Therefore, we note that by the $\mathfrak{P}$-adic addition formula

$$W(Q'-Q) = H(W(Q'), W(Q))$$

with a power series

$$H(X,Y) \in XL[[X,Y]] + YL[[X,Y]],$$

having $\mathfrak{p}$-integral coefficients. Therefore, we can write

$$g(W(Q'-Q)) = \sum_{n_1,n_2 \geq 0} b_{n_1 n_2} W(Q')^{n_1} W(Q)^{n_2}$$

with $\mathfrak{P}$-integral coefficients $b_{n_1 n_2} \in L$. Hence, the $c_{Q'}$ are of the form

$$c_{Q'} = \sum_{n_1 \geq 0} \left(\sum_{n,n_2 \geq 0} a_n b_{n_1 n_2} \sum_{Q \in G_m} W(Q)^{n+n_2} \right) W(Q')^{n_1},$$

which is a power series of $W(Q')$ having coefficients divisible by π^m according to Lemma 8.8.3. Furthermore, as in the proof of Lemma 8.8.1, we conclude that the coefficients are again in $\mathfrak{O}_{L_{\mathfrak{p}'}}$. Therefore, $\gamma\delta$ is in $\mathfrak{A}_{\mathfrak{p}'}$, and it follows that $\mathfrak{A}$ is a ring.

To prove the neutral element O to be in $\mathfrak{A}_{\mathfrak{p}'}$, we note that $h_O(0) = h_O(W(O)) = 0$, hence

$$f(X) := \frac{h_O(X)}{X} \in \mathfrak{O}_{L_{\mathfrak{p}'}}[X].$$

Therefore,

$$\gamma := \frac{1}{\pi^m} \sum_{Q \in G_m} f(W(Q))Q$$

defines an element in $\mathfrak{A}_{\mathfrak{p}'}$ satisfying

$$\gamma = \frac{h'_m(W(O))}{\pi^m} O.$$

This shows that O is in $\mathfrak{A}$ because by Lemma 8.8.1 the coefficient $\frac{h'_m(W(O))}{\pi^m}$ is in L and prime to $\mathfrak{p}$.

This proves the first assertion of Theorem 8.3.1. Next, we will show the second assertion: $\tilde{\mathfrak{O}}_P \circ \mathfrak{A} \subseteq \tilde{\mathfrak{O}}_P$. For the local components outside $\mathfrak{p}$, this is again trivial, so let $\mathfrak{p}'$ be a prime ideal of L above $\mathfrak{p}$, and let

$$\theta = (f(W(R))_{R\in P+G_m} \in \left(\tilde{\mathfrak{O}}_P\right)_{\mathfrak{p}'}, \qquad \gamma = \frac{1}{\pi^m}\sum_{Q\in G_m} g(W(Q))Q \in \mathfrak{A}_{\mathfrak{p}'}$$

with power series

$$f(X), g(X) \in \mathfrak{O}_{L_{\mathfrak{p}'}}[[X]].$$

Then

$$\theta\circ\gamma = \left(\frac{1}{\pi^m}\sum_{Q\in G_m} f(W(R+Q))g(W(Q))\right)_{R\in P+G_m}.$$

Writing $W(R+Q)$ as a power series of $W(R)$ and $W(Q)$, according to the addition formula, we obtain the following representation for the components of $\theta\circ\gamma$:

$$\frac{1}{\pi^m}\sum_{Q\in G_m} f(W(R+Q))g(W(Q)) = \frac{1}{\pi^m}\sum_{m,n\geq 0} d_{mn}\sum_{Q\in G_m} W(Q)^m W(R)^n$$

with $\mathfrak{p}'$-integral coefficients $d_{mn}\in L$. Lemma 8.8.3 now tells us that $\theta\circ\gamma$ lies in $\left(\tilde{\mathfrak{O}}_P\right)_{\mathfrak{p}'}$.

To prove the third assertion of Theorem 8.3.1, we start by computing the $\mathfrak{p}$-part of the discriminant of $\tilde{\mathfrak{O}}_P$ that, by definition of $\tilde{\mathfrak{O}}_P$, is equal to the $\mathfrak{p}$-part of the polynomial discriminant of

$$h_P(X) := \prod_{R\in P+G_m}(X - W(R)).$$

We proceed as in the proof of Lemma 8.8.1:

$$\begin{aligned} h_P'(W(R)) &= \prod_{Q'\neq O}(W(R) - W(R+Q')) \\ &= \prod_{Q'\neq O}(-W(Q') - W(R)W(Q')G(W(R),W(Q'))) \\ &= \pm\left(\prod_{Q'\neq O} W(Q')\right)\left(\prod_{Q'\neq O}(1 + W(R)G(W(R),W(Q')))\right). \end{aligned}$$

Therefore, the $\mathfrak{p}$-part of $h_P'(W(R))$ is equal to $\mathfrak{p}^m$, and hence the $\mathfrak{p}$-part of the discriminant must be $\prod_{R\in P+G_m} h_P'(W(R)) = \mathfrak{p}^{mp^m}$. We are now left with the proof of $\tilde{\mathfrak{O}}_P$ being unramified outside $\mathfrak{p}$.

If E has good reduction everywhere, then the field extensions of L occuring in the definition of $\mathbb{M}_P$ must be unramified outside $\mathfrak{p}$ since they are subfields of $L(G_{m+r})$.

If $E(L)$ contains points of order $\mathfrak{q}$ with $\frac{\mathfrak{q}}{\gcd(w_K,\mathfrak{q})}$ composite, then by Theorem 8.4.2 there exists an element $\Theta = (T(R))_{R\in P+G_m}$ in $\tilde{\mathfrak{O}}_P$ such that

$$\tilde{\mathfrak{O}}_P = \mathfrak{O}_L[\Theta] \quad \text{and} \quad discr(\mathfrak{O}_L[\Theta]) = \mathfrak{p}^{mp^m}.$$

This proves the third assertion of Theorem 8.3.1.

Proof of Theorem 8.5.1 For every prime ideal $\mathfrak{p}'$ in L we choose an integral power basis θ^i of $\left(\tilde{\mathfrak{O}}_P\right)_{\mathfrak{p}'}$ over $L_{\mathfrak{p}'}$. Then we have

$$\mathfrak{A}_{\mathfrak{p}'} = \begin{cases} \bigoplus_{i=0}^{p^m-1} \mathfrak{O}_{L_{\mathfrak{p}'}}\tau_i \quad \text{with} \quad \tau_i = \frac{1}{\pi^m}\sum_{Q\in G_m} \theta_Q^i Q, \quad \text{for} \quad \mathfrak{p}'|\mathfrak{p}, \\ \bigoplus_{i=0}^{p^m-1} \mathfrak{O}_{L_{\mathfrak{p}'}}\tau_i \quad \text{with} \quad \tau_i = \sum_{Q\in G_m} \theta_Q^i Q, \quad \text{for} \quad \mathfrak{p}'\nmid\mathfrak{p}. \end{cases}$$

Let ϵ be an element in $\tilde{\mathfrak{O}}_P$ satisfying the hypothesis of Theorem 8.5.1. Then

$$\left(\epsilon\circ\mathfrak{A}\right)_{\mathfrak{p}'} =$$

$$\begin{cases} \bigoplus_{i=0}^{p^m-1} \mathfrak{O}_{L_{\mathfrak{p}'}}\epsilon\circ\tau_i \quad \text{with} \quad \epsilon\circ\tau_i = \left(\frac{1}{\pi^m}\sum_{Q\in G_m} \theta_Q^i \epsilon_{R-Q}\right)_{R\in P+G_m} \\ \qquad \text{if} \quad \mathfrak{p}'|\mathfrak{p}, \\ \bigoplus_{i=0}^{p^m-1} \mathfrak{O}_{L_{\mathfrak{p}'}}\epsilon\circ\tau_i \quad \text{with} \quad \epsilon\circ\tau_i = \left(\sum_{Q\in G_m} \theta_Q^i \epsilon_{R-Q}\right)_{R\in P+G_m} \\ \qquad \text{if} \quad \mathfrak{p}'\nmid\mathfrak{p}. \end{cases}$$

Hence for $\mathfrak{p}' \mid \mathfrak{p}$ we have

$$\begin{aligned} discr\left(\epsilon\circ\mathfrak{A}\right)_{\mathfrak{p}'} &= \det\left(\frac{1}{\pi^m}\sum_{Q\in G_m} \theta_Q^i \epsilon_{P+Q'-Q}\right)^2_{i=0,\dots,p^m-1;\ Q'\in G_m} \\ &= \frac{1}{\pi^{2mp^m}} \det\left(\theta_Q^i\right)^2 \det\left(\epsilon_{P+Q'-Q}\right)^2_{Q',Q\in G_m} \end{aligned}$$

and therefore

$$\begin{aligned}\prod_{\mathfrak{p}'|\mathfrak{p}} discr\left(\epsilon\circ\mathfrak{A}\right)_{\mathfrak{p}'} &= \frac{1}{\pi^{2mp^m}} discr(\tilde{\mathfrak{O}}_O)_{\mathfrak{p}} \prod_{\chi\in\hat{G}_m} (\epsilon,\chi)^2 \\ &= \mathfrak{p}^{-2mp^m+mp^m+2mp^m} = \mathfrak{p}^{mp^m} = \left(discr\left(\tilde{\mathfrak{O}}_P\right)\right)_{\mathfrak{p}}.\end{aligned}$$

For $\mathfrak{p}'\nmid\mathfrak{p}$ the same computation yields

$$discr\left(\epsilon\circ\mathfrak{A}\right)_{\mathfrak{p}'} = discr(\tilde{\mathfrak{O}}_O)_{\mathfrak{p}'}.$$

If E has good reduction everywhere, then $\mathbb{M}_O$ is unramified outside $\mathfrak{p}$ and therefore $discr(\tilde{\mathfrak{O}}_O) = \mathfrak{p}^{mp^m}$. This implies that

$$discr(\epsilon\circ\mathfrak{A}) = \mathfrak{p}^{mp^m}$$

and then $\epsilon\circ\mathfrak{A} = \tilde{\mathfrak{O}}_P$.

If E only has good reduction above $\mathfrak{p}$, we can find a pair $\hat{x},\hat{y}$ of Weierstrass functions associated with $\mathfrak{m}$ defining an elliptic curve $\hat{E}$ over an extension $\hat{L}$ having good reduction everywhere and, replacing $\hat{L}$ by a further finite extension, we can achieve

$$\hat{L}(\hat{x},\hat{y}) = \hat{L}(x,y).$$

We identify the points of E and $\hat{E}$ according to their parametrisation by cosets $\xi+\mathfrak{m}$, and we can write

$$\mathfrak{O}_P(\hat{L}) = \epsilon\circ\mathfrak{A}(\hat{L}).$$

Observe that for a corresponding pair $P_1,P_2\in E(\mathbb{Q}^c)$; $\hat{P}_1,\hat{P}_2\in\hat{E}(\mathbb{Q}^c)$ and $\sigma\in\Omega_{\hat{L}}$, we have the equivalence

$$P_1^{\sigma} = P_2 \iff \hat{P}_1^{\sigma} = \hat{P}_2.$$

The uniformising parameter $\hat{W} = -\frac{\hat{x}}{\hat{y}}$ has the series expansion

$$\hat{W} = \sum_{n=1}^{\infty} a_n W^n$$

with $\mathfrak{p}$-integral coefficients in $\hat{L}$ and a $\mathfrak{p}$-unit a_1 as leading coefficient. Hence, in the local descriptions of $\tilde{\mathfrak{O}}_P(\hat{L})$ and $\tilde{\mathfrak{O}}_P(L)$ above $\mathfrak{p}$ we can use the same local parameter W:

$$(\tilde{\mathfrak{O}}_P(\hat{L}))_{\mathfrak{p}'} = \bigoplus_{i=0}^{p^m-1} \mathfrak{O}_{\hat{L}_{\mathfrak{p}'}}\,\theta^{(i)}, \qquad (\tilde{\mathfrak{O}}_P(L))_{\mathfrak{p}'} = \bigoplus_{i=0}^{p^m-1} \mathfrak{O}_{L_{\mathfrak{p}'}}\,\theta^{(i)},$$

$$\theta^{(i)} = \left(W(R)^i\right)_{R\in P+G_m}.$$

By intersecting the local components of $\mathbb{M}_P(L)$ with the local components of both sides of the equation $\mathfrak{O}_P(\hat{L}) = \epsilon \circ \mathfrak{A}(\hat{L})$, we obtain

$$\left(\tilde{\mathfrak{O}}_P(L)\right)_{\mathfrak{p}'} = \epsilon \circ \mathfrak{A}(L)_{\mathfrak{p}'} \quad \text{for all } \mathfrak{p}' \mid \mathfrak{p}.$$

For the other prime ideals $\mathfrak{p}'$ of L this equation is trivial since, then, the local components of $\tilde{\mathfrak{O}}_P(L)$ and $\tilde{\mathfrak{O}}_O(L)$ are the local components of the maximal order in $\mathfrak{O}_P(L)$ resp. $\mathfrak{O}_O(L)$. This proves that

$$\tilde{\mathfrak{O}}_P(L) = \epsilon \circ \mathfrak{A}(L).$$

Further, we can deduce from this equation that $\mathfrak{A}$ must be the associated order $\hat{\mathfrak{A}}$ of $\tilde{\mathfrak{O}}_P$ in $\mathbb{A}$, because clearly $\mathfrak{A} \subseteq \hat{\mathfrak{A}}$. This implies that

$$\tilde{\mathfrak{O}}_P = \epsilon \circ \mathfrak{A} = \epsilon \circ \hat{\mathfrak{A}},$$

and, since multiplication by ϵ is an automorphism of $\mathbb{C}^{p^m}$ because of

$$\det\left(\epsilon_{P+Q-Q'}\right)_{Q,Q' \in G_m} = \pm \prod_{\chi \in \hat{G}_m} (\epsilon, \chi) \neq 0,$$

it follows that

$$\mathfrak{A} = \hat{\mathfrak{A}}.$$

□

8.9 Proofs of Theorems 8.4.1, 8.4.2 and 8.5.2

For the proofs of the above-mentioned theorems, we need two special versions of Theorems 1.6.1 and 1.6.3.

Proposition 8.9.1 *We assume (8.5a) and (8.5b) to be given, and we suppose E to satisfy (8.6) with respect to an integral ideal $\mathfrak{q}$. Let $f \in \mathbb{C}(x, y)$ have the following property: there exist infinitely many prime ideals $\mathfrak{l}$ in K such that*

$$f(\xi) \in LK_{\mathfrak{q}\mathfrak{l}^n}, \text{ for all } \xi \in \mathfrak{l}^{-n}\mathfrak{m} \setminus \mathfrak{m},\ n \in \mathbb{N}.$$

Then $f \in L(x, y)$.

Proof Let $\mathfrak{l}$ be a fixed prime ideal with the above properties. Then obviously, there exists a null sequence $(z_n)_{n \geq n_0}$ with

$$z_n \in \mathfrak{l}^{-n}\mathfrak{m} \setminus \mathfrak{l}^{-(n-1)}\mathfrak{m}, \quad n \geq n_0$$

that does not contain poles of f. By Theorem 1.6.1, from the assumptions for f and in view of the general condition (8.7) there exists some $N \in \mathbb{N}$ with

$$f \in LK_{\mathfrak{q}\mathfrak{l}^N}(x, y).$$

To prove the proposition, we contend that almost all pairs $\mathfrak{l}, \mathfrak{l}'$ satisfy

$$LK_{\mathfrak{q}\mathfrak{l}^n} \cap LK_{\mathfrak{q}\mathfrak{l}'^n} = L \text{ for all } n \in \mathbb{N}.$$

In fact, this is true for any two prime ideals $\mathfrak{l}, \mathfrak{l}'$ of K unramified in L/K because the prime ideals of $K_{\mathfrak{q}}$ ramified in $K_{\mathfrak{q}\mathfrak{l}^n}$ for $K_{\mathfrak{q}\mathfrak{l}^n} \neq K_{\mathfrak{q}}$ are exactly the prime ideals above $\mathfrak{l}$, and, furthermore, these are totally ramified. This finishes the proof of proposition 8.9.1. □

Our next aim is the p-adic series expansion for singular values of elliptic functions from $L(x, y)$, where again we make the general assumption (8.5). According to Theorem 1.6.2 in a neighbourhood of zero we have the series

$$\begin{aligned} -\tfrac{1}{y(\xi)} &= W(\xi)^3 + \sum_{n\geq 4} A_n W(\xi)^n, \\ x(\xi) &= W(\xi)^{-2} + \sum_{n\geq -1} B_n W(\xi)^n, \\ y(\xi) &= -W(\xi)^{-3} + \sum_{n\geq -2} C_n W(\xi)^n \end{aligned} \tag{8.38}$$

with $\mathfrak{p}$-integral coefficients $A_n, B_n, C_n \in L$. This implies that every function $f \in L(x, y)$ can be written as

$$f(\xi) = \sum_{n=n_0}^{\infty} d_n W(\xi)^n \tag{8.39}$$

in a neighbourhood of zero with coefficients in L. The following proposition contains a criterion for the $\mathfrak{p}$-integrality of the coefficients d_n.

Proposition 8.9.2 *We assume (8.5) to be given, (a)–(c), and we let $f \in L(x, y)$ have the following properties:*

(i) *all poles of f are in $K \setminus \mathfrak{p}^{-\infty}\mathfrak{m}$,*
(ii) *$f(\xi)$ is $\mathfrak{p}$-integral for all $\xi \in \mathfrak{p}^{-\infty}\mathfrak{m}$.*

Then the coefficients d_n in (8.39) are $\mathfrak{p}$-integral in L, and $n_0 = 0$. In particular, the series in (8.39) is also $\mathfrak{P}$-adically convergent to $f(\xi)$ for $\xi \in \mathfrak{p}^{-\infty}\mathfrak{m}$ and, for every prime ideal $\mathfrak{P}$ above $\mathfrak{p}$ of $L(f(\xi), W(\xi))$.

Proof First $n_0 = 0$ because f has no pole at 0. We choose a null sequence $(z_n)_{n\geq 1}$ with $z_n \in \mathfrak{p}^{-n}\mathfrak{m} \setminus \mathfrak{p}^{n-1}\mathfrak{m}$. Then, by Theorem 1.6.3 the coefficients d_n are all in $\frac{1}{r}R_0$,

$$R_0 := \mathbb{Z}[a_1, .., a_6, x(\xi_1), .., x(\xi_m), w(z_1), .., w(z_s), s(z_1), .., s(z_n), f(z_1), .., f(z_N)],$$

where $\xi_1, \dots, \xi_m$ modulo $\mathfrak{m}$ are the poles of f and N is a sufficiently large natural number. According to Theorem 1.6.2 and because of assumptions (i) and (ii), the generators of R_0 are $\mathfrak{p}$-integers. Hence the d_n are all in $\frac{1}{r}\mathfrak{O}_\mathfrak{p}$, where $\mathfrak{O}_\mathfrak{p}$ denotes the ring of $\mathfrak{p}$-integers. Now, let $\mathfrak{p}'$ be a prime ideal in L above $\mathfrak{p}$, $\mathfrak{P}$ a prime ideal above $\mathfrak{p}'$ in $L(f(\xi), W(\xi))$, and let R be the valuation ring of $\mathfrak{P}$. Further, we choose an ascending chain of valuation rings R_n in $L(f(\xi), W(\xi), W(z_n))$ containing R. Then $\pi_n = W(z_n)$ satisfies the hypothesis of Theorem 1.6.7. Hence, the series is also $\mathfrak{P}$-adically convergent to $f(\xi)$, and the d_n are in R. More precisely, the d_n are $\mathfrak{p}'$-integral in $R \cap L$ and, since this is true for every prime ideal $\mathfrak{p}'$ over $\mathfrak{p}$, they are even $\mathfrak{p}$-integral. □

Proof of Theorem 8.4.1 A function satisfying the hypothesis of Theorem 8.4.1 also satisfies the hypothesis of Proposition 8.9.2. In particular, for $R = O$ it follows that $T(Q) = T(R+Q) - T(R)$ is integral for every $Q \in E[\mathfrak{p}^\infty]$. Therefore, the element defined in Theorem 8.4.1,

$$\theta := (T(R))_{R\in P+G_m},$$

is in $\tilde{\mathfrak{O}}_P$. For the discriminant of the polynomial

$$g_P(X) := \prod_{R\in P+G_m} (X - \theta_R)$$

we obtain

$$discr(g_P(X)) \sim \mathfrak{p}^{mp^m},$$

bearing in mind that by assumption $T(R) - T(R+Q) \sim \mathfrak{p}^{\frac{1}{\Phi(o(Q))}}$ for $Q \in G_m \setminus \{O\}$. Since this is a divisor of the discriminant of $\tilde{\mathfrak{O}}_P$, we finally obtain

$$\tilde{\mathfrak{O}}_P = \mathfrak{O}_L[\theta] \quad \text{and} \quad discr(\tilde{\mathfrak{O}}_P) = \mathfrak{p}^{mp^m}.$$

□

Proof of Theorem 8.4.2. Let T be the function defined in (8.10). By Proposition 8.7.6 (or following from Proposition 8.9.1) T is in $L(x, y)$.

To prove that T satisfies the other hypothesis of Theorem 8.4.1, we use the formula (7.1)

$$\mathcal{P}(\xi_1) - \mathcal{P}(\xi_2) = (-\epsilon)^e \prod_{i=0}^{e-1} \frac{\varphi(\xi_1 + \zeta^i \xi_2)\varphi(\xi_1 - \zeta^i \xi_2)}{\varphi(\xi_1)^2 \varphi(\zeta^i \xi_2)^2},$$

with $e = \frac{w_K}{2}$ and a generating element ζ for the group of roots of unity in K. Now let $Q = Q(\xi) \neq O$, $R = Q(\xi_1)$ be in $E[\mathfrak{p}^\infty]$. We set $\eta_1 := \xi_1 + \delta$. By the above formula we obtain

$$T(R+Q) - T(R) = (-\epsilon)^{e-1} \prod_{i=0}^{e-1} \frac{\varphi(\xi + (1+\zeta^i)\eta_1)\varphi(\xi + (1-\zeta^i)\eta_1)}{\varphi(\xi + \eta_1)^2 \varphi(\zeta^i \eta_1)^2}.$$

Herein

$$\varphi(\xi + (1 - \zeta^0)\eta_1) = \varphi(\xi) \sim \mathfrak{p}^{\frac{1}{\Phi(o(Q))}}$$

according to Theorem 4.3.1. All other factors are units, because η_1 is of composite order divisible by $\mathfrak{q}$ with $\frac{\mathfrak{q}}{\gcd(w_K, \mathfrak{q})}$ composite. This proves the factorisation

$$T(R+Q) - T(R) \sim \mathfrak{p}^{\frac{1}{\Phi(o(Q))}},$$

we wanted to show. □

Proof of Theorem 8.5.2. The transformation formula for h_γ in section 1.7 tells us that g is elliptic for $\mathfrak{m}$. More precisely, g is in $L(x,y)$, as we will show using Proposition 8.9.1. So let $\mathfrak{l}$ be an arbitrary prime ideal of K not dividing $\mathfrak{q}$ and $\xi \in \mathfrak{l}^{-n}\mathfrak{m}$, $n \in \mathbb{N}$. Then by the Reciprocity Law together with the transformation formula for the φ function we first obtain $g(\xi) \in K_{12N(\mathfrak{q}\mathfrak{l}^n)^2}$ and then

$$g(\xi)^{\sigma(\lambda)} = \cdots = g(\xi\lambda) = \cdots = g(\xi)$$

for every $\lambda \in \mathfrak{O}_K$ prime to $12N(\mathfrak{q}\mathfrak{l})$ satisfying $\lambda \equiv 1 \bmod \mathfrak{q}\mathfrak{l}^n$. Herein, $\sigma(\lambda)$ denotes the Frobenius automorphism corresponding to the ideal (λ). Hence $g(\xi)$ is in $K_{\mathfrak{q}\mathfrak{l}^n}$. Proposition 8.9.1 now implies that

$$g \in L(x,y).$$

Furthermore, by factoring φ-values and keeping in mind that by assumption the order $\mathfrak{q}$ of δ is composite, we obtain that the $g(\xi)$ are integral for all $\xi \in \mathfrak{p}^{-\infty}\mathfrak{m}$ and all prime ideals $\mathfrak{p}$ not dividing $\mathfrak{q}$. Therefore, using Proposition 8.9.1 we find that the element θ defined in Theorem 8.5.2 is in $\mathfrak{O}_P$.

Now we contend that the resolvents of the element defined in Theorem 8.5.2 are associated with $\mathfrak{p}^m$. Therefore, we write the resolvents in the form

$$(\theta, \chi) = \sum_{\xi \in \mathfrak{p}^{-m}\mathfrak{m} \bmod \mathfrak{m}} g(\xi_1 + \xi)\overline{\chi(\xi)},$$

with a parameter ξ_1 of $P = Q(\xi_1)$. By the resolvent formula of Theorem 1.7.2 we then have

$$\begin{aligned}(g(z|\mathfrak{L}), \chi) &:= \sum_{\substack{\xi \in \hat{\mathfrak{L}} \\ \xi \bmod \mathfrak{L}}} g(z + \xi|\mathfrak{L})\bar{\chi}(\xi) \\ &= \sqrt[12]{\frac{\Delta(\hat{\mathfrak{L}})}{\Delta(\mathfrak{L})}} C_{\chi\chi_0}(\delta + z) h_\gamma(\omega(\delta + z)|\mathfrak{L})^{-1},\end{aligned}$$

with the character $\chi_0(\xi) = e^{l_{\mathfrak{L}}((1-\omega)\xi, \gamma)}$,

$$C_{\chi\chi_0}(\delta + z) = e^{-\frac{1}{2}l_{\hat{\mathfrak{L}}}(\delta+z, \frac{\gamma}{n}+\mu_{\chi\chi_0})} \frac{\varphi(\gamma|\mathfrak{L})\varphi(\delta + z + \frac{\gamma}{n} + \mu_{\chi\chi_0}|\hat{\mathfrak{L}})}{\varphi(\delta + z|\hat{\mathfrak{L}})\varphi(\frac{\gamma}{n} + \mu_{\chi\chi_0}|\hat{\mathfrak{L}})},$$

$$\mathfrak{L} = \mathfrak{m}, \quad \hat{\mathfrak{L}} = \mathfrak{p}^{-m}\mathfrak{m}, \quad n = [\hat{\mathfrak{L}} : \mathfrak{L}] = N(\mathfrak{p}^m)$$

and elements

$$\mu_{\chi\chi_0} \in \frac{1}{n}\mathfrak{L}.$$

Theorem 4.2.2 implies that

$$\sqrt[12]{\frac{\Delta(\hat{\mathfrak{L}})}{\Delta(\mathfrak{L})}} \sim \mathfrak{p}^m,$$

and for $z = \xi_1$ all φ-values in the definition of $C_{\chi\chi_0}(\delta + z)$ and $h_\gamma(\omega(\delta + z)|\mathfrak{L})$ are units. This follows from Theorem 4.3.1, keeping in mind (8.9), (8.11) and (8.12). Finally, since the exponential factor in the definition of $C_{\chi\chi_0}(\delta+\xi_1)$ is a root of unity, the resolvent formula tells us that (θ, χ) has the factorisation $\mathfrak{p}^m$, as asserted. This finishes the proof of Theorem 8.5.2 under the hypothesis (8.9). The construction of a generator under the weaker hypothesis (8.8) is explained in the following proof of (8.18). □

Proof of (8.18). Using the $\mathfrak{p}$-adic series for θ,

$$\theta_R = \sum_{n=0}^{\infty} a_n W(R|\mathfrak{n})^n,$$

we obtain

$$(\lambda(\theta))_R = \sum_{R'\in\lambda^{-1}(R)} \theta_{R'} = \sum_{Q\in G_s^{\mathfrak{n}}} \theta_{R_0+Q} = \sum_{n=0}^{\infty} a_n \sum_{Q\in G_s^{\mathfrak{n}}} W(R_0+Q|\mathfrak{n})^n$$

with some $R_0 \in \lambda^{-1}(R)$. Now, applying the p-adic addition formula (8.3) to $W(R_0+Q|\mathfrak{n})$, Lemma 8.8.3 yields

$$\sum_{R'\in\lambda^{-1}(R)} \theta_{R'} \equiv 0 \bmod \mathfrak{p}^s.$$

To write $\lambda(\theta)$ as a power series of $W(R)$, $R \in \lambda(P) + G_{m-s}^{\mathfrak{m}}$, observe that $\sum_{Q\in G_s^{\mathfrak{n}}} W(z+\xi_Q|\mathfrak{n})$ is elliptic for $\mathfrak{m}$ and can be written as

$$\sum_{Q\in G_s^{\mathfrak{n}}} W(z+\xi_Q|\mathfrak{n}) = \sum_{k=0}^{\infty} b_k W(z|\mathfrak{m})^k$$

with coefficients b_k in $\hat{L}$ because of (8.17). Moreover, by Proposition 8.9.2 the b_k are $\mathfrak{p}$-integral. Now we arrive at the series we are aiming for by inserting $W(z|\mathfrak{n})$ into the expansion of $(\lambda(\theta))_R$.

With the above construction we obtain by

$$\Theta = \left(\frac{1}{\pi^s} \sum_{R'\in\lambda^{-1}(R)} g(R'|\mathfrak{m}\mathfrak{p}^s)\right)_{R\in\lambda(P)+G_s^{\mathfrak{m}}}$$

an element in $\mathfrak{O}_{\lambda(P)}^{\mathfrak{m}}(m-s)$, whose resolvents are associated with $\mathfrak{p}^{m-s}$. However, this does not end the proof since, according to our construction, the base field for $\mathfrak{O}_{\lambda(P)}^{\mathfrak{m}}(m-s)$ is $\hat{L}$. But applying Theorem 8.7.5 to the function

$$\tilde{g}(z) := \frac{1}{\pi^s} \sum_{Q'\in G_s^{\mathfrak{n}}} g(z+\xi_{Q'}|\mathfrak{m}\mathfrak{p}^s),$$

we find that in fact Θ is in $\mathfrak{O}_{\lambda(P)}^{\mathfrak{m}}(m-s)$ with the base field L. □

8.10 Proofs of Theorems 8.9.2 and 8.6.2

Proof of Theorem 8.9.2. First, we consider the case $\mathfrak{p}$ **odd** and $\mathfrak{q}$ satisfying (8.9). The assertions of Theorem 8.9.2 then follow from

Theorems 8.4.2 and 8.5.2, using the Fueter model E associated with $\mathfrak{m} = \mathfrak{O}_K$. Since the coefficients of E are in $K_{(4)}$, we can take

$$L = K_{4\mathfrak{q}\mathfrak{p}^r}.$$

as base field. By Reciprocity Law we find that

$$L(E[\mathfrak{q}]) \subseteq K_{4\mathfrak{q}}.$$

Hence, $E(L)$ contains a point of order $\mathfrak{q}$, and (8.6) is satisfied, so we can apply Theorems 8.4.2 and 8.5.2. We choose a point

$$P = Q(\xi_1) \text{ with } o(\xi_1) = \mathfrak{p}^{r+m}.$$

By Theorem 8.5.2 we then have

$$\tilde{\mathfrak{O}}_P = \theta \circ \mathfrak{A},$$

where θ and $\mathfrak{A}$ are explicitly given as values of g and T defined in (8.13) and (8.10):

$$\theta = (g(P+Q))_{Q \in G_m} = (g(\xi_1 + \xi))_{\xi \in \mathfrak{p}^{-m}/\mathfrak{O}_K},$$

$$\mathfrak{A} = \bigoplus_{i=0}^{p^m-1} \mathfrak{O}_{M_{4\mathfrak{q}}} \tau_i, \quad \tau_i = \frac{1}{\pi^m} \sum_{Q \in G_m} T(Q)^i Q,$$

$$T(Q(\xi)) := T(\xi), \quad \xi \in \mathfrak{p}^{-m}.$$

By Reciprocity Law $W(\xi_1) \in N_{4\mathfrak{q}}$ and

$$W(\xi_1)^{\sigma(1+\xi_0\xi)} = W(\xi_1 + \xi) \text{ for } \xi \in \mathfrak{p}^{-m}.$$

This implies that

$$\kappa : \mathbb{M}_P \to N_{4\mathfrak{q}}, \quad (a_{P+Q})_{Q \in G_m} \mapsto a_P,$$

is an isomorphism of $M_{4\mathfrak{q}}$-algebras. Further, according to (8.19), we have the isomorphism

$$\hat{\kappa} : G_m \to Gal(N_{4\mathfrak{q}}/M_{4\mathfrak{q}}), \quad Q(\xi) \mapsto \sigma_\xi := \sigma(1 + \xi_0 \xi),$$

that can be extended to $\mathfrak{A}$ by

$$\hat{\kappa}\left(\sum_{Q(\xi) \in G_m} f(T(Q(\xi)))Q(\xi)\right) := \sum_{\sigma_\xi \in G} f(T(\sigma_\xi))\sigma_\xi \text{ for } f(X) \in L[X].$$

This proves $\kappa(\tilde{\mathfrak{O}}_P)$ to be a $\hat{\kappa}(\mathfrak{A})$-module contained in $\mathfrak{O}_{N_{4\mathfrak{q}}}$ generated by θ_P:

$$\kappa(\tilde{\mathfrak{O}}_P) = \theta_P \circ \hat{\kappa}(\mathfrak{A}).$$

Furthermore,

$$d_{N_{4\mathfrak{q}}/M_{4\mathfrak{q}}}(\kappa(\tilde{\mathfrak{O}}_P)) = discr(\tilde{\mathfrak{O}}_P) = \mathfrak{p}^{mp^m},$$

and since the computation of the relative discriminant of $N_{4\mathfrak{q}}/M_{4\mathfrak{q}}$ via the conductor discriminant formula leads to the same result, we have proved that:

$$\mathfrak{O}_{N_{4\mathfrak{q}}} = \kappa(\tilde{\mathfrak{O}}_P), \quad \mathfrak{A}_{N_{4\mathfrak{q}}/M_{4\mathfrak{q}}} = \hat{\kappa}(\mathfrak{A})$$

and

$$\mathfrak{O}_{N_{4\mathfrak{q}}} = \theta_P \circ \mathfrak{A}_{N_{4\mathfrak{q}}/M_{4\mathfrak{q}}}.$$

To prove the last equation for the extension $N_\mathfrak{q}/M_\mathfrak{q}$ instead of $N_{4\mathfrak{q}}/M_{4\mathfrak{q}}$, note that $N_\mathfrak{q} \cap M_{4\mathfrak{q}} = M_\mathfrak{q}$. Hence, $Gal(N_\mathfrak{q}/M_\mathfrak{q})$ can be identified with $Gal(N_{4\mathfrak{q}}/M_{4\mathfrak{q}})$, and we denote both by G. Since θ_P lies in $N_\mathfrak{q}$, we obtain

$$\mathfrak{O}_{N_\mathfrak{q}} = \theta_P \circ ((\mathfrak{A}_{N_{4\mathfrak{q}}/M_{4\mathfrak{q}}} \cap N_\mathfrak{q}[G]))$$

by intersection of both sides in the above equation with $N_\mathfrak{q}$. Further, observing that, according to Theorems 8.4.1 and 8.4.2, $\mathfrak{A}$ is a free $\mathfrak{O}_{N_{4\mathfrak{q}}}$-module, it follows that

$$(\mathfrak{A}_{N_{4\mathfrak{q}}/M_{4\mathfrak{q}}} \cap N_\mathfrak{q}[G]) = \hat{\kappa}(\mathfrak{A} \cap N_\mathfrak{q}[G_m]) =$$

$$\hat{\kappa}\left(\frac{1}{\pi^m}\left\{\sum_{Q\in G_m} f(T(Q))Q \;\middle|\; f(X) \in \mathfrak{O}_{M_\mathfrak{q}}[X]\right\}\right).$$

This finishes the proof of the second and third assertions of Theorem 8.6.1 for an odd $\mathfrak{p}$ and $\mathfrak{q}$ satisfying (8.9). If $\mathfrak{q}$ only satisfies the weaker condition (8.8) and is not a divisor of 2, we obtain the second assertion of Theorem 8.6.1, by repeating the proof using the element ϑ' from (8.21) instead of θ_P.

For "$\mathfrak{p}$ even" we proceed in the same way using Deuring's instead of Fueter's model.

The first assertion of Theorem 8.6.1 follows easily from Theorem 7.4.1. □

Proof of Theorem 8.6.2. Keeping in mind that the element ϑ' in (8.21) lies in $\mathfrak{O}_{N_1}$ with resolvents associated with $\mathfrak{p}^m$, the assertion of Theorem 8.6.2 follows using same arguments as in the proof of Theorem 8.6.1. □

8.11 Analogy to the cyclotomic case

Similarly to the construction of integral bases we obtain a result on the Galois module structure for cyclotomic fields by taking a suitable limit in the basic resolvent formula. We consider two complex lattices

$$\mathfrak{L} = [\omega, 1] \subset \hat{\mathfrak{L}} = \left[\omega, \frac{1}{n}\right]$$

of index n. Further, in the resolvent formula of Theorem 1.7.2 we let z and γ be real, and we let the ξ be of the form

$$\xi = \frac{\nu}{n}, \quad \nu = 0, \ldots, n-1.$$

In this special case the resolvent formula of Theorem 1.7.2 can be written as

$$\sum_{\xi} \frac{\sigma^*(z+\xi+\gamma|\mathfrak{L})}{\sigma^*(z+\xi|\mathfrak{L})}\overline{\chi}(\xi) \;=\; e^{\frac{1}{2}l_{\hat{\mathfrak{L}}}(z,\mu_\chi)} \frac{\sigma^*(\gamma|\mathfrak{L})\sigma^*\left(z+\frac{\gamma}{n}+\mu_\chi\middle|\,\hat{\mathfrak{L}}\right)}{\sigma^*\left(\gamma\middle|\hat{\mathfrak{L}}\right)\sigma^*\left(\frac{\gamma}{n}+\mu_\chi\middle|\,\hat{\mathfrak{L}}\right)}. \tag{8.40}$$

Herein, we choose μ_χ as a coset representative of $\frac{1}{n}\mathfrak{L}$ modulo $\hat{\mathfrak{L}}$ of the form

$$\mu_\chi = u_\chi \frac{\omega}{n}, \quad u_\chi = 0, \ldots, n-1.$$

Now, keeping in mind the q-expansion of σ^* for arbitrary $z \in \mathbb{C}$,

$$\sigma^*\left(z \,\middle|\, \begin{matrix}\omega_1\\ \omega_2\end{matrix}\right) = \frac{\omega_2}{2\pi i} Q^{\frac{z_1}{\omega_2}} \left(Q^{\frac{1}{2}} - Q^{-\frac{1}{2}}\right) \prod_{n=1}^{\infty} \frac{(1-Qq^n)(1-Q^{-1}q^n)}{(1-q^n)^2}$$

with

$$Q = e^{2\pi i \frac{z}{\omega_2}}, \quad z = z_1\omega_1 + z_2\omega_2, \quad z_1, z_2 \in \mathbb{R}; \quad q = e^{2\pi i\omega}, \; \omega = \frac{\omega_1}{\omega_2},$$

by taking the limit for $\omega \to i\infty$ in (8.40) we obtain the relation

$$\sum_{\xi} \frac{d(z+\xi+\gamma)}{d(z+\xi)}\overline{\chi}(\xi) = n\zeta C_\chi \tag{8.41}$$

with a root of unity ζ and

$$C_\chi = \begin{cases} \frac{d(\gamma)}{d(nz)}, & \text{for } \chi \neq 1, \\ \frac{d(nz+\gamma)}{d(\gamma)}, & \text{for } \chi = 1, \end{cases} \tag{8.42}$$

where for short we use the notation

$$d(w) := e^{\pi i w} - e^{-\pi i w}.$$

For the arithmetical interpretation of this formula we consider, instead of an elliptic curve, the unit circle, defined over $L = \mathbb{Q}$:

$$E(\mathbb{C}) := \{(x, y) \in \mathbb{C}^2 \mid x^2 + y^2 = 1\}.$$

In order to elucidate the analogy, we use the same notations as used for elliptic curves in 8.1, 8.2 and 8.3. Via the bijection

$$\mathbb{C}/\mathbb{Z} \to E(\mathbb{C}), \quad \xi + \mathbb{Z} \mapsto Q(\xi) := (x(Q(\xi)), y(Q(\xi))) = (\cos(2\pi\xi), \sin(2\pi\xi)),$$

we define a group structure on $E(\mathbb{Q})$ by

$$Q(\xi_1) + Q(\xi_2) := Q(\xi_1 + \xi_2).$$

Then the n-torsion points are given by the n-th roots of unity

$$E[n] = \left\{ Q(\xi) \;\middle|\; \xi \in \frac{1}{n}\mathbb{Z}/\mathbb{Z} \right\}.$$

Further, the Galois group $\Omega_{\mathbb{Q}}$ of $\mathbb{Q}^c/\mathbb{Q}$ is acting in an obvious way on $E(\mathbb{Q}^c)$.

As uniformising parameter at the origin (neutral element) of $E(\mathbb{C})$ we choose

$$W(\xi) := e^{2\pi\xi} - 1 = x(Q(\xi)) + iy(Q(\xi)) - 1,$$

and we set

$$W(Q) := W(\xi), \text{ for } Q = Q(\xi).$$

If $Q = Q(\xi)$ is a torsion point of order n, then $W(Q)$ generates the n-th cyclotomic field, and for a prime power $n = p^e$ we have

$$W(Q) \sim p^{\Phi(p^e)},$$

where Φ is the Euler function in $\mathbb{Q}$. Further, similar to (8.3), we have the relation

$$W(Q_1 + Q_2) = W(Q_1) + W(Q_2) + W(Q_1)W(Q_2).$$

Now we choose a fixed prime number $p \neq 2$ and two natural numbers

$$m, r \geq 1.$$

For $s \in \mathbb{N}$ we set

$$G_s = E[p^s].$$

Given a point $P \in G_{r+m}$ we define the polynomial

$$h_P(X) := \prod_{R \in P + G_m} (X - W(R))$$

and then the algebra

$$\mathbb{M}_P := \mathbb{Q}[X]/(h_P(X)) = \{(f(W(R)))_{R \in P+G_m} \mid f \in \mathbb{Q}[X]\}$$

$$= \mathbb{Q}[\theta], \quad \theta = (W(R))_{R \in P+G_m}.$$

Since this is a direct sum of cyclotomic fields, we find, unlike in 8.3, that

$$\mathfrak{O}_P := \mathbb{Z}[\theta]$$

is the maximal order. Further, as in section 8.3 we define

$$\mathfrak{A} := \frac{1}{p^m} \left\{ \sum_{Q \in G_m} a_Q\, Q \;\middle|\; (a_Q)_{Q \in G_m} \in \mathfrak{O}_O \right\},$$

where $O = Q(0)$ denotes the neutral element in $E(\mathbb{C})$. Now, as in the proof of Theorem 8.3.1 defining the operation of $\mathfrak{A}$ on $\mathfrak{O}_P$ analogously to (8.4) we can prove:

Theorem 8.11.1

(i) $\mathfrak{A}$ *is a ring,*
(ii) $\mathfrak{O}_P \circ \mathfrak{A} \subseteq \mathfrak{O}_P$,
(iii) $discr(\mathfrak{O}_P) = discr(h_P(X)) = p^{mp^m}$.

In particular, the last assertion shows that every element $\theta \in \mathfrak{O}_P$ is a generating element of $\mathfrak{O}_P$ over $\mathfrak{A}$,

$$\mathfrak{O}_P = \theta \circ \mathfrak{A},$$

if for every character χ of G_m

$$\sum_{Q \in G_m} \theta_{P+Q} \overline{\chi}(Q) \sim p^m.$$

In the case $P = Q\left(\frac{1}{p^r+m}\right)$ such an element can be found via the resolvent formula (8.41) by setting

$$\theta = (\theta_R)_{R\in P+G_m} = (g(R))_{R\in P+G_m}$$

with the 1-periodic function

$$g(z) = \frac{d\left(z+\frac{1}{p^r}\right)}{d(z)} = \frac{e^{2\pi iz}-1}{e^{2\pi iz}e^{\frac{\pi i}{p^r}} - e^{-\frac{\pi i}{p^r}}}.$$

8.12 Generalisation to ring classes by Bettner and Bley

Most of the constructions in sections 8.4 and 8.5 carry over to ring classes and lead to similar results. However, the subgroup $G_{\mathfrak{p}^s}$ for a prime ideal power $\mathfrak{p}^s$, that is crucial for the definition of the objects to be studied, has to be modified. Therefore, we fix a proper ideal $\mathfrak{a}$ of the order $\mathfrak{O}_t$ of conductor $t > 1$ in an imaginary quadratic number field K and consider a Weierstrass model associated with $\mathfrak{a}$, defined over a finite extension L of K. The prime ideal power $\mathfrak{p}^s$ in the construction for $t = 1$ is replaced by a primary ideal $\mathfrak{q}$ of $\mathfrak{O}_t$, and we define

$$G_{\mathfrak{q}} := \{Q(\xi) \in E(\mathbb{Q}^c) \mid \mathfrak{q}\xi \subseteq \mathfrak{a}\}.$$

For $t = 1$ this is the old definition, because then $\mathfrak{q}$ is a prime ideal power. In order to use the $\mathfrak{p}$-adic power series expansions as in the case $t = 1$ it is necessary to modify $G_{\mathfrak{q}}$ by taking its subgroup

$$G_{\mathfrak{q}}^* := \{Q \in G_{\mathfrak{q}} \mid v_{\mathfrak{P}}(W(Q)) > 0 \text{ for all prime ideals } \mathfrak{P} \supseteq \mathfrak{p} \text{ of } \mathfrak{O}_1\}$$

instead, where $\mathfrak{p}$ denotes the prime ideal associated with $\mathfrak{q}$. Let p be the prime number in $\mathfrak{p}$, $p\mathbb{Z} = \mathfrak{p} \cap \mathbb{Z}$. Then, by factoring φ-values we obtain

$$\begin{array}{lll} G_{\mathfrak{q}}^* \subsetneqq G_{\mathfrak{q}} & \text{if} & p \mid t \text{ and } p \text{ is split in } K, \\ G_{\mathfrak{q}}^* = G_{\mathfrak{q}} & \text{otherwise.} & \end{array}$$

The general assumptions (8.5) for the definition of the integral objects in the case $t = 1$ are now generalised as follows:

General assumptions for $t > 1$ (8.43)

(a) K is a quadratic imaginary number field and $\mathfrak{O}_t$ the order of conductor t in K.

(b) E is an elliptic curve defined by an equation (8.1) of a pair of Weierstrass functions associated with an ideal $\mathfrak{a} \in \mathfrak{J}_t$ having coefficients in a finite extension L of the ring class field Ω_t. For $d = -3, -4$ we further assume L to contain the ring class field Ω_{3t} resp. Ω_{2t}.

(c) $\mathfrak{p}$ is a fixed prime ideal of $\mathfrak{O}_t$ of norm $p = N_t(\mathfrak{p}) := [\mathfrak{O}_t : \mathfrak{p}]$, we assume the coefficients of the equation (8.1) to be $\mathfrak{p}\mathfrak{O}_1$-integral and E to have good reduction above $\mathfrak{p}\mathfrak{O}_1$.

(d) Let $\mathfrak{q}_r$ be a primary ideal in $\mathfrak{O}_t$ or $\mathfrak{q}_r = \mathfrak{O}_t$ with

$$G^*_{\mathfrak{q}_r} \subseteq E(L),$$

(e) $\mathfrak{q}_m$ a primary ideal in $\mathfrak{O}_t$ and $P \in E(\mathbb{Q}^c)$ with

$$[\mathfrak{p}_m]P \in G^*_{\mathfrak{q}_r}.$$

The definition of $\mathbb{M}_P$, $\mathbb{A}$ and $\tilde{\mathfrak{O}}_P$, $\mathfrak{A}$ is completely analogous to the case $t = 1$, replacing the coset $P + G_{\mathfrak{p}^m}$ by the coset

$$P + G^*_{\mathfrak{q}_m}.$$

Further, in the definition of $\mathfrak{A}$ the power π^m is replaced by an element $\pi_m \in L$ with

$$\pi_m \sim \prod_{Q(\xi) \in G^*_{\mathfrak{q}_m} \setminus \{0\}} \varphi(\xi|\mathfrak{a}).$$

For $d_K \neq -3, -4$ the existence of such an element follows from the principal ideal theorem and similarly in the cases $d_K = -3, -4$ if we add the assumptions $\mathfrak{q}_m \in \mathfrak{J}_t$ and $G_{\mathfrak{q}_m} = G^*_{\mathfrak{q}_m}$.

To prove the existence of a Galois generating element for $\tilde{\mathfrak{O}}_P$ over $\mathfrak{A}$, we have, as in the case when $t = 1$, to assume that $E(L)$ contains enough torsion points. Therefore, let $\mathfrak{q}$ be an ideal of $\mathfrak{O}_t$, prime to $\mathfrak{p}$ with decomposition

$$\mathfrak{q} = \mathfrak{q}_1 \cdot \ldots \cdot \mathfrak{q}_s, \quad s \geq 2,$$

into a product of primary ideals, and for $s = 2$ we assume that

$$N(\sqrt{\mathfrak{q}_i}) \leq 3 \Rightarrow \mathfrak{q}_i \supsetneqq \sqrt{\mathfrak{q}_i}, \quad i = 1, 2 \tag{8.44}$$

and

$$\mathfrak{q} \nmid 2.$$

Then there exist elements $\gamma, \delta \in K \setminus \mathfrak{a}$ with $\mathfrak{q}\gamma, \mathfrak{q}\delta \subseteq \mathfrak{a}$, such that

$$o(\gamma|\mathfrak{a}),\ o(\delta|\mathfrak{a}),\ o(\gamma + \delta|\mathfrak{a}),\ o(\gamma + |G^*_{\mathfrak{q}_m}|\delta|\mathfrak{a}) \text{ are composite.} \tag{8.45}$$

Here the order $o(\lambda|\mathfrak{a})$ of an element $\lambda \in K \setminus \mathfrak{a}$ is defined to be the $\mathfrak{O}_t$-ideal

$$o(\lambda|\mathfrak{a}) := \{\xi \in \mathfrak{O}_t \mid \xi\lambda \in \mathfrak{a}\}.$$

The following theorem generalises Theorem 8.5.2.

Theorem 8.12.1 *Let $\mathfrak{q} \nmid 2$ be an integral ideal of $\mathfrak{O}_t$ prime to p, satisfying condition (8.44), and let $E(L)$ contain $G_{\mathfrak{q}}$. Let $\gamma, \delta \in K \setminus \mathfrak{a}$ with $\mathfrak{q}\gamma, \mathfrak{q}\delta \in \mathfrak{a}$ and (8.45). Further, let $\omega \in \mathfrak{O}_t$ satisfy the congruences*

$$\omega \equiv 0 \; mod \; \mathfrak{q}_m \quad and \quad \omega \equiv 1 \; mod \; p.$$

Let g be the function defined by γ, δ and ω in (8.13). Then

$$\theta \; := \; (g(R))_{R \in P + G^*_{\mathfrak{q}_m}}$$

is a Galois generator for $\tilde{\mathfrak{O}}_P$ over $\mathfrak{A}$.

As in the case for $t = 1$ we can skip the hypothesis $G_{\mathfrak{q}} \subseteq E(L)$, assuming that p does not divide w_K.

Theorem 8.12.2 *For $p \nmid w_K$ there exists an element $\theta \in \tilde{\mathfrak{O}}_P$ with*

$$\tilde{\mathfrak{O}}_P = \theta \circ \mathfrak{A}.$$

For the global construction of $\tilde{\mathfrak{O}}_P$ and $\mathfrak{A}$ we are using, as in 7.1 for $t = 1$, the normalisation $\mathcal{P}(z|\mathfrak{a})$ of the $\wp$ function and define:

$$T(z) := \mathcal{P}(z + \delta|\mathfrak{a}) - \mathcal{P}(\delta|\mathfrak{a}),$$

where $\delta \in K \setminus \mathfrak{a}$ has order $\mathfrak{q}$.

Theorem 8.12.3 *Assuming the hypothesis of Theorem 8.12.1 on δ and $\mathfrak{q}$, we have*

$$\tilde{\mathfrak{O}}_P = \mathfrak{O}_L \oplus \mathfrak{O}_L\theta \oplus \cdots \oplus \mathfrak{O}_L\theta^{N-1}$$

with

$$\theta = (T(R))_{R \in P + G^*_{\mathfrak{q}_m}}$$

and $N = |G^*_{\mathfrak{q}_m}|$. In particular,

$$\mathfrak{A} = \bigoplus_{i=0}^{N-1} \mathfrak{O}_L \tau_i \quad with \quad \tau_i = \frac{1}{\pi_m} \sum_{Q \in G^*_{\mathfrak{q}_m}} T(Q)^i Q.$$

Using Theorems 8.12.1, 8.12.2 and 8.12.3 we obtain similar results on the Galois module structure of the extensions

$$N_{\mathfrak{q}}/M_{\mathfrak{q}} = K_{t,\mathfrak{q}\mathfrak{q}_r\mathfrak{q}_m}/K_{t,\mathfrak{q}\mathfrak{q}_r}$$

with

$$\mathfrak{q}_m \mid \mathfrak{q}_r,$$

because by the last condition we have an isomorphism

$$\mathfrak{q}_m^{-1}/\mathfrak{O}_t \cong G(N_{\mathfrak{q}}/M_{\mathfrak{q}})$$

given by the map

$$\xi + \mathfrak{O}_t \mapsto \sigma_\xi := \sigma(1 - \xi_0\xi)$$

with some element $\xi_0 \in \mathfrak{q}\mathfrak{q}_m\mathfrak{q}_r \setminus \mathfrak{p}\mathfrak{q}_m\mathfrak{q}_r$ satisfying

$$\begin{aligned} \xi_0 &\equiv 0 \bmod 8 \quad \text{if} \quad \mathfrak{q} \nmid 2, \\ \xi_0 &\equiv 0 \bmod 9 \quad \text{if} \quad \mathfrak{q} \nmid 3. \end{aligned}$$

Assuming that

$$p \nmid t \ \text{ or } \ p \text{ not split in } K$$

and

$$d_K \neq -3, -4,$$

we obtain the following results:

Theorem 8.12.4 *Let $\delta \in \mathfrak{q}^{-1} \setminus \frac{1}{2}\mathfrak{O}_t$ have composite denominator $o(2\delta|\mathfrak{O}_t)$. We set*

$$T(z) := \mathcal{P}(z + \delta|\mathfrak{O}_t) - \mathcal{P}(\delta|\mathfrak{O}_t).$$

For $\xi_1 \in K \setminus \mathfrak{O}_t$ with $o(\xi_1|\mathfrak{O}_t) = \mathfrak{q}_m\mathfrak{q}_r$, $\xi_0\xi_1 \equiv 1 \bmod \mathfrak{q}_m$ and a $\pi_m \sim \mathfrak{q}_m\mathfrak{O}_K$ we then have

$$\mathfrak{O}_{N_{\mathfrak{q}}} = \mathfrak{O}_{M_{\mathfrak{q}}}[T(\xi_1)],$$

$$\mathfrak{A}_{N_{\mathfrak{q}}/M_{\mathfrak{q}}} = \frac{1}{\pi_m}\left\{ \sum_{\xi \in \mathfrak{q}_m^{-1} \bmod \mathfrak{O}_t} f(T(\xi))\sigma_\xi \,\middle|\, f(X) \in \mathfrak{O}_{M_{\mathfrak{q}}}[X] \right\}.$$

Theorem 8.12.5 *Let $\gamma, \delta \in \mathfrak{q}^{-1} \setminus \mathfrak{O}_t$ have the property of*

$$o(\gamma|\mathfrak{O}_t), o(\delta|\mathfrak{O}_t), o(\gamma + \delta|\mathfrak{O}_t), o(\gamma + N_t(\mathfrak{q}_m)\delta|\mathfrak{O}_t)$$

being composite and let $\omega \in \mathfrak{O}_t$ *satisfy*

$$\omega \equiv 0 \ mod\ \mathfrak{q}_m \quad and \quad \omega \equiv 1 \ mod\ p.$$

Let g *denote the function defined in (8.13) with* γ, δ *and* ω. *Then*

$$\mathfrak{O}_{N_\mathfrak{q}} = \vartheta \circ \mathfrak{A}_{N_\mathfrak{q}/M_\mathfrak{q}}$$

with

$$\vartheta = g(\xi_1),$$

where $\xi_1 \in K \setminus \mathfrak{O}_t$ *has order* $o(\xi_1|\mathfrak{O}_t) = \mathfrak{q}_m\mathfrak{q}_r$.

As in the case for $t = 1$, Theorem 8.12.5 implies:

Theorem 8.12.6 *For* $\mathfrak{p} \nmid 2$ *there exists* $\vartheta \in \mathfrak{O}_{N_1}$ *with*

$$\mathfrak{O}_{N_1} = \vartheta \circ \mathfrak{A}_{N_1/M_1}.$$

However, from the proof of Theorem 8.12.6 we do not obtain an explicit construction of $\mathfrak{A}_{N_1/M_1}$.

9
Berwick's congruences

In 1927 the English mathematician W.E.H. Berwick published an article on numerical computation of singular values of j including a series of conjectures concerning congruences satisfied by these singular values, without giving any idea of proof. In the framework of complex multiplication these congruences constitute a rather untypical result. Therefore, it was not surprising that the first proof of some of these congruences by Gross and Zagier (1985) did not use the classical methods of complex multiplication but was relying on counting isogenies of elliptic curves. Moreover, Gross and Zagier only treated congruences modulo powers of the primes $p = 2, 3, 5, 7$ and 11 without giving any hint on how similar congruences modulo powers of other primes could be found or deduced.

9.1 Bettner's results

A complete proof of Berwick's congruences is due to Stefan Bettner (2004). Essentially his proof uses division polynomials for $\wp$-values in combination with results on the denominators of these values, as will be outlined later. To state the results we use the following notation for an element x of an algebraic numberfield L, a prime number p and a rational number r by setting

$$v_p(x) = r, \quad v_p(x) \geq r$$

if

$$v_{\mathfrak{p}}(x) = r, \quad v_{\mathfrak{p}}(x) \geq r$$

for **all** prime ideals $\mathfrak{p}$ above p.

In the following let K be a quadratic imaginary number field of discriminant d and $\mathfrak{a}$ a proper ideal of $\mathfrak{O}_t$.

Theorem 9.1.1 *Let $t = 2^s m,\ 2 \nmid m$, Then*

$$v_2(j(\mathfrak{a})) \begin{cases} \geq 15 & \text{if 2 is inert in } K \text{ and } s = 0, \\ = 2^{3-s} & \text{if 2 is inert in } K \text{ and } s \geq 1, \\ = 3 \cdot 2^{1-s} & \text{if 2 is ramified in } K, \\ = 0 & \text{if 2 is split in } K. \end{cases}$$

Theorem 9.1.2 *Let $t = 3^s m,\ 3 \nmid m$, Then*

$$v_3(j(\mathfrak{a}) - 12^3) \begin{cases} \geq 6 & \text{if 3 is inert in } K \text{ and } s = 0, \\ = \frac{1}{2} 3^{2-s} & \text{if 3 is inert in } K \text{ and } s \geq 1, \\ = 3^{1-s} & \text{if 3 is ramified in } K, \\ = 0 & \text{if 3 is split in } K. \end{cases}$$

To state similar congruences for an arbitrary prime number p, we assume $d < -4$, and, given a natural number $N > 1$, we consider the modified division polynomial

$$\tilde{\Psi}_N(X) := \delta(\mathfrak{a})^N \Psi_N \left(\frac{X}{\delta(\mathfrak{a})} \right) = N \prod_{\substack{\xi \in \frac{1}{N}\mathfrak{a} \setminus \frac{1}{2}\mathfrak{a} \\ mod\ \mathfrak{a}}} (X - \mathcal{P}(\xi \mid \mathfrak{a})),$$

with the normalisation

$$\mathcal{P} = \delta(\mathfrak{a})\wp, \quad \delta(\mathfrak{a}) = \frac{\epsilon}{\sqrt[6]{\Delta}}$$

of the $\wp$ function as defined in 7.1.1. Writing

$$\tilde{\Psi}_N(X) = \sum_{\nu} a_\nu^{(N)}(\mathfrak{a}) X^\nu,$$

the coefficients are polynomials of

$$\tilde{g}_2(\mathfrak{a}) = \delta(\mathfrak{a})^2 g_2(\mathfrak{a}) = \frac{\epsilon^2 \gamma_2(\mathfrak{a})}{12^3} \quad \text{and} \quad \tilde{g}_3(\mathfrak{a}) = \delta(\mathfrak{a})^3 g_3(\mathfrak{a}) = \frac{\epsilon^3 \gamma_3(\mathfrak{a})}{6^3}.$$

More precisely, every coefficient is of the form

$$c_p \gamma_2^{\nu_2} \gamma_3^{\nu_3} h(j)$$

with a factor c_p prime to p, exponents $\nu_2 = 0, 1, 2;\ \nu_3 = 0, 1$ and a polynomial $h \in \mathbb{Z}[j]$. Congruences modulo a prime number are then obtained by writing the coefficients $a_\nu^{(p)}(\mathfrak{a})$ in terms of the values $\mathcal{P}(\xi \mid \mathfrak{a})$ and using the results of Theorem 7.3.2 about the factorisation of denominators. Berwick's congruences modulo 5, 7 and 11 are then obtained by considering

$$a^{(p)}_{\binom{p}{2}}(\mathfrak{a}).$$

This coefficient has the property that in the representation of

$$\tfrac{1}{p}\, a^{(p)}_{\binom{p}{2}}(\mathfrak{a})$$

as a symmetric function of $\mathcal{P}(\xi \mid \mathfrak{a})$-values there is exactly one summand having a maximal p-denominator. Therefore, this gives us the exact p-part of the denominator of $a^{(p)}_{\binom{p}{2}}(\mathfrak{a})$. In this way Bettner proves the following theorem:

Theorem 9.1.3 *Let $t = p^s m$, $3 \nmid m$, $p > 3$. Then*

$$v_p\left(a^{(p)}_{\binom{p}{2}}(\mathfrak{a})\right) \quad \begin{cases} \geq 1 & \text{if } p \text{ is inert in } K \text{ and } s = 0, \\ = \frac{1}{p+1}p^{1-s} & \text{if } p \text{ is inert in } K \text{ and } s \geq 1, \\ = \frac{1}{2}p^{-s} & \text{if } p \text{ is ramified in } K, \\ = 0 & \text{if } p \text{ is split in } K. \end{cases}$$

For $p = 5, 7, 11, 13, 17, 23, 29$ we quote from Bettner (2004)

$$a^{(p)}_{\binom{p}{2}}(\mathfrak{a}) = c_p \begin{cases} \gamma_2(\mathfrak{a}) & \text{if } p = 5, \\ \gamma_3(\mathfrak{a}) & \text{if } p = 7, \\ \gamma_2(\mathfrak{a})\gamma_3(\mathfrak{a}) & \text{if } p = 11, \\ (j(\mathfrak{a}) + 8) & \text{if } p = 13, \\ \gamma_2(\mathfrak{a})(j(\mathfrak{a}) + 9) & \text{if } p = 17, \\ \gamma_3(\mathfrak{a})(j(\mathfrak{a}) + 12) & \text{if } p = 19, \\ \gamma_2(\mathfrak{a})\gamma_3(\mathfrak{a})(j(\mathfrak{a}) + 4) & \text{if } p = 23, \\ \gamma_2(\mathfrak{a})(j(\mathfrak{a}) + 4)(j(\mathfrak{a}) + 27) & \text{if } p = 29, \end{cases}$$

and, using Theorem 9.1.3, we obtain congruences that include Berwick's conjectures.

9.2 Method of proof

The details of proof are rather complicated because, besides other things, they involve an explicit version of the factorisation of $\varphi(\xi|\mathfrak{a})$-values, in particular for proper ideals on non-maximal orders. To illustrate the basic idea, we will treat a simple case and assume that

$$t = 1 \ \text{ and } p \text{ inert in } K.$$

For numerical examples of the above congruences, we refer to the rich material presented by Gross and Zagier (1985).

$\underline{p=2}$:
In our case the equation Ψ_2 turns out to be of no use, so we consider

$$\tilde{\Psi}_4(X) =$$

$$4X^6 - 5\tilde{g}_2(\mathfrak{a})X^4 - 20\tilde{g}_3(\mathfrak{a})X^3 - \tfrac{5}{4}\tilde{g}_2^2(\mathfrak{a})X^2 - \tilde{g}_2(\mathfrak{a})\tilde{g}_3(\mathfrak{a})X - 2\tilde{g}_3(\mathfrak{a})^2 + \tfrac{1}{16}\tilde{g}_2(\mathfrak{a})^3,$$

which yields

$$\tfrac{-5}{4}\tilde{g}_2(\mathfrak{a}) = \sum_{1\le i<j\le 6} \mathcal{P}(\xi_i \mid \mathfrak{a})\mathcal{P}(\xi_j \mid \mathfrak{a}) = tr_{K_{(4)}/\Omega}\left(\mathcal{P}(\xi_1 \mid \mathfrak{a})\mathcal{P}(\xi_2 \mid \mathfrak{a})\right), \quad (9.1)$$

where $\pm\xi_i$ runs through a system of representatives for $\frac{1}{4}\mathfrak{a}$ mod $\mathfrak{a}$ with $\xi \notin \frac{1}{2}\mathfrak{a}$. Herein, the trace satisfies the congruence

$$tr_{K_{(4)}/\Omega}\left(\mathcal{P}(\xi_1 \mid \mathfrak{a})\mathcal{P}(\xi_2 \mid \mathfrak{a})\right) \equiv 0 \bmod 2 \quad (9.2)$$

in the ring of numbers integral for 2. Observe that the 2-part of the denominator of $\mathcal{P}(\xi_i \mid \mathfrak{a})$ is a divisor of $(2)^{\frac{1}{6}}$. This follows from Theorem 7.3.1 in combination with Theorem 7.3.2, according to which the number P in Theorem 7.3.1 is integral for 2 because 2 is inert in K. In our case the relative discriminant of $K_{(4)}/\Omega$ is $(2)^8$, and this implies the relative different to be

$$\vartheta_{K_{(4)}/\Omega} = (2)^{-\frac{8}{6}} = \frac{1}{2}(2)^{\frac{1}{3}},$$

which proves (9.2). By (9.1) and (9.2) we now obtain

$$v_2\left(\frac{\tilde{g}_2(\mathfrak{a})}{4}\right) \ge 1,$$

and, bearing in mind that $j(\mathfrak{a}) \sim 12^3\tilde{g}_2(\mathfrak{a})^3$, this implies that

$$2^{15} \mid j(\mathfrak{a}).$$

$\underline{p=3}$:
Writing the coefficient of X in the division polynomial

$$\tilde{\Psi}_3(X) = 3X^4 - \frac{3}{2}\tilde{g}_2(\mathfrak{a})X^2 - 3\tilde{g}_3(\mathfrak{a})X - \frac{1}{16}\tilde{g}_2(\mathfrak{a})^2$$

as a symmetric function we obtain the identity

$$\tilde{g}_3(\mathfrak{a}) = \sum_{1\le i<j<k\le 4} \mathcal{P}(\xi_i \mid \mathfrak{a})\mathcal{P}(\xi_j \mid \mathfrak{a})\mathcal{P}(\xi_k \mid \mathfrak{a})$$

$$= tr_{K_3/\Omega}\left(\mathcal{P}(\xi_1 \mid \mathfrak{a})\mathcal{P}(\xi_2 \mid \mathfrak{a})\mathcal{P}(\xi_3 \mid \mathfrak{a})\right),$$

where $\pm\xi_i, i = 1,2,3$, runs through a system of representatives of $\frac{1}{3}\mathfrak{a}$ modulo $\mathfrak{a}$, $\xi_i \notin \mathfrak{a}$. Since 3 is inert in K by assumption, we can conclude as for $p = 2$ that the numbers $\mathcal{P}(\xi_i \mid \mathfrak{a})$ have denominator $(3)^{\frac{1}{4}}$. This implies that

$$\tilde{g}_3(\mathfrak{a}) \in tr_{K_3/\Omega}\left(\vartheta_{K_3/\Omega}\right) \in \mathfrak{O}_\Omega,$$

and then

$$3^6 \mid (j(\mathfrak{a}) - 12^3)$$

because $j - 12^3 \sim 6^6\tilde{g}_3(\mathfrak{a})$.

<u>$p > 3$:</u>
In all other cases we conclude as for $p = 5$. Looking at the $\binom{5}{2}$-th coefficient of $\tilde{\Psi}_5(X)$, we obtain:

$$a^{(5)}_{\binom{5}{2}}(\mathfrak{a}) = -\frac{31}{2}\tilde{g}_2(\mathfrak{a}) = 5 \sum_{1\leq i<j\leq 12} \mathcal{P}(\xi_i \mid \mathfrak{a})\mathcal{P}(\xi_j \mid \mathfrak{a})$$

$$= 5 \cdot tr_{K_{(5)}/\Omega}\left(\mathcal{P}(\xi_1 \mid \mathfrak{a})\mathcal{P}(\xi_2 \mid \mathfrak{a})\right),$$

and, since the 5-part of the denominator of $\mathcal{P}(\xi_i \mid \mathfrak{a})$ is $(5)^{\frac{1}{12}}$, using the above conclusion via the different, we again obtain,

$$v_5\left(\tilde{g}_2(\mathfrak{a})\right) \geq 1,$$

which implies that

$$v_5(j(\mathfrak{a})) \geq 3.$$

10
Cryptographically relevant elliptic curves

The contents of this chapter arose from collaboration with Andreas Enge. Following an unpublished idea of Hasse (1933), independently rediscovered by H.M. Stark (1996) and refined by Rubin and Silverberg (2007) and Morain (2007), the reduction of elliptic curves with complex multiplication leads to elliptic curves over finite fields whose cardinality is explicitly predicted. Compared to the choice of elliptic curves at random this method is useful for primality tests and for cryptographic applications based on pairings as, for instance, described by Freeman, Scott and Teske (2006). In particular, this method can be applied to the construction of elliptic curves of prime cardinality.

The approaches in this chapter will on the one hand be to use suitable modular functions that accelerate the process of finding the reduced curve. On the other hand we will also study curves admitting reduction defined over fields of characteristic 2 and 3.

10.1 Reduction of the Weierstrass model

In 8.7.1 the Weierstrass model associated with a lattice $\mathfrak{L}$ has been defined by:

$$E: \quad y^2 = x^3 + a_4 x + a_6 \tag{10.1}$$

with

$$a_4 = -\frac{g_2(\mathfrak{L})}{4\sqrt[3]{\Delta(\mathfrak{L})}} = -\frac{1}{4 \cdot 12}\sqrt[3]{j(\mathfrak{L})} \tag{10.2}$$

and

$$a_6 = -\frac{g_3(\mathfrak{L})}{4\sqrt[2]{\Delta(\mathfrak{L})}} = -\frac{1}{4\cdot 6^3}\sqrt{j(\mathfrak{L})-12^3}. \tag{10.3}$$

For $\mathfrak{L}$ imaginary quadratic the a_i are integral outside 2 and 3, and, since the discriminant of E is equal to 1, this implies that E has good reduction outside 2 and 3. The uniformising functions of E are given by

$$x(z) := \frac{\wp(z \mid \mathfrak{L})}{\sqrt[6]{\Delta(\mathfrak{L})}}$$

and

$$y(z) := \frac{\wp'(z \mid \mathfrak{L})}{2\sqrt[4]{\Delta(\mathfrak{L})}} = \frac{\varphi(2z \mid \mathfrak{L})}{2\varphi(z \mid \mathfrak{L})^4}.$$

For the sake of simplicity we choose $\mathfrak{L}$ to be the maximal order $\mathfrak{O}$ of an imaginary quadratic number field K of discriminant d,

$$\mathfrak{O} := [\alpha, 1],$$

where, in view of the normalisation of the root of $\Delta(\mathfrak{O})$ the generating element α is chosen as follows:

$$\alpha = \begin{cases} \frac{3+\sqrt{d}}{2} & \text{if} \quad d \equiv 5 \bmod 8, \\ \frac{9+\sqrt{d}}{2} & \text{if} \quad d \equiv 1 \bmod 16, \\ \frac{3+\sqrt{d}}{2} & \text{if} \quad d \equiv 9 \bmod 16, \\ \frac{\sqrt{d}}{2} & \text{if} \quad d \equiv 0 \bmod 4. \end{cases}$$

Let $\mathfrak{p}$ be a prime ideal in K of norm $p \nmid 6$, whose ideal class has order e,

$$\mathfrak{p}^e \sim \pi = u\alpha + v, \quad u, v \in \mathbb{Z}.$$

The above Weierstrass model will now be modified so that it is defined over the fixed field of $\sigma(\pi)$. As we will explain later, there exists an element θ abelian over K and prime to p such that

$$\left[\theta^4\gamma_2(\alpha)\right]^{\sigma(\pi)} = \left[\theta^4\gamma_2(\alpha)\right] \text{ and } \left[\theta^6\gamma_3(\alpha)\right]^{\sigma(\pi)} = \left[\theta^6\gamma_3(\alpha)\right]. \tag{10.4}$$

The functions

$$\tilde{x}(\xi) := \theta^2 x(\xi), \quad \tilde{y}(\xi) := \theta^3 y(\xi)$$

then uniformise the elliptic curve

$$\tilde{E}: \quad Y^2 = X^3 + \tilde{a}_4 X + \tilde{a}_6, \text{ with } \tilde{a}_4 := \theta^4 a_4, \ \tilde{a}_6 := \theta^6 a_6, \tag{10.5}$$

which has good reduction above $\mathfrak{p}$ and is defined over the fixed field of $\sigma(\pi)$. Let $\mathfrak{P}$ be a prime ideal in $K^\gamma := \Omega_1(\tilde{a}_4, \tilde{a}_6)$ lying above $\mathfrak{p}$. Then, the reduction of the curve (10.5) modulo $\mathfrak{P}$ has coefficients in $\mathbb{F}_{p^e} = \mathfrak{O}_{K^\gamma}/\mathfrak{P}$. As we will see, replacing π by $-\pi$ if necessary, we can further achieve that the same is true for the torsion points $P = (\tilde{x}(\xi), \tilde{y}(\xi))$, $\xi \in K \setminus \mathfrak{O}$, of (10.5). The reduction of such a point P modulo a prime ideal $\mathfrak{P}'$ above $\mathfrak{P}$ of $K(P, \tilde{a}_4, \tilde{a}_6)$ is a point of $E(\mathbb{F}_{p^e})$ if

$$P^{\sigma(\pi)} = P. \tag{10.6}$$

In fact, by the Reciprocity Law we find that

$$\tilde{x}(\xi)^{\sigma(\pi)} = \tilde{x}(\pi\xi)$$

if $\mathfrak{p}$ does not divide the denominator of (ξ). The action of $\sigma(\pi)$ on $\tilde{y}(\xi)$ can be deduced from the proof of Theorem 6.8.7:

$$\tilde{y}(\xi)^{\sigma(\pi)} = \tilde{y}(\pi\xi)\epsilon(\pi, \alpha)^{-3}\theta^{3(\sigma(\pi)-1)}$$

with

$$\epsilon(\pi, \alpha)^3 := \begin{cases} \zeta_4^{(-uvn+1)p^e - uv - v} & \text{if} \quad 2 \nmid v, \\ \zeta_4^{sp^e + p^e - u + 1} & \text{if} \quad 2 | v, \end{cases}$$

where n and s denote the norm and the trace of α. Note that the equation of the curve implies that

$$\epsilon(\pi, \alpha)^{-3}\theta^{3(\sigma(\pi)-1)} = \pm 1,$$

that for $\mathfrak{p} \nmid 2$

$$\epsilon(\pi, \alpha)^3 = -\epsilon(-\pi, \alpha)^3 \tag{10.7}$$

and that by definition we have $\sigma(\pi) = \sigma(-\pi)$. Therefore, we can normalise π by the condition

$$\epsilon(\pi, \alpha)^{-3}\theta^{3(\sigma(\pi)-1)} = 1.$$

Then $P^{\sigma(\pi)} = P$ is equivalent to

$$\text{denominator of } (\xi) \ \mid (\pi - 1).$$

Let $E[(\pi - 1)]$ denote the group of $(\pi - 1)$-torsion points, and let $\mathfrak{P}'$ be a prime ideal above $\mathfrak{P}$ in the field generated by $\tilde{a}_4, \tilde{a}_6$ and the points of $E[(\pi - 1)]$. Then $E[(\pi - 1)]$ is mapped by reduction onto a subgroup U of $E(\mathbb{F}_{p^e})$, and since the reduction is injective on $E[(\pi - 1)]$ because of $\mathfrak{P}' \nmid (\pi - 1)$, this implies that $E(\mathbb{F}_{p^e})$ has a subgroup U of order

$$|U| = |E[(\pi - 1)]| = N(\pi - 1) = p^e + 1 - tr(\pi).$$

Herein

$$|tr(\pi)| \le 2\sqrt{p^e},$$

and by the Riemann hypothesis we have

$$|E(\mathbb{F}_{p^e})| \le p^e + 1 + 2\sqrt{p^e}.$$

Therefore, in view of $p \nmid 6$ we have $p^e > 3$, and we obtain

$$[E(\mathbb{F}_{p^e}) : U] \le \frac{p^e + 1 + 2\sqrt{p^e}}{p^e + 1 - 2\sqrt{p^e}} < 2.$$

Hence, $E(\mathbb{F}_{p^e}) = U$ and

$$|E(\mathbb{F}_{p^e})| = p^e + 1 - tr(\pi),$$

where, as above, π is normalised by $\epsilon(\pi, \alpha)^{-3}\theta^{3(\sigma(\pi)-1)} = 1$.

$\tilde{a}_4$ and $\tilde{a}_6$ may be determined modulo $\mathfrak{P}$ by the following steps:

(i) computation of θ^{12} modulo $\mathfrak{P}$,
(ii) computation of $\theta^6\gamma_3(\alpha)$ modulo $\mathfrak{P}$ (in some cases, explained below, this is a rational function of θ^{12}),
(iii) computation of $\theta^{12}j(\alpha) = (\theta^6\gamma_3(\alpha))^2 + 12^3\theta^{12}$ modulo $\mathfrak{P}$,
(iv) computation of a root $\sqrt[3]{\theta^{12}j(\alpha)}$ modulo $\mathfrak{P}$.

Of course, this determines $\overline{a}_4 = \tilde{a}_4$ mod $\mathfrak{P}$ only up to a third root of unity in $\mathbb{F}_{p^e}$, but all curves

$$Y^2 = X^3 + \rho\overline{a}_4X + \overline{a}_6, \qquad \rho \in \mathbb{F}_{p^e}, \rho^3 = 1, \tag{10.8}$$

are isogenous to each other over $\mathbb{F}_{p^e}$ and hence have cardinality

$$p^e + 1 - tr(\pi).$$

Remarks 10.1.1 To avoid the the computation of $\sqrt[3]{\theta^{12}j(\alpha)}$ modulo $\mathfrak{P}$, it is convenient to multiply the equation (10.5) by $\tilde{a}_4^{12}$. This yields the curve

$$E' : \quad Y^2 = X^3 + \tilde{a}_4^9X + \tilde{a}_6\tilde{a}_4^{12},$$

which is isogenous to $\tilde{E}$ over K^γ. It has good reduction above $\mathfrak{p}$, if $j(\mathfrak{O})$ is prime to $\mathfrak{p}$, which can be verified by Theorem 4.4.2. Then, instead of the third root, one has to compute the third power modulo $\mathfrak{P}$, which is easier.

If $p^e \equiv 1$ mod 4, then -1 is a square in $\mathbb{F}_{p^e}$. Therefore, all curves

$$Y^2 = X^3 + \rho\overline{a}_4X + (-1)^\nu\overline{a}_6, \qquad \rho \in \mathbb{F}_{p^e}, \rho^3 = 1, \nu = 0, 1, \tag{10.9}$$

are isogenous to each other over $\mathbb{F}_{p^e}$.

For $p^e \equiv -1 \bmod 4$ the curves in (10.9) for $\nu = 0, 1$ are not isogenous. To determine the cardinality for $\nu = 1$ we parametrise the curve by the functions $-\tilde{x}(z|\mathfrak{L}), i\tilde{y}(z|\mathfrak{L})$ and, concluding analogously, we find its cardinality to be

$$p^e + 1 + tr(\pi)$$

if π is normalised as above.

Construction of θ

In many cases $\theta = 1$ has the desired properties, as for instance, when d is prime to 6, since then $\gamma_2(\alpha)$ and $\gamma_3(\alpha)$ are in Ω_1. The choice $\theta = 1$ in (10.4) is also possible when $\pi \equiv 1 \bmod 6$ because we know $\gamma_2(\alpha)$ and $\gamma_3(\alpha)$ to be in K_6 for every fundamental discriminant d. In the remaining cases we have the following construction, which can also be generalised to non-fundamental discriminants.

According to section 2.4.3 the following functions are modular for $\Gamma_{\left(\begin{smallmatrix}1 & 0\\ 0 & n\end{smallmatrix}\right)}$, $n \in \mathbb{N}$:

$$h_2(\omega) := \left(\frac{\eta(\frac{\omega}{n})}{\eta(\omega)}\right)^8 \gamma_2(\omega)^{n-1} \text{ if } 3 \nmid n.$$
$$h_3(\omega) := \left(\frac{\eta(\frac{\omega}{n})}{\eta(\omega)}\right)^6 \gamma_3(\omega)^{\frac{n-1}{2}} \text{ if } 2 \nmid n,$$

In $\mathfrak{O}$ we choose primitive ideals $\mathfrak{n}_2, \mathfrak{n}_3$ of norm n_2, n_3 satisfying

$$n_2 \equiv 2 \bmod 3 \quad \text{and} \quad n_3 \equiv 3 \bmod 4, \quad \gcd(\mathfrak{n}_2, \overline{\mathfrak{n}}_3) = 1.$$

Two such ideals exist except for $d = -3, -4$, where $\theta = 1$ has the desired properties. Further, we set

$$\hat{\alpha} := \alpha + 6\mu,$$

with $\mu \in \mathbb{Z}$ such that $\hat{\alpha}, n_i$ is a basis of $\mathfrak{n}_i$:

$$\mathfrak{n} = [\hat{\alpha}, n_i].$$

Then, according to Theorem 6.6.4,

$$h_i(\hat{\alpha}) \in \Omega_1, \quad \text{for } i = 2, 3,$$

and it follows that a suitable product of

$$\theta_i := \frac{\eta\left(\frac{\hat{\alpha}}{n_i}\right)}{\eta\left(\hat{\alpha}\right)}, \quad i = 2, 3,$$

as, for instance,

$$\theta := \theta_2^8 \theta_3^3$$

has the desired properties since $\gamma_2(\hat{\alpha}) = \gamma_2(\alpha)$ and $\gamma_3(\hat{\alpha}) = \gamma_3(\alpha)$. Of course, when $\gamma_2(\alpha)$ resp. $\gamma_3(\alpha)$ are in Ω_1, we can simplify our construction by setting $\theta_2 := 1$ resp. $\theta_3 := 1$ in the definition of θ.

To normalise $\pi = u\alpha + v$ we need the action of $\sigma(\pi)$ on θ^3:

$$\theta_i^{3(\sigma(\pi)-1)} = \left(\frac{v}{n_i}\right) \zeta_4^{\frac{n_i-1}{2}\left[\frac{uvC}{n_i p^e} - u\left(\frac{v-uB}{p^e}(1-v^2)-v\right) - 3(v-1)u_1\right]}, \quad \zeta_4 = \sqrt{-1}.$$

Herein B and C are the coefficients of the primitive equation $X^2+BX+C=0$ of $\hat{\alpha}$, and u_1 is defined by the decomposition $u = 2^\lambda u_1,\ 2 \nmid u_1$.

We can also use the fact that $\sqrt{-1}\gamma_3(\alpha)$ is in Ω_1 for $d \equiv 4 \bmod 8$. In this case we can replace θ_3 by $\theta_3' = \sqrt[4]{-1} = \frac{1+\sqrt{-1}}{2}$.

Finally, we mention the special cases $n_3 = 3, 7$, where $\gamma_3(\alpha)$ can be computed via θ_3 because then, according to Example 2.10.3, we have the equations

$$x^4 + 18x^2 - \gamma_3(\alpha)x - 27 = 0 \quad \text{for } n_3 = 3 \tag{10.10}$$

and

$$x^8 + 7 \cdot 2x^6 + 7 \cdot 9x^4 + 7 \cdot 10x^2 - \gamma_3(\alpha)x - 7 = 0 \quad \text{for } n_3 = 7 \tag{10.11}$$

satisfied by $x = \theta_3^6$ resp. $x = \theta_3^2$.

Example 10.1.2 For $d = -248$ we have $\gamma_2(\alpha) \in \Omega_1$ and $\gamma_3(\alpha) \notin \Omega_1$. Considering the functions h_2 and h_3 with $n_2 = n_3 = 7$ we find that

$$\theta := \frac{\eta\left(\frac{\hat{\alpha}}{7}\right)}{\eta\left(\hat{\alpha}\right)}, \qquad \hat{\alpha} = 1980 + \sqrt{-62},$$

is a possible choice for θ. The power θ^4 is in Ω_1 and satisfies the equation

$$\begin{aligned} m(X) := X^8 &+ \left(-676 + 116\sqrt{-62}\right)X^7 + \left(-33118 + 6420\sqrt{-62}\right)X^6 \\ &+ \left(-759552 + 63920\sqrt{-62}\right)X^5 + \left(-4412757 - 333684\sqrt{-62}\right)X^4 \\ &+ \left(3945984 - 5648048\sqrt{-62}\right)X^3 + \left(68813474 - 16223748\sqrt{-62}\right)X^2 \\ &+ \left(56573660 - 15383252\sqrt{-62}\right)X + 2747137 + 643656\sqrt{-62} \end{aligned}$$

Now we choose the prime

$$p = 123456790743209877383 = \pi\overline{\pi}$$

with

$$\pi = 11111111139 + \sqrt{-62},$$

and, using the above formulae we find that

$$\epsilon(\pi, \alpha)^3 \theta^{3(\sigma(\pi)-1)} = 1.$$

The factorisation of $m(X)$ modulo (π) is given by

$$(X + 1149105037916946068 09)(X + 17485763858982704816)$$

$$\times(X + 5773378016543806 1637)(X + 4525422775233611858 2)$$

$$\times(X + 5798443105885732527 4)(X + 3933739211668551832 8)$$

$$\times(X + 1830041204395024323 6)(X + 1936386015279616066 7),$$

where for simplicity we have omitted the bars on the numbers. For the reduced curve to be constructed we choose the root

$$a := \overline{\theta^4} = -11491050379169460 6809.$$

Then, using (10.11), we find that

$$\begin{aligned}
\overline{\gamma_3(\alpha)\theta^6} &= a(a^4 + 14a^3 + 63a^2 + 70a - 7) = 5788686413212959 1728,\\
\overline{j(\alpha)\theta^{12}} &= (a(a^4 + 14a^3 + 63a^2 + 70a - 7))^2 + 12^3a^3\\
&= 8569078197658431129 2,\\
\overline{\gamma_2(\alpha)\theta^4} &= 502249127012118986.
\end{aligned}$$

This leads us to the curve

$$E: \quad y^2 = x^3 + 5657389890049177465 5x + 7537881787995681211 4$$

over $\mathbb{F}_p$ having cardinality

$$p + 1 - tr(\pi) = 12345679072098765510 6.$$

The considerations so far for $t = 1$ carry over almost word by word to the case $t > 1$ when the elliptic curve (10.1) is associated with a suborder $\mathfrak{O}_t$ in K. Instead of the prime ideal $\mathfrak{p}$ prime to 6 we have to consider a prime ideal coprime to $6t$ and to replace $\mathfrak{p}$ by the corresponding ring ideal $\mathfrak{p}_t$. Then e is the order of $\mathfrak{p}_t$ in the ring ideal class group modulo t and π is defined up to a unit factor by

$$\mathfrak{p}_t^e = \mathfrak{O}_t \pi.$$

10.2 Computation of $j(\mathfrak{O})$ modulo $\mathfrak{P}$

Let $\mathfrak{p}$ be a prime ideal of degree 1 and of norm p. Then the residue classes of $j(\mathfrak{O})$ modulo the different prime ideals $\mathfrak{P}$ in Ω_1 lying over $\mathfrak{p}$ can be obtained by factoring the minimal equation of $j(\mathfrak{O})$ over K modulo $\mathfrak{p}$, keeping in mind that the equation has coefficients in $\mathbb{Z}$. It is explicitly given by

$$m_{j(\mathfrak{O}),K}(X) = \prod_{A,B,C} \left(X - j\left(\frac{-B+\sqrt{d}}{2A} \right) \right),$$

where (A,B,C) runs through a system of representatives of the classes of quadratic forms of discriminant d. We take as example $d_K = -251$, in which case the choice $\theta = 1$ is possible. Then

$$\begin{aligned} &m_{j(\mathfrak{O}),K}(X) = \\ &X^7 + 412844619031530949836 8\,X^6 \\ &-6620418537314440399828077772 8\,X^5 \\ &+106200888027012610597600802840877465 6\,X^4 \\ &+7966552994949346594041401247164174172160\,X^3 \\ &+4161316087934374015778329997816103879709818 88\,X^2 \\ &-17919115457058418400843204272511348592207598714 88\,X \\ &+193758723946570326967205666068586405015246425240371 2. \end{aligned}$$

We choose the prime ideal of degree 1

$$\mathfrak{p} = (\pi), \pi = \frac{22222222221 + \sqrt{-251}}{2} \quad \text{having norm}$$

$$p = \pi\overline{\pi} = 123456790109876543273.$$

Then $\mathfrak{p}$ splits completely in the Hilbert class field, and by factoring of $m_{j(\mathfrak{O}),K}$ modulo p,

$$\begin{aligned} &m_{j(\mathfrak{O}),K} \equiv \\ &(X + 81666724111688493696)(X + 82896316073688673794) \\ &\cdot(X + 100882973722407244045)(X + 118936065735023292899) \\ &\cdot(X + 119357543134306544875)(X + 87317650069448975347) \\ &\cdot(X + 80598794392203062068) \bmod p, \end{aligned}$$

we find seven representatives of $j(\mathfrak{O})$ modulo the seven prime ideals above $\mathfrak{p}$. We take the first representative,

$$\bar{j}_1 = -81666724111688493696 = 41790065998188049577.$$

Since $p \equiv -1 \bmod 3$, we have only one possibility for $\bar{a}_4$:

$$\bar{a}_4 = 32267382838068516622$$

and two possibilities for $\bar{a}_6$:

$$\bar{a}_6 = \pm 116150345266117935436.$$

Since $p \equiv 1 \bmod 4$, the two elliptic curves arising from $\bar{j}_1$ are isogenous over $\mathbb{F}_p$ and have the same cardinality

$$p + 1 - tr(\pi) \ \text{ resp. } \ p + 1 + tr(\pi)$$

according to whether $\epsilon(\pi, \alpha)^3 = 1$ resp. $\epsilon(\pi, \alpha)^3 = -1$ is satisfied. For verification we write

$$\pi = \frac{22222222221 + \sqrt{-251}}{2} = \frac{3 + \sqrt{-251}}{2} + 11111111109,$$

so $u, v, n, s, p \equiv 1, 1, 1, -1, 1 \bmod 4$, hence $\epsilon(\pi, \alpha)^3 = \zeta_4^{(-uvn+1)p^e - uv - v} = -1$. Consequently all curves determined have cardinality

$$p + 1 + tr(\pi) = 123456790132098765495.$$

In conclusion, we obtained 14 elliptic curves having this cardinality, two for every j-invariant modulo $\mathfrak{P}$ that are isogenous to each other over $\mathbb{F}_p$.

For a prime ideal of degree 1 and of norm $p \equiv 2 \bmod 3$ we find two elliptic curves over $\mathbb{F}_p$ for every j-invariant modulo $\mathfrak{P}$. For $p \equiv 1 \bmod 4$ they are isogenous over $\mathbb{F}_p$ and, if $p \equiv -1 \bmod 4$, then they have different cardinality, which can be decided either by computing $\overline{\gamma_3(\alpha)}$ or by counting points.

However, with growing discriminants the coefficients of $m_{j(\mathfrak{O}),K}(X)$ become astronomically high. This has been shown in Enge and Morain (2002), Enge (2009) and Schoof (1991), so instead of j one uses modular functions g of level $N > 1$, whose singular values $g(\alpha)$ satisfy algebraic equations with rather small coefficients. Then, to compute $j(\alpha)$ modulo $\mathfrak{P}$, one uses the modular equation

$$\Phi(g, j) = 0$$

having coefficients in $\mathbb{Z}$ for suitably chosen g. In the sequel we will explain this method by some examples.

10.2.1 Schläfli–Weber functions

Let g be one of the functions f, f_1, f_2 defined in 6.4. Then, by the modular equation for g and j we see that j is a rational function of g. Let d be the discriminant of an imaginary quadratic number field K. First, we treat the case

$d \equiv 1$ mod 4: A zero of $m_{j(\mathfrak{O}),K}(X)$, generating the Hilbert class field over K, is given by the value $j(\alpha)$, $\alpha = \frac{3+\sqrt{d}}{2}$. To find a simpler generator, we use the identity

$$j = \frac{(f_2^{24}+16)^3}{f_2^{24}}.$$

Then, by the η-transformation formula we write $f_2(\alpha)$ as

$$f_2(\alpha) = \frac{e^{\frac{\pi i}{8}}\sqrt{2}}{f(\sqrt{d})},$$

and we end up with the relation

$$j(\alpha) = -\frac{\left(\left(\frac{\sqrt{2}}{f(\sqrt{d})}\right)^{24} + 16\right)^3}{\left(\frac{\sqrt{2}}{f(\sqrt{d})}\right)^{24}}. \tag{10.12}$$

Now, if some n-th power, $n \mid 24$, of $\frac{\sqrt{2}}{f(\sqrt{d})}$ is in the Hilbert class field we can compute $j(\alpha)$ modulo $\mathfrak{P}$ via $\left(\frac{\sqrt{2}}{f(\sqrt{d})}\right)^n$ modulo $\mathfrak{P}$. For illustration let $d \equiv 1 \bmod 8$ and $d \not\equiv 0 \bmod 3$. Then, according to Theorem 6.4.1, $\frac{f(\sqrt{d})}{\sqrt{2}}$ generates the Hilbert class field. For example, if $d = -247$, we find that

$$m_{\frac{f(\sqrt{d})}{\sqrt{2}},K}(X) = X^6 - 4X^5 - 7X^4 - 7X^3 - 6X^2 - 3X - 1,$$

which has the factorisation

$$(X + 2096108431151)(X + 205881311392)(X + 194833047732)\cdot$$
$$\cdot(X + 1715786557544)(X + 1032395073636)(X + 878555233535)$$

modulo $p = 3061779827497$. The zeros are representatives of $\frac{f(\sqrt{d})}{\sqrt{2}}$ modulo the prime ideals above $\mathfrak{p} = (\pi)$ in the Hilbert class field and yield the corresponding representatives of $j(\alpha)$ via the formula (10.12).

For $d \equiv 5 \bmod 8$, $d \not\equiv 0 \bmod 3$ the situation is a little more complicated. Then, $f(\sqrt{d})$ is a generating element of the ring class field modulo 2, which is of degree 3 over the Hilbert class field. Therefore, the minimal polynomial of $f(\sqrt{d})$ has degree $3h_K$, and for $\pi \not\equiv 1 \bmod 2$ the minimal polynomial of $f(\sqrt{d})$ has to be factored over $\mathbb{F}_{p^3}$. Unfortunately this happens in the case of cryptographic interest, when we want the elliptic curve over $\mathbb{F}_p$ to have prime cardinality.
In the case $d \equiv 4 \bmod 8$ we use the identity

$$j = \frac{(f^{24} - 16)^3}{f^{24}}$$

applied to $\alpha = \frac{\sqrt{d}}{2}$. However, the generators of the Hilbert class field obtained in this way are less simple than in the preceeding case, because by Theorem 6.4.1 the lowest power contained in the Hilbert class field is $f\left(\frac{\sqrt{d}}{2}\right)^4$. For example, in the case $d = -820$ we have:

$$m_{\frac{1}{2}f\left(\frac{\sqrt{d}}{2}\right)^4,K}(X) =$$

$$X^8 - 903\,X^7 + 1688\,X^6 - 315\,X^5 + 1806\,X^4 + 315\,X^3 + 1688\,X^2 + 903\,X + 1.$$

In the case $d \equiv 0 \bmod 8$ we use the identity

$$j = \frac{(f_1^{24} - 16)^3}{f_1^{24}}$$

for the argument $\frac{\sqrt{d}}{2}$ and the results of Theorem 6.4.1 for f_1.

10.2.2 Double η-quotients

As we have seen above, there are some cases when the Schläfli functions are not very suitable for the computation of $j(\mathfrak{O})$ modulo prime ideals. In many such cases the double η-quotients, used already for the construction of simple generators of ring class fields in Theorems 6.6.2 and 6.6.6, turn out to be quite effective. We choose two primes $q_1, q_2 \neq 2$ that are not inert in K, and we let $\mathfrak{q}_1, \mathfrak{q}_2$ be two prime ideals above q_1, q_2 satisfying

$$\mathfrak{q}_1 \neq \overline{\mathfrak{q}}_2.$$

Then we can choose $\alpha \in \mathbb{H}$ such that

$$\mathfrak{O} = [\alpha, 1], \quad \mathfrak{q}_1 = [\alpha, q_1], \quad \mathfrak{q}_2 = [\alpha, q_2], \quad \mathfrak{q}_1\mathfrak{q}_2 = [\alpha, q_1 q_2].$$

In addition, we can achieve

$$tr(\alpha) \equiv 0, 1 \bmod 4$$

and

$$tr(\alpha) \equiv 0 \bmod 3 \quad \text{if } q_1, q_2 \neq 3.$$

We define a function g by

$$g(\omega) = g_{q_1,q_2}(\omega) = \frac{\eta\left(\frac{\omega}{q_1}\right)\eta\left(\frac{\omega}{q_2}\right)}{\eta\left(\frac{\omega}{q_1 q_2}\right)\eta(\omega)}.$$

Applying Theorem 6.6.4 to the functions in 2.4.3 it follows that

$$g(\alpha)\left(\gamma_2(\alpha)\gamma_3(\alpha)\right)^{\frac{q_1-1}{2}\frac{q_2-1}{2}}$$

is in the Hilbert class field if $q_1, q_2 \neq 3$ or $q_1 = 3$ and $3 \mid (q_2 - 1)$. Clearly, the same holds for $g(\alpha)$ if $\frac{q_1-1}{2}\frac{q_2-1}{2} \equiv 0 \bmod 6$. To compute $j(\alpha) = j(\mathfrak{O})$ modulo a prime ideal we then need the modular equation $\Phi_{q_1,q_2}(X)$ of g over $\mathbb{C}_{U_6}$, which by Theorem 2.9.3 has coefficients in $\mathbb{Z}[\gamma_2, \gamma_3, X]$. We illustrate this procedure by the following example:

$$\begin{aligned}
\Phi_{5,7} = X^{48} &+ (-j + 708)X^{47} + (35j + 171402)X^{46} \\
&+ (-525j + 15185504)X^{45} + (4340j + 248865015)X^{44} \\
&+ (-20825j + 1763984952)X^{43} + (52507j + 6992359702)X^{42} \\
&+ (-22260j + 19325688804)X^{41} + (-243035j + 42055238451)X^{40} \\
&+ (596085j + 70108209360)X^{39} + (-272090j + 108345969504)X^{38} \\
&+ (-671132j + 121198179480)X^{37} + (969290j + 155029457048)X^{36} \\
&+ (-1612065j + 97918126080)X^{35} + (2493785j + 141722714700)X^{34} \\
&+ (647290j - 1509796288)X^{33} + (-3217739j + 108236157813)X^{32} \\
&+ (3033590j - 93954247716)X^{31} + (-5781615j + 91135898154)X^{30} \\
&+ (1744085j - 108382009680)X^{29} + (1645840j + 66862445601)X^{28} \\
&+ (-2260650j - 66642524048)X^{27} + (6807810j + 38019611082)X^{26} \\
&+ (-2737140j - 28638526644)X^{25} + (2182740j + 17438539150)X^{24} \\
&+ (-125335j - 8820058716)X^{23} + (-1729889j + 5404139562)X^{22} \\
&+ (1024275j - 1967888032)X^{21} + (-1121960j + 1183191681)X^{20} \\
&+ (395675j - 370697040)X^{19} + (-54915j + 103145994)X^{18} \\
&+ (15582j - 42145404)X^{17} + (34755j - 15703947)X^{16} \\
&+ (-6475j - 3186512)X^{15} + (1120j - 4585140)X^{14} \\
&+ (-176j + 1313040)X^{13} + (j^2 - 1486j - 38632)X^{12} \\
&+ (-7j + 399000)X^{11} + (-19j + 211104)X^{10} + (-9j + 6771)X^{8}
\end{aligned}$$

$$+(8j-6084)X^7+(7j-5258)X^6+(j-792)X^5-105X^4+16X^3$$
$$+42X^2+12X+1$$

We choose $d=-251$. Then 5 and 7 are split in $K=\mathbb{Q}(\sqrt{-251})$. Further, since $\frac{q_1-1}{2}\frac{q_2-1}{2}\equiv 0 \bmod 6$, we know that $g_{5,7}(\alpha)$ is in the Hilbert class field. Its minimal equation is

$$X^7+5X^6+12X^5-2X^4-6X^3+16X^2+5X+1.$$

As in section 10.2 we take

$$\mathfrak{p}=(\pi), \pi=\frac{22222222221+\sqrt{-251}}{2} \quad \text{having the norm}$$

$$p=\pi\overline{\pi}=123456790109876543273.$$

Then $\mathfrak{p}$ splits completely in the Hilbert class field, and the factorisation of the minimal equation of $g_{5,7}(\alpha)$ modulo $\mathfrak{p}$ is given by:

$$(X+16228912223456689 57)(X+35941263036181148886)\cdot$$
$$\cdot(X+59130329136279099 16)(X+21929643311898901652)\cdot$$
$$\cdot(X+50952364010769406202)(X+26163603504599516570)\cdot$$
$$\cdot(X+10439078222033053 4368).$$

We choose the zero $\overline{g_{5,7}(\alpha)}=-16228912223456689 57$ and for the corresponding value of j we obtain the equation

$$\begin{aligned}0&=\Phi_{5,7}(-16228912223456689 57,\overline{j})\\&=111416970477297685996\overline{j}+67496371020885195151\\&\quad+97730862037371935492\overline{j}^2\end{aligned}$$

with solutions

$$36139140040427567926, \quad 41790065998188049577.$$

In fact, $\overline{j}=41790065998188049577$ leads to the elliptic curves discovered in section 10.2.

10.2.3 Application of η-quotients in the ramified case

The double η-quotients considered in the last section have the property that their singular values $g(\alpha)$ are in $\mathbb{Q}(j(\alpha))$ and in most cases they are generators of this field. However, according to Theorem 6.7.1, the last property does not hold in the special case when q_1, q_2 and a third prime are ramified in $K=\mathbb{Q}(\alpha)$ and satisfy $\frac{q_1-1}{2}\frac{q_2-1}{2}\equiv 0 \bmod 2$.

Then $g(\alpha)$ is contained in a subfield over which $K(j(\alpha))$ has degree 2. Hence $g(\alpha)$ satisfies an equation of degree $\frac{h}{2}$, where h denotes the degree $[\mathbb{Q}(j(\alpha)) : \mathbb{Q}]$. In practice this reduces the amount of calculation needed to find the curves considerably. We illustrate this by two examples:

First example of minimal equations: we choose $d = -2555 = -5 \cdot 7 \cdot 73$ and we set $\alpha = \frac{3 \cdot 5 \cdot 7 + \sqrt{d}}{2}$. The class number is $h = 12$. Then $g(\alpha) = g_{5,7}(\alpha)$ generates an extension of degree 6 over $\mathbb{Q}$ with minimal equation

$$X^6 + 86X^5 - 574X^4 + 1972X^3 + 574X^2 + 86X - 1.$$

On the other hand, the classical algorithm using Weber's function $f(\sqrt{d})$ leads to the class polynomial of the (non-fundamental) discriminant $4d = -10220$ with class number $3h = 36$:

$$\begin{aligned}
&X^{36} - 740X^{35} - 5576X^{34} + 38864X^{33} + 217388X^{32} - 1641560X^{31} \\
&+2427368X^{30} + 5721552X^{29} - 26437184X^{28} + 34413984X^{27} \\
&+18585616X^{26} - 141474752X^{25} + 241459440X^{24} - 160215168X^{23} \\
&-166796160X^{22} + 583142528X^{21} - 783022976X^{20} \\
&+512563456X^{19} + 237373184X^{18} - 1107810304X^{17} \\
&+1465448960X^{16} - 895317504X^{15} - 233253376X^{14} \\
&+1030334464X^{13} - 1072218368X^{12} + 632040448X^{11} \\
&-71012352X^{10} - 339048448X^{9} + 389827584X^{8} - 192313344X^{7} \\
&+41510912X^{6} - 16437248X^{5} + 20930560X^{4} - 11321344X^{3} \\
&+2527232X^{2} - 180224X + 4096
\end{aligned}$$

The largest coefficient of this polynomial has 10 digits instead of four, so, to find an elliptic curve via Weber's function, it is necessary to factor a polynomial of degree six times as high. Moreover, if $\pi \not\equiv 1 \bmod 2$ the polynomial has to be factorised over $\mathbb{F}_{p^3}$ instead of $\mathbb{F}_p$.

Second example to compare running times: determination of an elliptic curve of prime cardinality over a finite field which is cryptographically secure implemented by A. Enge.

The fundamental discriminant $d = -1170195 = -3 \cdot 5 \cdot 13 \cdot 17 \cdot 353$ has class number $h = 208$. We choose $p_1 = 3$ and $p_2 = 13$. For the prime

$$
\begin{aligned}
p &= 3138550867693340381917894711603833208051177722232017843049 \\
&= 2^{191} + 586601
\end{aligned}
$$

the equation $4P = x^2 + 1170195y^2$ has a solution $(x, y) \in \mathbb{Z}^2$ and leads to an elliptic curve over $\mathbb{F}_P$ of cardinality

$$3138550867693340381917894711570630703183127382496658188029.$$

The following running times were measured on a Pentium III with 700 MHz.

For our procedure it is enough to calculate with 84 digits to obtain a minimal polynomial with the largest coefficient having 79 digits; the running time needed is 0.3 s. The zero

$$(g_{3,13})_1 = 944791451623554577477696101351863927447586313905349688290$$

of this polynomial of degree Grad 104 over $\mathbb{F}_p$ is obtained within 6.1 s.

The modular equation $\Phi(X, j) = 0$ satisfied by $g_{3,13}$ is given by:

$$
\begin{aligned}
& j^2 \quad X^{16} \\
+ \; & j \quad (-X^{55} + 39X^{54} - 663X^{53} + 6331X^{52} - 35763X^{51} \\
& +106392X^{50} - 18070X^{49} - 1082016X^{48} + 3516903X^{47} \\
& -1278901X^{46} - 18277116X^{45} + 40532700X^{44} \\
& +11574823X^{43} - 161476962X^{42} + 168751479X^{41} \\
& +230086922X^{40} - 617987682X^{39} + 137626281X^{38} \\
& +928366231X^{37} - 959457720X^{36} - 477589944X^{35} \\
& +1429130144X^{34} - 466517064X^{33} - 963208272X^{32} \\
& 909996295X^{31} + 158515461X^{30} - 607329720X^{29} \\
& +197238236X^{28} + 179445279X^{27} - 140684622X^{26} \\
& -6888479X^{25} + 37909092X^{24} - 8835450X^{23} \\
& -4070053X^{22} + 1885689X^{21} + 44928X^{20} \\
& -111436X^{19} + 9516X^{18} + 740X^{17} - 1486X^{16} - 49X^{15} \\
& +29X^{14} + 246X^{13} - 364X^{12} - 221X^{11} + 650X^{10} \\
& -221X^{9} - 364X^{8} + 247X^{7} + 26X^{6} - 52X^{5} + 13X^{4} - X^{3})+
\end{aligned}
$$

$$
\begin{aligned}
+\quad & (X^{56} + 704X^{55} + 168568X^{54} + 14498520X^{53} + 187807764X^{52} \\
& +744637296X^{51} - 6562036X^{50} - 3840625568X^{49} \\
& +1058251610X^{48} + 10302034600X^{47} + 4510900472X^{46} \\
& -34331690432X^{45} - 7097865034X^{44} + 84188024320X^{43} \\
& +546780176X^{42} - 154959173464X^{41} - 12359340101X^{40} \\
& +327081484064X^{39} - 49301838300X^{38} - 576339027576X^{37} \\
& +284363953068X^{36} + 735938431592X^{35} - 558265224452X^{34} \\
& -890017323520X^{33} + 977815434427X^{32} + 966995235128X^{31} \\
& -1755072840368X^{30} - 345165085024X^{29} + 2218368968890X^{28} \\
& -911733108784X^{27} - 1540031876048X^{26} + 1628026178168X^{25} \\
& +261124933147X^{24} - 1229692547200X^{23} + 462040501468X^{22} \\
& +441029439032X^{21} - 422841966612X^{20} - 7261052136X^{19} \\
& +163453863300X^{18} - 59787354976X^{17} - 26470898021X^{16} \\
& +24009911816X^{15} - 1731574864X^{14} - 3926472080X^{13} \\
& +1333660406X^{12} + 158103088X^{11} - 172600168X^{10} \\
& +25597000X^{9} + 5195450X^{8} - 2155088X^{7} + 177164X^{6} \\
& +39936X^{5} - 9996X^{4} + 600X^{3} + 88X^{2} - 16X + 1)
\end{aligned}
$$

Substituting $(g_{3,13})_1$ for X yields a quadratic equation for j. The two zeros in $\mathbb{F}_p$,

$$j_1 = 1737712759221672293462706430740701274216885226872539199796$$

and

$$j_2 = 2096167087313280259570847931701751304363626115642779297546,$$

are calculated in 0.01 s. They represent two possible j-invariants. The invariant j_1 leads to the elliptic curve $Y^2 = X^3 + aX + b$ with

$$
\begin{aligned}
a &= 1717086893498159407390388826164234412641707842768794182\,53 \\
b &= 6897868117069496766800601950241497735418885810463312180\,49
\end{aligned}
$$

having the desired cardinality.†

For the classical procedure using Weber's function the calculation must be performed with 650 digits to obtain a minimal polynomial whose

† This procedure is protected by patent no. 103 29 885 at the "Deutsches Patent- und Markenamt".

largest coefficient has 643 digits. The running time then is 14 s, and to determine a zero of this polynomial of degree 624 over $\mathbb{F}_{p^3}$, we need 382 s.

Therefore, in this case with cryptographically relevant parameters, the algorithm described is more than 60 times faster than the classical procedure using Weber's functions.

10.3 Reduction of the Fueter and Deuring models

In the previous subsections we had, due to bad reduction, to exclude the prime ideals above 2 and 3. To obtain similar results for 2 and 3, we consider the models of Fueter and Deuring.

For simplicity, we choose $\mathfrak{L}$ to be the maximal order $\mathfrak{O}$ of an imaginary quadratic number field K of discriminant d,

$$\mathfrak{O} := [\alpha, 1].$$

We consider a prime ideal $\mathfrak{p}$ of degree 1 and norm p not dividing 2 resp. not dividing 3, in particular $p = 3$ resp. $p = 2$. We let e be an exponent for $\mathfrak{p}$,

$$\mathfrak{p}^e \sim \pi$$

with a generator π that will be subject to a certain congruence, which can always be realised by taking e high enough. Further, in the definition of the two models, we will choose suitable torsion points ψ and κ, and, as described below, we can conclude that the reduction of the curve modulo a prime ideal of $K_{(\pi-1)}$ dividing $\mathfrak{p}$ yields a curve of cardinality

$$\#E(\mathbb{F}_{p^e}) = p^e + 1 - tr(\pi). \tag{10.13}$$

10.3.1 *Reduction of the Fueter model*

From Theorem 8.7.3 we know the coefficient $a_2(\psi)$ to be in K_8 and the coordinates $x(\xi)$, $y(\xi)$ of a point of order $(\pi - 1)$ in $K_{8(\pi-1)}$. For the cardinality of the reduced curve we are aiming for, we need more precise information.

Let $\lambda \in \mathfrak{O}$ be prime to 2 and $(\lambda - 1)\psi \in \mathfrak{O}$. Then by reciprocity

$$a_2(\psi)^{\sigma(\lambda)} = a_2(\psi).$$

If, in addition, $\lambda \equiv 1 \bmod (\pi - 1)$, we find that

$$x(\xi)^{\sigma(\lambda)} = e^{-4l(1,\lambda)N(\psi)}x(\xi),$$

$$y(\xi)^{\sigma(\lambda)} = e^{18l(1,\lambda)N(\psi)}y(\xi)$$

with $l = l_{\mathfrak{O}}$. Now we distinguish cases according to the order $o(\psi)$:

Case "$o(\psi) = \mathfrak{q}^2$, $2 = \mathfrak{q}\overline{\mathfrak{q}}$, $\mathfrak{q} \neq \overline{\mathfrak{q}}$":
To find a suitable ψ, we observe that in this case $\alpha \in \mathfrak{O}$ can be chosen with

$$\mathfrak{O} := [\alpha, 1] \quad \text{and} \quad \mathfrak{q}^\nu \;=\; [\alpha, 2^\nu], \quad \nu = 1, 2.$$

We set

$$\psi := \frac{\overline{\alpha}}{4}.$$

Then ψ has order $\mathfrak{q}^2$, so $a_2(\psi)$ is in $K_{\mathfrak{q}^2}$, hence

$$a_2(\psi) \in \Omega_1,$$

keeping in mind the formula $[K_{\mathfrak{q}^2} : \Omega_1] = \frac{1}{2}(N(\mathfrak{q}) - 1)N(\mathfrak{q})$. Further, we observe that $N(\psi)$ is in $\mathbb{Z}$, so that for π satisfying

$$\pi \equiv 1 \bmod \mathfrak{q}^2$$

the $(\pi - 1)$-torsions points have coordinates in $K_{(\pi-1)}$, and we can conclude that the reduction of the curve modulo a prime ideal of $K_{(\pi-1)}$ dividing $\mathfrak{p}$ yields a curve of cardinality (10.13).

Case "$o(\psi) = \mathfrak{q}^3$, $2 = \mathfrak{q}^2$":
In this case we can choose α such that

$$\mathfrak{O} := [\alpha, 1] \quad \text{and} \quad \mathfrak{q} \;=\; [\alpha, 2].$$

We set

$$\psi := \frac{\alpha}{4}.$$

Then ψ has order $\mathfrak{q}^3$, which implies that

$$a_2(\psi) \in K_{\mathfrak{q}^3}.$$

More precisely,

$$[K_{\mathfrak{q}^3} : \Omega_1] = 4,$$

and $a_2(\psi)$ is a generator for $K_{\mathfrak{q}^3}/\Omega_1$. In fact, according to the remarks in subsection 8.7.2, $a_2(\psi)$ is even a generator for $K_{\mathfrak{q}^3}/K$. Now we assume that

$$\pi \equiv 1 \bmod 4.$$

Then $K_{\mathfrak{q}^3}$ is contained in $K_{(\pi-1)}$, and, as in the preceeding case, we can conclude that the reduction of the curve modulo a prime ideal of $K_{(\pi-1)}$ yields a curve of cardinality (10.13).

Case "$o(\psi) = 4$":
In this case we set

$$\psi := \frac{1}{4}.$$

Then ψ has order 4 and $N(\psi) = \frac{1}{16}$, which implies that

$$a_2(\psi) \in K_4.$$

More precisely, by studying the Galois action, it transpires that the field generated by $a_2(\psi)$ over K is K_4. Now we assume that

$$\pi \equiv 1 \bmod 8.$$

Then K_4 is contained in $K_{(\pi-1)}$, and, observing $N(\psi) = \frac{1}{16}$, we find, as in the preceeding case, that the reduction of the curve modulo a prime ideal of $K_{(\pi-1)}$ yields a curve of cardinality (10.13).

Example 10.3.1 Let $d = -71$. Then 2 splits in K, $2 = \mathfrak{q}\overline{\mathfrak{q}}$, $\mathfrak{q} = \left[2, \frac{27+\sqrt{-71}}{2}\right]$ and $\kappa = \frac{27+\sqrt{-71}}{8}$ has denominator $\mathfrak{q}^2$. Further, 3 splits in K, $3 = \mathfrak{p}\overline{\mathfrak{p}}$ with $\mathfrak{p} = \left[3, \frac{1+\sqrt{-71}}{2}\right]$ having order 7 in the class group of K. We find that

$$\mathfrak{p}^7 = \pi\overline{\pi}, \quad \pi = -(46 + \sqrt{-71} \equiv 1 \bmod \mathfrak{q}^2.$$

In this case $4a_2(\kappa)$ is in Ω and satisfies the equation

$$\begin{aligned} &X^7 + \left(\frac{1107 - 119\sqrt{-71}}{2}\right) X^6 + \left(27611 + 273\sqrt{-71}\right) X^5 \\ &+ \left(258179 + 26313\sqrt{-71}\right) X^4 + \left(\frac{-3025469 + 464009\sqrt{-71}}{2}\right) X^3 \\ &+ \left(-20295609 - 478443\sqrt{-71}\right) X^2 + \left(\frac{338261 - 31579457\sqrt{-71}}{2}\right) X \\ &+ \frac{460015479 - 120288091\sqrt{-71}}{2}, \end{aligned}$$

having reduction

$$X^7 + X^6 + 2X^5 + 2X^4 + X^3 + 2X + 2 \text{ modulo } \mathfrak{p}.$$

Therefore, any root $\alpha \in \mathbb{F}_{3^7}$ of this last equation defines an elliptic curve

$$y^2 = x(4x^2 + \alpha x + 4)$$

over $\mathbb{F}_{3^{7n}}$, $n = 1, 2, \ldots$, having cardinality

$$3^{7n} + 1 - tr(\pi^n) = 2280,\ 477888, \ldots$$

10.3.2 Reduction of the Deuring model

We proceed completely analogously as for the Fueter model. From Theorem 8.7.3 we know the coefficient $a_1(\kappa)$ to be in K_9 and the coordinates $x(\xi)$, $y(\xi)$ of a point of order $(\pi - 1)$ in $K_{9(\pi-1)}$. For the cardinality of the reduced curve we are aiming for, we need more precise information.

Let $\lambda \in \mathfrak{O}$ be prime to 3 and $(\lambda - 1)\kappa \in \mathfrak{O}$. Then by reciprocity

$$a_1(\kappa)^{\sigma(\lambda)} = e^{2l(1,\lambda)N(\kappa)} a_1(\kappa)$$

with $l = l_{\mathfrak{O}}$. If, in addition, $\lambda \equiv 1 \bmod (\pi - 1)$, we find that

$$x(\xi)^{\sigma(\lambda)} = e^{l(1,\lambda)N(\kappa)} x(\xi),$$

$$y(\xi)^{\sigma(\lambda)} = e^{3l(1,\lambda)N(\kappa)} y(\xi).$$

Now we distinguish cases according to the order $o(\kappa)$:

Case "$o(\kappa) = \mathfrak{q},\ 3 = \mathfrak{q}\overline{\mathfrak{q}},\ \mathfrak{q} \neq \overline{\mathfrak{q}}$":
To find a suitable κ we observe that in this case $\alpha \in \mathfrak{O}$ can be chosen with

$$\mathfrak{O} := [\alpha, 1] \quad \text{and} \quad \mathfrak{q}^\nu = [\alpha, 3^\nu], \quad \nu = 1, 2.$$

We set

$$\kappa := \frac{\overline{\alpha}}{3}.$$

Then κ has order $\mathfrak{q}$ and $N(\kappa)$ is in $\mathbb{Z}$, which implies that $a_1(\kappa) \in K_{\mathfrak{q}}$, hence

$$a_1(\kappa) \in \Omega_1,$$

keeping in mind the formula $[K_{\mathfrak{q}} : \Omega_1] = \frac{1}{2}(N(\mathfrak{q}) - 1)$. Further, assuming that

$$\pi \equiv 1 \bmod \mathfrak{q},$$

the $(\pi - 1)$-torsion points have coordinates in $K_{(\pi-1)}$, and we can conclude that the reduction of the curve modulo a prime ideal of $K_{(\pi-1)}$ yields a curve of cardinality (10.13).

Case "$o(\kappa) = \mathfrak{q}$, $3 = \mathfrak{q}^2$":
In this case we can choose α such that

$$\mathfrak{O} := [\alpha, 1] \quad \text{and} \quad \mathfrak{q} = [\alpha, 3].$$

We set

$$\kappa := \frac{\alpha}{3}.$$

Then, κ has order $\mathfrak{q}$ and $N(\kappa)$ is in $\frac{1}{3}\mathbb{Z} \setminus \mathbb{Z}$, which implies that

$$a_1(\kappa) \in K_3 \setminus \Omega_1 \text{ and } a_1(\kappa)^3 \in \Omega_1.$$

More precisely,

$$[K_3 : \Omega_1] = 3,$$

and $a_1(\kappa)$ is a generator for K_3/Ω_1. In fact, according to the remarks in subsection 8.7.3, $a_1(\kappa)$ is even a generator for K_3/K. Now we assume that

$$\pi \equiv 1 \bmod 3.$$

Then K_3 is contained in $K_{(\pi-1)}$, and, as in the preceeding case, we can conclude that the reduction of the curve modulo a prime ideal of $K_{(\pi-1)}$ yields a curve of cardinality (10.13).

Case "$o(\kappa) = 3$":
In this case we set

$$\kappa := \frac{1}{3}.$$

Then, κ has order 3 and $N(\kappa) = \frac{1}{9}$, which implies that

$$a_1(\kappa) \in K_9 \setminus K_3 \text{ and } a_1(\kappa)^3 \in K_3.$$

More precisely, by studying the Galois action, it transpires that the field generated by $a_1(\kappa)$ over K is the fixed field of all Frobenius automorphisms $\sigma(\lambda)$ of K_9 with $\lambda \equiv 1 \bmod 3$ and $\lambda \equiv r \bmod 9$ for some $r \in \mathbb{Z}$. Now we assume that

$$\pi \equiv 1 \bmod 9.$$

Then K_9 is contained in $K_{(\pi-1)}$, and, as in the preceeding case, we can conclude that the reduction of the curve modulo a prime ideal of $K_{(\pi-1)}$ yields a curve of cardinality (10.13).

Example 10.3.2 Let $d = -71$. Then 3 splits in K, $3 = \mathfrak{q}\overline{\mathfrak{q}}$, $\mathfrak{q} = \left[3, \frac{1+\sqrt{-71}}{2}\right]$ and $\kappa = \frac{-1+\sqrt{-71}}{6}$ has denominator $\mathfrak{q}$. Further, 2 splits in K, $2 = \mathfrak{p}\overline{\mathfrak{p}}$ with $\mathfrak{p} = \left[2, \frac{1+\sqrt{-71}}{2}\right]$ having order 7 in the class group of K. We find that

$$\mathfrak{p}^7 = \pi\overline{\pi}, \quad \pi = \frac{21+\sqrt{-71}}{2} \equiv 1 \bmod \mathfrak{q}.$$

In this case $a_1(\kappa)$ is in Ω and and satisfies the equation

$$\begin{aligned} &X^7 + \left(-13+\sqrt{-71}\right)X^6 + \left(-100+\sqrt{-71}\right)X^5 \\ &+ \left(\frac{-273-51\sqrt{-71}}{2}\right)X^4 + \left(\frac{2247-195\sqrt{-71}}{2}\right)X^3 \\ &+ \left(\frac{11849-527\sqrt{-71}}{2}\right)X^2 + \left(\frac{25285+5\sqrt{-71}}{2}\right)X \\ &+8602-187\sqrt{-71} \end{aligned}$$

having reduction

$$X^7 + X^5 + X^4 + X^3 + 1 \quad \text{modulo } \mathfrak{p}.$$

Therefore, any root $\alpha \in \mathbb{F}_{2^7}$ of this last equation defines an elliptic curve

$$y^2 + \alpha xy + y = x^3$$

over $\mathbb{F}_{2^{7n}}$, $n = 1, 2, \ldots$, having cardinality

$$2^{7n} + 1 - tr(\pi^n) = 108,\ 16200,\ 2095956,\ 268434000, \ldots$$

11
The class number formulae of Curt Meyer

The ζ function of an abelian extension L of $\mathbb{Q}$ is known to have the decomposition

$$\zeta_L(s) = \zeta_{\mathbb{Q}} \prod_{\chi \neq 1} L(s, \chi),$$

where $\zeta_{\mathbb{Q}}$ is the ζ function of $\mathbb{Q}$. χ runs through all characters different from the principal character associated with the extension $L/\mathbb{Q}$, and $L(s, \chi)$ denotes the corresponding rational L-function. By evaluating this relation at $s = 1$ we then obtain the known class number formulae for cyclotomic fields. In particular, for a real field L the values $L(1, \chi)$ can be expressed in terms of cyclotomic units, thereby establishing a relation between cyclotomic units and class numbers of real cyclotomic fields.

As shown by Meyer (1957), a similar situation is given for the abelian extensions of a quadratic imaginary number field K and their subfields. The elliptic functions and modular functions, used in the previous chapters for the construction of class fields over K, also appear in the values at $s = 1$ of L-function s in K associated with abelian characters. Analogously to the cyclotomic case, this establishes a relation between class numbers and singular values of modular and elliptic functions, providing the possibility to investigate class numbers by studying singular values. Conversely, by this connection, the non-vanishing of L-function s at $s = 1$ can be used to prove certain properties of singular values, as in Theorem 6.6.1.

In the following we summarise the formulae needed from Meyer (1957) and provide the relations to the singular values of the η function and σ function that occured in the previous chapters. Let $\chi \neq 1$ be a primitive character of the ideal group of K with conductor $\mathfrak{f}_\chi$ and let

$$L(s,\chi) := \sum_{\mathfrak{g}} \frac{\chi(\mathfrak{g})}{N(\mathfrak{g})},$$

denote the corresponding L-function, where the summation is over all integral ideals $\mathfrak{g}$ of K. $N(\mathfrak{g})$ denotes the norm of $\mathfrak{g}$ and $\chi(\mathfrak{g})$ is set to be zero for $\mathfrak{g}$ not prime to $\mathfrak{f}_\chi$. To express this L-function at $s=1$ according to the results of C. Meyer by singular values of modular and elliptic functions, one has to distinguish between two (overlapping) cases. For ring class characters considered in the next section $L(1,\chi)$ can essentially be written in terms of singular values of Δ. The following section then deals with ray class characters of conductor different from (1). Here the role of Δ is taken over by division values of the normalised σ function.

11.1 L-Functions of ring class characters

In the following let $\chi \neq 1$ be a ring class character, i.e. there exists a natural number f, such that χ is trivial on $\mathfrak{U}_f$. Then it is easy to see that the conductor of χ must be rational, $\mathfrak{f}_\chi = (f_\chi)$ with a natural number f_χ, which is also called the conductor of χ. Due to the isomorphy in Theorem 3.1.7, χ can also be viewed as a character of $\mathfrak{R}_{f_\chi}$. With these notations C. Meyer has proved the following formula:

$$L(1,\chi) = \frac{2\pi}{f_\chi\sqrt{|d|}} A(\chi), \quad A(\chi) := \sum_{\mathfrak{k}\in\mathfrak{R}_{f_\chi}} -\overline{\chi}(\mathfrak{k}) \log D(\mathfrak{k}), \tag{11.1}$$

where the bar denotes complex conjugation. $D(\mathfrak{k})$ is the so-called **modulnormfunction** of the class $\mathfrak{k}$,

$$D(\mathfrak{k}) := \sqrt[24]{\delta(\mathfrak{a})^{12}\,|\Delta(\mathfrak{a})|^2},$$

defined by an arbitrary proper ideal $\mathfrak{a}$ of the class $\mathfrak{k}$ and the area $\delta(\mathfrak{a})$ of a fundamental domain for the lattice $\mathfrak{a}$. $\delta(\mathfrak{a})$ can be written as

$$\delta(\mathfrak{a}) = \left|\det\begin{pmatrix}\alpha_1 & \overline{\alpha_1}\\ \alpha_2 & \overline{\alpha_2}\end{pmatrix}\right|,$$

which easily implies the relation

$$\delta(\mathfrak{a}) = N(\mathfrak{a})\sqrt{|d|}$$

with the discriminant d of K.

We generalise the notation in (11.1) by setting

$$A_t^R(\chi) := \sum_{\mathfrak{k}\in\mathfrak{R}_t} \overline{\chi}(\mathfrak{k}) \log D(\mathfrak{k}) \tag{11.2}$$

for any not necessarily primitive character $\chi \neq 1$ of $\mathfrak{R}_t$. These sums play a crucial role in the generation of ring class fields by quotients of singular values of Δ. For this purpose we need the following theorem.

Theorem 11.1.1 *Let $\chi \neq 1$ be a character of $\mathfrak{R}_t$. Then $A_t^R(\chi) \neq 0$.*

Proof For $t = f_\chi$ the assertion follows directly from the class number formula for ring class fields, as we will see later. In all other cases we will show that $A_t^R(\chi)$ is a non-zero multiple of $A(\chi)$. For $t \in \mathbb{N}$ we denote by $\mathfrak{K}_t$ the ring divisor class group modulo t. Then for $\mathfrak{k} \in \mathfrak{K}_t$ the class zeta function of $\mathfrak{k}$ is defined by

$$\zeta_t^*(s|\mathfrak{k}) := \sum_{\mathfrak{g}\in\mathfrak{k}} \frac{1}{N(\mathfrak{g})}, \quad \Re(s) > 1,$$

where the summation is over all integral ideals $\mathfrak{g} \in \mathfrak{k}$. As in Meyer (1957) we write this function as

$$\zeta_t^*(s|\mathfrak{k}) = \frac{N(\mathfrak{c})^s}{w_t} \sum_{\substack{\gamma\in\mathfrak{c}_t \\ \gcd(\gamma,t)=1}} \frac{1}{N(\gamma)^s}$$

with an integral ideal $\mathfrak{c} \in \mathfrak{k}^{-1}$ and the number w_t of roots of unity in $\mathfrak{O}_t$. In order to use Kronecker's limit formula we must get rid of the condition $\gcd(\gamma, t) = 1$, so we set

$$\zeta_t(s|\mathfrak{k}) := \frac{N(\mathfrak{c})^s}{w_t} \sum_{\gamma\in\mathfrak{c}_t\setminus\{0\}} \frac{1}{N(\gamma)^s}, \quad \Re(s) > 1.$$

The two functions are related by

$$\zeta_t(s|\mathfrak{k}) = \sum_{t'|t} \frac{w_{t/t'}}{w_t t'^{2s}} \; \zeta_{t/t'}^*(s|\mathfrak{k}'),$$

where $\mathfrak{k}'$ denotes the class in $\mathfrak{K}_{t/t'}$ containing $\mathfrak{k}$. To prove the relation, note that $\gcd(\gamma, t) = t' \in \mathbb{N}$ since γ is in $\mathfrak{c}_t \subseteq \mathfrak{O}_t$, and we can write $\gamma = t'\gamma'$ with $\gamma' \in \mathfrak{c}_{t/t'}$. By summation we then obtain

$$\sum_{\mathfrak{k}\in\mathfrak{K}_t} \chi(\mathfrak{k})\zeta_t(s|\mathfrak{k}) = \sum_{t'|t} \frac{w_{t/t'}}{w_t t'^{2s}} \frac{|\mathfrak{K}_t|}{|\mathfrak{K}_{t/t'}|} \sum_{\mathfrak{k}'\in\mathfrak{K}_{t/t'}} \chi(\mathfrak{k}')\zeta_{t/t'}^*(s|\mathfrak{k}'). \tag{11.3}$$

Herein, as shown in Meyer (1957), the left-hand side at $s = 1$ is a non-zero multiple of $A_t^R(\chi)$ by Kronecker's limit formula, whereas on the right-hand side we have

$$\sum_{\mathfrak{k}' \in \mathfrak{K}_{t/t'}} \chi(\mathfrak{k}')\zeta^*_{t/t'}(s|\mathfrak{k}') = 0, \quad \text{if } f_\chi \nmid \frac{t}{t'}.$$

For $f_\chi \mid \frac{t}{t'}$ we find, using the Euler product expansion, that

$$\sum_{\mathfrak{k}' \in \mathfrak{K}_{t/t'}} \chi(\mathfrak{k}')\zeta^*_{t/t'}(s|\mathfrak{k}') = \sum_{\gcd(\mathfrak{g}, t/t')=1} \frac{\chi(\mathfrak{g})}{N(\mathfrak{g})} = \prod_{\mathfrak{p} | \frac{t}{t' f_\chi}} \left(1 - \frac{\chi(\mathfrak{p})}{N(\mathfrak{p})^s}\right)^{-1} L(s \mid \chi).$$

Therefore, (11.3) becomes

$$\sum_{\mathfrak{k} \in \mathfrak{K}_t} \chi(\mathfrak{k})\zeta_t(s|\mathfrak{k}) = \sum_{\substack{t'|t \\ f_\chi | \frac{t}{t'}}} \frac{w_{t/t'}}{w_t t'^{2s}} \frac{|\mathfrak{K}_t|}{|\mathfrak{K}_{t/t'}|} \prod_{\mathfrak{p} | \frac{t}{t' f_\chi}} \left(1 - \frac{\chi(\mathfrak{p})}{N(\mathfrak{p})^s}\right)^{-1} L(s \mid \chi). \quad (11.4)$$

The products on the right are all positive for $s = 1$ as can easily be seen by distinguishing the cases for $p = N(\mathfrak{p})$, p inert, ramified and split. In the first case $\chi(\mathfrak{p}) = 1$, in the second $\chi(\mathfrak{p}) = \pm 1$ and for $p = \mathfrak{p}\overline{\mathfrak{p}}$ the factors with $\mathfrak{p}$ and $\overline{\mathfrak{p}}$ are non-zero and complex conjugate, so in view of $L(1 \mid \chi) \neq 0$ and Kronecker's limit formula as mentioned above, (11.4) implies the assertion of our theorem. □

11.2 L-function s of ray class characters χ with $\mathfrak{f}_\chi \neq (1)$.

In the following let χ be a ray class character of conductor $\mathfrak{f}_\chi \neq (1)$. Then, according to Meyer (1957), the value at $s = 1$ of the L-function associated with χ can be written as

$$L(1, \chi) = \frac{2\pi\tau(\chi)}{N(\mathfrak{f}_\chi)\sqrt{|d|}} A(\chi), \quad A(\chi) := \sum_{\mathfrak{k} \in \mathfrak{K}_{\mathfrak{f}_\chi}} -\overline{\chi}(\mathfrak{k}) \log S(\mathfrak{k}),$$

where $\mathfrak{K}_{\mathfrak{f}_\chi} = \mathfrak{A}^{\mathfrak{f}_\chi}/\mathfrak{S}_{\mathfrak{f}_\chi}$ denotes the ray class group modulo $\mathfrak{f}_\chi$, and $S(\mathfrak{k})$ is defined by the normalised division value of the σ function:

$$S(\mathfrak{k}) := \left|\varphi(1|\mathfrak{f}_\chi\mathfrak{n}^{-1})\right|^2 = \left| \sqrt[12]{\Delta\left(\mathfrak{f}_\chi\mathfrak{n}^{-1}\right)}\ \sigma^*(1|\mathfrak{f}_\chi\mathfrak{n}^{-1})\right|^2.$$

Herein $\mathfrak{n}$ is an arbitrary integral ideal of the ray class $\mathfrak{k}$. By $\tau(\chi)$ we denote the Gaussian sum

$$\tau(\chi) := \sum_{\mathfrak{x}} \chi(\mathfrak{x}) \sum_{\substack{\xi \bmod^+ \frac{1}{\sqrt{d}} \\ \xi \sim \frac{\mathfrak{x}}{\mathfrak{f}_\chi \sqrt{d}}}} e^{2\pi i\, tr(\xi)},$$

where $\mathfrak{x}$ runs through a system of integral ideals that are representatives for the ray classes modulo $\mathfrak{f}_\chi$ contained in the absolute ideal class of $\mathfrak{f}_\chi$. The absolute value of $\tau(\chi)$ is given by

$$|\tau(\chi)| = \sqrt{N(\mathfrak{f}_\chi)}.$$

As in the preceeding section, given an integral ideal $\mathfrak{f}$ divisible by $\mathfrak{f}_\chi$, we consider the sum

$$A_{\mathfrak{f}}^S(\chi) := \sum_{\mathfrak{k} \in \mathfrak{K}_{\mathfrak{f}}} \overline{\chi}(\mathfrak{k}) \log S(\mathfrak{k}), \tag{11.5}$$

where in the definition of $S(\mathfrak{k})$ the conductor $\mathfrak{f}_\chi$ of χ is replaced by $\mathfrak{f}$,

$$S(\mathfrak{k}) := \left|\varphi(1|\mathfrak{f}\mathfrak{n}^{-1})\right|^2 = \left| \sqrt[12]{\Delta\left(\mathfrak{f}\mathfrak{n}^{-1}\right)}\ \sigma^*(1|\mathfrak{f}\mathfrak{n}^{-1})\right|^2,$$

and where $\mathfrak{n}$ denotes an integral ideal of the ray class $\mathfrak{k}$ modulo $\mathfrak{f}$. As for ring class characters we then express $A_{\mathfrak{f}}^S(\chi)$ by $A_{\mathfrak{f}_\chi}^S(\chi)$ using the relation between division values of the normalised σ function from Theorem 1.9.3. For two lattices $\hat{\mathfrak{L}} \supset \mathfrak{L}$ this relation yields

$$\prod_{\substack{\xi \in \hat{\mathfrak{L}} \\ \xi \bmod \mathfrak{L}}} \left|\varphi(z+\xi|\mathfrak{L})\right| = \left|\varphi(z|\hat{\mathfrak{L}})\right|.$$

In particular, for $\mathfrak{L} = \mathfrak{f}\mathfrak{p}\mathfrak{n}^{-1}$ and $\hat{\mathfrak{L}} = \mathfrak{f}\mathfrak{n}^{-1}$ with an integral ideal $\mathfrak{f}$, a prime ideal $\mathfrak{p}$ and an integral ideal $\mathfrak{n}$ prime to $\mathfrak{f}\mathfrak{p}$, the above formula, for $z=1$, gives us for every class $\mathfrak{k}_{\mathfrak{f}} \in \mathfrak{K}_{\mathfrak{f}}$ the relation

$$\prod_{\substack{\mathfrak{k}_{\mathfrak{f}\mathfrak{p}} \in \mathfrak{K}_{\mathfrak{f}\mathfrak{p}} \\ \mathfrak{k}_{\mathfrak{f}\mathfrak{p}} \subset \mathfrak{k}_{\mathfrak{f}}.}} S(\mathfrak{k}_{\mathfrak{f}\mathfrak{p}})^{\frac{w_{\mathfrak{f}}}{w_{\mathfrak{f}\mathfrak{p}}}} = \begin{cases} S(\mathfrak{k}_{\mathfrak{f}}) & \text{if } \mathfrak{p}|\mathfrak{f}, \\ \dfrac{S(\mathfrak{k}_{\mathfrak{f}})}{S(\mathfrak{k}_{\mathfrak{f}}\mathfrak{c}^{-1})} & \text{if } \mathfrak{p}\nmid\mathfrak{f}, \end{cases}$$

with an integral ideal $\mathfrak{c}$ in the ray class modulo $\mathfrak{f}$ of $\mathfrak{p}^{-1}$. As usual $w_{\mathfrak{f}}$ resp. $w_{\mathfrak{f}\mathfrak{p}}$ denotes the number of roots of unity in K that are congruent to 1 modulo $\mathfrak{f}$ resp. modulo $\mathfrak{f}\mathfrak{p}$. The relation implies the following theorem:

Theorem 11.2.1 *Let $\chi \neq 1$ be a character of $\mathfrak{K}_{\mathfrak{f}}$. Then*

$$A_{\mathfrak{f}}^S(\chi) = \frac{w_{\mathfrak{f}_\chi}}{w_{\mathfrak{f}}} \left(\prod_{\mathfrak{p} \mid \frac{\mathfrak{f}}{\mathfrak{f}_\chi}} (1 - \bar{\chi}(\mathfrak{p})) \right) A(\chi).$$

In contrast to ring classes the product in the formula of Theorem 11.2.1 can vanish, a fact we have to live with, unfortunately.

11.3 Class number formulae

In the following let K be a quadratic imaginary number field, and let L be an abelian extension of K. The ζ function of L has the product decomposition

$$\zeta_L(s) = \zeta_K(s) \prod_{\chi \neq 1} L(s, \chi),$$

with the ζ function of K. χ runs through all characters of the ideal group of K different from the principal character that belongs to the extension L/K, and $L(s, \chi)$ denotes the L-function in K associated with such a character.

Taking residues at $s = 1$ on both sides, we obtain the class number formula

$$\frac{w_K h_L}{w_L h_K} R_L = \prod_{\chi \neq 1} |A(\chi)| \tag{11.6}$$

with the class numbers h_L and h_K of L and K, the regulator R_L and the number w_L resp. w_K of roots of unity L resp. K and the sums defined in (11.2) resp. (11.5):

$$A(\chi) := \begin{cases} A_1^R(\chi), & \text{if} \quad \mathfrak{f}_\chi = (1), \\ A_{\mathfrak{f}_\chi}^S(\chi), & \text{if} \quad \mathfrak{f}_\chi \neq (1). \end{cases}$$

Herein $A_1(\chi)$ is essentially defined by singular values of Δ whereas $A_{\mathfrak{f}_\chi}(\chi)$ is defined by division values of the normalised σ function φ. For a ring class character of conductor different from (1) we can also write

$$A(\chi) = A_t^S(\chi), \quad \mathfrak{f}_\chi = (t),$$

keeping in mind that the conductor of a ring class character is rational. Therefore, in this case, $A(\chi)$ can also be expressed by singular values of φ.

(11.6) is derived from the formulae

$$\lim_{s\to 1}(s-1)\zeta_L(s) = \frac{(2\pi)^n}{w_L\sqrt{|d_L|}} R_L h_L, \quad \lim_{s\to 1}(s-1)\zeta_K(s) = \frac{2\pi}{w_K\sqrt{|d|}} h_K,$$

with the discriminant d_L of L and $n = [L:\mathbb{Q}]$ and, furthermore, using the relation

$$d_L = \prod_{\chi\neq 1} N(\mathfrak{f}_\chi)\ \ d^n$$

following from the conductor-discriminant theorem and the tower formula for the discriminant.

For the ζ function of a subfield M of an abelian extension of K that does not contain K,

$$K \nsubseteq M \subset L := MK, \quad L/K \text{ abelian},$$

one has a similar decomposition into a product of L-functions in K involving the maximal subextension M_0 abelian over $\mathbb{Q}$, too. The class number formula obtained by taking values at $s = 1$ is given by

$$\frac{w_{M_0} h_M}{w_M h_{M_0}} \frac{R_M}{R_{M_0}} = \prod_{\chi\in\mathcal{H}} |A(\chi)|\,. \tag{11.7}$$

Herein $\mathcal{H}$ is a system of representatives for the characters of L/K satisfying $\chi \neq \chi^\tau$ with respect to the equivalence relation

$$\chi' \sim \chi :\Longleftrightarrow (\chi' = \chi \text{ oder } \chi' = \chi^\tau).$$

12

Arithmetic interpretation of class number formulae

The class number formulae of the last section have been studied by several authors, for instance Robert (1973), Kersey (1980), Nakamula (1981–1985), Hayashi (1983–1986), Hajir (1993), and Limmer (1994). In this chapter we follow and refine the ideas of Schertz(1974, 1977, 1978). The aim is to express the product of the $A(\chi)$ in (11.6), up to a factor c_L, as the regulator of a unit group U_0 in L generated by elliptic units. Then (11.6) can be read as index formula

$$c_L \frac{w_K h_L}{w_L h_K} = \frac{R_L(U_0)}{R_L} = [E : U_0],$$

in which the unit group E of L and the index $[E : U_0]$ has to be understood as the index of the corresponding subgroups in the logarithmic space. For subfields M of abelian extensions of K not containing K,

$$K \not\subseteq M \subset L := MK, \quad L/K \text{ abelian},$$

we will derive a similar formula from (11.7).

The next section provides the tools needed for transformation of the class number formulae we are aiming for.

12.1 Group-theoretical lemmas for the case $L \supseteq K$

Let $\mathfrak{K}$ be a finite abelian group and $\mathfrak{U}$ a subgroup of index

$$n = [\mathfrak{K} : \mathfrak{U}].$$

Let $\mathbb{X}$ denote the group of characters of $\mathfrak{K}$ that are trivial on $\mathfrak{U}$. Further, let

$$u : \mathfrak{K} \to \mathbb{R}$$

be a map and $\delta \in \mathbb{R}$ satisfying:

$$u(\mathfrak{k}) \text{ only depends on } \mathfrak{k}\mathfrak{U}, \tag{12.1}$$

$$\sum_{\mathfrak{h} \bmod \mathfrak{U}} (u(\mathfrak{k}\mathfrak{h}) - \delta u(\mathfrak{h})) = 0 \text{ for all } \mathfrak{k} \in \mathfrak{K}. \tag{12.2}$$

For $\chi \in \mathbb{X}$ we set:

$$A(\chi) := \sum_{\mathfrak{k} \bmod \mathfrak{U}} \chi(\mathfrak{k}) u(\mathfrak{k})$$

and define a quadratic matrix by

$$\begin{aligned} U := \quad & \left(u(\mathfrak{k}\mathfrak{h}^{-1}) - \delta u(\mathfrak{h}^{-1})\right) \\ & \text{row index: } \mathfrak{k} \bmod \mathfrak{U}, \quad \mathfrak{k} \notin \mathfrak{U}, \\ & \text{column index: } \mathfrak{h} \bmod \mathfrak{U}, \quad \mathfrak{h} \notin \mathfrak{U}. \end{aligned} \tag{12.3}$$

For the transformation of (11.6) for subfields of the Hilbert class field we need:

Theorem 12.1.1 *The matrix U in (12.3) has determinant*

$$|\det(U)| = \left| \prod_{\chi \neq 1} A(\chi) \right|.$$

Proof We multiply U from the left by

$$\begin{aligned} X = \quad & (\chi(\mathfrak{k})) \\ & \text{row index: } \chi \in \mathbb{X} \setminus \{1\}, \\ & \text{column index: } \mathfrak{k} \bmod \mathfrak{U}, \quad \mathfrak{h} \notin \mathfrak{U}. \end{aligned} \tag{12.4}$$

To compute XU note the relation $\sum\limits_{\mathfrak{k} \bmod \mathfrak{U}} \chi(\mathfrak{k}) = 0$ for $\chi \in \mathbb{X} \setminus \{1\}$. Then

$$\begin{aligned} \sum_{\substack{\mathfrak{k} \bmod \mathfrak{U} \\ \mathfrak{k} \notin \mathfrak{U}}} \chi(\mathfrak{k})(u(\mathfrak{k}\mathfrak{h}^{-1}) - \delta u(\mathfrak{h}^{-1})) &= \sum_{\mathfrak{k} \bmod \mathfrak{U}} \chi(\mathfrak{k}) u(\mathfrak{k}\mathfrak{h}^{-1}) \\ &= \chi(\mathfrak{h}) \sum_{\mathfrak{k} \bmod \mathfrak{U}} \chi(\mathfrak{k}) u(\mathfrak{k}). \end{aligned}$$

This shows that

$$\det(X)\det(U) = \det(X) \prod_{\chi \neq 1} A(\chi).$$

Therefore, for the proof of Theorem 12.1.1 we are left with the proof of $\det(X) \neq 0$. Using the above relation for characters, we find that

$$X\,{}^t\overline{X} = (\xi_{ij})_{i,j=1,\dots,n}, \qquad \xi_{ij} = \begin{cases} n-1, & \text{if } i=j, \\ -1, & \text{if } i \neq j, \end{cases}$$

and by elementary row operations we obtain

$$|\det(X)|^2 = \det(X\,{}^t\overline{X}) = n^{n-2}. \tag{12.5}$$

This completes the proof of Theorem 12.1.1. □

The transformation of (11.6) in the general case is much more difficult because the sums in (11.6) depend on the conductor of the character, which implies that more complicated matrices than in Theorem 12.1.1 come into play. For a given character $\chi \in \mathbb{X}$ we consider the subgroup

$$\mathfrak{U}_{\tilde{\chi}} := \{\mathfrak{k} \in \mathfrak{K} \mid \chi(\mathfrak{k}) = 1\}$$

only depending on the Frobenius equivalence class $\tilde{\chi}$. With every such class $\tilde{\chi}$ we associate a map

$$u_{\tilde{\chi}} : \mathfrak{K} \to \mathbb{R}$$

and a number $\delta_{\tilde{\chi}} \in \mathbb{R}$ having the following properties:

$$u_{\tilde{\chi}}(\mathfrak{k}) \text{ only depends on } \mathfrak{k}\mathfrak{U}_{\tilde{\chi}}, \tag{12.6}$$

$$\sum_{\mathfrak{h} \bmod \mathfrak{U}} \left(u_{\tilde{\chi}}(\mathfrak{k}\mathfrak{h}) - \delta_{\tilde{\chi}} u_{\tilde{\chi}}(\mathfrak{h})\right) = 0 \ \text{ for all } \mathfrak{k} \in \mathfrak{K}. \tag{12.7}$$

For $\chi \in \mathbb{X}$ we set:

$$A(\chi) := \sum_{\mathfrak{k} \bmod \mathfrak{U}_{\tilde{\chi}}} \chi(\mathfrak{k}) u_{\tilde{\chi}}(\mathfrak{k}).$$

Further, we associate every class $\tilde{\chi}$ with a real matrix $\Gamma_{\tilde{\chi}}$,

$$\begin{aligned} &\Gamma_{\tilde{\chi}} = \left(\gamma_{i_{\tilde{\chi}},\mathfrak{k}}(\tilde{\chi})\right), \\ &\text{row index: } i_{\tilde{\chi}} = 1, \dots, \varphi(n_{\tilde{\chi}}), \\ &\text{column index: } \mathfrak{k} \bmod \mathfrak{U}_{\tilde{\chi}}, \end{aligned} \tag{12.8}$$

where $n_{\tilde{\chi}}$ denotes the order of χ and φ the Euler function. We define a quadratic matrix by

$$U = \left(\sum_{\mathfrak{k} \bmod \mathfrak{U}_{\tilde{\chi}}} \gamma_{i_{\tilde{\chi}},\mathfrak{k}}(\tilde{\chi}) \left[u_{\tilde{\chi}}(\mathfrak{k}\mathfrak{h}^{-1}) - \delta_{\tilde{\chi}} u_{\tilde{\chi}}(\mathfrak{h}^{-1}) \right] \right),$$

row index: $(\tilde{\chi}, i_{\tilde{\chi}})$, $\tilde{\chi}$ runs through Frobenius classes $\tilde{\chi} \neq \tilde{1}$ of $\mathbb{X}$ and $i_{\tilde{\chi}}$ the numbers $1, \ldots, \varphi(n_{\tilde{\chi}})$, (12.9)
column index: $\mathfrak{h}$ runs through a system of representatives modulo $\mathfrak{U}$ with $\mathfrak{h} \notin \mathfrak{U}$.

Now we can state the theorem needed for the transformation of (11.6) in the general case:

Theorem 12.1.2 *Let U be the matrix defined in (12.9). Then*

$$|\det(U)| = C(\mathfrak{K}/\mathfrak{U}, \Gamma, \delta) \left| \prod_{\chi \in \mathbb{X} \setminus \{1\}} A(\overline{\chi}) \right|$$

with

$$C(\mathfrak{K}/\mathfrak{U}, \Gamma, \delta) = n^{-\frac{n}{2}} \prod_{\tilde{\chi} \neq \tilde{1}} |\det(g(\tilde{\chi}, \Gamma_{\tilde{\chi}}, \delta_{\tilde{\chi}}))| \left(\frac{n}{n_{\tilde{\chi}}} \right)^{\varphi(n_{\tilde{\chi}})}$$

and

$$g(\tilde{\chi}, \Gamma_{\tilde{\chi}}, \delta_{\tilde{\chi}}) = \left(\sum_{\mathfrak{k} \bmod \mathfrak{U}_{\tilde{\chi}}} \gamma_{i_{\tilde{\chi}},\mathfrak{k}}(\tilde{\chi})(\chi^{\mu}(\mathfrak{k}) - \delta_{\tilde{\chi}}) \right)$$

row index: $i_{\tilde{\chi}} = 1, \ldots, \varphi(n_{\tilde{\chi}})$,
column index: μ runs through a system of prime residues modulo $n_{\tilde{\chi}}$

The bar in $\overline{\chi}$ denotes complex conjugation.

Before proving Theorem 12.1.2 we state and prove:

Theorem 12.1.3 *For $\chi \in \mathbb{X} \setminus \{1\}$ and $\delta_{\tilde{\chi}} \in \{0, 1\}$ the coefficients $\gamma_{i_{\tilde{\chi}},\mathfrak{k}}$ of the matrix $\Gamma_{\tilde{\chi}}$ can be chosen in $\mathbb{Z}$ such that*

$$|\det(g(\tilde{\chi}, \Gamma_{\tilde{\chi}}, \delta_{\tilde{\chi}}))| = \sqrt{|d_{n_{\tilde{\chi}}}|}\ \Phi_{n_{\tilde{\chi}}}(1)^{\delta_{\tilde{\chi}}}$$

$$= \begin{cases} \sqrt{|d_{n_{\tilde{\chi}}}|} & \textit{if } n_{\tilde{\chi}} \textit{ is composite or if } \delta_{\tilde{\chi}} = 0, \\ \sqrt{|d_{n_{\tilde{\chi}}}|}\ p & \textit{if } n_{\tilde{\chi}} = p^k \textit{ is a prime power and } \delta_{\tilde{\chi}} = 1. \end{cases}$$

Herein $d_{n_{\tilde{\chi}}}$ denotes the discriminant of the $n_{\tilde{\chi}}$-th cyclotomic field and $\Phi_{n_{\tilde{\chi}}}$ the $n_{\tilde{\chi}}$-th cyclotomic polynomial.

Proof For $\delta_{\tilde{\chi}} = 0$ the numbers $\chi(\mathfrak{k}), \mathfrak{k} \in \mathfrak{K}$, are generators over $\mathbb{Z}$ of the maximal order $\mathfrak{O}_{n_{\tilde{\chi}}}$ in the $n_{\tilde{\chi}}$-th cyclotomic field. Hence $\gamma_{i_{\tilde{\chi}},\mathfrak{k}} \in \mathbb{Z}$ can be chosen such that

$$\sum_{\mathfrak{k} \bmod \mathfrak{U}_{\tilde{\chi}}} \gamma_{i_{\tilde{\chi}},\mathfrak{k}}(\tilde{\chi})(\chi(\mathfrak{k}) - \delta_{\tilde{\chi}}), \quad \mathfrak{k} \bmod \mathfrak{U}_{\tilde{\chi}}, \tag{12.10}$$

is a $\mathbb{Z}$-basis of $\mathfrak{O}_{n_{\tilde{\chi}}}$. The same holds for $\delta_{\tilde{\chi}} = 1$ if $n_{\tilde{\chi}}$ is composite because then one of the numbers $\chi(\mathfrak{k}) - 1$ is a unit. This implies the assertion of Theorem 12.1.3 in the first case. However, if $n_{\tilde{\chi}} = p^k$ is a prime power and $\delta_{\tilde{\chi}} = 1$, then the numbers $\chi(\mathfrak{k}) - \delta_{\tilde{\chi}}, \mathfrak{k} \in \mathfrak{K}$, are generators over $\mathbb{Z}$ of the prime ideal $\mathfrak{p}$ of norm p in the p^k-th cyclotomic field. Then, choosing $\gamma_{i_{\tilde{\chi}},\mathfrak{k}} \in \mathbb{Z}$ such that the numbers in (12.10) are a basis for $\mathfrak{p}$, we obtain the assertion of Theorem 12.1.3 in case 2. □

To prove Theorem 12.1.2 we further need the following three lemmas.

Lemma 12.1.4 *For $\chi, \psi \in \mathbb{X}$ and $\psi \notin <\chi>$ we have*

$$\sum_{\mathfrak{h} \bmod \mathfrak{U}} \psi(\mathfrak{h}) u_{\tilde{\chi}}(\mathfrak{h}) = 0.$$

Proof $\psi \notin <\chi>$ implies that $\psi | \mathfrak{U}_{\tilde{\chi}} \neq 1$. So

$$\sum_{\substack{\mathfrak{h} \bmod \mathfrak{U} \\ \mathfrak{h} \in \mathfrak{U}_{\tilde{\chi}}}} \psi(\mathfrak{h}) = 0$$

and, using (12.6), we obtain

$$\sum_{\mathfrak{h} \bmod \mathfrak{U}} \psi(\mathfrak{h}) u_{\tilde{\chi}}(\mathfrak{h}) = \sum_{\mathfrak{h}_1 \bmod \mathfrak{U}_{\tilde{\chi}}} \psi(\mathfrak{h}_1) \left(\sum_{\substack{\mathfrak{h}_2 \bmod \mathfrak{U} \\ \mathfrak{h}_2 \in \mathfrak{U}_{\tilde{\chi}}}} \psi(\mathfrak{h}_2) \right) u_{\tilde{\chi}}(\mathfrak{h}_1) = 0,$$

as asserted. □

The next two lemmas can be proved by obvious elementary row transformations.

Lemma 12.1.5 *Let* $e_i \in \mathbb{R}$, $\mathfrak{a}_i \in \mathbb{C}^l$, $i = 1, \ldots, l$, $e_o \neq 0$ *and* $\sum_{i=0}^{l} e_i \mathfrak{a}_i = 0$. *Then*

$$\det(e_1(\mathfrak{a}_1 - \mathfrak{a}_0), \ldots, e_l(\mathfrak{a}_l - \mathfrak{a}_0)) = \left(\sum_{i=0}^{l} \frac{e_i}{e_0}\right) \det(e_1\mathfrak{a}_1, \ldots, e_l\mathfrak{a}_l).$$

Lemma 12.1.6 *Let* $\alpha_\nu, \beta_\mu, \gamma_\mu \in \mathbb{C}$; $\nu, \mu = 1, \ldots, l$, $\gamma_\mu \neq 0$, $\mu = 1, \ldots, l$, *and let* $\delta_{\nu\mu}$ *denote the Kronecker symbol. Then*

$$\det(\alpha_\nu \beta_\mu + \delta_{\nu\mu}\gamma_\mu) = \left(\prod_{\mu=1}^{l} \gamma_\mu\right)\left(1 + \sum_{\mu=1}^{l} \frac{\alpha_\mu \beta_\mu}{\gamma_\mu}\right).$$

Proof of Theorem 12.1.2: Using Lemma 12.1.5 we transform the determinant of U:

$$|\det(U)| = \frac{1}{n} |\det(U_0)| \quad \text{with} \tag{12.11}$$

$$U_0 = \left(\sum_{\mathfrak{k} \bmod \mathfrak{U}_{\tilde{\chi}}} \gamma_{i_{\tilde{\chi}},\mathfrak{k}}(\tilde{\chi})\left(\left[u_{\tilde{\chi}}\left(\mathfrak{k}\mathfrak{h}^{-1}\right) - \delta_{\tilde{\chi}} u_{\tilde{\chi}}\left(\mathfrak{h}^{-1}\right)\right] - \left[u_{\tilde{\chi}}(\mathfrak{k}) - \delta_{\tilde{\chi}} u_{\tilde{\chi}}(\mathfrak{e})\right]\right)\right),$$

row index:	$(\tilde{\chi}, i_{\tilde{\chi}})$, $\tilde{\chi}$ runs through all classes
	$\tilde{\chi} \neq \tilde{1}$ of $\mathbb{X}$ and $i_{\tilde{\chi}}$ the numbers $1, \ldots, \varphi(n_{\tilde{\chi}})$,
column index:	$\mathfrak{h}$ runs through a system of residues
	modulo $\mathfrak{U}$ with $\mathfrak{h} \notin \mathfrak{U}$,

where $\mathfrak{e}$ denotes the neutral element in $\mathfrak{K}$. Unlike as in the proof of Theorem 12.1.1 we now multiply U_0 from the right by the matrix $X = (\psi(\mathfrak{h}))_{\psi,\mathfrak{h}}$ defined in (12.4), and we obtain

$$U_0 X = \left(\xi_{(\tilde{\chi}, i_{\tilde{\chi}}),\psi}\right) \quad \text{with} \quad \xi_{(\tilde{\chi}, i_{\tilde{\chi}}),\psi} =$$

$$\sum_{\substack{\mathfrak{h} \bmod \mathfrak{U} \\ \mathfrak{h} \notin \mathfrak{U}}} \left(\sum_{\mathfrak{k} \bmod \mathfrak{U}_{\tilde{\chi}}} \gamma_{i_{\tilde{\chi}},\mathfrak{k}}(\tilde{\chi})\psi(\mathfrak{h})\left(\left[u_{\tilde{\chi}}\left(\mathfrak{k}\mathfrak{h}^{-1}\right) - \delta_{\tilde{\chi}} u_{\tilde{\chi}}\left(\mathfrak{h}^{-1}\right)\right] - \left[u_{\tilde{\chi}}(\mathfrak{k}) - \delta_{\tilde{\chi}} u_{\tilde{\chi}}(\mathfrak{e})\right]\right)\right). \tag{12.12}$$

Since for $\mathfrak{h} = \mathfrak{e}$ the summand in (12.12) vanishes, we can extend the outer summation over a full system of residues $\mathfrak{h}$ modulo $\mathfrak{U}$, and, keeping in mind $\sum_{\mathfrak{h} \bmod \mathfrak{U}} \psi(\mathfrak{h}) = 0$ for $\psi \in \mathbb{X} \setminus \{1\}$, the matrix in (12.12) becomes

$$\xi_{(\tilde{\chi}, i_{\tilde{\chi}}), \psi} = \sum_{\mathfrak{h} \bmod \mathfrak{U}} \gamma_{i_{\tilde{\chi}}, \mathfrak{k}}(\tilde{\chi}) \sum_{\mathfrak{k} \bmod \mathfrak{U}_{\tilde{\chi}}} \psi(\mathfrak{h}) \left[u_{\tilde{\chi}} \left(\mathfrak{k}\mathfrak{h}^{-1} \right) - \delta_{\tilde{\chi}} u_{\tilde{\chi}} \left(\mathfrak{h}^{-1} \right) \right]$$

$$= \sum_{\mathfrak{h} \bmod \mathfrak{U}} \gamma_{i_{\tilde{\chi}}, \mathfrak{k}}(\tilde{\chi})(\psi(\mathfrak{k}) - \delta_{\tilde{\chi}}) \sum_{\mathfrak{k} \bmod \mathfrak{U}_{\tilde{\chi}}} \overline{\psi}(\mathfrak{h}) u_{\tilde{\chi}}(\mathfrak{h}).$$

In view of Lemma 12.1.4, this shows that:

$$\xi_{(\tilde{\chi}, i_{\tilde{\chi}}), \psi} = 0 \quad \text{if } \psi \notin < \chi > .$$

Arranging the rows in $U_0 X$ according to classes such that the class of ψ^ν with $\nu \neq 1$, $\nu | n_{\tilde{\psi}}$ follows the class of ψ, and, combining the rows belonging to the same $\tilde{\chi}$, we obtain a matrix which has along the diagonal the quadratic matrices

$U_{\tilde{\chi}} = \left(\xi_{(\tilde{\chi}, i_{\tilde{\chi}}), \chi^\mu} \right)$
row index: $i_{\tilde{\chi}} = 1, \ldots, \varphi(n_{\tilde{\chi}})$,
column index: μ runs through a system of prime residues modulo $n_{\tilde{\chi}}$.

Since all entries below these matrices in $U_0 X$ vanish, the determinant of $U_0 X$ equals the product of determinants of the $U_{\tilde{\chi}}$:

$$|\det(U_0 X)|$$

$$= \left| \prod_{\tilde{\chi} \neq \tilde{1}} \left\{ |\det(g(\tilde{\chi}, \Gamma_{\tilde{\chi}}))| \left(\frac{n}{n_{\tilde{\chi}}} \right)^{\varphi(n_{\tilde{\chi}})} \prod_{\substack{\mu \bmod n_{\tilde{\chi}} \\ (\mu, n_{\tilde{\chi}}) = 1}} \left(\sum_{\mathfrak{h} \bmod \mathfrak{U}_{\tilde{\chi}}} \overline{\chi}^\mu(\mathfrak{h}) u_{\tilde{\chi}}(\mathfrak{h}) \right) \right\} \right|$$

$$= \left| \left(\prod_{\tilde{\chi} \neq \tilde{1}} |\det(g(\tilde{\chi}, \Gamma_{\tilde{\chi}}))| \left(\frac{n}{n_{\tilde{\chi}}} \right)^{\varphi(n_{\tilde{\chi}})} \right) \prod_{\chi \in \mathbb{X} \setminus \{1\}} \left(\sum_{\mathfrak{h} \bmod \mathfrak{U}_{\tilde{\chi}}} \chi(\mathfrak{h}) u_{\tilde{\chi}}(\mathfrak{h}) \right) \right| .$$

Using (12.11) and (12.5) this implies the formula in Theorem 12.1.2. □

12.2 Applications of Theorems 12.1.1, 12.1.2

In the following we let

- K be the field of rational numbers or a quadratic imaginary number field,
- $\mathfrak{f}$ an integral ideal in K,
- $\mathfrak{K} = \mathfrak{K}_\mathfrak{f}$ the ray class group modulo $\mathfrak{f}$ in K,
- $\mathfrak{U}$ a subgroup of $\mathfrak{K}$,

- $\mathbb{X}$ the group of characters χ of $\mathfrak{K}$ with $\chi|\mathfrak{U} = 1$,
- $K_{\mathfrak{f}}$ the ray class field modulo $\mathfrak{f}$ over K,
- $\sigma : \mathfrak{K}_{\mathfrak{f}} \to G(K_{\mathfrak{f}}|K)$ the isomorphism from class field theory mapping $\mathfrak{K}_{\mathfrak{f}}$ onto the Galois group of $K_{\mathfrak{f}}$ over K,
- L the fixed field of $\sigma(\mathfrak{U})$,
- h_L and h_K the class numbers of L and K,
- R_L and r_L the regulator and unit rank of L.

With these notations we assume L to have a class number formula that can be transformed using Theorems 12.1.1 or 12.1.2 as we will describe below.

12.2.1 Application of Theorem 12.1.1

We assume L to have a class number given by the formula

$$c_L \frac{h_L}{h_K} = \frac{1}{R_L} \left| \prod_{\chi \in \mathbb{X} \setminus \{1\}} \left(\sum_{\mathfrak{h} \bmod \mathfrak{U}} \overline{\chi}(\mathfrak{h}) u(\mathfrak{h}) \right) \right| \tag{12.13}$$

with a number $c_L \in \mathbb{R}$ and a map $u \colon \mathfrak{K}_{\mathfrak{f}} \to \mathbb{R}$ satisfying (12.1) and (12.2) with some $\delta \in \mathbb{R}$. Further, we assume the unit rank r_L of L to be

$$r_L = \left| \mathfrak{K}_{\mathfrak{f}} / \mathfrak{U} \right| - 1,$$

the existence of units

$$\epsilon(\mathfrak{k}), \quad \mathfrak{k} \in \mathfrak{K}_{\mathfrak{f}} \bmod \mathfrak{U}, \quad \mathfrak{k} \notin \mathfrak{U},$$

in L with

$$\log \left| \epsilon(\mathfrak{k})^{\sigma(\mathfrak{h})} \right|^{e(\mathfrak{h})} = u(\mathfrak{k}\mathfrak{h}^{-1}) - \delta u(\mathfrak{h}^{-1})$$

and exponents $e(\mathfrak{h})$, that are equal to 1 or 2 depending on whether $\sigma(\mathfrak{h})$ is real or complex. Theorem 12.1.1 then implies that the regulator of the system of units $\epsilon(\mathfrak{k})$ is equal to the product on the right-hand side in (12.13), so (12.13) can be written as

$$c_L \frac{h_L}{h_K} = [E_L : U_0] \tag{12.14}$$

with the unit group

$$U_0 = < \{ \epsilon(\mathfrak{k}) \mid \mathfrak{k} \in \mathfrak{K}_{\mathfrak{f}} \setminus \mathfrak{U} \} > . \tag{12.15}$$

Herein the index is to be understood as the index of the corresponding subgroups in the logarithmic space.

12.2.2 Application of Theorem 12.1.2

We assume that we have a class number formula for L of the form

$$c_L \frac{h_L}{h_K} = \frac{1}{R_L} \left| \prod_{\chi \in \mathbb{X}\setminus\{1\}} \left(\sum_{\mathfrak{h} \bmod \mathfrak{U}_{\tilde{\chi}}} \overline{\chi}(\mathfrak{h}) u_{\tilde{\chi}}(\mathfrak{h}) \right) \right| \tag{12.16}$$

with some $c_L \in \mathbb{R}$ and maps $u_{\tilde{\chi}} : \mathfrak{K}_{\mathfrak{f}} \to \mathbb{R}$ depending only on the class $\tilde{\chi}$ of χ that satisfy conditions (12.6) and (12.7) for some $\delta\tilde{\chi} \in \mathbb{R}$. Further, let

$$r_L = |\mathfrak{K}_{\mathfrak{f}}/\mathfrak{U}| - 1,$$

and for every class $\tilde{\chi}$, $\chi \in \mathbb{X} \setminus \{1\}$ we assume the existence of units

$$\epsilon_{i_{\tilde{\chi}}}(\tilde{\chi}), \quad i_{\tilde{\chi}} = 1, \ldots, \varphi(n_{\tilde{\chi}}), \tag{12.17}$$

in L such that the equation

$$\log \left| \epsilon_{i_{\tilde{\chi}}}(\tilde{\chi})^{\sigma(\mathfrak{h})} \right|^{e(\mathfrak{h})} = \sum_{\mathfrak{k} \bmod \mathfrak{U}_{\tilde{\chi}}} \gamma_{i_{\tilde{\chi}},\mathfrak{k}}(\tilde{\chi}) \left(u_{\tilde{\chi}}(\mathfrak{k}\mathfrak{h}^{-1}) - \delta_{\tilde{\chi}} u_{\tilde{\chi}}(\mathfrak{h}^{-1}) \right)$$

with some coefficients $\gamma_{i_{\tilde{\chi}},\mathfrak{k}}(\tilde{\chi}) \in \mathbb{R}$ holds for all $\mathfrak{h} \in \mathfrak{K}_{\mathfrak{f}}$. The exponents $e(\mathfrak{h})$ are 1 or 2 depending on whether $\sigma(\mathfrak{h})$ is real or complex. φ denotes Euler's function. The union of the systems in (12.17) for all $\tilde{\chi}$, $\chi \in \mathbb{X} \setminus \{1\}$, is then a system of r_L units with a regulator that is equal to the product on the right-hand side in (12.16) times the factor $C(\mathfrak{K}_{\mathfrak{f}}/\mathfrak{U}, \Gamma, \delta)$. If this factor is not vanishing, the system of units must be independent and the class number formula (12.16) becomes

$$c_L C(\mathfrak{K}_{\mathfrak{f}}/\mathfrak{U}, \Gamma, \delta) \frac{h_L}{h_K} = [E_L : U_0], \tag{12.18}$$

with the product

$$U_0 = \prod_{\tilde{\chi} \neq \tilde{1}} U_{\tilde{\chi}} \tag{12.19}$$

of unit groups

$$U_{\tilde{\chi}} = \left\langle \{ \epsilon_{i_{\tilde{\chi}}}(\tilde{\chi}) \, | i_{\tilde{\chi}} = 1, \ldots, \varphi(n_{\tilde{\chi}}) \} \right\rangle$$

of rank $\varphi(n_{\tilde{\chi}})$, the product being direct modulo roots of unity.

12.3 Class number formulae for $\Omega \supseteq L \supseteq K$

First, let $\Omega = K_1$ be the Hilbert class field of K and $\mathfrak{K}$ the ideal class group of K. By $h = |\mathfrak{K}|$ we denote the class number of K. To every $\mathfrak{k} \in \mathfrak{K}$ we associate a unit in Ω in the following way: we choose an arbitrary ideal $\mathfrak{a}$ in $\mathfrak{k}$. Since $\mathfrak{a}^h$ is principal, we have

$$\mathfrak{a}^h = \mathfrak{O}\alpha$$

with some $\alpha \in K$ and the maximal order $\mathfrak{O}$ of K. Then, according to Theorem 4.2.2,

$$\epsilon(\mathfrak{k}) := \left(\frac{\Delta(\mathfrak{a})}{\Delta(\mathfrak{O})}\right)^h \alpha^{12} \tag{12.20}$$

defines a unit in Ω. In fact, $\epsilon(\mathfrak{k})$ only depends on $\mathfrak{k}$ and not on $\mathfrak{a}$, which easily follows from homogeneity of Δ. Further, keeping in mind that the unit group of K is a subgroup of the group of 12-th roots of unity, we see that α^{12} is uniquely defined by $\mathfrak{a}$. The action of the Galois group of Ω/K is given, according to Theorem 6.6.1, by

$$\epsilon(\mathfrak{k})^{\sigma(\mathfrak{h})} = \frac{\epsilon(\mathfrak{k}\mathfrak{h}^{-1})}{\epsilon(\mathfrak{h}^{-1})}, \tag{12.21}$$

and this implies that

$$\log\left|\epsilon(\mathfrak{k})^{\sigma(\mathfrak{h})}\right| = u(\mathfrak{k}\mathfrak{h}^{-1}) - u(\mathfrak{h}^{-1})$$

with

$$u(\mathfrak{k}) = \log|\epsilon(\mathfrak{k})|^2.$$

Further, by definition of $\epsilon(\mathfrak{k})$,

$$\left|\epsilon(\mathfrak{k})\right|^2 = \left|\frac{\Delta(\mathfrak{a})N(\mathfrak{a})^{12}}{\Delta(\mathfrak{O})}\right|^{2h} = \left(\frac{D(\mathfrak{k})}{D(\mathfrak{e})}\right)^{24h}.$$

Using this relation together with the character relation $\sum\limits_{\mathfrak{k}\in\mathfrak{K}} \chi(\mathfrak{k}) \log D(\mathfrak{e}) = 0$ for $\chi \in \mathbb{X} \setminus \{1\}$, the class number formula (11.6) for Ω can be written as

$$(24h)^{h-1}\frac{w_K}{w_\Omega}\frac{h_\Omega}{h} = \left|\prod_{\chi\in\mathbb{X}\setminus\{1\}} \left(\sum_{\mathfrak{k}\in\mathfrak{K}} \chi(\mathfrak{k})u(\mathfrak{k})\right)\right|. \tag{12.22}$$

Since the map u satisfies conditions (12.1), (12.2) and, keeping in mind that Ω is totally imaginary with unit rank

$$r_\Omega = h - 1,$$

we can, as described in 12.2.1 transform the formula (12.22) into

$$(24h)^{h-1} \frac{w_K}{w_\Omega} \frac{h_\Omega}{h} = \left[E_\Omega : U_{ell}^{(0)}\right]. \tag{12.23}$$

Herein E_Ω denotes the unit group of Ω and

$$U_{ell}^{(0)} = \left\langle \{\epsilon(\mathfrak{k}) \mid \mathfrak{k} \in \mathfrak{K}\} \right\rangle.$$

Note that $\epsilon(\mathfrak{e}) = 1$ so that $U_{ell}^{(0)}$ is generated by $|\mathfrak{K}| - 1 = r_\Omega$ units.

To eliminate the factors in (12.23) on the left-hand side of h_Ω we will replace the subgroup $U_{ell}^{(0)}$ by a suitable other subgroup of E_Ω.

Lemma 12.3.1 *Let $U_{ell}^{(1)}$ be the subgroup of $U_{ell}^{(0)}$ generated by units of the form*

$$\prod_{\mathfrak{k} \in \mathfrak{K}} \epsilon(\mathfrak{k})^{x_\mathfrak{k}} \quad \text{with} \quad x_\mathfrak{k} \in \mathbb{Z}, \quad \prod_{\mathfrak{k} \in \mathfrak{K}} \mathfrak{k}^{x_\mathfrak{k}} = \mathfrak{e}.$$

Then

$$\left[U_{ell}^{(0)} : U_{ell}^{(1)}\right] = h,$$

and for

$$U_{ell}^{(2)} = \left(U_{ell}^{(1)}\right)^{\frac{1}{24h}} \bigcap \Omega$$

we have

$$\left[U_{ell}^{(2)} : U_{ell}^{(1)}\right] = (24h)^{h-1} \frac{w_K}{w_\Omega}.$$

Using Lemma 12.3.1 we can now deduce from (12.23):

Theorem 12.3.2 *Let $U_{ell}^{(2)}$ be the unit group defined in Lemma 12.3.1. Then*

$$h_\Omega = \left[E_\Omega : U_{ell}^{(2)}\right].$$

Proof of Lemma 12.3.1 To prove the first assertion, we define the epimorphism

$$\kappa : \mathbb{Z}^{h-1} \to \mathfrak{K}, \qquad (x_\mathfrak{k}) \mapsto \prod_{\mathfrak{k} \in \mathfrak{K} \setminus \{\mathfrak{e}\}} \mathfrak{k}^{x_\mathfrak{k}}.$$

Then

$$[\mathbb{Z}^{h-1} : \ker \kappa] = |\mathfrak{K}| = h,$$

and since the $\epsilon(\mathfrak{k})$, $\mathfrak{k} \in \mathfrak{K} \setminus \{\mathfrak{e}\}$, are independent, this implies the first assertion of the lemma.

For the proof of the second assertion, note that $U_{ell}^{(1)}$ is generated by the units

$$\epsilon(\mathfrak{k}, \mathfrak{h}) := \frac{\epsilon(\mathfrak{k})\epsilon(\mathfrak{h})}{\epsilon(\mathfrak{k}\mathfrak{h})} \tag{12.24}$$

which can be written as

$$\epsilon(\mathfrak{k}, \mathfrak{h}) = \left(\frac{\eta\left(\frac{\alpha}{p}\right)\eta\left(\frac{\alpha}{q}\right)}{\eta\left(\frac{\alpha}{pq}\right)\eta\left(\alpha\right)} \right)^{24h}. \tag{12.25}$$

Herein $\alpha \in \mathfrak{O} \cap \mathbb{H}$ is chosen such that $\mathfrak{p} = [\alpha, p]$ and $\mathfrak{q} = [\alpha, q]$ are prime ideals of degree 1 in $\mathfrak{k}$ resp. $\mathfrak{h}$ of norm p resp. q and coprime to 6. Further, we can assume that $[\alpha, pq] = \mathfrak{p}\mathfrak{q}$, and α can always be chosen such that its trace is divisible by 3. According to Theorem 6.6.4 the η-quotient in (12.25) satisfies

$$\frac{\eta\left(\frac{\alpha}{p}\right)\eta\left(\frac{\alpha}{q}\right)}{\eta\left(\frac{\alpha}{pq}\right)\eta\left(\alpha\right)} \left(\gamma_2(\alpha)\gamma_3(\alpha)\right)^{\frac{p-1}{2}\frac{q-1}{2}} \in \Omega. \tag{12.26}$$

Now we have to distinguish cases according to the number of roots of unity contained in Ω. As we know by Theorem 6.1.4, Ω can at most contain the fourth and third roots of unity. First, we assume Ω to contain neither the fourth nor the third roots of unity. Then $\sqrt{-3}$ and $\sqrt{-4}$ are not in Ω, hence by Theorem 3.3.1 there exists a prime ideal of norm $\equiv -1 \bmod 3$ and $\equiv -1 \bmod 4$ in every ideal class. Since in this case $\gamma_2(\alpha)$ and $\gamma_3(\alpha)$ are non-zero elements in Ω, (12.25) and (12.26) imply that $U_{ell}^{(1)}$ is generated by $24h$-th powers of elements from E_Ω. Hence

$$\left[U_{ell}^{(2)} : U_{ell}^{(1)}\right] = (24h)^{h-1}.$$

Further, since in this case $w_K = 2$ and $w_\Omega = 2$, we have proved the second assertion of the lemma in the case $w_\Omega = 2$.

If Ω contains the third, but not the fourth roots of unity, we can still find a prime ideal of degree 1 and norm $\equiv 1 \bmod 4$ in every ideal class, but, according to Theorem 3.3.1, either all prime ideals not dividing 3 in an ideal class are of norm $\equiv 1 \bmod 3$ or are all of norm $\equiv -1 \bmod 3$.

Theorem 6.1.4 implies the discriminant of K to be divisible by 3. Therefore, the singular value $\gamma_2(\alpha)$ in (12.26) generates an extension H_3 of degree 3 over Ω. Excluding the trivial case $K = \mathbb{Q}(\sqrt{-3})$, we can now, by Theorem 3.3.1, choose an ideal class $\mathfrak{k}_0$, in which all prime ideals not dividing 3 have norm $\equiv -1 \bmod 3$. Then (12.26) tells us that

$$\epsilon_0 = \sqrt[24h]{\epsilon(\mathfrak{k}_0, \mathfrak{k}_0)} := \frac{\eta\left(\frac{\alpha_0}{p_0}\right)\eta\left(\frac{\alpha_0}{p_0}\right)}{\eta\left(\frac{\alpha_0}{p_0^2}\right)\eta\left(\alpha_0\right)}$$

generates H_3 over Ω and further, that for two ideal classes $\mathfrak{k}, \mathfrak{h}$ we always have

$$\sqrt[24h]{\epsilon(\mathfrak{k}, \mathfrak{h})}\epsilon_0^{\nu} \in \Omega$$

for some $\nu \in \{0, 1, 2\}$. This implies that

$$\left[U_{ell}^{(2)} : U_{ell}^{(1)}\right] = \frac{(24h)^{h-1}}{3},$$

which proves the assertion in the case $w_\Omega = 6$ because we have $w_K = 2$.

If Ω contains the fourth but not the third roots of unity, we conclude analogously. Excluding the trivial case $K = \mathbb{Q}(\sqrt{-4})$, we find that

$$\left[U_{ell}^{(2)} : U_{ell}^{(1)}\right] = \frac{(24h)^{h-1}}{2} \quad \text{and} \quad \frac{w_K}{w_\Omega} = \frac{1}{2},$$

and if Ω contains the third and the fourth roots of unity:

$$\left[U_{ell}^{(2)} : U_{ell}^{(1)}\right] = \frac{(24h)^{h-1}}{6} \quad \text{and} \quad \frac{w_K}{w_\Omega} = \frac{1}{6}.$$

This proves the second assertion of our lemma in the last two cases. □

A formula for the extensions L of K contained in Ω analogous to that of Theorem 12.3.2 can be derived by taking the relative norms

$$\epsilon_L(\mathfrak{k}) := \mathbf{N}_{\Omega/L}(\epsilon(\mathfrak{k})) \tag{12.27}$$

instead of $\epsilon(\mathfrak{k})$. Then as in (12.21)

$$\epsilon_L(\mathfrak{k})^{\sigma(\mathfrak{h})} = \frac{\epsilon_L(\mathfrak{k}\mathfrak{h}^{-1})}{\epsilon_L(\mathfrak{h}^{-1})}, \tag{12.28}$$

and this implies that

$$\log\left|\epsilon_L(\mathfrak{k})^{\sigma(\mathfrak{h})}\right| = u(\mathfrak{k}\mathfrak{h}^{-1}) - u(\mathfrak{h}^{-1})$$

with

$$u(\mathfrak{k}) = \log |\epsilon_L(\mathfrak{k})|^2.$$

Further, by definition of $\epsilon_L(\mathfrak{k})$

$$|\epsilon_L(\mathfrak{k})|^2 = \prod_{\mathfrak{h}\in\mathfrak{U}} \left(\frac{D(\mathfrak{k}\mathfrak{h})}{D(\mathfrak{h})}\right)^{24h},$$

where $\mathfrak{U}$ denotes the subgroup of $\mathfrak{K}$ associated with L. Let $\mathbb{X}$ be the group of characters of $\mathfrak{K}/\mathfrak{U}$. Then, keeping in mind the relation $\sum_{\mathfrak{k} \bmod \mathfrak{U}} \chi(\mathfrak{k})$ $\log D(\mathfrak{h}) = 0$ for $\chi \in \mathbb{X} \setminus \{1\}$, the class number formula of L in (11.6) can be written in the form

$$(24h)^{[L:K]-1} \frac{w_K}{w_L} \frac{h_L}{h} = \left| \prod_{\chi\in\mathbb{X}\setminus\{1\}} \left(\sum_{\mathfrak{k} \bmod \mathfrak{U}} \chi(\mathfrak{k}) u(\mathfrak{k}) \right) \right|. \tag{12.29}$$

Analogously to Lemma 12.3.1, we successively define the subgroups

$$U^{(0)}_{ell,L} := \left\langle \{\epsilon_L(\mathfrak{k}) \mid \mathfrak{k} \in \mathfrak{K} \bmod \mathfrak{U}\} \right\rangle,$$

$$U^{(1)}_{ell,L} := \left\{ \prod_{\mathfrak{k} \bmod \mathfrak{U}} \epsilon_L(\mathfrak{k})^{x_\mathfrak{k}} \mid \prod_{\mathfrak{k} \bmod \mathfrak{U}} \mathfrak{k}^{x_\mathfrak{k}} \in \mathfrak{U} \right\}.$$

Then, there exists a subgroup

$$U^{(2)}_{ell,L} \subseteq \left(U^{(1)}_{ell,L}\right)^{\frac{1}{24h}} \bigcap L$$

with

$$\left[U^{(2)}_{ell,L} : U^{(1)}_{ell,L}\right] = \frac{24h^{[L:K]-1}}{w_L/w_K},$$

and the same conclusions as in the proof of Theorem 12.3.2 yield the following result first obtained by Kersey (1980):

Theorem 12.3.3 *Let L be a subfield of Ω, $K \subseteq L \subseteq \Omega$. Then*

$$\tfrac{1}{[\Omega:L]} h_L = \left[E_L : U^{(2)}_{ell,L}\right].$$

The reason for the factor $\frac{1}{[\Omega:L]} = \frac{[L:K]}{h}$ missing in Theorem 12.3.2 is because, unlike Lemma 12.3.1, we here have

$$\left[U^{(1)}_{ell,L} : U^{(0)}_{ell,L}\right] = [\mathfrak{K} : \mathfrak{U}] = [L : K].$$

12.4 Class number formulae for $K_{\mathfrak{f}} \supseteq L \supseteq K$

In the following let L be an arbitrary abelian extension of K. Then L is a subfield of $K_{\mathfrak{f}}$ for a suitable integral ideal $\mathfrak{f}$ in K. Since in this general case the sums $A(\chi)$ on the right-hand side of (11.6) are defined with possibly **different** conductors, the units needed for the transformation of (11.6) cannot be constructed as for the subfields of the Hilbert class field. Therefore, the following construction is more complicated. On the other hand, as we will see later, the units obtained in this way allow relations between class numbers of different subfields to be discovered.

Let $\mathfrak{U}$ be the subgroup of $\mathfrak{K}_{\mathfrak{f}}$ associated with L and $\mathbb{X}$ the subgroup of characters of $\mathfrak{K}_{\mathfrak{f}}$ with $\chi|\mathfrak{U} = 1$. For $\chi \in \mathbb{X}$ let

$\tilde{\chi} := \{\psi \in <\chi> \mid <\psi> = <\chi>\}$ the class of χ,
$n_{\tilde{\chi}}$ the order of χ,
$\mathfrak{f}_{\tilde{\chi}}$ the conductor of χ,
$\mathfrak{U}_{\tilde{\chi}} = \{\mathfrak{k} \in \mathfrak{K}_{\mathfrak{f}} \mid \chi(\mathfrak{k}) = 1\}$ and
$K_{\tilde{\chi}}$ the subfield of $K_{\mathfrak{f}_{\tilde{\chi}}}$ corresponding to $\mathfrak{U}_{\tilde{\chi}}$
(which is also a subfield of L).

We construct a group of units as a product of groups

$$U^{(0)}_{ell,\mathbb{X}} = \prod_{\tilde{\chi} \neq \tilde{1}} U^{(0)}_{ell,\tilde{\chi}}, \tag{12.30}$$

where the factors $U^{(0)}_{ell,\tilde{\chi}}$ only depend on $\tilde{\chi}$ and not on L. We have to distinguish the two cases $\mathfrak{f}_{\tilde{\chi}} = (1)$ and $\mathfrak{f}_{\tilde{\chi}} \neq 1$. For $\mathfrak{f}_{\tilde{\chi}} \neq (1)$ we have two possibilities of construction if χ is a ring class character. Therefore, in the following, besides L, we fix a ring class field

$$\Omega_t \subseteq K_{\mathfrak{f}},$$

and for the construction of units we distinguish between those which are and are not characters of Ω_t/K.

For "χ a ring class character of Ω_t/K" we define, generalising (12.20),

$$\epsilon(\tilde{\chi}, \mathfrak{k}) := \mathbf{N}_{K_{\mathfrak{f}_{\tilde{\chi}}}/K_{\tilde{\chi}}}\left(\epsilon_{f_{\tilde{\chi}}}(\mathfrak{k})\right) = \prod_{\mathfrak{h} \in \mathfrak{U}_{\tilde{\chi}}} \frac{\epsilon_{f_{\tilde{\chi}}}(\mathfrak{k}\mathfrak{h})}{\epsilon_{f_{\tilde{\chi}}}(\mathfrak{h})}, \tag{12.31}$$

where

$$\epsilon_{f_{\tilde{\chi}}}(\mathfrak{k}) = \left(\frac{\Delta(\mathfrak{a}_{f_{\tilde{\chi}}})}{\Delta(\mathfrak{O}_{f_{\tilde{\chi}}})}\right)^{h_{f_{\tilde{\chi}}}} \alpha^{12}, \qquad \text{with } \mathfrak{a} \in \mathfrak{k} \in \mathfrak{K}_{\mathfrak{f}}.$$

Herein $f_{\tilde{\chi}}$ denotes the (necessarily rational) conductor of χ and $h_{f_{\tilde{\chi}}}$ the ring class number modulo $f_{\tilde{\chi}}$. $\mathfrak{a}_{f_{\tilde{\chi}}}$ is the ring ideal of $\mathfrak{O}_{f_{\tilde{\chi}}}$ associated with $\mathfrak{a}$ and α a generator of the principal ideal $\mathfrak{a}_{f_{\tilde{\chi}}}^{h_{f_{\tilde{\chi}}}}$.

For "χ not a ring class character of $\Omega_{\mathfrak{t}}/K$" we have $\mathfrak{f}_{\tilde{\chi}} \neq (1)$. Then, for $\mathfrak{a} \in \mathfrak{k} \in \mathfrak{K}_{\mathfrak{f}}$ the singular value $\Phi_{\mathfrak{f}_{\tilde{\chi}}}(\mathfrak{a})$, essentially defined by the σ function, only depends on $\mathfrak{k}$ and more precisely only on the ray class modulo $\mathfrak{f}_{\tilde{\chi}}$ that contains $\mathfrak{k}$. We set

$$\Phi_{\mathfrak{f}_{\tilde{\chi}}}(\mathfrak{k}) := \Phi_{\mathfrak{f}_{\tilde{\chi}}}(\mathfrak{a})$$

and then we obtain a unit by

$$\epsilon(\tilde{\chi}, \mathfrak{k}) := \mathbf{N}_{K_{\mathfrak{f}_{\tilde{\chi}}}/K_{\tilde{\chi}}}\left(\frac{\Phi_{\mathfrak{f}_{\tilde{\chi}}}(\mathfrak{k})}{\Phi_{\mathfrak{f}_{\tilde{\chi}}}(\mathfrak{e})}\right) = \prod_{\mathfrak{h}\in\mathfrak{U}_{\tilde{\chi}}} \frac{\Phi_{\mathfrak{f}_{\tilde{\chi}}}(\mathfrak{k}\mathfrak{h})}{\Phi_{\mathfrak{f}_{\tilde{\chi}}}(\mathfrak{h})}, \quad \mathfrak{k} \in \mathfrak{K}_{\mathfrak{f}}, \tag{12.32}$$

where $\mathfrak{e}$ denotes the principal class in $\mathfrak{K}_{\mathfrak{f}}$.

As we will discuss later, there is a further way of construction yielding larger subgroups if $\mathfrak{f}_{\tilde{\chi}}$ is composite.

With the units $\epsilon(\tilde{\chi}, \mathfrak{k})$ from (12.31) and (12.32) we define

$$\epsilon_{i_{\tilde{\chi}}}(\tilde{\chi}) = \prod_{\substack{\mathfrak{k}\in\mathfrak{K}_{\mathfrak{f}} \\ \mathfrak{k} \bmod \mathfrak{U}_{\tilde{\chi}}}} \epsilon(\tilde{\chi}, \mathfrak{k})^{\gamma_{i_{\tilde{\chi}},\mathfrak{k}}(\tilde{\chi})}, \quad i_{\tilde{\chi}} = 1, \ldots, \varphi(n_{\tilde{\chi}}),$$

where φ denotes the Euler function, and the exponents $\gamma_{i_{\tilde{\chi}},\mathfrak{k}}(\tilde{\chi}) \in \mathbb{Z}$ are chosen in such a way that the matrix defined in (12.8) has a non-zero determinant. Then we set

$$U_{ell,\tilde{\chi}}^{(0)} :=< \{\epsilon_{i_{\tilde{\chi}}}(\tilde{\chi}) \mid i_{\tilde{\chi}} = 1, \ldots, \varphi(n_{\tilde{\chi}})\} > . \tag{12.33}$$

The action of the Galois group of $K_{\mathfrak{f}}/K$ on $\epsilon(\mathfrak{k})$ is deduced from Theorem 6.6.1:

$$\epsilon(\tilde{\chi}, \mathfrak{k})^{\sigma(\mathfrak{h})} = \frac{\epsilon(\tilde{\chi}, \mathfrak{k}\mathfrak{h}^{-1})}{\epsilon(\tilde{\chi}, \mathfrak{h}^{-1})}, \quad \mathfrak{h} \in \mathfrak{K}_{\mathfrak{f}},$$

and by definition of $\epsilon_{i_{\tilde{\chi}}}(\tilde{\chi})$ we have

$$\log \left|\epsilon_{i_{\tilde{\chi}}}(\tilde{\chi})^{\sigma(\mathfrak{h})}\right|^{e(\mathfrak{h})} = \sum_{\mathfrak{k} \bmod \mathfrak{U}_{\tilde{\chi}}} \gamma_{i_{\tilde{\chi}},\mathfrak{k}}(\tilde{\chi}) \left(u_{\tilde{\chi}}(\mathfrak{k}\mathfrak{h}^{-1}) - u_{\tilde{\chi}}(\mathfrak{h}^{-1})\right)$$

with

$$u_{\tilde{\chi}}(\mathfrak{k}) = \sum_{\substack{\mathfrak{h} \in \mathfrak{K}_{\mathfrak{f}_{\tilde{\chi}}} \\ \mathfrak{h} \bmod \mathfrak{U}_{\tilde{\chi}}}} \log \left(D(\mathfrak{k}_{f_{\tilde{\chi}}} \mathfrak{h}_{f_{\tilde{\chi}}}) \right)^{24 h_{f_{\tilde{\chi}}}} \tag{12.34}$$

for a ring class character mod t,

where $\mathfrak{k}_{f_{\tilde{\chi}}}$ and $\mathfrak{h}_{f_{\tilde{\chi}}}$ denote the ring ideal classes modulo $f_{\tilde{\chi}}$ belonging to $\mathfrak{k}$ and $\mathfrak{h}$. In the other cases we set

$$u_{\tilde{\chi}}(\mathfrak{k}) = \sum_{\substack{\mathfrak{h} \in \mathfrak{K}_{\mathfrak{f}_{\tilde{\chi}}} \\ \mathfrak{h} \bmod \mathfrak{U}_{\tilde{\chi}}}} \log \left(S(\mathfrak{k}\mathfrak{h}) \right)^{12 f_{\tilde{\chi}}}, \tag{12.35}$$

where $f_{\tilde{\chi}}$ denotes the smallest natural number in $\mathfrak{f}_{\tilde{\chi}}$. Now the class number formula for L in (11.6) can be written as

$$\begin{aligned} &\left(\prod_{\tilde{\chi}} (24 h_{f_{\tilde{\chi}}})^{n_{\tilde{\chi}}} \right) \left(\prod_{\tilde{\chi}} (e_{f_{\tilde{\chi}}})^{n_{\tilde{\chi}}} \right) \frac{w_K}{w_L} \frac{h_\Omega}{h} R_L \\ &= \left| \prod_{\chi \in \mathbb{X} \setminus \{1\}} \left(\sum_{\substack{\mathfrak{k} \in \mathfrak{K}_{\mathfrak{f}} \\ \mathfrak{k} \bmod \mathfrak{U}_{\tilde{\chi}}}} \chi(\mathfrak{k}) u_{\tilde{\chi}}(\mathfrak{k}) \right) \right|. \end{aligned} \tag{12.36}$$

Herein, the first product on the left-hand side is over all classes of characters of L/K that are characters of Ω_t/K, and the second is over the remaining classes of characters of L/K. Now $u_{\tilde{\chi}}$ satisfies (12.6) and (12.7) and

$$r_L = \frac{[L : \mathbb{Q}]}{2} - 1 = |\mathbb{X}| - 1$$

because L is totally imaginary. Therefore, as explained in 12.2.2, we can transform (12.36) into

$$\begin{aligned} &\left(\prod_{\tilde{\chi}} (24 h_{f_{\tilde{\chi}}})^{n_{\tilde{\chi}}} \right) \left(\prod_{\tilde{\chi}} (e_{f_{\tilde{\chi}}})^{n_{\tilde{\chi}}} \right) C(\mathfrak{K}_{\mathfrak{f}}/\mathfrak{U}, \Gamma, 1) \frac{w_K}{w_L} \frac{h_L}{h} \\ &= \left[E_L : U^{(0)}_{ell,\mathbb{X}} \right] \end{aligned} \tag{12.37}$$

with the unit group E_L of L and the subgroup $U^{(0)}_{ell,\mathbb{X}}$ from (12.30), whose factors are defined in (12.33). $C(\mathfrak{K}_{\mathfrak{f}}/\mathfrak{U}, \Gamma, 1)$ is the factor from Theorem

12.1.2. Further, the number of generators in the definition of $U^{(0)}_{ell,\mathbb{X}}$ being equal to the unit rank of L, implies the product

$$U^{(0)}_{ell,\mathbb{X}} = \prod_{\tilde{\chi}\neq\tilde{1}} U^{(0)}_{ell,\tilde{\chi}}$$

to be direct modulo roots of unity.

Later, as for the unramified extensions L/K the factors $24h$ and $e_{f_{\tilde{\chi}}}$ in (12.37) will be reduced by modification of $U^{(0)}_{ell,\mathbb{X}}$. Before doing that, we will first make use of the fact that the factors $U^{(0)}_{ell,\tilde{\chi}}$ only depend on $\tilde{\chi}$, so let

$$K \subseteq L_1, \dots, L_s \subseteq L \subseteq K_{\mathfrak{f}}$$

be a system of subextensions $L_1, \dots, L_s$, linearly disjointed over K,

$$L_i \cap (L_1 \cdot \ldots \cdot L_{i-1} \cdot L_{i+1} \cdot \ldots \cdot L_s) = K, \quad i = 1, \dots, s.$$

Then, the groups of characters $\mathbb{X}_i$ of L_i/K satisfy

$$\mathbb{X}_i \cap \mathbb{X}_j = \{1\} \text{ for } i \neq j.$$

Hence, the partial product of factors in $U^{(0)}_{ell,\mathbb{X}}$ corresponding to the L_i is direct modulo roots of unity. Therefore, by combination of (12.37) and the corresponding class number formulae for the L_i, we obtain

$$C_1C_2C_3 \frac{w_K}{w_L} \frac{w_{L_1}\cdots w_{L_s}}{w_K^s} \frac{h_L h^{s-1}}{h_{L_1}\cdots h_{L_s}} = \left[E_L : (E_{L_1} \cdot \ldots \cdot E_{L_s} \cdot U^{(0)}_{ell,\mathbb{X}^c})\right], \quad (12.38)$$

where h_{L_i} and E_{L_i} denote class numbers and unit groups of the L_i. Further, we have

$$U^{(0)}_{ell,\mathbb{X}^c} = \prod_{\substack{\tilde{\chi} \\ \chi\in\mathbb{X}^c}} U^{(0)}_{ell,\tilde{\chi}} \quad \text{with } \mathbb{X}^c = \mathbb{X} \setminus (\mathbb{X}_1 \cup \dots \cup \mathbb{X}_s \cup \{1\})$$

and the factors

$$C_1 = \left(\prod_{\chi\in\mathbb{X}^c_{ring}} (24h_{f_{\tilde{\chi}}})\right),$$

$$C_2 = \left(\prod_{\chi\in\mathbb{X}^c_{ray}} (e_{f_{\tilde{\chi}}})\right), \quad (12.39)$$

$$C_3 = \frac{C(\mathfrak{K}_{\mathfrak{f}}/\mathfrak{U},\Gamma,1)}{C(\mathfrak{K}_{\mathfrak{f}}/\mathfrak{U}_1,\Gamma_1,1)\cdots C(\mathfrak{K}_{\mathfrak{f}}/\mathfrak{U}_s,\Gamma_s,1)},$$

that are defined according to the decomposition

$$\mathbb{X}^c = \mathbb{X}^c_{ring} \uplus \mathbb{X}^c_{ray}$$

with $\mathbb{X}^c_{ring} = \{\chi \in \mathbb{X}^c \mid \chi \text{ character of } \Omega_t/K\}$, $\mathbb{X}^c_{ray} = \mathbb{X}^c \setminus \mathbb{X}^c_{ring}$.

C_1 and C_2 can be reduced as follows: decompose $U^{(0)}_{ell,\mathbb{X}^c}$ according to the above decomposition of X^c into the product

$$U^{(0)}_{ell,\mathbb{X}^c} = U^{(0)}_{ell,ring}\, U^{(0)}_{ell,ray}.$$

To reduce $\mathbf{C_1}$ we write

$$U^{(0)}_{ell,ring} = \prod_{t'|t} U^{(0)}_{ell,t'},$$

where $U^{(0)}_{ell,t'}$, $t' \mid t$, is generated by the units $\epsilon_{i_{\tilde{\chi}}}(\tilde{\chi})$ with χ being a ring class character in $\mathbb{X}^c$ of conductor t'. If $\mathbb{X}^c_{ring}$ contains a character of conductor t', we define

$$U^{(1)}_{ell,t'} :=$$

$$\left\{ \prod_{\tilde{\chi},i_{\tilde{\chi}}} \epsilon_{i_{\tilde{\chi}}}(\tilde{\chi})^{x_{i_{\tilde{\chi}}}(\tilde{\chi})} \,\middle|\, x_{i_{\tilde{\chi}}}(\tilde{\chi}) \in \mathbb{Z},\ \prod_{\tilde{\chi},i_{\tilde{\chi}}} \left(\prod_{\mathfrak{k}_{\tilde{\chi}} \bmod \mathfrak{U}_{L_{\Omega_{t'}}}} \mathfrak{k}_{\tilde{\chi}}^{\gamma_{i_{\tilde{\chi}},\mathfrak{k}_{\tilde{\chi}}}} \right)^{x_{i_{\tilde{\chi}}}(\tilde{\chi})} \in \mathfrak{U}_{L_{\Omega_{t'}}} \right\},$$

where the product is taken over all classes of characters $\chi \in \mathbb{X}^c_{ring}$ of conductor t'. $L_{\Omega_{t'}}$ denotes the field $L \cap \Omega_{t'}$ and $\mathfrak{U}_{L_{\Omega_{t'}}}$ the subgroup of $\mathfrak{K}_{\mathfrak{f}}$ associated with $L_{\Omega_{t'}}/K$. Immediately by definition of $U^{(1)}_{ell,t'}$ we obtain

$$\left[U^{(0)}_{ell,t'} : U^{(1)}_{ell,t'}\right] = \left[L_{\Omega_{t'}} : K\right].$$

The elements of $U^{(1)}_{ell,t'}$ are products of relative norms with respect to $\Omega_{t'}/L_{\Omega_{t'}}$ of $24h_{t'}$-th powers of elements

$$\epsilon_{p,q} = \frac{\eta\left(\frac{\alpha}{p}\right)\eta\left(\frac{\alpha}{q}\right)}{\eta\left(\frac{\alpha}{pq}\right)\eta(\alpha)} \tag{12.40}$$

with prime ideals $\mathfrak{p}, \mathfrak{q}$ not dividing $6\mathfrak{f}$, of degree 1 and norm p, q with associated ring ideals $\mathfrak{p}_{t'} = [\alpha, p], \mathfrak{q}_{t'} = [\alpha, q], \mathfrak{p}_{t'}\mathfrak{q}_{t'} = [\alpha, pq]$, where $\alpha \in \mathbb{H}$ then has the property that $\mathfrak{O}_t = [\alpha, 1]$. Similarly to Lemma 12.3.1 we define

$$U^{(2)}_{ell,t'} := \left(U^{(1)}_{ell,t'}\right)^{\frac{1}{24h_{t'}}} \bigcap L_{\Omega_{t'}}. \tag{12.41}$$

As in the proof of Theorem 12.3.2, keeping in mind that

$$\epsilon_{p,q}\left(\gamma_2(\alpha)\gamma_3(\alpha)\right)^{\frac{p-1}{2}\frac{q-1}{2}} \in \Omega_{t'},$$

we obtain the equality

$$\left[U^{(2)}_{ell,t'} : U^{(1)}_{ell,t'}\right] = \frac{(24h_{t'})^{m_{t'}}}{n_{t'}},$$

where $m_{t'}$ denotes the rank of $U^{(1)}_{ell,t'}$ and $n_{t'} \in \mathbb{N}$ is a divisor of

$$\begin{array}{lll} 2 & \text{if} & \sqrt{-3} \notin L_{\Omega_{t'}} \setminus K, \\ 3 & \text{if} & \sqrt{-1} \notin L_{\Omega_{t'}} \setminus K, \\ 6, & \text{otherwise.} & \end{array}$$

Now, replacing $U^{(0)}_{ell,t'}$ by $U^{(2)}_{ell,t'}$ in (12.38) reduces C_1 to

$$c_1 = \prod_{t'|t} n_{t'} \left[L_{\Omega_{t'}} : K\right]. \tag{12.42}$$

To reduce $\mathbf{C_2}$ we proceed similarly as for C_1. For a given ideal dividing $\mathfrak{f}$,

$$\mathfrak{t} \mid \mathfrak{f}$$

we consider the subgroup

$$U^{(0)}_{ell,\mathfrak{t}} := \prod_{\tilde{\chi}}{}' U_{ell,\tilde{\chi}},$$

where the product is over all classes of characters $\chi \in \mathbb{X}^c_{ray}$ with $\mathfrak{f}_{\tilde{\chi}} = \mathfrak{t}$. To modify the subgroup $U^{(0)}_{ell,\mathfrak{t}}$ contained in

$$L_{\mathfrak{t}} := L \cap K_{\mathfrak{t}}$$

we will use Theorem 6.8.7. We write

$$\mathfrak{t} = t_1 \mathfrak{t}_2$$

with $t_1 \in \mathbb{N}$ and a primitive ideal $\mathfrak{t}_2$ of norm t_2. Then

$$t = t_1 t_2 = \min(\mathfrak{t} \cap \mathbb{N}).$$

Further, we write

$$t_2 = t_2^* t_2^{**},$$

where t_2^* denotes the ramified and t_2^{**} the split part of t_2. For a given prime p we note the two properties with a primitive p^ν-th root of unity ζ_{p^ν}.

$$L_{\mathfrak{t}} \cap K(\zeta_{p^\nu}) \neq K \quad \text{for a} \quad \nu \geq 1. \tag{12.43}$$

If this condition is **not** satisfied, then for every automorphism σ of $L_{\mathfrak{t}}/K$ and for every $\nu \geq 1$ there exists a continuation of σ on $L_\nu = L(\zeta_{p^\nu})$ leaving ζ_{p^ν} fixed. Keeping in mind Theorem 3.3.2, this implies that in every coset of $\mathfrak{K}_{\mathfrak{f}}$ modulo $\mathfrak{U}_{\tilde{\chi}}$, $\mathfrak{f}_{\tilde{\chi}} = \mathfrak{t}$, there is an ideal class $\mathfrak{k}$ and in $\mathfrak{k}$ an integral ideal $\mathfrak{a}$ coprime to $2\mathfrak{f}$ such that

$$N(\mathfrak{a}) \equiv 1 \bmod p^\nu. \tag{12.44}$$

Now we define

$$f_{L_{\mathfrak{t}}} := \prod_{\substack{p \\ p \text{ satisfies } (12.43)}} p^{\nu_p(\mathrm{lcm}(2^{\alpha_{\mathfrak{t}}} 3^{\beta_{\mathfrak{t}}}, 2t_1t_2^*))} \text{ and } f_{L_{\mathfrak{t}}}^c := \frac{\mathrm{lcm}(2^{\alpha_{\mathfrak{t}}} 3^{\beta_{\mathfrak{t}}}, 2t_1t_2^*)}{f_{L_{\mathfrak{t}}}},$$

where $\nu_p(\mathrm{lcm}(2^{\alpha_{\mathfrak{t}}} 3^{\beta_{\mathfrak{t}}}, 2t_1t_2^*))$ denotes the p-exponent of $\mathrm{lcm}(2^{\alpha_{\mathfrak{t}}} 3^{\beta_{\mathfrak{t}}}, 2t_1t_2^*)$ and

$$\alpha_{\mathfrak{t}} = \begin{cases} 1 & \text{if} \quad 2|d \text{ and } 2 \nmid \mathfrak{t}, \\ 0 & \text{otherwise}, \end{cases}$$
$$\beta_{\mathfrak{t}} = \begin{cases} 1 & \text{if} \quad 3|d \text{ and } 3 \nmid \mathfrak{t}, \\ 0 & \text{otherwise}. \end{cases}$$

To modify $U_{ell,\mathfrak{t}}^{(0)}$, we write the relative norm with respect to $L_{\mathfrak{t}}/K_{\tilde{\chi}}$ of an element θ as

$$\mathbf{N}_{L_{\mathfrak{t}}/K_{\tilde{\chi}}}(\theta) = \theta^{\gamma_{\tilde{\chi}}}, \quad \gamma_{\tilde{\chi}} = \sigma(\mathfrak{b}_1) + \cdots + \sigma(\mathfrak{b}_k),$$

with suitable Frobenius automorphisms $\sigma(\mathfrak{b}_i)$ of $K_{12f^2/K}$. Using the notation of Theorem 6.8.2, the generating units in the definition of $U_{ell,\tilde{\chi}}^{(0)}$ can then be written in the form

$$\epsilon_{i_{\tilde{\chi}}}(\tilde{\chi}) = \mathbf{N}_{K_{\mathfrak{t}}/L_{\mathfrak{t}}}\left(\Phi_{\mathfrak{t}}(\mathfrak{e})^{\left\{ \sum_{\mathfrak{k}_{\tilde{\chi}}} (\sigma(\mathfrak{k}_{\tilde{\chi}})-1)\gamma_{i_{\tilde{\chi}},\mathfrak{k}_{\tilde{\chi}}}(\tilde{\chi})\gamma_{\tilde{\chi}} \right\}} \right).$$

with

$$\Phi_{\mathfrak{t}}(\mathfrak{e}) = \varphi\left(\xi \middle| \begin{matrix} \alpha \\ 1 \end{matrix}\right)^{e_t}, \quad e_t = \mathrm{lcm}(12, t),$$

and some $\xi \in K \setminus \mathfrak{t}$ satisfying $\mathfrak{t}\xi \subseteq \mathfrak{O}$. Now we define $U^{(1)}_{ell,\mathfrak{t}}$ to be the set of all products

$$\prod_{i_{\tilde{\chi}},\tilde{\chi}} \epsilon_{i_{\tilde{\chi}}}(\tilde{\chi})^{x(i_{\tilde{\chi}},\tilde{\chi})}, \qquad x(i_{\tilde{\chi}},\tilde{\chi}) \in \mathbb{Z},$$

with exponents $x(i_{\tilde{\chi}},\tilde{\chi})$ satisfying the congruence

$$\sum_{i_{\tilde{\chi}}\tilde{\chi}} \left[\sum_{\mathfrak{k}_{\tilde{\chi}}} (N(\mathfrak{a}_{\tilde{\chi}}) - 1)\gamma_{i_{\tilde{\chi}},\mathfrak{k}_{\tilde{\chi}}}(\tilde{\chi}) N(\gamma_{\tilde{\chi}}) \right] x(i_{\tilde{\chi}},\tilde{\chi}) \tag{12.45}$$

$$\equiv 0 \bmod \operatorname{lcm}(4^{\alpha_{\mathfrak{t}}}3^{\beta_{\mathfrak{t}}}, 2t_1t_2^*)$$

and representatives $\mathfrak{a}_{\tilde{\chi}} \in \mathfrak{k}_{\tilde{\chi}}$. From the above considerations it is clear that the classes $\mathfrak{k}_{\tilde{\chi}}$ and representatives $\mathfrak{a}_{\tilde{\chi}}$ can be chosen in such a way that the $\mathfrak{a}_{\tilde{\chi}}$ are integral with norms satisfying

$$N(\mathfrak{a}_{\tilde{\chi}}) \equiv 1 \bmod f^c_{L_{\mathfrak{t}}}, \quad N(\mathfrak{a}_{\tilde{\chi}}) \equiv 1 \bmod 2.$$

From (12.45) we now obtain

$$\left[U^{(0)}_{ell,\mathfrak{t}} : U^{(1)}_{ell,\mathfrak{t}}\right] \mid f_{L_{\mathfrak{t}}},$$

and by Theorem 6.8.7 all elements in $U^{(1)}_{ell,\mathfrak{t}}$ must be e_t-th powers of elements from $L_{\mathfrak{t}}$. Hence for

$$U^{(2)}_{ell,\mathfrak{t}} := \left(U^{(1)}_{ell,\mathfrak{t}}\right)^{\frac{1}{e_t}} \bigcap L_{\mathfrak{t}} \tag{12.46}$$

we have the index formula

$$\left[\left(U^{(2)}_{ell,\mathfrak{t}}\right)^{\frac{1}{e_t}} : U^{(2)}_{ell,\mathfrak{t}}\right] = e_t^{rank\left(U^{(0)}_{ell,\mathfrak{t}}\right)}.$$

Now, replacing in (12.38) the subgroups $U^{(0)}_{ell,\mathfrak{t}}$ by $U^{(2)}_{ell,\mathfrak{t}}$, the factor C_2 becomes

$$c_2 = {\prod_{\mathfrak{t}|\mathfrak{f}}}' f_{L_{\mathfrak{t}}}, \tag{12.47}$$

where the product is over those divisors $\mathfrak{t}$ of $\mathfrak{f}$ that occur as conductors of characters in $\mathbb{X}^c_{ray}$.

We summarise our results:

Theorem 12.4.1 *Let L be a subfield of $K_{\mathfrak{f}}$, $K \subseteq L \subseteq K_{\mathfrak{f}}$, and $L_1, \ldots, L_s$ a system of subextensions of L/K that is linearly disjoint over K,*

$$L_i \cap (L_1 \cdot \ldots \cdot L_{i-1} \cdot L_{i+1} \cdot \ldots \cdot L_s) = K, \quad i = 1, \ldots, s.$$

For every class $\tilde{\chi}$, $\chi \in \mathbb{X}$, we choose a matrix $\Gamma_{\tilde{\chi}}$ with coefficients in $\mathbb{Z}$ such that the matrix $g(\tilde{\chi}, \Gamma_{\tilde{\chi}}, 1)$ as defined in Theorem 12.1.2 has non-zero determinant. Moreover, in the following we will assume the exponents $\gamma_{i_{\tilde{\chi}}, \mathfrak{k}}(\tilde{\chi})$ to be chosen according to Theorem 12.1.3. By Γ we denote the system of these matrices and by Γ_i the subsystem associated with L_i. Further, let

$$U^{(2)}_{ell,\mathbb{X}^c} = \left(\prod_{t'|t}{}' U^{(2)}_{ell,t'}\right)\left(\prod_{\mathfrak{t}|\mathfrak{f}}{}' U^{(2)}_{ell,\mathfrak{t}}\right)$$

denote the subgroup of the unit group of L obtained by modification of $U^{(0)}_{ell,\mathbb{X}^c}$. With this subgroup we obtain from (12.38)

$$C\,\frac{w_K}{w_L}\,\frac{w_{L_1}\cdots w_{L_s}}{w_K^s}\,\frac{h_L}{h_K}\,\frac{h_K^s}{h_{L_1}\cdots h_{L_s}} = \left[E_L : (E_{L_1} \cdot \ldots \cdot E_{L_s} \cdot U^{(2)}_{ell,\mathbb{X}^c})\right]$$

with the class numbers $h_{\ldots}$, unit groups $E_{\ldots}$ and numbers of roots of unity $w_{\ldots}$ of the corresponding fields and the factor

$$C = c_1 c_2 C_3,$$

that is composed by the factors defined in (12.42), (12.47) and (12.39):

$$\begin{aligned} c_1 &= \prod_{t'|t}{}' n_{t'}\left[L_{\Omega_{t'}} : K\right], \\ c_2 &= \prod_{\mathfrak{t}|\mathfrak{f}}{}' f_{L_{\mathfrak{t}}}, \end{aligned}$$

$$C_3 = \frac{c(\mathfrak{K}_{\mathfrak{f}}/\mathfrak{U},\Gamma,1)}{c(\mathfrak{K}_{\mathfrak{f}}/\mathfrak{U}_1,\Gamma_1,1)\cdot\ldots\cdot c(\mathfrak{K}_{\mathfrak{f}}/\mathfrak{U}_s,\Gamma_s,1)}.$$

12.4.1 Application of the formulae from 12.4

12.4.1.1 Divisibility between class numbers

An interesting application of Theorem 12.4.1 is the derivation of divisibility relations between class numbers. Of course, for this purpose the determination of the constant C is important. The following two remarks are useful:

(i) Choosing the matrices $\Gamma_{\tilde{\chi}}$ according to Theorem 12.1.3 yields

$$C(\mathfrak{K}_\mathfrak{f}/\mathfrak{U},\Gamma,1) = n^{\frac{n}{2}} \prod_{\tilde{\chi}\neq\tilde{1}} \frac{\sqrt{|d_{n_{\tilde{\chi}}}|}}{n_{\tilde{\chi}}^{\varphi(n_{\tilde{\chi}})}},$$

where $d_{n_{\tilde{\chi}}}$ is the discriminant of the $n_{\tilde{\chi}}$-th cyclotomic field. With this choice we know from Hasse (1952) that:

(ii) $C(\mathfrak{K}_\mathfrak{f}/\mathfrak{U},\Gamma,1) \in \mathbb{N}$ and

$$C(\mathfrak{K}_\mathfrak{f}/\mathfrak{U},\Gamma,1) = 1 \iff \mathfrak{K}_\mathfrak{f}/\mathfrak{U} \quad \text{cyclic.}$$

For special types of field the general results of Theorem 12.4.1 can be made more precise, as will be explained in the following example.

Example 12.4.2 Let $L = L_6 = \mathbb{Q}(\sqrt{-3}, \sqrt[6]{a})$, $a \in \mathbb{Z}$, be the normal field of the $\mathbb{Q}(\sqrt[6]{a})$ having degree 6 over $\mathbb{Q}$, and let $L_2 = \mathbb{Q}(\sqrt{-3}, \sqrt{a})$, $L_3 = \mathbb{Q}(\sqrt{-3}, \sqrt[3]{a})$. Then L_6, L_2, L_3 are abelian over $K = \mathbb{Q}(\sqrt{-3})$. More precisely, L is a subfield of a ring class field Ω_f, so we choose $t = f$ for the construction of $U^{(2)}_{ell,\mathbb{X}^c}$. Then $C = c_1 = 6$ in Theorem 12.4.1, all numbers of roots of unity are equal to 6, and $h_K = 1$. Hence

$$6\,\frac{h_{L_6}}{h_{L_2}h_{L_3}} = \left[E_{L_6} : E_{L_2}E_{L_3}U^{(2)}_{ell,\mathbb{X}^c}\right].$$

Modifying the construction of $U^{(2)}_{ell,\mathbb{X}^c}$ we obtain the formula

$$9\,\frac{h_{L_6}}{h_{L_2}h_{L_3}} = \left[E_{L_6} : E_{L_2}E_{L_3}\tilde{U}^{(2)}_{ell,\mathbb{X}^c}\right],$$

showing that the quotient $\frac{h_{L_6}}{h_{L_2}h_{L_3}}$ has a denominator dividing 3.

Proof The first formula is immediate by Theorem 12.4.1. To prove the second formula we start with the formula (12.38). In our case it tells us that

$$(24f_{\tilde{\chi}})^2\,3\,\frac{h_{L_6}}{h_{L_2}h_{L_3}} = \left[E_{L_6} : E_{L_2}E_{L_3}U^{(0)}_{ell,\tilde{\chi}}\right],$$

where χ denotes the generating character of L_6/K. $U^{(0)}_{ell,\tilde{\chi}}$ has rank 2 and is defined by

$$\tilde{U}^{(0)}_{ell,\tilde{\chi}} := \left\langle \epsilon_{\tilde{\chi}}(\mathfrak{k}^2), \epsilon_{\tilde{\chi}}(\mathfrak{k}^4)\right\rangle$$

with a class $\mathfrak{k} \in \mathfrak{K}_\mathfrak{f}$ representing a generator for $\mathfrak{K}_\mathfrak{f}/\mathfrak{U}_6$. Here, we have $|\det(g(\tilde{\chi},\Gamma_{\tilde{\chi}}))| = 3\sqrt{3} = 3\sqrt{|d_6|}\Phi_6(1)$, hence $C_3 = 3$ in (12.38). Further, we define

$$\tilde{U}^{(1)}_{ell,\tilde{\chi}} := \left\langle \frac{\epsilon_{\tilde{\chi}}(\mathfrak{k}^2)^2}{\epsilon_{\tilde{\chi}}(\mathfrak{k}^4)}, \frac{\epsilon_{\tilde{\chi}}(\mathfrak{k}^4)^2}{\epsilon_{\tilde{\chi}}(\mathfrak{k}^2)} \right\rangle.$$

Since $(\mathfrak{k}^2)^2\mathfrak{k}^{-4}$ and $(\mathfrak{k}^4)^2\mathfrak{k}^{-2}$ are in $\mathfrak{U}_6$, the two generators of $\tilde{U}^{(1)}_{ell,\tilde{\chi}}$ are $24f_{\tilde{\chi}}$-th powers in L_6. This implies the second formula in Example 12.4.2 with

$$\tilde{U}^{(2)}_{ell,\tilde{\chi}} := \left(\tilde{U}^{(1)}_{ell,\tilde{\chi}}\right)^{\frac{1}{24h_{f_{\tilde{\chi}}}}} \bigcap L,$$

keeping in mind that

$$\left[\tilde{U}^{(0)}_{ell,\tilde{\chi}} : \tilde{U}^{(1)}_{ell,\tilde{\chi}}\right] = 3.$$

□

Remark 12.4.3 If, in particular, $L = K_{\mathfrak{f}}$, then the subgroup

$$U^{(2)}_{ell,ray} := \prod_{\mathfrak{t}|\mathfrak{f}}{}' U^{(2)}_{ell,\mathfrak{t}}$$

in Theorem 12.4.1 can be replaced by a larger subgroup $\hat{U}^{(2)}_{ell,ray}$ thereby reducing c_2 to

$$\hat{c}_2 = f_{K_{\mathfrak{f}}}.$$

The construction of $\hat{U}^{(2)}_{ell,ray}$ is done in two steps. First, we consider the subgroup $\hat{U}^{(1)}_{ell,ray}$ consisting of units of the form

$$\prod_{i_{\tilde{\chi}},\tilde{\chi}} \epsilon_{i_{\tilde{\chi}}}(\tilde{\chi})^{x(i_{\tilde{\chi}},\tilde{\chi})},$$

where the product is taken over **all** classes $\tilde{\chi}$ of characters in $\mathbb{X}^c$ that are no ring class characters modulo t and where the exponents $x(i_{\tilde{\chi}},\tilde{\chi}) \in \mathbb{Z}$ satisfy

$$\sum_{i_{\tilde{\chi}}\tilde{\chi}} \left[\sum_{\mathfrak{k}_{\tilde{\chi}}} (N(\mathfrak{a}_{\tilde{\chi}})-1)\gamma_{i_{\tilde{\chi}},\mathfrak{k}_{\tilde{\chi}}}(\tilde{\chi})N(\hat{\gamma}_{\tilde{\chi}})\right] \frac{e_f}{e_{f_{\tilde{\chi}}}} x(i_{\tilde{\chi}},\tilde{\chi}) \equiv 0 \bmod \operatorname{lcm}(4^{\alpha_{\mathfrak{f}}} 3^{\beta_{\mathfrak{f}}}, 2f_1 f_2^*).$$

Herein $\hat{\gamma}_{\tilde{\chi}}$ is defined by

$$\hat{\gamma}_{\tilde{\chi}} = \sigma(\mathfrak{b}_1) + \cdots + \sigma(\mathfrak{b}_k),$$

with suitable Frobenius automorphisms $\sigma(\mathfrak{b}_i)$ of $K_{12f^2/K}$ such that

$$\mathbf{N}_{K_{\mathfrak{f}_{\tilde{\chi}}}/K_{\tilde{\chi}}}(\theta) = \theta^{\hat{\gamma}_{\tilde{\chi}}}$$

holds for all elements $\theta \in K_{\mathfrak{f}_{\tilde{\chi}}}$. Then, with

$$\hat{U}^{(2)}_{ell,ray} := \left(\hat{U}^{(1)}_{ell,ray}\right)^{\frac{1}{e_f}} \bigcap K_{\mathfrak{f}}$$

we obtain by Theorem 6.8.7 the formula

$$\left[\hat{U}^{(2)}_{ell,ray} : \hat{U}^{(1)}_{ell,ray}\right] = \prod_{\chi \in \mathbb{X}^c_{ray}} e_{f_{\tilde{\chi}}}.$$

Now, replacing the subgroup $U^{(2)}_{ell,\mathbb{X}^c}$ in Theorem 12.4.1 by

$$\hat{U}^{(2)}_{ell,\mathbb{X}^c} = \left(\prod_{t'|t}{}' U^{(2)}_{ell,t'}\right) \hat{U}^{(2)}_{ell,ray},$$

we obtain the formula from Theorem 12.4.1 with c_2 replaced by $\hat{c}_2$.

An application of this remark is:

Example 12.4.4 Let $K = \mathbb{Q}(\sqrt{d})$, $d = d_K < 0$, $\gcd(d,6) = 1$, and let $\mathfrak{f}$ be an integral odd ideal in K. Let $L = K_{\mathfrak{f}}$, and $L_1 = \Omega$ the Hilbert class field of K. Further, we assume $G(K_{\mathfrak{f}}/K)$ to be **cyclic**. In the construction of $\hat{U}^{(2)}_{ell,\mathbb{X}^c}$ we take $t = 1$. Then $C = 1$ in the formula of Theorem 12.4.1 modified according to Remark 12.4.3 and

$$\frac{h_{K_{\mathfrak{f}}}}{h_\Omega} = \left[E_{K_{\mathfrak{f}}} : E_\Omega \hat{U}^{(2)}_{ell,\mathbb{X}^c}\right].$$

Proof To prove the formula in Example 12.4.4 the factors c_1, $\hat{c}_2$ and the numbers of roots of unity in Ω and $K_{\mathfrak{f}}$ have to be computed. We start by determining the roots of unity in $K_{\mathfrak{f}} \setminus \Omega$, so let p^ν prime power with $\nu \geq 1$ for $p \neq 2$ resp. $\nu \geq 2$ for $p = 2$ and let

$$\zeta = \zeta_{p^\nu}$$

be a primitive p^ν-th root of unity. We contend that

$$\zeta \in K_{\mathfrak{f}} \setminus \Omega \Longrightarrow p^\nu \mid f_1 f_2^*. \tag{12.48}$$

First, note that $\zeta \in K_{\mathfrak{f}} \cap \overline{K_{\mathfrak{f}}} = K_{\mathfrak{f}} \cap K_{\overline{\mathfrak{f}}} = K_{f_1 \mathfrak{f}_2^*}$. Further, on the one hand $\Omega(\zeta)/\Omega$ is unramified above p, since Ω is the maximal unramified extension of K and since the discriminant of $\Omega(\zeta)/\Omega$ is a divisor of a power of p^ν. On the other hand the extension $K_{f_1\mathfrak{f}_2^*}/\Omega$ is unramified outside $f_1 f_2^*$. Hence, p must divide $f_1 f_2^*$. We write

$$f_1\mathfrak{f}_2^* = f_1[\alpha, f_2^*]$$

and keep in mind that ζ is invariant under all Frobenius automorphisms $\sigma(1+\omega)$, $\omega \in f_1\mathfrak{f}_2^*$. In view of Theorem 3.3.2 this implies that

$$N(1+\omega) \equiv 1 \bmod p^\nu \quad \text{for all } \omega \in f_1\mathfrak{f}_2^*.$$

Choosing $\omega = \pm f_1 f_2^*$ in particular, we find that

$$f_1 f_2^*(2 \pm f_1 f_2^*) \equiv 0 \bmod p^\nu$$

and further, by addition, that

$$4 f_1 f_2^* \equiv 0 \bmod p^\nu.$$

As by assumption $\mathfrak{f}$ and hence $f_1 f_2^*$ are odd and since p is a divisor of $f_1 f_2^*$ as shown above, p must be odd, and (12.48) follows.

Conversely, $K_{\mathfrak{f}}$ contains all $f_1 f_2^*$-th roots of unity, because all $\omega \in \mathfrak{f}$ satisfy

$$N(1+\omega) \equiv 1 \bmod f_1 f_2^*.$$

Hence, the number of roots of unity in $K_{\mathfrak{f}}$ is

$$w_{K_{\mathfrak{f}}} = \operatorname{lcm}(w_\Omega, f_1 f_2^*).$$

To determine w_Ω note that, according to Theorem 6.1.4, the maximal abelian subextension of Ω is a product of quadratic fields, which implies that only the 2-nd, 4-th, 6-th or 12-th roots of unity can be contained in Ω. More precisely, because of $\gcd(d,6) = 1$, Theorem 6.1.4 tells us that

$$w_\Omega = 2.$$

Further, in this case

$$\hat{c}_2 = f_{K_{\mathfrak{f}}} = f_1 f_2^*$$

and by choice of t

$$c_1 = 1.$$

The class number formula in Example 12.4.4 now follows from Remark 12.4.3. □

12.4.1.2 Divisibility of class numbers by divisors of the field degree

In the last section the unit groups were constructed by quotients of invariants $\Phi_{\mathfrak{f}_{\tilde{\chi}}}(\mathfrak{k})$ applying Theorem 12.1.2 with $\delta_{\tilde{\chi}} = 1$. In the following, we assume $\mathfrak{f}_{\tilde{\chi}}$ to be composite, in which case the invariants $\Phi_{\mathfrak{f}_{\tilde{\chi}}}(\mathfrak{k})$ themselves are units. Then, the application of Theorem 12.1.2 with $\delta_{\tilde{\chi}} = 0$ yields class number formulae showing that the class numbers of certain fields are divisible by divisors of the field degree.

In the "case of ray classes", we therefore modify the construction of units different from (12.32) as follows:

If $\mathbf{n}_{\tilde{\chi}}$ **is a prime power and** $\mathfrak{f}_{\tilde{\chi}}$ **composite,** we set

$$\epsilon(\tilde{\chi}, \mathfrak{k}) := \mathbf{N}_{K_{\mathfrak{f}_{\tilde{\chi}}}/K_{\tilde{\chi}}}\left(\Phi_{\mathfrak{f}_{\tilde{\chi}}}(\mathfrak{k})\right) = \prod_{\mathfrak{h} \in \mathfrak{U}_{\tilde{\chi}}} \Phi_{\mathfrak{f}_{\tilde{\chi}}}(\mathfrak{k}\mathfrak{h}), \quad \mathfrak{k} \in \mathfrak{K}_{\mathfrak{f}}. \tag{12.49}$$

Modifying the construction of $U^{(0)}_{ell,\mathbb{X}^c}$ by $U^{(0)'}_{ell,\mathbb{X}^c}$ in this way the formula (12.38) becomes

$$C_1 C_2 \tilde{C}_3 \frac{w_K}{w_L} \frac{w_{L_1} \cdots w_{L_s}}{w_K^s} \frac{h_L h^{s-1}}{h_{L_1} \cdots h_{L_s}} = \left[E_L : (E_{L_1} \cdot \ldots \cdot E_{L_s} \cdot U^{(0)'}_{ell,\mathbb{X}^c})\right]. \tag{12.50}$$

The factors C_3 in (12.38) and C_3' are related by

$$\tilde{C}_3 = \frac{C_3}{N}$$

with

$$N = \prod_{\chi}{}' p_\chi,$$

where χ runs through all characters $\mathbb{X}^c_{ray}$ with composite conductor and an order which is a prime power $p_\chi^{a_\chi}$. Further, if L/K is **cyclic**, then $C_3 = 1$, and by (12.50) we obtain divisibility properties for the product of class numbers contained in the formula if the remaining factors are coprime to N. For example, in this way we can prove the following result:

Theorem 12.4.5 *Let K be a quadratic imaginary number field and n a natural number without multiple prime divisors and let* $\gcd(n, 6h_K) = 1$. *Then, there exist infinitely many dihedral fields L with $L \supseteq K$, $[L : Q] = 2n$ and*

$$n \mid h_L.$$

Proof Let $n = p_1 \cdot \ldots \cdot p_r$ with pairwise different primes p_i. For $i = 1, \ldots, r$ we construct infinitely many ring class characters χ_{p_i} of order

p_i and composite conductor. Then, the extension L/K corresponding to $\mathbb{X} =< \chi_{p_1}, \dots, \chi_{p_r} >$ will have the desired properties.

First, for every p_i there are infinitely many primes q_i with

$$p_i \mid h_{q_i} \quad \text{and} \quad q_i \neq p_i,$$

where h_{q_i} denotes the ring class number modulo q_i. This is immediate by the formula for h_{q_i} in Theorem 3.1.9. Hence, there exists a ring class character χ modulo q_i of order p_i and

$$f_\chi = q_i$$

since, in view of $p_i \nmid h_K$, the conductor of χ cannot be 1. With two such ring class characters χ, χ' modulo q_i resp. modulo q_i' with different primes q_i, q_i' we set

$$\chi_i := \chi\chi'.$$

Then obviously

$$n_{\tilde{\chi}_i} = p_i \quad \text{and} \quad f_{\tilde{\chi}_i} = q_i q_i'.$$

Now let $\mathbb{X} =< \chi_{p_1}, \dots, \chi_{p_r} >$ be one of these infinitely many character groups and L the associated extension of K. Then L is Galois over $\mathbb{Q}$ as a subfield of a ring class field and the Galois group of $L/\mathbb{Q}$ is the semi-direct product of the Galois group G of L/K and $< \tau >$, with the complex conjugation τ:

$$\sigma\tau = \tau\sigma^{-1} \text{ for all } \sigma \in G.$$

Further, G must be cyclic since $|G| = n$ has no multiple prime factors, so L is a dihedral field. From the class number formula (12.50) we can deduce the divisibility

$$n \mid c_1 c_2 \frac{h_L}{h_K},$$

which implies that $n|h_L$ because c_1 and c_2 are integers relatively prime to n. This is because they only have prime divisors dividing $2 \cdot 3 \cdot h_K \cdot q_i \cdot q_i'$.

□

12.5 Group-theoretical lemmas for $M \not\supseteq K$

Let $\mathfrak{K}$ be a finite abelian group and $\mathfrak{U}$ a subgroup of $\mathfrak{K}$ of index

$$[\mathfrak{K} : \mathfrak{U}] = n.$$

Further, let $\mathfrak{k} \mapsto \overline{\mathfrak{k}}$ be an automorphism on $\mathfrak{K}$ with

$$\overline{\overline{\mathfrak{k}}} = \mathfrak{k} \text{ for all } \mathfrak{k} \in \mathfrak{K} \quad \text{and} \quad \overline{\mathfrak{U}} = \mathfrak{U}. \tag{12.51}$$

We fix an element $\mathfrak{k}_0 \in \mathfrak{K}$ having the property

$$\mathfrak{k}_0 \overline{\mathfrak{k}}_0 \in \mathfrak{U}. \tag{12.52}$$

Then

$$\mathfrak{h}'\mathfrak{U} \sim \mathfrak{h}\mathfrak{U} \;:\Longleftrightarrow\; (\, \mathfrak{h}'\mathfrak{U} = \mathfrak{h}\mathfrak{U} \text{ or } \mathfrak{h}'\mathfrak{U} = \mathfrak{k}_0 \overline{\mathfrak{h}}\mathfrak{U} \,) \tag{12.53}$$

defines an equivalence relation on $\mathfrak{K}/\mathfrak{U}$. In the following, let $\mathfrak{H}$ denote a system of elements $\mathfrak{h} \in \mathfrak{K}$ such that the cosets $\mathfrak{h}\mathfrak{U}$ form a system of representatives with respect to (12.53). We assume $\mathfrak{H}$ to contain the neutral element $\mathfrak{e}$ of $\mathfrak{K}$, and we define

$$\mathfrak{H}' := \mathfrak{H} \setminus \{\mathfrak{e}\}$$

and

$$e(\mathfrak{h}) := \begin{cases} 2 & \text{if} \quad \mathfrak{h}\mathfrak{U} \neq \mathfrak{k}_0 \overline{\mathfrak{h}}\mathfrak{U}, \\ 1 & \text{if} \quad \mathfrak{h}\mathfrak{U} = \mathfrak{k}_0 \overline{\mathfrak{h}}\mathfrak{U}. \end{cases}$$

Then

$$\sum_{\mathfrak{h} \in \mathfrak{H}} e(\mathfrak{h}) = [\mathfrak{K} : \mathfrak{U}]. \tag{12.54}$$

Let $\mathbb{X}$ denote the subgroup of characters χ of $\mathfrak{K}$ with $\chi \mid \mathfrak{U} = 1$. Keeping in mind $\overline{\mathfrak{U}} = \mathfrak{U}$, it is clear that for $\chi \in \mathbb{X}$ we obtain a new character by defining

$$\chi^{\tau}(\mathfrak{k}) := \chi(\overline{\mathfrak{k}}).$$

We set

$$\mathbb{X}^{(0)} := \{\chi \in \mathbb{X} \mid \chi^{\tau} = \chi\},$$

$$\mathbb{X}^{(l)} := \{\chi \in \mathbb{X}^{(0)} \mid \chi(\mathfrak{k}_0) = l\}, \quad l = \pm 1.$$

Lemma 12.5.1

$$\mathbb{X}^{(0)} = \begin{cases} \mathbb{X}^{(1)} & \textit{if} \qquad \mathfrak{k}_0 \overline{\mathfrak{h}}\mathfrak{U} = \mathfrak{h}\mathfrak{U} \textit{ for some } \mathfrak{h} \in \mathfrak{K}, \\ \mathbb{X}^{(1)} \uplus \mathbb{X}^{(-1)} & \textit{otherwise}. \end{cases}$$

In the second case $|\mathbb{X}^{(1)}| = |\mathbb{X}^{(-1)}|$.

Proof First, let $\mathfrak{k}_0\bar{\mathfrak{h}}\mathfrak{U} = \mathfrak{h}\mathfrak{U}$, hence $\mathfrak{k}_0\mathfrak{U} = \bar{\mathfrak{h}}^{-1}\mathfrak{h}\mathfrak{U}$ for some $\mathfrak{h} \in \mathfrak{K}$. Then

$$\chi(\mathfrak{k}_0) = \chi(\bar{\mathfrak{h}}^{-1}\mathfrak{h}) = \chi(\bar{\mathfrak{h}})^{-1}\chi(\mathfrak{h}) = \chi(\mathfrak{h})^{-1}\chi(\mathfrak{h}) = 1$$

for $\chi \in \mathbb{X}^{(0)}$, so $\mathbb{X}^{(0)} = \mathbb{X}^{(1)}$, as asserted. To prove the remaining assertions, we consider the endomomorphism $g: \ \mathfrak{h} \mapsto \bar{\mathfrak{h}}^{-1}\mathfrak{h}$ of $\mathfrak{K}$. Then we can write

$$\mathbb{X}^{(0)} = \{\chi \in \mathbb{X} \mid (\chi \mid g(\mathfrak{K})\mathfrak{U}\ ) = 1\}. \tag{12.55}$$

If $\mathfrak{k}_0\bar{\mathfrak{h}}\mathfrak{U} \neq \mathfrak{h}\mathfrak{U}$ for all $\mathfrak{h} \in \mathfrak{K}$, then obviously $\mathfrak{k}_0 \notin g(\mathfrak{K})\mathfrak{U}$. Hence, there exists a $\chi_0 \in \mathbb{X}^{(0)}$ with $\chi_0(\mathfrak{k}_0) \neq 1$. On the other hand

$$\chi_0(\mathfrak{k}_0)^2 = \chi(\mathfrak{k}_0)\chi^{\tau}(\mathfrak{k}_0) = \chi_0(\mathfrak{k}_0\overline{\mathfrak{k}_0}) = 1.$$

Thus, $\chi_0(\mathfrak{k}_0) = -1$, and it follows that

$$\mathbb{X}^{(0)} = \mathbb{X}^{(1)} \uplus \mathbb{X}^{(-1)} \quad \text{and} \quad \mathbb{X}^{(-1)} = \chi_0\mathbb{X}^{(1)},$$

which implies the remaining assertions of the lemma. □

On $\mathbb{X} \setminus (\mathbb{X}^{(-1)} \cup \{1\})$ we consider the equivalence relation

$$\chi' \sim \chi \ :\Longleftrightarrow \quad (\ \chi' = \chi \ \text{ or } \ \chi' = \chi^{\tau}\ ),$$

and we let $\mathcal{H}$ be a system of representatives for $\sim$. Then:

Lemma 12.5.2 $$|\mathfrak{H}'| = |\mathcal{H}|.$$

Proof We have $|\mathbb{X}| = n$, and we set $n_0 := |\mathbb{X}^{(0)}|$. First, we consider the case of $\mathfrak{k}_0\bar{\mathfrak{h}}\mathfrak{U} \neq \mathfrak{h}\mathfrak{U}$ for all $\mathfrak{h} \in \mathfrak{K}$. Then $e(\mathfrak{h}) = 2$ for all $\mathfrak{h} \in \mathfrak{K}$. In view of (12.54) it follows that

$$|\mathfrak{H}'| = \frac{n}{2} - 1$$

and further, by Lemma 12.5.1

$$|\mathcal{H}| = \frac{n - n_0}{2} + \frac{n_0}{2} - 1 = \frac{n}{2} - 1.$$

If $\mathfrak{k}_0\bar{\mathfrak{h}}\mathfrak{U} = \mathfrak{h}\mathfrak{U}$ for a $\mathfrak{h} \in \mathfrak{K}$, then $\mathfrak{k}_0 \in (\ker g)\mathfrak{U}$ and, in view of (12.55), this implies that

$$\begin{aligned} &|\{\mathfrak{h} \in \mathfrak{H} \mid e(\mathfrak{h}) = 1\}| = |\{\mathfrak{h}\mathfrak{U} \in \mathfrak{K}/\mathfrak{U} \mid g(\mathfrak{h})\mathfrak{U} = \mathfrak{k}_0\mathfrak{U}\}| \\ &= |\{\mathfrak{h}\mathfrak{U} \in \mathfrak{K}/\mathfrak{U} \mid g(\mathfrak{h})\mathfrak{U} = \mathfrak{U}\}| \\ &= [(\ker g\ \mathfrak{U}) : \mathfrak{U}] = [\mathfrak{K} : g(\mathfrak{K})\mathfrak{U}] = |\mathbb{X}^{(0)}| = n_0. \end{aligned}$$

Now we have

$$|\mathfrak{H}'| = |\mathfrak{H}| - 1 = \frac{n - n_0}{2} + n_0 - 1 = \frac{n + n_0}{2} - 1,$$

and by Lemma 12.5.1 we obtain

$$|\mathcal{H}| = \frac{n - n_0}{2} + n_0 - 1 = \frac{n + n_0}{2} - 1.$$

□

Similarly to section 12.1 we consider a map

$$u : \mathfrak{K} \to \mathbb{R}$$

and a number $\delta \in \mathbb{R}$ satisfying the following conditions:

$$u(\mathfrak{k}) \text{ only depends on } \mathfrak{k}\mathfrak{U}, \tag{12.56}$$

$$\sum_{\mathfrak{h} \bmod \mathfrak{U}} \big(u(\mathfrak{k}\mathfrak{h}) - \delta u(\mathfrak{h})\big) = 0 \quad \text{for all } \mathfrak{k} \in \mathfrak{K}. \tag{12.57}$$

For $\chi \in \mathbb{X}$ we set:

$$A(\chi) := \sum_{\mathfrak{k} \bmod \mathfrak{U}} \chi(\mathfrak{k}) u(\mathfrak{k})$$

and we assume that

$$A(\chi^{\tau}) = A(\chi) \quad \text{for all } \chi \in \mathbb{X}. \tag{12.58}$$

We define the quadratic matrix by

$$V := \Big(e(\mathfrak{h})(v(\mathfrak{k}, \mathfrak{h}, \delta) + v(\mathfrak{k}, \mathfrak{k}_0\overline{\mathfrak{h}}, \delta))\Big) \qquad \begin{array}{l} \text{row index: } \mathfrak{k} \in \mathfrak{H}', \\ \text{column index: } \mathfrak{h} \in \mathfrak{H}'. \end{array} \tag{12.59}$$

with

$$v(\mathfrak{k}, \mathfrak{h}, \delta) = u(\mathfrak{k}\mathfrak{h}^{-1}) - \delta u(\mathfrak{h}^{-1}).$$

Theorem 12.5.3 *The matrix V in (12.59) has the determinant*

$$|\det(V)| = 2^{|\mathcal{H}|} \left| \prod_{\chi \in \mathcal{H}} A(\chi) \right|.$$

Proof We multiply V by

$$Y = \Big(e(\mathfrak{k})(\ \chi(\mathfrak{k}) + \chi(\mathfrak{k}_0\overline{\mathfrak{k}})\ )\Big) \qquad \begin{array}{l}\text{row index: } \chi \in \mathcal{H},\\ \text{column index: } \mathfrak{k} \in \mathfrak{H}'.\end{array}$$

Then the elements in YV are

$$\xi_{\chi,\mathfrak{h}} = \sum_{\mathfrak{k}\in\mathfrak{H}'} e(\mathfrak{k})(\ \chi(\mathfrak{k}) + \chi(\mathfrak{k}_0\overline{\mathfrak{k}})\ )(e(\mathfrak{h})(v(\mathfrak{k},\mathfrak{h},\delta) + v(\mathfrak{k},\mathfrak{k}_0\overline{\mathfrak{h}},\delta))). \quad (12.60)$$

Herein, the summation can be extended over a full system $\mathfrak{H}$ because

$$v(\mathfrak{k},\mathfrak{h},\delta) + v(\mathfrak{k},\mathfrak{k}_0\overline{\mathfrak{h}},\delta) = (u(\mathfrak{k}\mathfrak{h}^{-1}) - \delta u(\mathfrak{h}^{-1})) + (u(\mathfrak{k}_0\overline{\mathfrak{k}}\mathfrak{h}^{-1}) - \delta u(\mathfrak{k}_0\mathfrak{h}^{-1}))$$

vanishes for $\mathfrak{k} = \mathfrak{e}$. Further, keeping in mind the relation

$$\sum_{\mathfrak{k}\in\mathfrak{H}} e(\mathfrak{k})(\ \chi(\mathfrak{k}) + \chi(\mathfrak{k}_0\overline{\mathfrak{k}})\ ) = 2 \sum_{\mathfrak{k} \bmod \mathfrak{U}} \chi(\mathfrak{k}) = 0, \quad (12.61)$$

we can omit in (12.60) the summands not depending on $\mathfrak{k}$. Then, observing (12.52) and (12.58), we obtain

$$\begin{aligned}\xi_{\chi,\mathfrak{h}} &= \sum_{\mathfrak{k}\in\mathfrak{H}} e(\mathfrak{h})e(\mathfrak{k})[(\chi(\mathfrak{k})u(\mathfrak{k}\mathfrak{h}^{-1}) + \chi(\mathfrak{k}_0\overline{\mathfrak{k}})u(\mathfrak{k}_0\overline{\mathfrak{k}}\mathfrak{h}^{-1}))\\ &\qquad + (\chi(\mathfrak{k})u(\mathfrak{k}_0\overline{\mathfrak{k}}\mathfrak{h}^{-1}) + \chi(\mathfrak{k}_0\overline{\mathfrak{k}})u(\mathfrak{k}\mathfrak{h}^{-1}))]\\ &= \sum_{\mathfrak{k}\in\mathfrak{H}} e(\mathfrak{h})e(\mathfrak{k})[(\chi(\mathfrak{k})u(\mathfrak{k}\mathfrak{h}^{-1}) + \chi(\mathfrak{k}_0\overline{\mathfrak{k}})u(\mathfrak{k}_0\overline{\mathfrak{k}}\mathfrak{h}^{-1}))\\ &\qquad + (\chi(\mathfrak{k})u(\mathfrak{k}_0\overline{\mathfrak{k}}\mathfrak{h}^{-1}) + \chi(\mathfrak{k}_0\overline{\mathfrak{k}})u(\mathfrak{k}_0\overline{(\mathfrak{k}_0\overline{\mathfrak{k}})}\mathfrak{h}^{-1}))]\\ &= \sum_{\mathfrak{k} \bmod \mathfrak{U}} 2e(\mathfrak{h})[\chi(\mathfrak{k})u(\mathfrak{k}\mathfrak{h}^{-1}) + \chi(\mathfrak{k})u(\mathfrak{k}_0\overline{\mathfrak{k}}\mathfrak{h}^{-1})]\\ &= 2e(\mathfrak{h})[\chi(\mathfrak{h})A(\chi) + \chi(\mathfrak{k}_0\overline{\mathfrak{h}})A(\chi^{\tau})]\\ &= 2e(\mathfrak{h})(\chi(\mathfrak{h}) + \chi(\mathfrak{k}_0\overline{\mathfrak{h}}))A(\chi).\end{aligned}$$

This implies that

$$|\det(YV)| = 2^{|\mathcal{H}|}\left|\det(Y)\prod_{\chi\in\mathcal{H}} A(\chi)\right|. \quad (12.62)$$

Hence, to prove Theorem 12.5.3, we have to show that Y has a non-zero determinant. Therefore, we compute YY' with

$$Y' = \left(\overline{\psi}(\mathfrak{k}) + \overline{\psi}(\mathfrak{k}_0\overline{\mathfrak{k}}) \right) \qquad \begin{array}{l}\text{row index: } \mathfrak{k} \in \mathfrak{H}',\\ \text{column index: } \psi \in \mathcal{H}.\end{array}$$

The elements in YY' are

$$\left\{ \sum_{\mathfrak{k}\in\mathfrak{H}} e(\mathfrak{k}) \left[((\chi\overline{\psi})(\mathfrak{k}) + (\chi\overline{\psi})(\mathfrak{k}_0\overline{\mathfrak{k}})) \;+\; \overline{\psi}(\mathfrak{k}_0)((\chi\overline{\psi}^{\tau})(\mathfrak{k}) + (\chi\overline{\psi}^{\tau})(\mathfrak{k}_0\overline{\mathfrak{k}}))\right] \right\}$$
$$-e(\mathfrak{e})(1+\chi(\mathfrak{k}_0))\overline{(1+\psi(\mathfrak{k}_0))}$$

$$= \left\{ 2 \sum_{\mathfrak{k} \bmod \mathfrak{U}} \left[(\chi\overline{\psi})(\mathfrak{k}) \;+\; \overline{\psi}(\mathfrak{k}_0)(\chi\overline{\psi}^{\tau})(\mathfrak{k}) \right] \right\}$$
$$-e(\mathfrak{e})(1+\chi(\mathfrak{k}_0))\overline{(1+\psi(\mathfrak{k}_0))}$$

$$= \begin{cases} 2n(1+\overline{\chi}(\mathfrak{k}_0)) - |1+\chi(\mathfrak{k}_0)|^2 e(\mathfrak{e}) & \text{if} \quad \chi = \psi \neq \psi^{\tau}, \\ 2n - |1+\underline{\chi(\mathfrak{k}_0)|^2 e(\mathfrak{e})} & \text{if} \quad \chi = \psi \neq \psi^{\tau}, \\ -(1+\chi(\mathfrak{k}_0))\overline{(1+\psi(\mathfrak{k}_0))} e(\mathfrak{e}) & \text{if} \quad \chi \neq \psi. \end{cases}$$

$$= \alpha_\chi \beta_\psi + \delta_{\chi,\psi}\gamma_\chi$$

with

$$\alpha_\chi = -(1+\chi(\mathfrak{k}_0)), \quad \beta_\psi = \overline{(1+\psi(\mathfrak{k}_0))} e(\mathfrak{e}),$$
$$\gamma_\chi = \begin{cases} 2n(1+\chi(\mathfrak{k}_0)) & \text{if} \quad \chi = \chi^{\tau}, \\ 2n & \text{if} \quad \chi \neq \chi^{\tau}. \end{cases}$$

Using Lemma 12.1.6 we now obtain

$$|\det(YY')|$$
$$= (2n)^{|\mathcal{H}|} \left| \prod_{\substack{\chi\in\mathcal{H}\\ \chi=\chi^{\tau}}} (1+\chi(\mathfrak{k}_0)) \right| \times \tag{12.63}$$
$$\times \left| 1 - \sum_{\substack{\chi\in\mathcal{H}\\ \chi=\chi^{\tau}}} \frac{(1+\overline{\chi}(\mathfrak{k}_0))e(\mathfrak{e})}{2n} - \sum_{\substack{\chi\in\mathcal{H}\\ \chi\neq\chi^{\tau}}} \frac{|1+\chi(\mathfrak{k}_0)|^2 e(\mathfrak{e})}{2n} \right|.$$

For $\chi = \chi^\tau$ we have $\chi(\mathfrak{k}_0) = 1$ by definition of $\mathcal{H}$. Therefore (12.63) becomes

$$2^{|\mathcal{H}|+|\{\chi\in\mathcal{H}|\chi^\tau=\chi\}|} n^{|\mathcal{H}|} \left| \left\{ 1 - \sum_{\substack{\chi\in\mathcal{H}\\ \chi=\chi^\tau}} \frac{e(\mathfrak{e})}{n} - \sum_{\substack{\chi\in\mathcal{H}\\ \chi\neq\chi^\tau}} \frac{|1+\chi(\mathfrak{k}_0)|^2 e(\mathfrak{e})}{2n} \right\} \right|.$$

Herein, the expression in curly braces can be expressed in the following way

$$\begin{aligned}\{\ldots\} &= 1 - \frac{e(\mathfrak{e})}{4n}\left[\sum_{\substack{\chi\in\mathcal{H}\\ \chi=\chi^\tau}} 4 - \sum_{\substack{\chi\in\mathcal{H}\\ \chi\neq\chi^\tau}} 2|1+\chi(\mathfrak{k}_0)|^2\right] \\ &= 1 - \frac{e(\mathfrak{e})}{4n}\left[\left(\sum_{\chi\in\mathbb{X}} |1+\chi(\mathfrak{k}_0)|^2\right) - 4\right].\end{aligned}$$

Applying the relation

$$\sum_{g\in G} |1+\lambda(g)|^2 = \sum_{g\in G} (2+\lambda(g)+\lambda(g^{-1})) = \begin{cases} 2|G| & \text{if} \quad \lambda\neq 1, \\ 4|G| & \text{if} \quad \lambda = 1 \end{cases}$$

holding for any character λ of a finite abelian group G to $\lambda = (\chi \mapsto \chi(\mathfrak{k}_0))$, we find that

$$\{\ldots\} = \frac{e(\mathfrak{e})}{n}.$$

Thus

$$|\det(YY')| = 2^{|\mathcal{H}|+|\{\chi\in\mathcal{H}|\chi^\tau=\chi\}|} n^{|\mathcal{H}|-1} e(\mathfrak{e}).$$

In particular, $|\det(Y)| \neq 0$. More precisely, we have

$$|\det(Y)| = 2^{\frac{1}{2}(|\mathcal{H}|+|\{\chi\in\mathcal{H}|\chi^\tau=\chi\}|-|\{\mathfrak{h}\in\mathfrak{H}|e(\mathfrak{h})=2\}|)} n^{\frac{1}{2}(|\mathcal{H}|-1)} e(\mathfrak{e}). \quad (12.64)$$

□

Later, for transformations of class number formulae, we will need, as in the first part of this chapter, a more complicated analogue of Theorem 12.5.3. Therefore, given a character $\chi \in \mathbb{X}$, we consider the subgroup

$$\mathfrak{U}_{\tilde{\chi}} := \{\mathfrak{k} \in \mathfrak{K} \mid \chi(\mathfrak{k}) = 1\}$$

depending only on the Frobenius class $\tilde{\chi}$. We associate with χ a map

$$u_{\tilde{\chi}} : \mathfrak{K} \to \mathbb{R}$$

and a $\delta_{\tilde{\chi}} \in \mathbb{R}$, assuming the following properties to be given:

$$u_{\tilde{\chi}}(\mathfrak{k}) \text{ only depends on } \mathfrak{k}\mathfrak{U}_{\tilde{\chi}}, \tag{12.65}$$

$$\sum_{\mathfrak{h} \bmod \mathfrak{U}} \left(u_{\tilde{\chi}}(\mathfrak{k}\mathfrak{h}) - \delta_{\tilde{\chi}} u_{\tilde{\chi}}(\mathfrak{h})\right) = 0 \text{ for all } \mathfrak{k} \in \mathfrak{K}. \tag{12.66}$$

For $\chi \in \mathbb{X}$ we define:

$$A(\chi) := \sum_{\mathfrak{k} \bmod \mathfrak{U}_{\tilde{\chi}}} \chi(\mathfrak{k}) u_{\tilde{\chi}}(\mathfrak{k}),$$

and we assume that

$$A(\chi) = A(\chi^{\tau}) \tag{12.67}$$

for all $\chi \in \mathbb{X}$. Further, the above system $\mathcal{H}$ is chosen with the following properties:

$$\begin{aligned} &(\tilde{\chi} \subseteq \mathcal{H} \text{ and } \tilde{\chi}^{\tau} \cap \mathcal{H} = \emptyset) \text{ or } (\tilde{\chi}^{\tau} \subseteq \mathcal{H} && \text{and } \tilde{\chi} \cap \mathcal{H} = \emptyset) \\ & && \text{if } \tilde{\chi}^{\tau} \neq \tilde{\chi}, \\ &(\mathcal{H} \cap \tilde{\chi}) \uplus (\mathcal{H} \cap \tilde{\chi})^{\tau} = \tilde{\chi} && \text{if } \chi^{\tau} \neq \chi,\ \tilde{\chi}^{\tau} = \tilde{\chi}. \end{aligned} \tag{12.68}$$

It is easy to see that such a choice of $\mathcal{H}$ is always possible. Further, to every class $\tilde{\chi}$ we associate a real matrix $\Lambda_{\tilde{\chi}}$,

$$\begin{aligned} &\Lambda_{\tilde{\chi}} = \left(\lambda_{i_{\tilde{\chi}},\mathfrak{k}}(\tilde{\chi})\right), \\ &\text{row index: } i_{\tilde{\chi}} = 1, \ldots, |\mathcal{H} \cap \tilde{\chi}|, \\ &\text{column index: } \mathfrak{k} \bmod \mathfrak{U}_{\tilde{\chi}}. \end{aligned} \tag{12.69}$$

Then, we define the quadratic matrix

$$\begin{aligned} V = &\left(\sum_{\mathfrak{k} \bmod \mathfrak{U}_{\tilde{\chi}}} \lambda_{i_{\tilde{\chi}},\mathfrak{k}}(\tilde{\chi}) e(\mathfrak{h}) \left[v_{\tilde{\chi}}(\mathfrak{k},\mathfrak{h}) + v_{\tilde{\chi}}(\mathfrak{k},\mathfrak{k}_0\overline{\mathfrak{h}})\right]\right), \\ &\text{row index: } (\tilde{\chi}, i_{\tilde{\chi}}),\ \tilde{\chi} \text{ runs through all classes with } \chi \in \mathcal{H} \\ &\text{and } i_{\tilde{\chi}} \text{ the numbers } 1, \ldots, |\mathcal{H} \cap \tilde{\chi}|, \\ &\text{column index: } \mathfrak{h} \text{ runs though } \mathfrak{H}', \end{aligned} \tag{12.70}$$

where, for abbreviation

$$v_{\tilde{\chi}}(\mathfrak{k},\mathfrak{h}) := u_{\tilde{\chi}}(\mathfrak{k}\mathfrak{h}^{-1}) - \delta_{\tilde{\chi}} u_{\tilde{\chi}}(\mathfrak{h}^{-1}).$$

Theorem 12.5.4 *We assume (12.65) – (12.68). Then*

$$|\det(V)| = d(\mathfrak{K}/\mathfrak{U}, \mathfrak{k}_0, \Lambda) \left| \prod_{\chi \in \mathcal{H}} A(\chi) \right|,$$

$$d(\mathfrak{K}/\mathfrak{U}, \mathfrak{k}_0, \Lambda) = n^{-\frac{|\mathcal{H}|+1}{2}} 2^{\frac{a}{2}} \prod_{\substack{\tilde{\chi} \\ \chi \in \mathcal{H}}} |\det(l(\tilde{\chi}, \Lambda_{\tilde{\chi}}, \delta_{\tilde{\chi}}))| \left(\frac{n}{n_{\tilde{\chi}}}\right)^{|\mathcal{H} \cap \tilde{\chi}|}$$

with $a = |\mathcal{H}| - |\{\chi \in \mathcal{H} | \chi^\tau = \chi\}| + |\{\mathfrak{h} \in \mathfrak{H} | e(\mathfrak{h}) = 2\}|$ *and* $l(\tilde{\chi}, \Lambda_{\tilde{\chi}}, \delta_{\tilde{\chi}}) = (l_{i_{\tilde{\chi}},\mu}),\ l_{i_{\tilde{\chi}},\mu} =$

$$\begin{cases} \sum_{\mathfrak{k} \bmod \mathfrak{U}_{\tilde{\chi}}} \lambda_{i_{\tilde{\chi}},\mathfrak{k}}(\tilde{\chi})(\chi^\mu(\mathfrak{k}) - \delta_{\tilde{\chi}}) & \text{if} \quad \tilde{\chi}^\tau \neq \tilde{\chi}, \\ \sum_{\mathfrak{k} \bmod \mathfrak{U}_{\tilde{\chi}}} \lambda_{i_{\tilde{\chi}},\mathfrak{k}}(\tilde{\chi})\{(\chi^\mu(\mathfrak{k}) - \delta_{\tilde{\chi}}) + \chi^\mu(\mathfrak{k}_0)((\chi^\tau)^\mu(\mathfrak{k}) - \delta_{\tilde{\chi}})\} & \\ & \text{if} \quad \tilde{\chi}^\tau = \tilde{\chi} \end{cases}$$

row index: $i_{\tilde{\chi}} = 1, \ldots, |\mathcal{H} \cap \tilde{\chi}|$,
column index: μ *runs through a system modulo* $n_{\tilde{\chi}}$*, so that* χ^μ *are the different characters in* $\mathcal{H} \cap \tilde{\chi}$.

Proof Due to the assumptions (12.66) and (12.54), Lemma 12.1.5 implies that

$$|\det(V)| = \tfrac{e(\mathfrak{e})}{n} |\det(V_0)| \quad \text{with} \quad V_0 =$$

$$\left(\sum_{\mathfrak{k} \bmod \mathfrak{U}_{\tilde{\chi}}} \lambda_{i_{\tilde{\chi}},\mathfrak{k}}(\tilde{\chi}) e(\mathfrak{h}) \left\{ \left[v_{\tilde{\chi}}(\mathfrak{k}, \mathfrak{h}) + v_{\tilde{\chi}}(\mathfrak{k}, \mathfrak{k}_0 \overline{\mathfrak{h}}) \right] - \left[v_{\tilde{\chi}}(\mathfrak{k}, \mathfrak{e}) + v_{\tilde{\chi}}(\mathfrak{k}, \mathfrak{k}_0) \right] \right\} \right), \tag{12.71}$$

row index: $(\tilde{\chi}, i_{\tilde{\chi}})$, column index: $\mathfrak{h}$.

Multiplying V_0 from the right by

$$Y_0 = \left(\psi(\mathfrak{h}) + \psi(\mathfrak{k}_0 \overline{\mathfrak{h}}) \right),$$

row index: $\mathfrak{h} \in \mathfrak{H}'$,
column index: $\psi \in \mathcal{H}$,

we obtain

$$V_0Y_0 = \left(\xi_{(\tilde{\chi},i_{\tilde{\chi}}),\psi}\right) \text{ with } \xi_{(\tilde{\chi},i_{\tilde{\chi}}),\psi} =$$

$$\sum_{\mathfrak{h}\in\mathfrak{H}'}\ \sum_{\mathfrak{k} \bmod \mathfrak{U}_{\tilde{\chi}}} \lambda_{i_{\tilde{\chi}},\mathfrak{k}}(\tilde{\chi})e(\mathfrak{h})(\psi(\mathfrak{h})+\psi(\mathfrak{k}_0\overline{\mathfrak{h}}))\left\{[v_{\tilde{\chi}}(\mathfrak{k},\mathfrak{h})+v_{\tilde{\chi}}(\mathfrak{k},\mathfrak{k}_0\overline{\mathfrak{h}})]\right.$$
$$\left. - [v_{\tilde{\chi}}(\mathfrak{k},\mathfrak{e})+v_{\tilde{\chi}}(\mathfrak{k},\mathfrak{k}_0)]\right\}.$$

Since herein for $\mathfrak{h} = \mathfrak{e}$ the summand vanishes, the sumation can be extended over $\mathfrak{h} \in \mathfrak{H} = \mathfrak{H}' \cup \{\mathfrak{e}\}$ and, keeping in mind the relation (12.61) and the definition of $v(.,.)$, we find that

$$\xi_{(\tilde{\chi},i_{\tilde{\chi}}),\psi} =$$
$$2\sum_{\mathfrak{k} \bmod \mathfrak{U}_{\tilde{\chi}}} \lambda_{i_{\tilde{\chi}},\mathfrak{k}}(\tilde{\chi})\left\{(\psi(\mathfrak{k})-\delta_{\tilde{\chi}})\sum_{\mathfrak{h} \bmod \mathfrak{U}} \overline{\psi}(\mathfrak{h})u_{\tilde{\chi}}(\mathfrak{h})\right.$$
$$\left. + \psi(\mathfrak{k}_0)(\psi^{\tau}(\mathfrak{k})-\delta_{\tilde{\chi}})\sum_{\mathfrak{h} \bmod \mathfrak{U}} \overline{\psi}^{\tau}(\mathfrak{h})u_{\tilde{\chi}}(\mathfrak{h})\right\}.$$

In view of Lemma 12.1.4 this shows that

$$\xi_{(\tilde{\chi},i_{\tilde{\chi}}),\psi} = 0 \quad \text{for} \quad \psi \notin <\chi> \cup <\chi^{\tau}>. \tag{12.72}$$

Further, as in the proof of Theorem 12.1.2, using (12.67) and Lemma 12.1.4, we obtain

$$\xi_{(\tilde{\chi},i_{\tilde{\chi}}),\psi} =$$
$$\begin{cases} \frac{2n}{n_{\tilde{\chi}}}\sum_{\mathfrak{k} \bmod \mathfrak{U}_{\tilde{\chi}}} \lambda_{i_{\tilde{\chi}},\mathfrak{k}}(\tilde{\chi})(\psi(\mathfrak{k})-\delta_{\tilde{\chi}})A(\psi) & \text{if } \tilde{\chi}=\tilde{\psi}\neq\tilde{\psi}^{\tau}, \\ \frac{2n}{n_{\tilde{\chi}}}\sum_{\mathfrak{k} \bmod \mathfrak{U}_{\tilde{\chi}}} \lambda_{i_{\tilde{\chi}},\mathfrak{k}}(\tilde{\chi})\left\{(\psi(\mathfrak{k})-\delta_{\tilde{\chi}})+\psi(\mathfrak{k}_0)(\psi(\mathfrak{k}) \right. & \left. -\delta_{\tilde{\chi}})\right\}A(\psi) \\ & \text{if } \tilde{\chi}=\tilde{\psi}=\tilde{\psi}^{\tau}. \end{cases}$$

Now arrange the rows of V_0Y_0 according to classes such that the implication

$$"\tilde{\chi}' \text{ preceeds } \tilde{\chi}" \Longrightarrow (<\chi'> \ \nsubseteq \ <\chi> \text{ and } n_{\tilde{\chi}'} \geq n_{\tilde{\chi}})$$

holds and collect the rows belonging to the same $\tilde{\chi}$. Accordingly, collect the rows with indices ψ in the same class and arrange them corresponding to the rows. Then, V_0Y_0 is a matrix having the quadratic matrices along the diagonal

$$V_{\tilde{\chi}} = \left(\xi_{(\tilde{\chi}, i_{\tilde{\chi}}), \chi^{\mu}}\right),$$

row index: $i_{\tilde{\chi}} = 1, \dots |\mathcal{H} \cap \tilde{\chi}|$,
column index: μ runs though a system modulo $n_{\tilde{\chi}}$, so that χ^{μ} are the different characters in $\mathcal{H} \cap \tilde{\chi}$.

Below these quadratic matrices $V_0 Y_0$ has only zeros. To prove that $V_0 Y_0$ has the asserted form, we have to show for $\psi, \chi \in \mathcal{H}$

$$"\tilde{\psi} \text{ preceeds } \tilde{\chi}" \Longrightarrow \psi \notin <\chi> \cup <\chi^{\tau}>,$$

keeping in mind (12.72). If "$\tilde{\psi}$ preceeds $\tilde{\chi}$", then $\psi \notin <\chi>$ and $n_{\tilde{\psi}} \geq n_{\tilde{\chi}}$. If ψ was in $<\chi^{\tau}>$, we would have $n_{\tilde{\psi}} \leq n_{\tilde{\chi}^{\tau}} = n_{\tilde{\chi}}$, hence $n_{\tilde{\psi}} = n_{\tilde{\chi}}$ and so $\psi \in \tilde{\chi}^{\tau}$. Since $\psi \notin <\chi>$ this would imply that $\tilde{\chi} \neq \tilde{\chi}^{\tau}$, and because $\chi \in \mathcal{H}$, we could conclude in view of property (12.68) of $\mathcal{H}$ that $\tilde{\chi}^{\tau} \cap \mathcal{H} = \emptyset$, which is impossible because $\psi \in \mathcal{H}$ and $\psi \in \tilde{\chi}^{\tau}$. Hence $\psi \notin <\chi^{\tau}>$ and $\psi \notin <\chi> \cup <\chi^{\tau}>$. The determinant of $V_0 Y_0$ is therefore equal to the product of the subdeterminants of the $V_{\tilde{\chi}}$, and, determining the determinant of Y_0 by (12.64), we can prove the formula of Theorem 12.5.4. □

For later applications we prove the following theorem about determinants $\det(l(\tilde{\chi}, \Lambda_{\tilde{\chi}}, \delta_{\tilde{\chi}}))$ in Theorem 12.5.4.

Theorem 12.5.5 *The elements of the matrices $\Lambda_{\tilde{\chi}}$ can be chosen in $\mathbb{Z}$ such that for $\delta_{\tilde{\chi}} = 0, 1$ the absolute value of the determinant $l_{\tilde{\chi}} := |\det(l(\tilde{\chi}, \Lambda_{\tilde{\chi}}, \delta_{\tilde{\chi}}))|$ in Theorem 12.5.4 satisfies the following divisibility properties:*

(i) $l_{\tilde{\chi}} \mid \sqrt{|d_{n_{\tilde{\chi}}}|}\ \Phi_{n_{\tilde{\chi}}}(1)$ *if* $\tilde{\chi} \neq \tilde{\chi}^{\tau}$,

(ii) $l_{\tilde{\chi}} \mid (4n_{\tilde{\chi}})^{\varphi(n_{\tilde{\chi}})/2} \sqrt[4]{|d_{n_{\tilde{\chi}}}|}\ \Phi_{n_{\tilde{\chi}}}(1)$ *if* $\tilde{\chi} = \tilde{\chi}^{\tau}, \chi \neq \chi^{\tau}, \chi(\mathfrak{k}_0) \neq -1$,

(iii) $l_{\tilde{\chi}} \mid (2n_{\tilde{\chi}})^{\varphi(n_{\tilde{\chi}})/2} \sqrt[4]{|d_{n_{\tilde{\chi}}}|}$ *if* $\tilde{\chi} = \tilde{\chi}^{\tau}, \chi \neq \chi^{\tau}, \chi(\mathfrak{k}_0) = -1$,

(iv) $l_{\tilde{\chi}} \mid (2n_{\tilde{\chi}})^{\varphi(n_{\tilde{\chi}})} \sqrt{|d_{n_{\tilde{\chi}}}|}\ \Phi_{n_{\tilde{\chi}}}(1)$ *if* $\chi = \chi^{\tau}, \chi(\mathfrak{k}_0) \neq -1$.

Herein, $d_{n_{\tilde{\chi}}}$ is the discriminant of the $n_{\tilde{\chi}}$-th cyclotomic field, $\Phi_{n_{\tilde{\chi}}}(X)$ the $n_{\tilde{\chi}}$-th cyclotomic polynomial and φ Euler's function.

Proof We consider the $\mathbb{Z}$-module $\mathfrak{v}$ generated by the numbers

$$\chi(\mathfrak{k}) - \delta_{\tilde{\chi}}, \mathfrak{k} \bmod \mathfrak{U}_{\tilde{\chi}}, \text{ resp. } (\chi(\mathfrak{k}) - \delta_{\tilde{\chi}}) + \chi(\mathfrak{k}_0)(\chi^{\tau}(\mathfrak{k}) - \delta_{\tilde{\chi}}), \mathfrak{k} \bmod \mathfrak{U}_{\tilde{\chi}},$$

and construct in $\mathfrak{v}$ a basis of an ideal in the $n_{\tilde{\chi}}$-th cyclotomic field $\mathbb{Q}_{n_{\tilde{\chi}}}$ or in a subfield of $\mathbb{Q}_{n_{\tilde{\chi}}}^{(0)}$ over which $\mathbb{Q}_{n_{\tilde{\chi}}}$ has degree 2. The above assertions are then obtained by computing the discriminant of this ideal, which makes it necessary to distinguish cases. Of course, for special cases even sharper divisibility properties can be proved. First, we consider the case $\delta_{\tilde{\chi}} = 1$.

(i) For $\chi \in \mathcal{H}$ with $\tilde{\chi} \neq \tilde{\chi}^{\tau}$ we know by (12.68) that $\tilde{\chi} \cap \mathcal{H} = \tilde{\chi}$. Hence, the matrix $l(\tilde{\chi}, \Lambda_{\tilde{\chi}}, \delta_{\tilde{\chi}})$ is equal to the matrix $g(\tilde{\chi}, \Lambda_{\tilde{\chi}}, \delta_{\tilde{\chi}})$ in Theorem 12.1.3, so the assertion in (i) follows from Theorem 12.1.3.

(ii) For $\tilde{\chi} = \tilde{\chi}^{\tau}$, $\chi \neq \chi^{\tau}$, we have $\chi^{\tau} = \chi^{\nu_0}$ with $\nu_0 \in \mathbb{Z}$, and, keeping in mind $(\chi^{\tau})^{\tau} = \chi$, it follows that:

$$\nu_0^2 \equiv 1 \bmod n_{\tilde{\chi}}, \quad \nu_0 \not\equiv 1 \bmod n_{\tilde{\chi}}. \tag{12.73}$$

In this case $n_{\tilde{\chi}} \neq 2$, and if ζ denotes a primitive $n_{\tilde{\chi}}$-th root of unity, then $\sigma_{\nu_0} = (\zeta \mapsto \zeta^{\nu_0})$ defines a quadratic automorphism of $\mathbb{Q}_{n_{\tilde{\chi}}}$. Further,

$$\mathfrak{v} = \langle\ \{(\zeta^{\nu} - 1) + \zeta_0(\zeta^{\nu\nu_0} - 1) \mid \nu \in \mathbb{Z}\}\ \rangle \quad \text{with} \quad \zeta_0 = \chi(\mathfrak{k}_0),$$

and with

$$\xi_{\nu} := (\zeta^{\nu} - 1) + \zeta_0(\zeta^{\nu\nu_0} - 1),$$

we have

$$(1 + \zeta_0)[(\zeta^{\nu} - 1) + (\zeta^{\nu\nu_0} - 1)] = \xi_{\nu} + \xi_{\nu\nu_0} \in \mathfrak{v}.$$

Therefore, we consider the $\mathbb{Z}$-module

$$\mathfrak{v}_0 := \langle\ \{(\zeta^{\nu} - 1) + (\zeta^{\nu\nu_0} - 1) \mid \nu \in \mathbb{Z}\}\ \rangle$$

and the inclusion

$$2(\mathfrak{a} \cap \mathfrak{O}_{n_{\tilde{\chi}}}^{(0)}) \subseteq \mathfrak{v}_0,$$

where $\mathfrak{O}_{n_{\tilde{\chi}}}^{(0)}$ denotes the ring of integers in the fixed field of σ_{ν_0} and $\mathfrak{a} = (\zeta - 1)\mathfrak{O}_{n_{\tilde{\chi}}}$ the ideal generated by $(\zeta - 1)$ in $\mathbb{Q}_{n_{\tilde{\chi}}}$. Hence, there exist $\lambda_{i,\nu} \in \mathbb{Z}$ such that

$$\rho_i := \sum_{\nu \bmod n_{\tilde{\chi}}} \lambda_{i,\nu} \left[(\zeta^{\nu} - 1) + \zeta_0(\zeta^{\nu\nu_0} - 1)\right], \quad i = 1, \ldots, \frac{\varphi(n_{\tilde{\chi}})}{2},$$

is a $\mathbb{Z}$-basis of $(1+\zeta_0)2(\mathfrak{a}\cap\mathfrak{O}^{(0)}_{n_{\tilde{\chi}}})$. Let T be a subsystem of residues modulo $n_{\tilde{\chi}}$ with

$$\tilde{\chi}\cap\mathcal{H}=\{\chi^{\mu}\mid\mu\in T\}.$$

Then the property (12.68) of $\mathcal{H}$ implies the automorphisms $\sigma_\mu=(\zeta\mapsto\zeta^\mu)$, $\mu\in T$, to induce the different isomorphisms on the fixed field of σ_{ν_0}, so we have

$$discr\left(2(\mathfrak{a}\cap\mathfrak{O}^{(0)}_{n_{\tilde{\chi}}})\right)=\left(\det\left(\left(\frac{\rho_i}{1+\zeta_0}\right)^{\sigma_\mu}\right)_{1=1,\dots,\varphi(n_{\tilde{\chi}})/2;\ \mu\in T}\right)^2,$$

and this implies that

$$\left|\det\left(\rho_i^{\sigma_\mu}\right)\right|=\left|\prod_{\mu\in T}(1+\zeta_0)\right|2^{\varphi(n_{\tilde{\chi}})/2}\sqrt{\left|discr\left(\mathfrak{a}\cap\mathfrak{O}^{(0)}_{n_{\tilde{\chi}}}\right)\right|}. \quad (12.74)$$

By assumption $\zeta_0\neq-1$, hence $-\zeta_0$ is a root of unity of order n', $1<n'|2n_{\tilde{\chi}}$. This shows that

$$\prod_{\mu\in T}(1+\zeta_0)\,\Big|\,2^{\varphi(n_{\tilde{\chi}})/2}. \quad (12.75)$$

Further, we find that

$$\begin{aligned}&\sqrt{\left|discr\left(\mathfrak{a}\cap\mathfrak{O}^{(0)}_{n_{\tilde{\chi}}}\right)\right|}\\&=\mathbf{N}_{\mathbb{Q}^{(0)}_{n_{\tilde{\chi}}}/\mathbb{Q}}\left(\mathfrak{a}\cap\mathfrak{O}^{(0)}_{n_{\tilde{\chi}}}\right)\sqrt{\left|discr\left(\mathfrak{O}^{(0)}_{n_{\tilde{\chi}}}\right)\right|}\ \Big|\ \mathbf{N}_{\mathbb{Q}^{(0)}_{n_{\tilde{\chi}}}/\mathbb{Q}}(\mathfrak{a})\sqrt[4]{|d_{n_{\tilde{\chi}}}|} \quad (12.76)\\&=\Phi_{n_{\tilde{\chi}}}(1)\sqrt[4]{|d_{n_{\tilde{\chi}}}|},\end{aligned}$$

keeping in mind that by the tower formula the square of the discriminant of $\mathbb{Q}^{(0)}_{n_{\tilde{\chi}}}$ is a divisor of the discriminant of $\mathbb{Q}_{n_{\tilde{\chi}}}$. Now, choosing $\lambda_{i_{\tilde{\chi}},\mathfrak{k}}(\tilde{\chi}):=\lambda_{i_{\tilde{\chi}},\nu}$ if $\chi(\mathfrak{k})=\zeta^\nu$, then $l_{\tilde{\chi}}$ equals the left-hand side in (12.74), and the divisibility in (ii) follows from (12.75) and (12.76).

(iii) For the proof of (iii) we can conclude as for (ii). First, $\chi^\tau=\chi^{\nu_0}$ with a ν_0 satisfying (12.73) and $n_{\tilde{\chi}}\neq 2$. Further,

$$\mathfrak{v}=\langle\ \{\zeta^\nu-\zeta^{\nu\nu_0}\mid\nu\in\mathbb{Z}\}\ \rangle.$$

In view of

$$(\zeta-\zeta^{\nu_0})(\zeta^\nu+\zeta^{\nu\nu_0})(\zeta^{\nu+\nu_0}-\zeta^{\nu_0(\nu+\nu_0)})-(\zeta^{\nu+1}-\zeta^{\nu_0(\nu+1)})\in\mathfrak{v},$$

we consider the module

$$\mathfrak{v}_0 = \langle\ \{\zeta^\nu + \zeta^{\nu\nu_0} \mid \nu \in \mathbb{Z}\}\ \rangle,$$

which obviously ($\nu = 0$) satisfies

$$2\mathfrak{O}^{(0)}_{n_{\tilde{\chi}}} \subseteq \mathfrak{v}_0.$$

Hence, there exists $\lambda_{i_{\tilde{\chi}},\nu} \in \mathbb{Z}$, so that

$$\rho_i := \sum_{\nu \bmod n_{\tilde{\chi}}} \lambda_{i,\nu}(\zeta^\nu - \zeta^{\nu\nu_0}), \quad i = 1, \ldots, \frac{\varphi(n_{\tilde{\chi}})}{2},$$

is a basis of $(\zeta^{\nu_0} - \zeta)2\mathfrak{O}^{(0)}_{n_{\tilde{\chi}}}$. As in case (ii) this implies, by setting $\lambda_{i_{\tilde{\chi}},\mathfrak{k}}(\tilde{\chi}) = \lambda_{i_{\tilde{\chi}},\nu}$ for $\chi(\mathfrak{k}) = \zeta^\nu$, that:

$$l_{\tilde{\chi}} = \left|\det\left(\rho_i^{\sigma_\mu}\right)\right| = \left|\prod_{\mu\in T}(\zeta^{\nu_0\mu} - \zeta^\mu)\right| 2^{\varphi(n_{\tilde{\chi}})/2}\sqrt{\left|discr\left(\mathfrak{O}^{(0)}_{n_{\tilde{\chi}}}\right)\right|}$$

and then

$$l_{\tilde{\chi}} \mid n_{\tilde{\chi}}^{\varphi(n_{\tilde{\chi}})/2} 2^{\varphi(n_{\tilde{\chi}})/2} \sqrt[4]{|d_{n_{\tilde{\chi}}}|},$$

thereby proving (iii).

(iv) If $\chi = \chi^\tau$, $\chi(\mathfrak{k}_0) \neq -1$, then necessarily $\mathcal{H} \cap \tilde{\chi} = \tilde{\chi}$, which implies that

$$l_{\tilde{\chi}} = \mathbf{N}_{\mathbb{Q}_{n_{\tilde{\chi}}}/\mathbb{Q}}(1 + \zeta_0)\left|\det(g(\tilde{\chi}, \Lambda_{\tilde{\chi}}, 1))\right|, \quad \zeta_0 = \chi(\mathfrak{k}_0),$$

with the matrix $g(\tilde{\chi}, \Lambda_{\tilde{\chi}}, 1)$ from Theorem 12.1.3. Since $\zeta_0 \neq -1$, we have as in case (ii)

$$\mathbf{N}_{\mathbb{Q}_{n_{\tilde{\chi}}}/\mathbb{Q}}(1 + \zeta_0) \mid (2n_{\tilde{\chi}})^{\varphi(n_{\tilde{\chi}})},$$

and then the assertion follows, as in case (ii).

If $\delta_{\tilde{\chi}} = 0$, we can proceed analogously. □

12.6 The Galois group of MK/K

Let K be a quadratic imaginary number field and M a number field with

$$K \nsubseteq M$$

and the property that

$$L := MK \quad \text{is abelian over} \quad K.$$

Then, for some $f \in \mathbb{N}$

$$L \subseteq K_{(f)},$$

and $K_{(f)}/\mathbb{Q}$ is Galois because $f \in \mathbb{N}$. We have

$$G(K_{(f)}/\mathbb{Q}) = \sigma(\mathfrak{K}_{(f)}) \uplus \sigma(\mathfrak{K}_{(f)})\tau,$$

where τ denotes complex conjugation and σ the canonical isomorphism between $\mathfrak{K}_{(f)}$ and $G(K_{(f)}/K)$. Further, we have the relation

$$\sigma(\mathfrak{h})\tau = \tau\sigma(\overline{\mathfrak{h}}), \quad \mathfrak{h} \in \mathfrak{K}_{(f)}, \tag{12.77}$$

with complex conjugation of ray classes

$$\mathfrak{h} \mapsto \overline{\mathfrak{h}},$$

which, in view of $f \in \mathbb{N}$ defines an automorphism of $\mathfrak{K}_{(f)}$ satisfying

$$\overline{\overline{\mathfrak{h}}} = \mathfrak{h}.$$

Let $\mathfrak{U}$ be the subgroup of $\mathfrak{K}_{(f)}$ corresponding to the extension L of K, and for $\mathfrak{h} \in \mathfrak{K}_{(f)}$ we set

$$\kappa(\mathfrak{h}) := \sigma(\mathfrak{h})|L.$$

Then $\kappa(\mathfrak{h})$ depends on $\mathfrak{h}$ only modulo $\mathfrak{U}$, and

$$G(L/M) = \{1, \kappa(\mathfrak{k}_0)\}$$

with a class $\mathfrak{k}_0 \in K_{(f)}$ that is uniquely determined modulo $\mathfrak{U}$. By $(\kappa(\mathfrak{k}_0)\tau)^2 = 1$ and (12.77) we see $1 = (\sigma(\mathfrak{k}_0)\tau)^2|L = \sigma(\mathfrak{k}_0\overline{\mathfrak{k}}_0)|L$, hence

$$\mathfrak{k}_0\overline{\mathfrak{k}}_0 \in \mathfrak{U}. \tag{12.78}$$

For $\mathfrak{h} \in \mathfrak{U}$ we have $\kappa(\mathfrak{k}_0\mathfrak{h}) = \kappa(\mathfrak{k}_0)$, so $(\mathfrak{k}_0\mathfrak{h})\overline{\mathfrak{k}_0\mathfrak{h})} \in \mathfrak{U}$ and

$$\overline{\mathfrak{U}} = \mathfrak{U}. \tag{12.79}$$

This implies that

$$L/\mathbb{Q} \quad \text{Galois}.$$

With the map $\mathfrak{h} \mapsto \overline{\mathfrak{h}}$ and the relation (12.52) the hypothesis (12.51) and (12.52) for $\mathfrak{K}_{(f)}$ and $\mathfrak{U}$ are satisfied. Therefore, with a system $\mathfrak{H}$ of representatives for the equivalence relation (12.53) we have the decomposition

$$
\begin{aligned}
G(L/\mathbb{Q}) &= \biguplus_{\mathfrak{h} \bmod \mathfrak{U}} \{1, \kappa(\mathfrak{k}_0)\tau\}\kappa(\mathfrak{h}) = \biguplus_{\mathfrak{h} \bmod \mathfrak{U}} \{\kappa(\mathfrak{h}), \kappa(\mathfrak{k}_0\overline{\mathfrak{h}})\tau\} \\
&= \left(\biguplus_{\substack{\mathfrak{h}\in\mathfrak{H} \\ e(\mathfrak{h})=1}} \{\kappa(\mathfrak{h}), \kappa(\mathfrak{h})\tau\} \right) \\
&\quad \uplus \left(\biguplus_{\substack{\mathfrak{h}\in\mathfrak{H} \\ e(\mathfrak{h})=2}} \left(\{\kappa(\mathfrak{h}), \kappa(\mathfrak{k}_0\overline{\mathfrak{h}})\tau\} \uplus \{\kappa(\mathfrak{k}_0\overline{\mathfrak{h}}), \kappa(\mathfrak{h})\tau\} \right) \right),
\end{aligned}
$$

where the isomorphisms in braces have the same restriction on M. Hence, the different isomorphisms of $M/\mathbb{Q}$ are given by

$$
\begin{aligned}
Iso(M/\mathbb{Q}) = \; & \{(\sigma(\mathfrak{h})|M) \mid \mathfrak{h} \in \mathfrak{H}, e(\mathfrak{h}) = 1\} \\
& \uplus \biguplus_{\substack{\mathfrak{h}\in\mathfrak{H} \\ e(\mathfrak{h})=2}} \left\{(\sigma(\mathfrak{h})|M), (\sigma(\mathfrak{k}_0\overline{\mathfrak{h}})|M)\right\},
\end{aligned}
$$

where, according to our deduction, the restrictions of

$$\sigma(\mathfrak{h}), \quad \mathfrak{h} \in \mathfrak{H}, e(\mathfrak{h}) = 1,$$

are the real isomorphisms of $M/\mathbb{Q}$. Now, choosing the system $\mathfrak{H}$ with $\mathfrak{e} \in \mathfrak{H}$ and setting $\mathfrak{H}' = \mathfrak{H} \setminus \{\mathfrak{e}\}$, the unit rank of M is

$$r_M = |\mathfrak{H}'|,$$

and the regulator of a system $\epsilon_1, \ldots, \epsilon_{r_M}$ of units in M is equal to

$$R_M(\epsilon_1, \ldots, \epsilon_{r_M}) = \left| \det \left(\log \left| \epsilon_i^{\sigma(\mathfrak{h})} \right|^{e(\mathfrak{h})} \right)_{i=1,\ldots,r_M \;;\mathfrak{h}\in\mathfrak{H}'} \right|. \tag{12.80}$$

12.7 Class number formulae for $\Omega \supset M \not\supset K$

Keeping the notations of the preceeding section 12.6, we construct a system of units in M by taking relative norms with respect to L/M of the units $\epsilon_L(\mathfrak{k})$ defined in (12.27) using singular values of Δ. We set

$$\vartheta_M(\mathfrak{k}) := \mathbf{N}_{L/M}(\epsilon_L(\mathfrak{k})) = \epsilon_L(\mathfrak{k})^{1+\sigma(\mathfrak{k}_0)\tau} = \frac{\epsilon_L(\mathfrak{k})\epsilon_L(\overline{\mathfrak{k}}\mathfrak{k}_0)}{\epsilon_L(\mathfrak{k}_0)}$$

and define

$$V_{ell,M}^{(0)} := \left\langle \{\vartheta_M(\mathfrak{k}) \mid \mathfrak{k} \bmod \mathfrak{U}\} \right\rangle.$$

Since $\vartheta_M(\mathfrak{e}) = 1$ and since $\vartheta_M(\mathfrak{k})$ only depends on the equivalence class of $\mathfrak{k}$ with respect to (12.53), the rank of $V_{ell,M}^{(0)}$ is at most $|\mathfrak{H}'| = r_M$. In fact, we will show that it is equal to r_M by computing the regulator of the system

$$\vartheta_M(\mathfrak{k}), \quad \mathfrak{k} \in \mathfrak{H}', \tag{12.81}$$

of units $\vartheta_M(\mathfrak{k})$ different from 1. First,

$$\log \left| \vartheta(\mathfrak{k})^{\sigma(\mathfrak{h})} \right|^{e(\mathfrak{h})} = \frac{e(\mathfrak{h})}{2} \left(v(\mathfrak{k}, \mathfrak{h}, 1) + v(\mathfrak{k}, \mathfrak{k}_0 \overline{\mathfrak{h}}, 1) \right) \quad \text{for } \mathfrak{h}, \mathfrak{k} \in \mathfrak{H}'$$

with

$$v(\mathfrak{k}, \mathfrak{h}, 1) = u(\mathfrak{k}\mathfrak{h}^{-1}) - u(\mathfrak{h}^{-1})$$

and

$$u(\mathfrak{k}) = \log |\epsilon_L(\mathfrak{k})|^2 = \log \left(\prod_{\mathfrak{h} \in \mathfrak{U}} \left(\frac{D(\mathfrak{k}\mathfrak{h})}{D(\mathfrak{h})} \right)^{24h} \right).$$

By formula (12.80) and Theorem 12.5.3 we find that

$$R_M \left(V_{ell,M}^{(0)} \right) = \left| \prod_{\chi \in \mathcal{H}} A(\chi) \right| \tag{12.82}$$

with

$$A(\chi) = 24h \sum_{\mathfrak{k} \in \mathfrak{K}} \chi(\mathfrak{k}) \log D(\mathfrak{k}).$$

Therefore, since $A(\chi) \neq 0$, this implies that the system (12.81) has rank r_M. To write down (12.82) for $M = M_0$, observe that $\mathbb{X}^{(0)}$ is the character group of M_0K/K and $\mathcal{H} \cap \mathbb{X}^{(0)} = \mathbb{X}^{(1)} \setminus \{1\}$.Then

$$R_{M_0} \left(V_{ell,M_0}^{(0)} \right) = \left| \prod_{\chi \in \mathbb{X}^{(1)} \setminus \{1\}} A(\chi) \right|. \tag{12.83}$$

Further, according to (11.7), we have the class number formula

$$(24h)^{r_M - r_{M_0}} \frac{h_M}{h_{M_0}} \frac{R_M}{R_{M_0}} = \left| \prod_{\substack{\chi \in \mathcal{H} \\ \chi^\tau \neq \chi}} A(\chi) \right|.$$

Combining (12.82) and (12.83) yields the formula

$$(24h)^{r_M - r_{M_0}} \frac{h_M}{h_{M_0}} = \frac{R_{M_0}}{R_{M_0} \left(V_{ell,M_0}^{(0)} \right)} \frac{R_{M_0} \left(V_{ell,M}^{(0)} \right)}{R_M},$$

which can be written as

$$(24h)^{r_M - r_{M_0}} \frac{h_M}{h_{M_0}} = \frac{\left[E_M : V_{ell,M}^{(0)}\right]}{\left[E_{M_0} : V_{ell,M_0}^{(0)}\right]}, \tag{12.84}$$

where the indices are to be understood modulo roots of unity. To simplify the right-hand side in (12.84) we will show now that a subgroup of $V_{ell,M_0}^{(0)}$ is a direct summand of $V_{ell,M}^{(0)}$. Therefore, we choose a system $\mathfrak{H}_0$ of $\mathfrak{K}/\mathfrak{U}_0$ of representatives for the equivalence relation (12.53) with $\mathfrak{e} \in \mathfrak{H}_0$. Then, since $\mathfrak{U}_0 \supseteq \mathfrak{U}$, the elements of $\mathfrak{H}_0$ are in inequivalent classes modulo $\mathfrak{U}$ so that $\mathfrak{H}_0$ can be completed to a system $\mathfrak{H}$ for cosets of modulo $\mathfrak{U}$ with respect to (12.53):

$$\mathfrak{H} \supseteq \mathfrak{H}_0 \ni \mathfrak{e}.$$

In the system of independent units,

$$\vartheta_M(\mathfrak{k}), \mathfrak{k} \in \mathfrak{H}',$$

generating $V_{ell,M}^{(0)}$, we now replace the units $\vartheta_M(\mathfrak{k})$, $\mathfrak{k} \in \mathfrak{H}_0'$, by their relative norms to M_0 that generate $V_{ell,M_0}^{(0)}$. They can be written as

$$\begin{aligned} \vartheta_{M_0}(\mathfrak{k}) &= \left[\mathbf{N}_{\Omega/L_0}\left(\epsilon(\mathfrak{k})\right)\right]^{1+\sigma(\mathfrak{k}_0)\tau} \\ &= \prod_{\mathfrak{h} \in \mathfrak{U}_0 \bmod \mathfrak{U}} \left[\mathbf{N}_{\Omega/L}\left(\frac{\epsilon(\mathfrak{k}\mathfrak{h})}{\epsilon(\mathfrak{h})}\right)\right]^{1+\sigma(\mathfrak{k}_0)\tau} \\ &= \vartheta_M(\mathfrak{k}) \prod_{\substack{\mathfrak{h} \in \mathfrak{U}_0 \bmod \mathfrak{U} \\ \mathfrak{h} \notin \mathfrak{U}}} \frac{\vartheta_M(\mathfrak{k}\mathfrak{h})}{\vartheta_M(\mathfrak{h})}. \end{aligned} \tag{12.85}$$

The factors $\vartheta_M(\mathfrak{h})$ in the product on the right-hand side are all different from $\vartheta_M(\mathfrak{k}), \mathfrak{k} \in \mathfrak{H}_0'$. A factor $\vartheta_M(\mathfrak{k}\mathfrak{h}), \mathfrak{k} \in \mathfrak{H}_0'$ is equal to a factor $\vartheta_M(\mathfrak{k}'), \mathfrak{k}' \in \mathfrak{H}_0'$ if and only if

$$\mathfrak{k}'\mathfrak{U} = \mathfrak{k}\mathfrak{U} \quad \text{and} \quad \mathfrak{h}\mathfrak{U} = \mathfrak{k}_0\mathfrak{k}^{-2}\mathfrak{U}.$$

Consequently, the product

$$\hat{V}_{ell,M}^{(0)} := V_{ell,M_0}^{(0)} \times V_{ell,M^c}^{(0)},$$

which is direct modulo roots of unity with

$$V_{ell,M^c}^{(0)} := \left\langle \left\{ \vartheta_M(\mathfrak{k}) \mid \mathfrak{k} \in \mathfrak{H} \setminus \mathfrak{H}_0 \right\} \right\rangle,$$

contains $V^{(0)}_{ell,M}$, and further,

$$\left[\hat{V}^{(0)}_{ell,M} : V^{(0)}_{ell,M}\right] = 2^{|\{\mathfrak{k}\in\mathfrak{H}'_0 | \mathfrak{k}_0\mathfrak{k}^{-2}\notin\mathfrak{U}\}|}.$$

In view of this, the formula (12.84) becomes

$$2^{-|\{\mathfrak{k}\in\mathfrak{H}'_0 | \mathfrak{k}_0\mathfrak{k}^{-2}\notin\mathfrak{U}\}|}(24h)^{r_M - r_{M_0}} \frac{h_M}{h_{M_0}} = \left[E_M : E_{M_0} \times V^{(0)}_{ell,M^c}\right]. \quad (12.86)$$

To eliminate the factor $(24h)^{r_M - r_{M_0}}$, we proceed as in section 12.3 and define

$$V^{(1)}_{ell,M^c} := \left\{ \prod_{\mathfrak{k}\in\mathfrak{H}\setminus\mathfrak{H}_0} \vartheta_M(\mathfrak{k})^{x_\mathfrak{k}} \,\middle|\, x_\mathfrak{k} \in \mathbb{Z}, \prod_{\mathfrak{k}\in\mathfrak{H}\setminus\mathfrak{H}_0} \mathfrak{k}^{x_\mathfrak{k}} = \mathfrak{e} \right\}.$$

Then

$$\left[V^{(0)}_{ell,M^c} : V^{(1)}_{ell,M^c}\right] = [\mathfrak{K} : \mathfrak{U}] = [L : K] = [M : \mathbb{Q}].$$

Further, using the method of section 12.3, one can define a subgroup

$$V^{(2)}_{ell,M^c} \subseteq \left(V^{(1)}_{ell,M^c}\right)^{\frac{1}{24h}} \bigcap M$$

satisfying

$$\left[V^{(2)}_{ell,M^c} : V^{(1)}_{ell,M^c}\right] = \frac{(24h)^{r_M - r_{M_0}}}{e}$$

with

$$e = \begin{cases} 1 & \text{if} \quad \sqrt{-1}, \sqrt{-3} \notin L \setminus K, \\ 2 & \text{if} \quad \sqrt{-1} \in L \setminus K, \sqrt{-3} \notin L \setminus K, \\ 3 & \text{if} \quad \sqrt{-3} \in L \setminus K, \sqrt{-1} \notin L \setminus K. \end{cases}$$

Theorem 12.7.1 *Let M be a subfield of Ω, $\Omega \supset M \not\supseteq K$, and let M_0 be the maximal abelian subfield of M. Then, with the above notation*

$$2^{-|\{\mathfrak{k}\in\mathfrak{H}'_0 | \mathfrak{k}_0\mathfrak{k}^{-2}\notin\mathfrak{U}\}|} e[M : \mathbb{Q}] \frac{h_M}{h_{M_0}} = \left[E_M : E_{M_0} \times V^{(2)}_{ell,M^c}\right].$$

12.8 Class number formulae for $K_{\mathfrak{f}} \supset M \not\supseteq K$

Keeping the notations of section 12.6, we construct unit groups in M by taking relative norms of the units $\epsilon(\tilde{\chi}, \mathfrak{k})$ defined in section 12.4 (12.31) and (12.32). We set

$$\vartheta(\tilde{\chi}, \mathfrak{k}) := \mathbf{N}_{L/M}(\epsilon(\tilde{\chi}, \mathfrak{k})) = \epsilon(\tilde{\chi}, \mathfrak{k})^{1+\sigma(\mathfrak{k}_0)\tau} = \frac{\epsilon(\tilde{\chi}, \mathfrak{k})\epsilon(\tilde{\chi}, \overline{\mathfrak{k}}\mathfrak{k}_0)}{\epsilon(\tilde{\chi}, \mathfrak{k}_0)}$$

for a class $\tilde{\chi}$ in the character group $\mathbb{X}$ of $L = MK/K$ and an ideal class $\mathfrak{k} \in \mathfrak{K}_{(f)}$. Then we define

$$\vartheta_{i_{\tilde{\chi}}}(\tilde{\chi}) := \prod_{\mathfrak{k} \bmod \mathfrak{U}_{\tilde{\chi}}} \vartheta(\tilde{\chi}, \mathfrak{k})^{\lambda_{i_{\tilde{\chi}},\mathfrak{k}}(\tilde{\chi})}, \qquad i_{\tilde{\chi}} = 1, \ldots, |\tilde{\chi} \cap \mathcal{H}|,$$

where the exponents $\lambda_{i_{\tilde{\chi}},\mathfrak{k}}(\tilde{\chi}) \in \mathbb{Z}$ have to be chosen such that $\det(l(\tilde{\chi}, \Lambda_{\tilde{\chi}})) \neq 0$, which can be achieved, for instance, by defining them as in Theorem 12.5.5. Further, we assume that the system $\mathcal{H}$ satisfies the condition (12.68). For every class $\tilde{\chi}$ we set

$$V^{(0)}_{ell,\tilde{\chi}} := \langle \{ \vartheta_{i_{\tilde{\chi}}}(\tilde{\chi}) \mid i_{\tilde{\chi}} = 1, \ldots, |\tilde{\chi} \cap \mathcal{H}| \} \rangle,$$

and then we define

$$V^{(0)}_{ell,\mathbb{X}\setminus\mathbb{X}^{(0)}} := \prod_{\substack{\tilde{\chi} \\ \chi \in \mathbb{X}\setminus\mathbb{X}^{(0)}}} V^{(0)}_{ell,\tilde{\chi}},$$

$$V^{(0)}_{ell,\mathbb{X}^{(1)}} := \prod_{\substack{\tilde{\chi} \\ \chi \in \mathbb{X}^{(1)}\setminus\{1\}}} V^{(0)}_{ell,\tilde{\chi}}.$$

The number of generators in the definition of $V^{(0)}_{ell,\mathbb{X}^{(1)}} V^{(0)}_{ell,\mathbb{X}\setminus\mathbb{X}^{(0)}}$ is equal to r_M, and their regulator is the determinant with entries

$$\log \left| \vartheta_{i_{\tilde{\chi}}}(\tilde{\chi})^{\sigma(\mathfrak{h})} \right|^{e(\mathfrak{h})} = \frac{1}{2} \sum_{\mathfrak{k} \bmod \mathfrak{U}_{\tilde{\chi}}} \lambda_{i_{\tilde{\chi}},\mathfrak{k}}(\tilde{\chi})\, e(\mathfrak{h}) \left(v_{\tilde{\chi}}(\mathfrak{k}, \mathfrak{h}) + v_{\tilde{\chi}}(\mathfrak{k}, \mathfrak{k}_0 \overline{\mathfrak{h}}) \right)$$

where

$$v_{\tilde{\chi}}(\mathfrak{k}, \mathfrak{h}) = u_{\tilde{\chi}}(\mathfrak{k}\mathfrak{h}^{-1}) - u_{\tilde{\chi}}(\mathfrak{h}^{-1}),$$

with $u_{\tilde{\chi}}$ is defined in 12.4 (12.34) and (12.35). Using Theorem 12.5.4, the regulator of $V^{(0)}_{ell,\mathbb{X}^{(1)}} V^{(0)}_{ell,\mathbb{X}\setminus\mathbb{X}^{(0)}}$ is given by

$$R_M(V^{(0)}_{ell,\mathbb{X}^{(1)}} V^{(0)}_{ell,\mathbb{X}\setminus\mathbb{X}^{(0)}}) = 2^{-r_M} d(\mathfrak{K}_{(f)}/\mathfrak{U}, \mathfrak{k}_0, \Lambda) \left| \prod_{\chi \in \mathcal{H}} A(\chi) \right| \qquad (12.87)$$

with

$$A(\chi) := \sum_{\mathfrak{k} \bmod \mathfrak{U}_{\tilde{\chi}}} \overline{\chi}(\mathfrak{k}) u_{\tilde{\chi}}(\mathfrak{k}).$$

In particular, this implies that the regulator of $V^{(0)}_{ell,\mathbb{X}^{(1)}} V^{(0)}_{ell,\mathbb{X}\setminus\mathbb{X}^{(0)}}$ is non-zero and the product of the two unit groups is direct modulo roots of unity. We note (12.87) for $M = M_0$:

$$R_{M_0}(V^{(0)}_{ell,\mathbb{X}^{(1)}}) = 2^{-r_{M_0}} d(\mathfrak{K}_{(f)}/\mathfrak{U}_0, \mathfrak{k}_0, \Lambda) \left| \prod_{\chi \in \mathbb{X}^{(1)} \setminus \{1\}} A(\chi) \right|. \quad (12.88)$$

Now (11.7) can be written as

$$\underbrace{\left(\prod_{\substack{\tilde{\chi} \\ \chi \in \mathcal{H} \setminus \mathbb{X}^{(0)}}} (24 h_{f_{\tilde{\chi}}})^{|\tilde{\chi} \cap \mathcal{H}|} \right) \left(\prod_{\substack{\tilde{\chi} \\ \chi \in \mathcal{H} \setminus \mathbb{X}^{(0)}}} (e_{f_{\tilde{\chi}}})^{|\tilde{\chi} \cap \mathcal{H}|} \right) \frac{h_M}{h_{M_0}} \frac{R_M}{R_{M_0}}}_{=:C} \quad (12.89)$$

$$= \left| \prod_{\substack{\chi \in \mathcal{H} \\ \chi^\tau \neq \chi}} A(\chi) \right|.$$

Herein, the first product on the left-hand side is over all classes of characters of $\mathcal{H} \setminus \mathbb{X}^{(0)}$ that are charcters of Ω_t/K and the second over the remaining characters of $\mathcal{H} \setminus \mathbb{X}^{(0)}$. Now, combining (12.87), (12.88) and (12.89), we obtain, similarly to section 12.7,

$$C\, 2^{-r_M + r_{M_0}} \frac{d(\mathfrak{K}_{(f)}/\mathfrak{U}, \mathfrak{k}_0, \Lambda)}{d(\mathfrak{K}_{(f)}/\mathfrak{U}_0, \mathfrak{k}_0, \Lambda)} \frac{h_M}{h_{M_0}} = [E_M : E_{M_0} V^{(0)}_{ell,\mathbb{X}\setminus\mathbb{X}^{(0)}}]. \quad (12.90)$$

To prove Theorem 12.4.1, we first use the fact that the unit $V^{(0)}_{ell,\mathbb{X}\setminus\mathbb{X}^{(0)}}$ only depends on the class of characters $\chi \in \mathcal{H} \setminus \mathbb{X}^{(0)}$. We consider a system

$$M_1, \ldots, M_s \subseteq M$$

of subfields of M that, together with the maximal abelian subfield M_0 of M, is linearly disjoint over $\mathbb{Q}$:

$$M_i \cap (M_0 \cdot \ldots \cdot M_{i-1} \cdot M_{i+1} \cdot \ldots \cdot M_s) = \mathbb{Q}, \quad i = 0, \ldots, s.$$

For the character groups $\mathbb{X}_i$ of $L_i = M_i K/K$ we then have

$$\mathbb{X}_i \setminus \{1\} \subseteq \mathbb{X} \setminus \mathbb{X}^{(0)}, \quad i = 1, \ldots s,$$

and

$$\mathbb{X}_i \cap \mathbb{X}_j = \{1\} \quad \text{for } i \neq j.$$

Considering the fact that the maximal abelian subfields of the $M_i, i = 1, \ldots s$, are equal to $\mathbb{Q}$, we write down the class number formula (12.90) for the $M_i, i = 1, \ldots, s$. Then, in combining these formulae with (12.90), we obtain the relation

$$C_1' C_2' C_3' \frac{h_M}{h_{M_0} \cdot \ldots \cdot h_{M_s}} = \left[E_M : E_{M_0} \cdot \ldots \cdot E_{M_s} V^{(0)}_{ell,\mathbb{X}^c} \right] \tag{12.91}$$

with

$$V^{(0)}_{ell,\mathbb{X}^c} = \prod_{\substack{\tilde{\chi} \\ \chi \in \mathbb{X}^c}} V^{(0)}_{ell,\tilde{\chi}}, \qquad \mathbb{X}^c := \mathbb{X} \setminus (\mathbb{X}^{(0)} \cup \mathbb{X}_1 \cup \cdots \cup \mathbb{X}_s),$$

$$C_1' = \prod_{\substack{\tilde{\chi} \\ \chi \in \mathcal{H} \cap \mathbb{X}^c_{ring}}} (24 h_{f_{\tilde{\chi}}})^{|\tilde{\chi} \cap \mathcal{H}|},$$

$$C_2' = \prod_{\substack{\tilde{\chi} \\ \chi \in \mathcal{H} \cap \mathbb{X}^c_{ray}}} (e_{f_{\tilde{\chi}}})^{|\tilde{\chi} \cap \mathcal{H}|},$$

where $\mathbb{X}^c_{ring}$ denotes the set of ring class characters modulo t in $\mathbb{X}^c$ and $\mathbb{X}^c_{ray} := \mathbb{X}^c \setminus \mathbb{X}^c_{ring}$ and

$$C_3' = 2^{-r_M + (r_{M_0} + \cdots + r_{M_s})} \frac{d(\mathfrak{K}_{(f)}/\mathfrak{U}, \mathfrak{k}_0, \Lambda)}{d(\mathfrak{K}_{(f)}/\mathfrak{U}_0, \mathfrak{k}_0, \Lambda) \cdot \cdot d(\mathfrak{K}_{(f)}/\mathfrak{U}_s, \mathfrak{k}_0, \Lambda)}. \tag{12.92}$$

To reduce the factors C_1' and C_2' we write $V^{(0)}_{ell,\mathbb{X}^c}$ as the product

$$V^{(0)}_{ell,\mathbb{X}^c_{ring}} V^{(0)}_{ell,\mathbb{X}^c_{ray}}$$

and, as for the reduction of C_1, C_2 in (12.39) we construct subgroups

$$V^{(1)}_{ell,\mathbb{X}^c_{ring}} \subseteq V^{(0)}_{ell,\mathbb{X}^c_{ring}}$$

and

$$V^{(1)}_{ell,\mathbb{X}^c_{ray}} \subseteq V^{(0)}_{ell,\mathbb{X}^c_{ray}}.$$

Further, by taking radicals

$$V^{(2)}_{ell,\mathbb{X}^c_{ring}} \supseteq V^{(1)}_{ell,\mathbb{X}^c_{ring}}$$

and

$$V^{(2)}_{ell,\mathbb{X}^c_{ray}} \supseteq V^{(1)}_{ell,\mathbb{X}^c_{ray}}$$

The construction of the last subgroups is completely analogous to the construction of $U^{(i)}_{ell,\mathbb{X}^c_{ring}}$ and $U^{(i)}_{ell,\mathbb{X}^c_{ray}}$, $i = 1, 2$, in the proof of Theorem 12.4.1. In the defining equations one has to replace the units $\epsilon_{i_{\tilde{\chi}}}(\tilde{\chi})$ by their relative norms $\vartheta_{i_{\tilde{\chi}}}(\tilde{\chi})$ and the exponents $\gamma_{i_{\tilde{\chi}},\mathfrak{k}_{\tilde{\chi}}}(\tilde{\chi})$ by the exponents $\lambda_{i_{\tilde{\chi}},\mathfrak{k}_{\tilde{\chi}}}(\tilde{\chi})$. Then, replacing the unit groups in (12.91) by the subgroups $V^{(2)}_{ell,\mathbb{X}^c_{ring}}$ and $V^{(2)}_{ell,\mathbb{X}^c_{ray}}$, we obtain, as in Theorem 12.4.1, a formula where the factors C'_1 and C'_2 are replaced by c_1, c_2 defined in (12.42) and (12.47).

We summarise our result in:

Theorem 12.8.1 *Let M be a subfield of $K_{\mathfrak{f}}, \mathfrak{f} = (f), f \in \mathbb{N}$, with $M \not\supseteq K$, and let M_0 be its maximal abelian subfield. Further, let $M_1, \ldots, M_s$ be a system of subfields of M, which together with M_0 are linearly independent over $\mathbb{Q}$:*

$$M_i \cap (M_0 \cdot \ldots \cdot M_{i-1} \cdot M_{i+1} \cdot \ldots \cdot M_s) = K, \quad i = 0, 1, \ldots, s.$$

Let $\mathbb{X}$ denote the character group of $L = MK/K$ and $\mathbb{X}_i$ the subgroups of $\mathbb{X}$ associated with $L_i = M_iK/K$. We set

$$\mathbb{X}^c := \mathbb{X} \setminus (\mathbb{X}_0 \cup \cdots \cup \mathbb{X}_s,$$

and we choose, as described in section 12.5, a system $\mathcal{H}$ satisfying (12.68). Further, we associate with every class $\tilde{\chi}$, $\chi \in \mathcal{H} \cap \mathbb{X}^c$, a matrix $\Lambda_{\tilde{\chi}}$ with coefficients in $\mathbb{Z}$ such that the matrices $l(\tilde{\chi}, \Lambda_{\tilde{\chi}}, 1)$, defined according to Theorem 12.5.4, have a non-zero determinant. Moreover, in the following, we will assume that exponents $\lambda_{i_{\tilde{\chi}},\mathfrak{k}}(\tilde{\chi})$ are chosen according to Theorem 12.5.5. By Λ we denote the system of these matrices and by Λ_i the subsystem associated with M_i, $i \geq 1$. Further, let

$$V^{(2)}_{ell,\mathbb{X}^c} = \left(\prod_{t'|t}{}' V^{(2)}_{ell,t'}\right)\left(\prod_{\mathfrak{t}|\mathfrak{f}}{}' V^{(2)}_{ell,\mathfrak{t}}\right)$$

be the subgroup of the unit group of M obtained, as described above, by modification of $V^{(0)}_{ell,\mathbb{X}^c}$. Then by (12.91) we obtain the formula

$$C' \frac{h_M}{h_{M_0} \cdot \ldots \cdot h_{M_s}} = \left[E_M : E_{M_0} \cdot \ldots \cdot E_{M_s} V^{(2)}_{ell,\mathbb{X}^c}\right]$$

with the class numbers $h_{..}$ and unit groups $E_{..}$ of the fields in play and the product

$$C' = c_1 c_2 C'_3,$$

of factors defined in (12.42), (12.47), (12.92):

$$c_1 = \prod_{t'|t}{}' n_{t'} \left[L_{\Omega_{t'}} : K\right],$$

$$c_2 = \prod_{\mathfrak{t}|\mathfrak{f}}{}' f_{L_{\mathfrak{t}}},$$

$$C_3' = 2^{-r_M+(r_{M_0}+\cdots+r_{M_s})} \frac{d(\mathfrak{K}_{(f)}/\mathfrak{U}, \mathfrak{k}_0, \Lambda)}{d(\mathfrak{K}_{(f)}/\mathfrak{U}_0, \mathfrak{k}_0, \Lambda_0) \cdot \ldots \cdot d(\mathfrak{K}_{(f)}/\mathfrak{U}_s, \mathfrak{k}_0, \Lambda_s)}.$$

Herein, $\mathfrak{U}$ and $\mathfrak{U}_i$ denote the subgroups of $\mathfrak{K}_\mathfrak{f}$ associated with L/K and L_i/K, and $\mathfrak{k}_0 \in \mathfrak{K}_\mathfrak{f}$ is a class having the property that $\sigma(\mathfrak{k}_0)\tau$ generates the Galois group of the quadratic extension L/M.

12.8.1 Applications of the class number formulae in 12.8

12.8.1.1 Divisibility relations between class numbers

To illustrate Theorem 12.8.1, we choose the pure field of degree 6 from Example 12.4.2 and its two subfields of degree 2 and 3:

$$M_6 = \mathbb{Q}(\sqrt[6]{a}), \quad M_2 = \mathbb{Q}(\sqrt{a}), \quad M_3 = \mathbb{Q}(\sqrt[3]{a}).$$

Note that M_2 is the maximal abelian subfield of M_6, which is linearly disjoint to M_3 over $\mathbb{Q}$. Therefore, with $K = \mathbb{Q}(\sqrt{-3})$ we can apply Theorem 12.8.1, and with a suitable choice of the matrices $\Lambda_{\tilde{\chi}}$ we obtain:

Example 12.8.2 Let $M_6 = \mathbb{Q}(\sqrt[6]{a})$, $a \in \mathbb{Z}$, a pure field of degree 6, and let $M_2 = \mathbb{Q}(\sqrt{a})$, $M_3 = \mathbb{Q}(\sqrt[3]{a})$ be the two non-trivial subfields. Further, we assume that $M_2 \neq \mathbb{Q}(\sqrt{\pm 3}), \mathbb{Q}(\sqrt{-4})$. Then M_6 satisfies the hypothesis of Theorem 12.8.1 with $K = \mathbb{Q}(\sqrt{-3})$, and it is a subfield of some ring class field Ω_f. Let $\mathfrak{U}$ denote the subgroup of $\mathfrak{K}_{(f)}$ associated with M_6K/K, then by normalising the root $\sqrt[3]{a}$, we can achieve that

$$o(\mathfrak{k}_0\mathfrak{U}) = \begin{cases} 2 & \text{for} \quad a < 0, \\ 1 & \text{for} \quad a > 0. \end{cases}$$

We choose $t = f$ as in Theorem 12.4.2, and we obtain

$$6\,\frac{h_{M_6}}{h_{M_2}h_{M_3}} = \left[E_{M_6} : E_{M_2}E_{M_3}V^{(2)}_{ell,\mathbb{X}^c}\right].$$

By modification of $V^{(2)}_{ell,\mathbb{X}^c}$ we can, similarly to the proof of Example 12.4.2, derive

$$\frac{h_{M_6}}{h_{M_2}h_{M_3}} = \left[E_{M_6} : E_{M_2}E_{M_3}\tilde{V}^{(2)}_{ell,\mathbb{X}^c}\right].$$

Of course, Example 12.8.2 can be generalised, by taking, for instance, instead of M_6, the maximal real subfield of a dihedral field of degree $2pq$ with two primes p, q. In these cases the factor C' is no longer equal to 1 or 3 but is a product of powers of p and q with exponents growing with p and q.

Proof of example 12.8.2 First, in our case we have $c_1 = 6, c_2 = 1$, and the factor C_3' in Theorem 12.8.1 is

$$C_3' = \begin{cases} \frac{1}{\sqrt{3}} l_{\tilde{\chi}} & \text{if} \quad a < 0, \\ l_{\tilde{\chi}} & \text{if} \quad a > 0, \end{cases}$$

with the generating character χ of the cyclic extension M_6K/K and the factor $l_{\tilde{\chi}}$ depending on the construction of $V^{(0)}_{ell,\tilde{\chi}}$. Let $\mathfrak{U}$ denote the subgroup of $\mathfrak{K}_{(f)}$ associated with M_6K/K, and let $\mathfrak{k}$ be a class in $\mathfrak{K}_{(f)}$ with $\mathfrak{K}_{(f)}/\mathfrak{U} =< \mathfrak{k}\mathfrak{U} >$. Then $V^{(0)}_{ell,\mathfrak{X}^c} = V^{(0)}_{ell,\tilde{\chi}}$ has rank 1. We set

$$V^{(0)}_{ell,\tilde{\chi}} = \langle \vartheta_{\tilde{\chi}}(\mathfrak{k}) \rangle .$$

Then

$$l_{\tilde{\chi}} = |(\chi(\mathfrak{k}) - 1) + \chi(\mathfrak{k}_0)(\overline{\chi}(\mathfrak{k}) - 1)|,$$

and by definition of $\mathfrak{k}_0$ we can achieve

$$\chi(\mathfrak{k}_0) = \begin{cases} -1 & \text{if} \quad a < 0, \\ 1 & \text{if} \quad a > 0. \end{cases}$$

For $a > 0$ this is immediate. For $a < 0$, since M_6 is totally imaginary, $\mathfrak{k}_0$ cannot be in $\mathfrak{U}$. For a conjugate field $M_6^{\sigma(\mathfrak{h})}$, $\mathfrak{h} \in \mathfrak{K}_{(f)}$, the class $\mathfrak{k}_0$ has to be replaced by $\mathfrak{k}_0\mathfrak{h}^2$, so that we can achieve $o(\mathfrak{k}_0\mathfrak{U}) = 2$. Then $\chi(\mathfrak{k})$ is a primitive 6-th root of unity, hence

$$l_{\tilde{\chi}} = \begin{cases} \sqrt{3} & \text{if } a < 0, \\ 1 & \text{if } a > 0. \end{cases} \tag{12.93}$$

This implies the first formula in Example 12.8.2. To prove the second formula we choose

$$V'^{(0)}_{ell,\tilde{\chi}} = \left\langle \vartheta_{\tilde{\chi}}(\mathfrak{k}^2)^3 \vartheta_{\tilde{\chi}}(\mathfrak{k}^3)^{-2} \right\rangle = \left\langle \mathbf{N}_{M_6K/M_6} \left(\epsilon_{\tilde{\chi}}(\mathfrak{k}^2)^3 \epsilon_{\tilde{\chi}}(\mathfrak{k}^3)^{-2} \right) \right\rangle .$$

Then (12.93) still holds, and, in addition, we know that

$$\vartheta_{\tilde{\chi}}(\mathfrak{k}^2)^3 \vartheta_{\tilde{\chi}}(\mathfrak{k}^3)^{-2} = \mathbf{N}_{M_6K/M_6} \left(\epsilon_{\tilde{\chi}}(\mathfrak{k}^2)^3 \epsilon_{\tilde{\chi}}(\mathfrak{k}^3)^{-2} \right)$$

is the $24h_{f_{\tilde{\chi}}}$-th power of an element in M_6 because

$$(\mathfrak{k}^2)^3(\mathfrak{k}^3)^{-2} \in \mathfrak{U} \quad \text{and} \quad \sqrt{-3}, \sqrt{-4} \notin M_6K \setminus K.$$

Hence the first formula holds without the factor 6, after replacing $V^{(2)}_{ell,\tilde{\chi}}$ by $\tilde{V}^{(2)}_{ell,\mathbb{X}^c} := V'^{(0)}_{ell,\tilde{\chi}}$. □

12.8.1.2 Divisibility of class numbers by divisors of the field degree

By Theorem 12.8.1 the results for dihedral fields in Theorem 12.4.5 can also be proved for the class numbers of their maximal real subfields. Therefore, the definition of the units $\epsilon_{\tilde{\chi}}(\mathfrak{k})$ in section 12.4.1.2 will be modified for composite conductors $\mathfrak{f}_{\tilde{\chi}}$ by defining the units $\vartheta_{\tilde{\chi}}(\mathfrak{k})$. With the modified units the class number formula (12.91) becomes

$$C_1'C_2'\tilde{C}_3' \frac{h_M}{h_{M_0} \cdot \ldots \cdot h_{M_s}} = \left[E_M : E_{M_0} \cdot \ldots \cdot E_{M_s} V^{(0)}_{ell,\mathbb{X}^c}\right]'. \qquad (12.94)$$

The factors $\tilde{C}_3'$ and C_3' in (12.91) are related by

$$\tilde{C}_3' = \frac{C_3'}{N}$$

with

$$N = \prod_{\chi}{}' p_{\chi},$$

where the product is overall χ in $\mathbb{X}^c_{ray} \cap \mathcal{H}$ with composite conductor and an order, which is a prime power $p_{\chi}^{a_{\chi}}$. Further, we need a more precise version of Theorem 12.5.5, which is easily obtained by the conclusions in the proof of Theorem 12.5.5.

Theorem 12.8.3 *Let $\mathfrak{k}_0 = \mathfrak{e}$ and $\delta_{\tilde{\chi}} \in \{0,1\}$. Then the coefficients $\lambda_{i_{\tilde{\chi}},\mathfrak{k}}(\tilde{\chi})$ of the matrices $\Lambda_{\tilde{\chi}}$ in $\mathbb{Z}$ can be chosen such that*

$$l_{\tilde{\chi}} := |\det(l(\tilde{\chi}, \Lambda_{\tilde{\chi}}, \delta_{\tilde{\chi}}))| = 2^{|\tilde{\chi} \cap \mathcal{H}|} \sqrt{|d^{(0)}_{n_{\tilde{\chi}}}|}\, \Phi_{n_{\tilde{\chi}}}(1)^{\delta_{\tilde{\chi}}} \quad \text{for } \tilde{\chi}^{\tau} = \tilde{\chi}.$$

Herein, $d^{(0)}_{n_{\tilde{\chi}}}$ denotes the discriminant of the maximal real subfield of the $n_{\tilde{\chi}}$-th cyclotomic field and $\Phi_{n_{\chi}}(X)$ the $n_{\tilde{\chi}}$-th cyclotomic polynomial.

Theorem 12.8.4 *Let K be a quadratic imaginary number field, n a natural number without multiple prime divisor, and we assume that* $\gcd(n, 6h_K) = 1$. *Then the maximal real subfield M of the infinite* *number of dihedral fields L in Theorem 12.4.5 with $L \supseteq K$ of degree $[L:Q] = 2n$ has a class number divisible by n:*

$$n \mid h_M.$$

Proof The assertion is obtained by the formula (12.94) for the field M considered in Theorem 12.8.4:

$$C_1' C_2' \tilde{C}_3' h_M = [E_M : V^{(0)}_{ell,\mathbb{X}^c}].$$

Note that on the one hand by assumption we have $[M:\mathbb{Q}] = [L:K] = n \equiv 1 \bmod 2$, and that on the other hand M_0 must be equal to $\mathbb{Q}$, since M_0 is an abelian subfield of a ring class field, which implies that its degree is a power of 2. As in the proof of Theorem 12.4.5 the factors C_1' and C_2' are integral and coprime to n. The assertion then follows by computing the rational number $\tilde{C}_3'$, that turns out to have denominator n. Since by construction MK is a ring class field of odd degree over K, it follows that the system $\mathcal{H}$ occuring in the class number formula of M is also a system of representatives for the equivalence relation

$$\chi' \sim \chi : \Longleftrightarrow (\chi' = \chi \text{ or } \chi' = \chi^{-1}).$$

Therefore, by definition

$$\tilde{C}_3' = d(\mathfrak{K}_{(f)}/\mathfrak{U}, \Lambda) = n^{-\frac{n+1}{2}} 2^a \prod_{\tilde{\chi} \neq \tilde{1}} l_{\tilde{\chi}} \left(\frac{n}{n_{\tilde{\chi}}}\right)^{\frac{\varphi(n_{\tilde{\chi}})}{2}}.$$

This can be simplified, using the relation $\sum_{\tilde{\chi} \neq \tilde{1}} \varphi(n_{\tilde{\chi}}) = n - 1$:

$$\tilde{C}_3' = \frac{1}{n} 2^a n^{\frac{n}{4}} n^{\frac{1}{4}} \prod_{\tilde{\chi} \neq \tilde{1}} \frac{l_{\tilde{\chi}}}{n_{\tilde{\chi}}^{\varphi(n_{\tilde{\chi}})/2}}.$$

Since M is real, we can take $\mathfrak{k}_0 = \mathfrak{e}$, and, choosing the coefficients of the matrices $\Lambda_{\tilde{\chi}}$ according to Theorem 12.8.3, we obtain

$$l_{\tilde{\chi}} = 2^{\varphi(n_{\tilde{\chi}})/2} \sqrt{\left| d^{(0)}_{n_{\tilde{\chi}}} \right|}\ \Phi_{n_{\tilde{\chi}}}(1)^{\delta_{\tilde{\chi}}}.$$

As can be easily seen, we have the relation

$$|d_m| = \Phi_m(1) \left| d^{(0)}_m \right|^2 \quad \text{for } m \in \mathbb{N}, m \equiv 1 \bmod 2,$$

whereby $\tilde{C}_3'$ can be transformed to

$$\tilde{C}_3' = \frac{1}{n} \prod_{\tilde{\chi} \neq \tilde{1}} \Phi_{n_{\tilde{\chi}}}(1)^{\delta_{\tilde{\chi}}} \, 2^{a+\frac{n-1}{2}} \left(n^{\frac{1}{4}} \prod_{\tilde{\chi} \neq \tilde{1}} \Phi_{n_{\tilde{\chi}}}(1)^{-\frac{1}{4}} \right) \left(n^{\frac{n}{4}} \prod_{\tilde{\chi} \neq \tilde{1}} \frac{\sqrt[4]{|d_{n_{\tilde{\chi}}}|}}{n^{\varphi(n_{\tilde{\chi}})/2}} \right).$$

Herein, the last but one factor is equal to 1 because $n = \prod_{\tilde{\chi} \neq \tilde{1}} \Phi_{n_{\tilde{\chi}}}(1)$ and similarly the last factor because $\mathbb{X}$ is cyclic in our case. Therefore,

$$\tilde{C}_3' = \frac{1}{n} \prod_{\tilde{\chi} \neq \tilde{1}} \Phi_{n_{\tilde{\chi}}}(1)^{\delta_{\tilde{\chi}}} \, 2^{a+\frac{n-1}{2}}.$$

This implies the assertion of Theorem 12.8.4 because in the modified construction of $V_{ell,\mathbb{X}^c}^{(0)}$ all $\delta_{\tilde{\chi}}$ are equal to zero. □

Remark 12.8.5 The class number formulae of section 12.8 together with an estimate of the regulator can also be used for the simultaneous calculation of the class number and unit group of the fields considered. This has been realised, for instance, by Nakamula(1982, 1985) for cubic, quartic and sextic fields.

References

A. Agboola, Torsion points on elliptic curves and Galois module structure, *Invent. Math.* **123** (1996), 105–122.

E.J. Gòmez Ayala and R. Schertz, Eine Bemerkung zur Galoismodulstruktur in Strahlklassenkörpern über imaginär-quadratischen Zahlkörpern, *J. Number Theory* **44**, No 1 (1993), 41–46.

W.E. Berwick, Modular invariants expressible in terms of quadratic and cubic irrationalities, *Proceedings of the London Mathematical Society* **28**, Ser. 2 (1927), 53–69.

S. Bettner, Beweis der Kongruenzen von Berwick sowie deren Verallgemeinerung und weitere Anwendungen von Torsionspunkten auf elliptischen Kurven. Augsburger Schriften zur Mathematik, *Phys. Inform.* 8 (2004).

S. Bettner and R. Schertz, Lower powers of elliptic units, *J. Théor. Nombres Bordeaux* **13** (2001).

B.J. Birch, Weber's class invariants, *Mathematika* **16** (1969), 283–294.

W. Bley, Konstruktion von Ganzheitsbasen in abelschen Erweiterungen imaginär-quadratischer Zahlkörper, *J. Number Theory* **46**, (1994), 334–371.

W. Bley, Galois module structure and elliptic functions, *J. Number Theory* 52 (1995), 216–242.

J. Cassels, A note on the division polynomials of $\wp(u)$, *Proceedings of the Cambridge Philosophical Society* **45** Pt 2 (1949), 167–172.

Ph. Cassou-Noguès and M.J. Taylor, *Elliptic Functions and Rings of Integers*, Birkhäuser (1987).

J. Cougnard and V. Fleckinger, Sur la monogénéité de l'anneau des entiers de certain corps de rayon, *Manu. Math.* **63** (1989), 365–376.

M. Deuring, Die Klassenkörper der Komplexen Multiplikation, Enzyklopädie der *mathematischen Wissenschaften* Band I, 2. Teil, Heft 10, Teil II (1958).

D. Dorman, Prime factorization of singular moduli, thesis, Brown (1984).

A. Enge and F. Morain, Comparing invariants for class fields of imaginary quadratic fields, *Springer Lecture Notes in Computer Science* **2369** (2002).

A. Enge, The complexity of class polynomial computation via floating point approximations, *Math. Comput.* **78** (2009).

A. Enge and R. Schertz, Constructing elliptic curves over finite fields using double eta-quotients, *J. Théor. Nombres Bordeaux* **16** (2004), 555–568.

A. Enge and R. Schertz, Constructing elliptic curves from modular curves of positive genus, *J. Théor. Nombres Bordeaux* (2004).

A. Enge and R. Schertz, Modular curves of composite level, *Acta Arithmetica* **118**, No 2 (2005), 129–141.

A. Enge and R. Schertz, Doppelte Eta-Quotienten im verzweigten Fall, preprint.

V. Fleckinger, Modèle de Deuring et monogénéité des anneaux d'entiers des corps de rayon d'un corps quadratique imaginaire dans le cas "3 ramifié", Publ. Sci. Besancon, (1987–8).

D. Freeman, M. Scott and E. Teske, A taxonomy of pairing-friendly elliptic curves (2006), Cryptology ePrint Archive 2006/372, http://eprint.iacr.org/2006/372/; to appear in *J. Crypt.*

R. Fricke, *Die elliptischen Funktionen und ihre Anwendungen*, I, II Teubner (1916, 1922).

R. Fueter, Vorlesungen über die singulären Moduln und die Komplexe Multiplikation elliptischer Funktionen I, II, Teubner (1924, 1927).

B.H. Gross and D.B. Zagier, On singular moduli, *J. für die Reine und Angewandte Math.* **355** (1985), 191–220.

F. Hajir, Unramified elliptic units, Ph.D. thesis, Massachusetts Institute of Technology, Dept. of Mathematics (1993).

F. Halter-Koch, Geschlechtertheorie der Ringklassenkörper, *J. für die Reine und Angewandte Math.* **250** (1971), 107–108.

H. Hasse, Neue Begründung der Komplexen Multiplikation I, II, *J. für die Reine und Angewandte Math.* **157**, **165**, (1927, 1931), 115–139, 64–88.

H. Hasse, Zum Hauptidealsatz der komplexen Multiplikation, *Monatshefte für Math. und Phys.* **38**, (1931), 315–322.

H. Hasse, *Nicht veröffentlichte Durchführung des Beweises für die Riemannsche Vermutung in Kongruenzfunktionenkörpern vom Geschlecht 1 mittels der Methoden aus der Komplexen Multiplikation*, Marburg (1933) (scientific estate).

H. Hasse, *Über die Klassenzahl abelscher Zahlkörper*, Berlin, Akademie-Verlag (1952).

H. Hayashi, On elliptic units and class number of a certain non-galois number field, *Manu. Math.* **45**, No 1 (February, 1983), 13–27.

H. Hayashi, On elliptic units and class number of certain dihedral extensions of degree $2l$, *Acta Arith.* **XLIV** (1984), 35–45.

H. Hayashi, On index formulas of Siegel units in a ring class field, *Acta Arith.* **XLVII** (1986).

K. Heegner, Diophantische Analysis und Modulfunktionen, *Math. Zeitschrift* **56** (1952), 227–253.

D. Kersey, The index of modular units in complex multiplication, Yale thesis (1980).

M. Klebel, Zur Theorie der Potenzganzheitsbasen bei relativ galoisschen Zahlkörpern, Dissertation, Augsburg (1995).

S. Lang, *Elliptic Functions*, Addison-Wesley (1970).

S. Lang, *Elliptic Curves Diophantine Analysis*, Springer-Verlag (1978).

H.W. Leopoldt, *Über Einheitengruppe und Klassenzahl reeller Zahlkörper*, Abh. der Deutschen Akad. Wiss. Berlin, Akademie-Verlag, Berlin (1954).

H.W. Leopoldt, Über die Hauptordnung der ganzen Elemente eines abelschen Zahlkörpers, *J. für die Reine und Angewandte Mathematik* 209 (1962), 54–71.

R. Limmer, Konstruktion von vollständigen Einheitengruppen mit Hilfe der Weierstrass'schen σ- und $\wp$-Funktion, *Augsburger Mathematisch-Naturwissenschaftliche Schriften* Bd 2 (1994).

C. Meyer, *Die Berechnung der Klassenzahl abelscher Körper über quadratischen Zahlkörpern*, Akademie Verlag, Berlin (1957).

C. Meyer, Über einige Anwendungen Dedekindscher Summen, *J. für die Reine und Angewandte Mathematik* **198**, (1957), 143–203.

C. Meyer, Bemerkungen zum Satz von Heegner-Stark über die imaginäquadratischen Zahlkörper mit der Klassenzahl Eins, *J. für die Reine und Angewandte Mathematik* **242** (1970), 179–214.

C. Meyer, Zur Theorie und Praxis der elliptischen Einheiten, *Ann. Uni. Sarveniensis Series Math.* **6** No 2 (1995).

F. Morain, Computing the cardinality of CM elliptic curves using torsion points. *J. Théor. Nombres Bordeaux* **19**, No 3 (2007), 663–681.

K. Nakamula, Class number calculation and elliptic unit. I, II, III. *Proceedings of Japan Academy*, **57A**, 56–59, 117–120, 363–366 (1981).

K. Nakamula, Class number calculation of a cubic field from the elliptic unit, J. Reine Angew. Math. **331** (1982), 114–123.

K. Nakamula, Calculation of the class numbers and fundamental units of abelian extensions over imaginary quadratic fields from approximate values of elliptic units, *J. Math. Soc. Jap.* **37**, No 2 (1985).

K. Nakamula, Class number calculation of a quartic field from the elliptic unit, *Acta Arith.* **45** (1985) 215–227.

K. Nakamula, Class number calculation of a sextic field from the elliptic unit, *Acta Arith.* **45** (1985) 229–247.

G. Pappas, On torsion line bundles and torsion points on abelian varieties, *Invent. Math.* **133** (1998), 193–225.

K. Ramachandra, Some applications of Kronecker's limit formulae, *Ann. Math.* **236** (1969), 104–148.

K. Ramachandra, On the class number of relative abelian fields, *J. Reine Angewandte Math.* **236** (1969), 1–10.

I. Reiner, *Maximal Orders*, Academic Press, London (1975).

G. Robert, Unités elliptiques, *Bul. Soc. Math. France, Mémoire* No 36 (1973).

K. Rubin and A. Silverberg, Choosing the correct elliptic curve, preprint, http://eprint.iacr.org/2007/253.

K. Rubin and A. Silverberg, Point counting on reductions of CM elliptic curves, preprint, http://arxiv.org/abs/0706.3711.

R. Schertz, Über die Klassenzahl gewisser nicht galoisscher Körper 6-ten Grades, *Hamburger Abh.* **42** (1974), 217–227.

R. Schertz, Die Klassenzahl der Teilkörper abelscher Erweiterungen imaginärquadratischer Zahlkörper, Teil I, *J. Reine Angew. Math.* **295** (1977), 151–168, Teil II, *J. Reine Angew. Math.* 296, (1977), 58–79.

R. Schertz, Zur Theorie der Ringklassenkörper über imaginär-quadratischen Zahlkörpern, *J. Number Theory* **10**, No 1 (1978), 70–82.

R. Schertz, Teilkörper abelscher Erweiterungen imaginär-quadratischer Zahlkörper, deren Klassenzahl durch Primteiler des Körpergrades teilbar ist, *J. Reine Angew. Math.* **302** (1978), 59–69.

R. Schertz, Niedere Potenzen von Ringklasseneinheiten, Research Institute for Mathematical Sciences Kokyuroku 603, Kyoto Univ, Kyoto Japan (1986), 21–34.

R. Schertz, Konstruktion von Ganzheitsbasen in Strahlklassenkörpern über imaginär-quadratischen Zahlkörpern, *J. Reine Angew. Math.* **398** (1989), 105–129.

R. Schertz, Zur expliziten Berechnung von Ganzheitsbasen in Strahlklassenkörpern über einem imaginär-quadratischen Zahlkörper, *J. Number Theory* **34**, No 1, (1990), 41–53.

R. Schertz, Über die Nenner normierter Teilwerte der Weierstraßschen $\wp$-Funktion, *J. Number Theory* **34**, No 2, (1990), 229–234.

R. Schertz, Galoismodulstruktur und elliptische Funktionen, *J. Number Theory* **39**, No 3, (1991), 285–326.

R. Schertz, An elliptic resolvent, *J. Number Theory* **77** (1999), 97–121.

R. Schertz, Weber's class invariants revisited, *J. Théorie Nombres Bordeaux* **14** (2002), 325–343.

R. Schertz, Global construction of associated orders in complex multiplication, *J. Number Theory* **111** (2005) 197–226.

R. Schertz, On the gerneralized principal ideal theorem of complex multiplication, *J. Théorie Nombres Bordeaux* **18** (2006), 683–691.

B. Schöneberg, *Elliptic Modular Functions*, Grundlehren der Math. Wiss. Bd. 203 Springer (1974).

R.Schoof, The exponents of the groups of points on the reductions of an elliptic curve, *Arithmetic Algebraic Geometry*, Birkhäuser, Boston (1991).

G. Shimura, *Arithmetic Theory of Automorphic Functions*, Princeton University Press, Princeton, New Jersey (1971).

C.L. Siegel, Zum Beweise des Starkschen Satzes, *Invent. Math.* **5** (1968), 180–191.

J. Silverman, The arithmetic of elliptic curves, *Graduate Text in Mathematics* **106** (1985).

H. Söhngen, Zur Komplexen Multiplikation, *Math. Ann.* **111** (1935), 102–328.

A. Srivastav and M.J. Taylor, Elliptic curves with complex multiplication and Galois module structure, *Invent. Math.* **99** (1990), 165–184.

H.M. Stark, A complete determination of the complex quadratic fields of class number one. *Michigan Math. J.* **14** (1967) 1–27.

H.M. Stark, On the 'Gap' in a theorem of Heegner, *J. Number Theory* **1** (1969), 16–27.

H.M. Stark, L-functions at s = 1. IV. First derivatives at s = 0, *Adv. Math.*, **35**, No 3 (1980), 197–235.

H.M. Stark, Counting points on CM elliptic curves, *Rocky Mountain J. Math.* 26 (1996), 1115–1138.

M.J. Taylor, Mordell-Weil Groups and the Galois Module Structure of Rings of Integers, *Illinois J. Math.* **32** No 3, (1988), 428–452.

M. Verant, Monogénéité de l'anneau des entiers de certains corps de classes de corps quadratiques, *L'U.F.R. des sciences et techniques de l'Université de Franche-Comté* **179** (1990), 66–102.

H. Weber, *Lehrbuch der Algebra III, Neudruck der 2. Auflage*, Chelsea, New York (1908).

A. Weng, A class of hyperelliptic CM-curves of genus three, *J. Ramanujan Math. Soc.* **16**, No 4 (2001), 339–372.

A. Weng, Constructing hyperelliptic curves of genus 2 suitable for cryptography, *Math. Comput.* **72** No 241 (2003), 435–458.

Index of Notation

Index

www.ingramcontent.com/pod-product-compliance
Lightning Source LLC
LaVergne TN
LVHW101457200926
848483LV00001B/146
* 9 7 8 0 5 2 1 7 6 6 6 8 5 *